TECHNICAL DEVELOPMENT
OF TELEVISION

TECHNICAL DEVELOPMENT OF TELEVISION

Edited With an Introduction by
George Shiers

ARNO PRESS

A New York Times Company

New York / 1977

Editorial Supervision: MARIE STARECK

————◆————

(4)

Reprint Edition 1977 by Arno Press Inc.

Introduction and arrangement
Copyright © 1977 by George Shiers

HISTORICAL STUDIES IN TELECOMMUNICATIONS
ISBN for complete set: 0-405-07756-4
See last pages of this volume for titles.

Manufactured in the United States of America

————◆————

Library of Congress Cataloging in Publication Data
Main entry under title:

Technical development of television.

   (Historical studies in telecommunications)
   Reprint of articles published between 1911 and 1970
by various publishers.
   Includes bibliographical references.
   1. Television--History--Addresses, essays, lectures.
I. Shiers, George. II. Series.
TK6637.T4 1976   621.388   75-23902
ISBN 0-405-07761-0

ACKNOWLEDGMENTS

"Early Schemes for Television" by George Shiers, © 1970 by The Institute of Electrical and Electronics Engineers, Inc. was reprinted by permission of The Institute of Electrical and Electronics Engineers, Inc. and George Shiers, from *IEEE Spectrum,* 1970, Volume 7, Number 5, pp. 24-34.

"History of Modern Television" by P. K. Gorokhov, © 1961 by The Institute of Electrical and Electronics Engineers, Inc. was reprinted by permission of The Institute of Electrical and Electronics Engineers, Inc., from *Radio Engineering,* English edition of *Radiotekhnika,* 1961, Volume 16, Number 6, pp. 71-80.

"Twenty Five Years' Change in Television" by J. C. Wilson, "John L. Baird: The Founder of British Television" by J. D. Percy, "Television in America To-Day" by A. Dinsdale and "The Birth of a High Definition Television System" by S. J. Preston, were reprinted by permission of The Royal Television Society.

"Television: An Account of the Development and General Principles of Television As Illustrated By A Special Exhibition Held at the Science Museum—June-September, 1937" edited by G. R. M. Garratt and G. Parr, and "Report of the Television Committee" were reprinted by permission of the Controller of Her Britannic Majesty's Stationery Office.

"Creating An Industry" by Robert C. Bitting, Jr., *Journal of the SMPTE,* Volume 74, New York, Nov. 1965 © 1965 by the Society of Motion Picture and Television Engineers, Inc., "The Evolution of Modern Television" by Axel G. Jensen, *Journal of the SMPTE,* Volume 63, New York, Nov. 1954, © 1954 by the Society of Motion Picture and Television Engineers, Inc., "A Short History of Television Recording" by Albert Abramson, *Journal of the SMPTE,* Volume 64, New York, February 1955, © 1955 by the Society of Motion Picture and Television Engineers, Inc., and "A Short History of Television Recording" Part II, by Albert Abramson, *Journal of the SMPTE,* Volume 82, New York, March 1973, © 1973 by the Society of Motion Picture and Television Engineers, Inc., were reprinted by permission of the Society of Motion Picture and Television Engineers, 862 Scarsdale Avenue, Scarsdale, New York 10583.

INTRODUCTION

When it became possible to hear through an electric wire, the question of being able to see by electricity immediately attracted a number of would-be inventors. Since the telephone was a beautiful analogue of the human ear, all that was needed was a mechanism that would similarly serve as a far-seeing eye. By 1884 about a dozen schemes for "electric telescopes" had been proposed, all using selenium[1] as the optical pickup element. Apart from this universal feature, these early proposals were as diverse in their technical make-up as were the national origins of the proponents.

If one looks beyond the technical fantasies in most of these early schemes [1]*, the basic features of the problem of distant electric vision were even then fairly clear, as were the alternative arrangements. One approach employed mosaic assemblies with multiwire connections, the other consisted of single elements and synchronized scanning motions. A mosaic of selenium "cells" at the transmitter, with each cell connected by separate wires to the respective incandescent elements at the receiver, offered the simplest technical arrangement. The obvious drawback of multiple wires was offset by the fact that they eliminated the need for complex scanning means and synchronized motions. In the other approach, the technical challenge presented by the otherwise simpler single-line circuits was met, theoretically at least, by the adoption of various synchronizing mechanisms then available in telegraphic practices.

Even more than in most lines of engineering, the infant art of seeing by electricity had a mixed parentage. One direct line of descent is traceable from Alexander Bain's automatic copying

* Numbers in brackets refer to papers in this collection.

telegraph[2] through subsequent developments of facsimile telegraphy.[3] But less than half of the early schemes were concerned solely with television, the others being facsimile systems with direct reception of moving images little more than a tentative and intriguing suggestion. By 1884 almost all of the hitherto insuperable problems were solved by means of whirling perforated disks. This fundamental scanning system was embodied in the first television patent,[4] applied for by Paul Nipkow, a 23-year-old student of physics in Berlin.

The general problem was clearly a current topic in 1890, when an editorial[5] in The Electrician put the question, "Shall we ever see by electricity?" One answer to the receiver problem lay in the future: "It is possible to conceive of some as yet uninvented glow lamp of extraordinary delicacy that may serve the purpose." And, along the same prophetic lines: "Perhaps one day some sort of electro-optic action may be discovered...which may place the problem on a wholly different and simpler basis." The "electro-optic action" had already been discovered, however, since Julius Elster and Hans Geitel introduced the first phototube that year.[6] As for the "uninvented glow lamp," a potential answer came seven years later when Ferdinand Braun introduced the cathode-ray indicator tube.[7] During the quarter century from 1884, more than a score of different mechanical schemes appeared, including early proposals for two-way television and television in color [3].

The Braun tube opened a new line of development during the first decade of the present century. In 1907, Boris L. Rosing in St. Petersburg (Leningrad) devised a hybrid system of mirror drums and a cathode-ray tube receiver [2].[8] An all-electric scheme[9] was first proposed the following year by Alan A. Campbell Swinton in response to correspondence[10] on the problems and possibilities of picture transmission and electric vision. Swinton's familiarity with cathode rays and X rays, as well as his experience in electrical engineering, led to his more detailed plans for

the all-electric system which he described [18] in 1911. The
term "television," first used[11] in 1900, was then coming into use.

The popular rise of radio broadcasting after 1920 was mani-
fested by a flood of literature on the new medium, accompanied by
an undercurrent of interest in facsimile picture transmission and
radio movies,[12] the latter promoted by the work of C. Francis
Jenkins. His attempts at developing radio vision [13][13] were
limited, however, largely to the transmission of shadowgraphs and
motion picture film.[14]

By 1924 the armamentarium of radio technology, allied with
some of the traditional methods inherited from the 1880 period,
was being applied by enthusiasts in several countries in the
attempts to transmit moving images. In April that year, Swinton
released his definitive blueprint[15] for an all-electric system,
including the circuits for radio transmission, now a practicality.

The inventive dreams of nearly 50 years were finally justified early in 1926 when John L. Baird [11] gave the first public
demonstration[16] of true television with the aid of Nipkow's disks.
During the next three years Baird probed the frontier of the art
with a series of innovations: phonovision, noctovision, stereoscopic television, color television, long-distance transmission
over land lines, and the bridging of the Atlantic with television
signals.[17]

In the meantime, while an increasing number of inventors
attempted to develop their own apparatus, several large American
companies [5,12] took up the technical challenge.[18] Television
was clearly perceived through corporate eyes as a possible companion of the telephone, and also as an entertainment medium that
belonged in the province of broadcast radio. Public demonstrations and printed accounts revealed the team approach in solving
engineering problems and also exposed the limits of advanced technology that still prevented television from being a commercial
reality.

Technical and public interest in television reached a peak in 1928. By then it was truly an international affair and often the center of controversy, flavored with highly polarized views on personalities, corporate activities, national merits, methods, practicalities and possibilities.[19] In spite of the barrage of words and pictures in newspapers and magazines, the layman saw little proof of real television being close at hand. The tiny, erratic, neon-hued images, though novel, held no more than momentary interest to anyone except enthusiastic amateurs. Nevertheless, demonstrations of pictures by radio in many quarters proved that a miracle had indeed been accomplished and that radio television was a fact.

Perhaps after all, "real" television was an insoluble problem, at least with mechanical apparatus. On the other hand, it seemed entirely possible that television would soon leap ahead, in the way that silent films had recently blossomed into the talkies almost overnight. This hopeful outlook was supported by a constant stream of reports on new developments, often embellished with publicity prose. Part of this optimism took more tangible forms, however, with the birth of the first technical society devoted to television,[20] the publication of the first technical magazine[21] and of several books for the layman.[22]

While mechanical systems were being improved, demonstrated, promoted and exploited during the closing years of the 1920s,[23] research on cathode-ray systems gathered momentum, primarily through the work of Philo T. Farnsworth,[24] Vladimir K. Zworykin [14],[25] and Manfred von Ardenne.[26]

Interest in electromechanical and hybrid solutions reached a peak during the Great Depression; despite the widespread economic troubles, a wide range of apparatus was demonstrated and exhibited in America, England, and Germany [15,16].[27] Engineering achievement reached a high level during these transitional years (1930-33), and new methods were introduced. Big screens for theater television and tele-talkies,[28] zone and multichannel systems,[29] and intermediate film processes[30] were displayed in exhibitions and cinemas to stimulate public support and capture business interests.

Limited public service on an experimental basis continued in several countries throughout these years. But the mass market for home receivers remained elusive, as it had been ever since 1927. Thus, in 1933, the outlook for "real television" was still a conjectural issue. Nonetheless, with the picture tube creeping into favor and with continual rumors about the "magic eye" that seemed to be the key to a major solution, television began to appear as an inevitable extension of aural broadcasting. As research on many fronts transformed the cathode-ray tube from a laboratory device into a practical receiver component—albeit small, expensive, short-lived, and still a technical novelty—the picture tube began to appear to be the future "window on the world."[31]

Limitations on the picture definition obtainable by mechanical means, along with channel restrictions in the overcrowded broadcast band, gave impetus to the movement toward electronic solutions [8] and the use of the then relatively new ultra-short waves below 10 meters. By late 1933 the electronic camera tube existed in three experimental forms; Farnsworth's[32] Image Dissector [19], Zworykin's[33] Iconoscope [20], and the Emitron, developed in England by Electric and Musical Industries Ltd. [25].

Electron-beam techniques and allied radio developments progressed rapidly during the next few years.[34] Meanwhile, several research and manufacturing organizations explored other possibilities with electronic[35] and opto-mechanical[36] systems. Even though all-electronic solutions were clearly at hand in 1935, other technical approaches continued to flourish until the end of the decade [7]. Contemporary accounts show the increasing tempo of development and rising interest in Germany [17][37] and Russia,[38] as well as general progress in all countries.[39]

During the early thirties the idea of television broadcasting on a regional scale was accepted as a certainty. Several years of experimental broadcasting in England, Europe and the United States had opened the way to the future. New factors now demanded attention; rival claims of proponents, the economic base, allotments in the radio frequency spectrum, methods of regulatory control, social implications and possible technological advances. All these new problems pushed the older technical questions into the background. It was no longer a matter of being able to see by electricity, but rather a question of how it should be done.

The rival claims of several industrial concerns competing for restricted broadcasting facilities in Britain led to the formation of the Television Committee in London in May 1934. The Committee Report [21], published in January 1935, established the framework for a high-definition public service[40] which began on a regular schedule in November 1936 [4,22].[41] Following trials on an alternate basis, Baird's mixed systems were dropped in February 1937 in favor of EMI's all-electronic system [24].[42] After nearly three years of steady progress,[43] the service came to an abrupt end on the eve of World War II.

For a variety of reasons, the same trend in the United States took a more protracted course. Though the problem of technical standards had been studied as early as 1928-9,[44] and often in succeeding years, the matter was not fully dealt with until 1940. After prolonged field tests,[45] an informal commercial public service was inaugurated by RCA at the opening of the World's Fair in New York in May 1939.[46] At that time, technical advances pointed in several different directions. Faced with a discord of rival claims and opinions, the Federal Communications Commission avoided any decision on standards until engineers and manufacturers found an acceptable technical basis for a public service. From July 1940, the National Television System Committee studied every conceivable aspect of the problem and delivered

a voluminous report [23] in March 1941. Following FCC approval
in July, this material became the blueprint for American television [6] which, as in Britain, did not flower until the war was
over.[47]

The British television service was resumed in June 1946 on
the prewar basis. In addition to the technical advances accumulated during the war years, postwar television had the advantage
of improved camera tubes,[48] notably the image orthicon (1946),
C.P.S. Emitron (1947) and the vidicon (1950). With regular
broadcasting [9] established and millions of receivers [10] in
use in the United States at the end of the 1940s, and a comparable boom in Britain, the metamorphosis of television from one
of the greatest technical challenges in history to an essential
public service was complete.[49]

Color television, the great achievement of the 1950s, was
indeed a technical triumph.[50] Baird,[51] and Ives [28][52] experimented with color in the late 1920s, reviving ambitious schemes
for color that had been proposed from 1880 onward.[53] Baird
renewed his work in color ten years later[54] and continued independent development throughout the war.[55] The Columbia Broadcasting System demonstrated a semi-electronic high-definition
color system [29] in 1940.[56] Just after the war, RCA introduced
another semi-electronic color system[57] and also an all-electronic
system.[58] By 1950, color was the new frontier in television
research and a matter of public concern in the United States [30].
Rapid progress soon carried color television over the technical
threshold.[59] Once again the NTSC went to work and formulated
standards for a color system compatible with the monochrome
standards.

Following the adoption of these standards[60] in 1953, color
television entered a reluctant market in the U.S. Several different designs of display tubes appeared during these years as
possible alternatives to the basic tricolor shadowmask picture
tube.[61] The introduction of video tape recording for color as

well as monochrome was another major event of this period [26,27].[62] After years of slow commercial progress, color television in the United States came into its own with total color programming during the mid-sixties. Experimental transmissions based on an adaptation of the NTSC system had previously been made in Britain.[63] Other versions of the compatible system were developed in Europe,[64] and the BBC began the first regular color service with the PAL system in 1967.

The papers in this anthology record some of the highlights of progress from the simple selenium proposals of the late 1870s to advanced color systems of recent years. Companion documents and other contemporary sources, given in the following references and book list, provide further sampling of historical material. In a larger sense, the collection can be regarded as a "centennial" testimonial to inventors, engineers, physicists, financiers, promoters and others, as well as manufacturers, professional groups and legislators, whose efforts made "distant electric vision" not only a reality but a worldwide institution.

George Shiers

References

1. W. Smith, "Effect of Light on Selenium During the Passage of an Electric Current," Journal of the Society of Telegraph Engineers 2:31 (February 12, 1873). See also Nature 7:303 (February 20, 1873); and American Journal of Science 5:301 (January-June, 1873).

2. A. Bain, "Electric Time-Pieces and Telegraphs," British Patent No. 9745 (November 27, 1843).

3. For papers on the development of facsimile and its relationship to television, see G. Shiers (ed.) Facsimile: An Historical Anthology (New York: Arno Press, 1977) one of the anthologies in this series.

4. P. Nipkow, (Electric Telescope), German Patent No. 30,105 (January 6, 1884). Nipkow's disk was the basic component in many subsequent proposals.

See, for instance, L. Le Pontois, "The Telectroscope," Scientific American Supplement 35:546,547 (June 10, 1893); "The Dussaud Teleoscope," ibid, 46:18793 (July 2, 1898); "The Senlecq Telectroscope, An Apparatus for Electrical Vision," ibid, 64:372,373 (December 14, 1907); H. Sutton, "Tele-Photography," Telegraphic Journal and Electrical Review 27:549-551 (November 7, 1890). See also J.V.L. Hogan, "The Early Days of Television," Journal of the SMPTE 63:169-173 (November 1954).

5. "Seeing by Electricity," Electrician 24:448-450 (March 7, 1890).

6. J. Elster, and H. Geitel, "On the Use of Sodium Amalgam in Photoelectric Experiments," Annalen der Physik 41:161-176 (1890).

7. K.F. Braun, "A Method to Demonstrate and Study the Time Sequence of Variable Currents," Annalen der Physik 60:552-559 (February 1897).

8. B. Rosing, "New or Improved Method of Electrically Transmitting to a Distance Real Optical Images and Apparatus Therefor," British Patent No. 27,570 (December 13, 1907). R. Grimshaw, "The Telegraphic Eye," Scientific American 104:335,336 (April 1, 1911); 105:574 (December 23, 1911). The Braun tube had been used for pictorial reproduction earlier: M. Dieckmann and G. Glage, "A Method for the Transmission of Written Material and Line Drawings by Means of Cathode Ray Tubes," German Patent No. 190,102 (September 12, 1906). See also "Cathode Rays in Telephotography." Scientific American Supplement 71:163 (March 18, 1911).

9. A.A.C. Swinton, "Distant Electric Vision," Nature 78:151 (June 18, 1908).

10. S. Bidwell, "Telegraphic Photography and Electric Vision," Nature 78:105,106 (June 4, 1908).

11. Congrés International d'Electricité (Paris: Gauthier-Villars, 1903). C. Perskyi, report on television, pp. 54-56. Congress held August 18-25, 1900.

12. W. Davis, "The New Radio Movies," Popular Radio 4:436-443 (December 1923); H. Gernsback, "Radio Vision," Radio News 5:681,823 (December 1923).

13. G.L. Bidwell, "Television Arrives," QST 9:9-14 (July 1925); C.A. Herndon, "Motion Pictures by Ether Waves," Popular Radio 8:107-113 (August 1925); W.B. Arvin, "See With Your Radio," Radio News 7:278,384-387 (September 1925).

14. See book list, Jenkins, 1925, 1929.

15. A.A.C. Swinton, "The Possibilities of Television," <u>Wireless World</u> 14:51-56 (April 9), 82-84 (April 16), 114-118 (April 23, 1924).

16. "The Televisor. Successful Test of New Apparatus," <u>The Times</u> 9c (January 28, 1926). Demonstration to members of the Royal Institution at 22 Frith Street, London, Tuesday, January 26. For earlier work, see J.L. Baird, "An Account of some Experiments in Television," <u>Wireless World</u> 14:153-155 (May 7, 1924); J.L. Baird, "Television, Description of the Baird System by the Inventor," <u>ibid</u>, 15:533-535 (January 21, 1925); J.L. Baird, "Television, or Seeing by Wireless," <u>Discovery</u> 6:142,143 (April 1925); P.R. Bird, "Wireless Television, a Review of the Baird System," <u>Popular Wireless</u>, pages 622,623 (May 23, 1925).

17. J.L. Baird, "Television," <u>Journal of Scientific Instruments</u> 4:138-143 (February 1927); J.L. Baird, "Television and Black Light," <u>Popular Radio</u> 11:447-451,498 (May 1927); A. Dinsdale, "Television Sees in Darkness and Records Its Impressions," <u>Radio News</u> 8:1422,1423,1490-1492 (June 1927); E.T. Jones, "Television" (Note on London to Glasgow test), <u>Nature</u> 119:896 (June 18, 1927); A. Dinsdale, "Seeing Across the Atlantic Ocean," <u>Radio News</u> 9:1232,1233 (May 1928); R.F. Tiltman, "Television in Natural Colors Demonstrated," <u>ibid</u>, 10:320,374 (October 1928); R.F. Tiltman, "How 'Stereoscopic' Television is Shown," <u>ibid</u>, 10:418,419 (November 1928).

18. For General Electric, see E.F.W. Alexanderson, "Radio Photography and Television," <u>Radio News</u> 8:944,945,1030-1034 (February 1927), and <u>General Electric Review</u> 30:78-84 (February 1927); G.C.B. Rowe, "Television Comes to the Home," <u>Radio News</u> 9:1098-1100,1156 (April 1928); R. Hertzberg, "Television Makes the Radio Drama Possible," <u>ibid</u>, 10:524-527,587,588 (December 1928). For Bell Telephone Laboratories, see item 12 in this volume and the four companion papers, <u>Bell System Technical Journal</u> 6:560-652 (October 1927). For Westinghouse, see "Radio 'Movies' from KDKA," <u>Radio News</u> 10:416,417 (November 1928).
For RCA, see item 5 in this volume and "Large Television Images Broadcast by R.C.A.," <u>Radio News</u> 10:1121 (June 1929).

19. "Television and the Public," <u>Electrician</u> 100:296 (March 16, 1928); "Radio Television" (Radiotorial comment), <u>Radio</u> 10:9 (June 1928); "The

Progress of Television," _Electrician_ 101:379 (October 5, 1928); A.A.C. Swinton, "Television: Past and Future," _Discovery_ 9:337-339 (November 1928); R.F. Tiltman, "The Entertainment Value of Television Today!" _Television_ 1:13-16 (November 1928).

20. The Television Society, founded at the annual meeting of the British Association for the Advancement of Science, Leeds, September 1927. "The Birth of the Television Society," _Television_ 1:16 (March 1928).

21. _Television_ began monthly appearance in London with the March 1928 issue.

22. See book list, pre-1930 titles.

23. For American, see H.E. Rhodes, "Television—Its Progress To-day," _Radio Broadcast_ 13:331-333 (October 1928).
For British, see L.W. Corbett, "What Prospects of Television Abroad?" _Radio Broadcast_ 14:11-13 (November 1928).
For German, see A. Dinsdale, "Television at the Berlin Radio Exhibition," _Television_ 2:379-389 (October 1929).

24. P.T. Farnsworth and H.R. Lubcke, "Transmission of Television Images," _Radio_ 11:36,85,86 (December 1929); P.T. Farnsworth, "An Electrical Scanning System for Television," _Radio Industries_ 5:386-389 (November 1930); A. Dinsdale, "Television by Cathode Ray: The New Farnsworth System," _Wireless World_ 28:286-288 (March 18, 1931); P.T. Farnsworth, "Scanning with an Electric Pencil," _Television News_ 1:48-51,74 (March-April 1931); A.H. Halloran, "Scanning Without a Disc," _Radio News_ 12:998,999,1015 (May 1931).

25. V.K. Zworykin, "Television Through a Crystal Globe," _Radio News_ 11:905,949,954 (April 1930).

26. A. Neuburger, "A New Cathode Ray Tube for Television," _Television_ 3:99 (April 1930); A. Gradenwitz, "Cathode Ray Television: Manfred von Ardenne's Recent Work," _ibid_ 4:132,133 (June 1931); M. von Ardenne, "The Cathode Ray Tube Method of Television," _Journal of the Television Society_ 1:71-74 (1931).

27. See, for instance: "Practical Television is Now Here!" _Radio News_ 12:26-29 (July 1930); E.H. Felix, "Television Advances from Peephole to Screen," _ibid_, 12:228-230,268,269 (September 1930); W.G.W. Mitchell, "Developments in Television," _Journal of the Royal Society of Arts_ 79:616-642

(May 22, 1931); A. Gradenwitz, "Television at the Berlin Radio Exhibition," _Television_ 4:310-312,318 (October 1931); E.H. Traub, "The German Post Office Cathode Ray Television System," _Journal of the Television Society_ 1:75-81 (1931); H.J.B. Chapple, "Recent Advances in Television," _ibid_, 1:106-112 (1932); P. Shmakov, "Television in the U.S.S.R.," _ibid_, 1:126-130 (1932); "Television Apparatus: Demonstration by the Marconi Company During the British Association Meeting at York," _Electrician_ 109:311,312 (September 9, 1932); A.P. Peck, "Sight and Sound on One Wave," _Television_ 5:240-242 (September 1932); E.H. Traub, "Television at the 1932 Berlin Radio Exhibition," _Journal of the Television Society_ 1:155-166 (1932); A. Church, "Recent Developments in Television," _Nature_ 132:502-505 (September 30, 1933); E.H. Traub, "Television at the 1933 Berlin Radio Exhibition," _Journal of the Television Society_ 1:273-285 (1933).

28. "A Startling Development in Screen Television," _Television_ 3:255,256 (August 1930); "Baird Screen Television: The Coliseum Triumph," _ibid_, 3:292-295 (September 1930); D.R. Campbell, "The First Public Tele-Talkie Demonstration at the Coliseum," _ibid_, 3:304-306 (September 1930); A. Dinsdale, "Television Turns to Projection," _Radio Engineering_ 11:19,20 (September 1931).

29. H.E. Ives, "Multi-Channel Television Apparatus," _Bell System Technical Journal_ 10:33-45 (January 1931); "A New Television System," _Wireless World_ 28:38,39 (January 14, 1931); "Zone Television," _Television_ 3:476,477 (February 1931); L. Bruchiss, "Multi-plex System of Television Recently Introduced in England," _Radio Industries_ 5:554,555 (March 1931); C.O. Browne, "Multi-Channel Television," _Journal of the IEE_ 70:340-353 (March 1932).

30. G. Schubert, "The Television System of the Fernseh Company Using an Intermediate Film," _Fernsehen und Tonfilm_ 3:129-134 (July 1932); E.H. Traub, "Television at the 1932 Berlin Radio Exhibition," _Journal of the Television Society_ 1:165,166 (1932), and 1:275,276 (1933).

31. A. Dinsdale, "De-Bunking Television," _Radio News_ 12:592-594,660 (January 1931); "New Research on Cathode Rays," _Television_ 6:47,48 (February 1933); G. Parr and T.W. Price, "The Application of the Cathode Ray Tube to Television," _Journal of the Television Society_ 1:323-328 (1934); E.H. Traub, "Television at the Berlin Radio Exhibition, 1934," _ibid_, 1:341-351 (1934); "Television in Germany: Preponderance of Cathode Ray Equipment at Berlin Radio Show," _Electrician_ 113:506 (October 19, 1934); G. Parr and T.W. Price, "Recent

Improvements in Cathode Ray Tubes for Television," Journal of the Television
Society 2:15-18 (1935).

32. A.H. Brolly, "Television by Electronic Methods," Electrical Engineering 53:1153-1160 (August 1934); P.T. Farnsworth, "An Electron Multiplier,"
Electronics 7:242,243 (August 1934); R.C. Hergenrother, "The Farnsworth Electronic Television System," Journal of the Television Society 1:384-387 (1934).

33. "Zworykin's Iconoscope," Electronics 6:188 (July 1933); "The Icono-
scope: America's Latest Television Favourite," Wireless World 33:197
(September 1, 1933); V.K. Zworykin, "Television with Cathode-Ray Tubes: The
Iconoscope," Journal of the IEE 73:437-451 (October 1933), and 74:276,277
(March 1934); V.K. Zworykin and others (three papers on an experimental tele-
vision system), Proceedings of the IRE 21:1655-1706 (December 1933), and
E.W. Engstrom and others (four papers on an experimental television system),
ibid, 22:1241-1294 (November 1934).

34. (Farnsworth) S. Kaufman, "Demonstrates High-Definition Television,"
Radio News 17:265,308 (November 1935); "Farnsworth Television," ibid, 17:330,
331,375 (December 1935); P.T. Farnsworth, "An Improved Television Camera,"
Radio-Craft 8:92,113 (August 1936); C.C. Larson and B.C. Gardner, "The Image
Dissector," Electronics 12:24-27,50 (October 1939).
(Zworykin and RCA) V.K. Zworykin, "Iconoscopes and Kinescopes in Television,"
RCA Review 1:60-84 (July 1936); V.K. Zworykin and others, "Theory and Perform-
ance of the Iconoscope," Journal of the IEE 82:105-114,561,562 (January,
May 1938); H. Iams and others, "The Image Iconoscope," Proceedings of the IRE
27:541-547 (September 1939); A. Rose and H. Iams, "The Orthicon, a Television
Pick-up Tube," RCA Review 4:186-199 (October 1939).
(E.M.I.) "Super Emitron Camera," Wireless World 41:497-498 (November 18,
1937); J.D. McGee and H.G. Lubszynski, "E.M.I. Cathode-Ray Television Transmission Tubes," Journal of the IEE 84:468-482 (April 1939).

35. E.H. Traub, "New Television System: Thun's Principle of Variable Speed
Applied to Cathode Ray by von Ardenne," Journal of the Television Society
1:177-182 (1933); "New Television System: Velocity and Intensity Modulation
Combined," Wireless World 34:88-91 (February 9, 1934); L.H. Bedford and
O.S. Puckle, "A Velocity Modulation Television System," Journal of the IEE
75:63-92 (July 1934).

36. "Cinema Television," Journal of the Television Society 2:310-312 (1937); L.E.C. Hughes, "Television: The Scophony Receiving System," Electrician 120:515,516 (April 22, 1938); H.W. Lee, "The Scophony Television Receiver," Nature 142:59-62 (July 9, 1938); D.M. Robinson and others (four papers on the Scophony system), Proceedings of the IRE 27:483-500 (August 1939).

37. E.H. Traub, "Television at the Berlin Radio Exhibition," Journal of the Television Society 2:53-61 (1935), 2:181-187,188-191 (1936), 2:289-297 (1937).

38. P. Shmakov, "The Development of Television in the U.S.S.R.," Journal of the Television Society 2:97-105 (1936).

39. R.D. Washburne and W.E. Schrage, "World-Wide Television," Radio-Craft 7:76-80,123-126 (August 1935); A.W. Cruse, "The Status of Television in Europe," Electrical Engineering 54:966-969 (September 1935); "Special Television Number," Radio-Craft 8:69-127 (August 1936); M.P. Wilder, "Television in Europe," Electronics 10:13-15 (September 1937); "Radio Progress During 1937: Part IV, Report by the Technical Committee on Television and Facsimile," Proceedings of the IRE 26:298-301 (March 1938), 27:174-176 (March 1939), 28:120-124 (March 1940).

40. "High Definition Television Service in England," Journal of the Television Society 2:34-43 (June 1935); "Television Transmissions: Details of the Baird and Marconi-E.M.I. Systems," Wireless World 37:371-373 (October 4, 1935); "The London Television Transmitter, First Description of the Marconi-E.M.I. Installation," Wireless World 38:103,104 (January 31, 1936); "The London Television Station at Alexandra Palace," and "Technical Details of the Television Equipment Supplied by Baird Television Ltd. to the British Broadcasting Corporation at Alexandra Palace," and "Marconi-E.M.I. Television Equipment at the Alexandra Palace," Journal of the Television Society 2:156-160,161-168,169-176 (1936).

41. "B.B.C. Television Service," Electrician 117:563 (November 6, 1936).

42. T.C. MacNamara and D.C. Birkinshaw, "The London Television Service with Station Layout," and A.D. Blumlein and others, "The Marconi-E.M.I. Television System," Journal of the IEE 83:729-757,758-792,793-801 (discussion on both) (December 1938), also 85:271-279 (August 1939).

43. "Television Exhibition," Journal of the Television Society 2:265-273 (1936), and item 4 in this volume; "Radiolympia, 1937," ibid, 2:280-284 (1937); E.H. Traub, "English and Continental Television," ibid, 2:457-468 (1938); H.L. Kirke, "Recent Progress in Television," Journal of the Royal Society of Arts 87:302-327 (February 3, 1939).

44. J. Weinberger and others, "The Selection of Standards for Commercial Radio Television," Proceedings of the IRE 17:1584-1594 (September 1929); A.R. Murray, "R.M.A. Complete Television Standards," Electronics 11:28,29,55 (July 1938); Federal Communications Commission, "First Report of Television Committee," "Second Report, etc.," (May 22, November 15, 1939).

45. L.M. Clement and E.W. Engstrom, "RCA Television Field Tests," RCA Review 1:32-40 (July 1936); R.R. Beal, "Equipment Used in the Current RCA Television Field Tests," ibid, 1:36-48 (January 1937).

46. "Television I: A $13,000,000 'If'," Fortune 19:52-59,168-182 (April 1939); D.H. Castle, "A Television Demonstration System for the New York World's Fair," RCA Review 4:6-13 (July 1939).

47. G.R. Town, "Television," Electrical Engineering 59:313-322 (August 1940); and "Progress in Television," ibid, 66:580-590 (June 1947); "The Evolution of Television: 1927-1943. As Reported in the Annual Reports of the Federal Radio Commission and the Federal Communications Commission," Journal of Broadcasting 4:199-240 (Summer 1960); "The Evolution of Television: 1944-1948. As Reported in the Annual Reports of the Federal Communications Commission," ibid, 5:23-37 (Winter 1960-61); "RCA's Television," Fortune 38:80-85,194-204 (September 1948).

48. K. Schlesinger and E.G. Ramberg, "Beam-Deflection and Photo Devices," Proceedings of the IRE 50:991-1005 (May 1962).

49. L. Laden, "Television in Great Britain," Radio News 33:32-34,84,86 (January 1945); G. Parr, "British Television," Discovery 10:212-218 (July 1949); D.G. Fink, "Television Broadcasting in the United States, 1927-1950," Proceedings of the IRE 39:116-123 (February 1951); "British Contribution to Television," Proceedings of the IEE 99: Part IIIA (May 1952). See also E.L.E. Pawley, "B.B.C. Television 1939-1960; a Review of Progress," ibid, 108B:375-397 (July 1961), and "35 Years of World Television," International Broadcast Engineer, Supplement (January 1972).

50. F. Bello, "Color TV: Who'll Buy a Triumph?" Fortune 52:136-139,201-206 (November 1955).

51. R.F. Tiltman, "Television in Natural Colors Demonstrated," Radio News 10:320,374 (October 1928).

52. H.E. Ives and A.L. Johnsrud, "Television in Colors by a Beam Scanning Method," Journal of the Optical Society of America 20:11-22 (January 1930).

53. Some examples of early color proposals are: M. Leblanc, "Étude Sur La Transmission Électrique des Impressions Lumineuses," La Lumiere Electrique 2:477-481 (1880); J. Szczepanik, "Methods and Apparatus for Reproducing Pictures and the like at a Distance by Means of Electricity," British Patent No. 5031 (1897); A.C. and L.S. Andersen, "Improvements in Apparatus for Electrically Transmitting Images of Natural Objects to a Distance," British Patent No. 30,188 (1909). See also item 2 in this volume.

54. "Television in Colour," Electrician 120:197 (February 18, 1938); "Baird Colour Television," Television and Short Wave World 11:151,152 (March 1938); F.W. Marchant, "New Baird Colour Television System," ibid, 12:541,542 (September 1939).

55. "Baird High-Definition Colour Television," Journal of the Television Society 3:171-174 (1941); "Television in Colour and Stereoscopic Relief," ibid, 3:225,226 (1941); G. Parr, "Review of Progress in Colour Television," ibid, 3:251-256 (1942); "New Baird Tube Gives Television in Color," Electronics 17:190,194,198 (October 1944).

56. P.C. Goldmark and others, "Color Television, Part I," Proceedings of the IRE 30:162-182 (April 1942), and Part II, 31:465-478 (September 1943); R.W. Ehrlich, "Color Television," Radio News 34:32-34,130-138 (July 1945).

57. R.D. Kell and others, "An Experimental Color Television System," RCA Review 7:141-154 (June 1946).

58. RCA Laboratories, "Simultaneous All-Electronic Color Television; A Progress Report," RCA Review 7:459-468 (December 1946), and "A Six-Megacycle Compatible High-Definition Color Television System; A Report," ibid, 10:504-524 (December 1949).

59. "Color Television Issue," Proceedings of the IRE 39:1123-1360 (October 1951). Includes: H.B. Law, "A Three-Gun Shadow-Mask Color Kinescope,"

(pp.1186-1194), P.C. Goldmark and others, "Color Television—U.S.A. Standard," (pp.1288-1313), and 18 companion papers.

60. D.G. Fink, "NTSC Color Television Standards," Electronics 26:138-150 (December 1953). "Second Color Television Issue," Proceedings of the IRE 42:1-348 (January 1954). Includes five introductory papers, NTSC signal specifications, field test, color standards, 41 companion papers and a supplementary bibliography.

61. D. Gabor, "Colour TV," Endeavour 21:25-34 (January 1962).

62. J.W. Wentworth, "The Technology of Television Program Production and Recording," Proceedings of the IRE 50:830-836 (May 1962).

63. G.N. Patchett, "Colour Television," Journal of the British IRE 16:591-620 (November 1956); W.N. Sproson and others, "The BBC Colour Television Tests: An Appraisal of Results," BBC Engineering Monograph, No. 18 (May 1958); I.R. Atkins and others, "A New Survey of the BBC Experimental Colour Transmissions," BBC Engineering Monograph, No. 32 (October 1960); J. Redmond, "Television Broadcasting 1960-70: BBC 625-line Services and the Introduction of Colour," IEE Reviews, Proceedings of the IEE 117:1469-1488, Special Issue (August 1970).

64. R. Theile, "Compatible Color Transmission and European Modulation Methods," IEEE Spectrum 3:60-68 (June 1966).

Chronological List of Selected Books on Television

Jenkins, C.F. Vision by Radio, Radio Photographs, Radio Photograms. Washington: Jenkins Laboratories, 1925.

Baker, T.T. Wireless Pictures and Television. A Practical Description of the Telegraphy of Pictures, Photographs and Visual Images. London: Constable, 1926.

Dinsdale, A. Television. London: Pitman, 1926 (revised edition, Television Press, 1928).

Secor, H.W., and J.H. Kraus. All About Television, vol. 1, No. 1. New York: Experimenter Publishing Co., 1927, vol. 1, No. 2, 1928.

Tiltman, R.F. Television for the Home: The Wonders of Seeing by Wireless. London: Hutchinson, 1927.

Larner, E.T. Practical Television. London: Benn; New York: D. Van Nostrand, 1928 (revised 1929).

Jenkins, C.F. Radiomovies, Radiovision, Television. Washington: Jenkins Laboratories, 1929.

Sheldon, H.H., and E.N. Grisewood. Television: Present Methods of Picture Transmission . New York: D. Van Nostrand, 1929.

Yates, R.F. ABC of Television, or Seeing by Radio. New York: Henley, 1929.

Benson, T.W. Fundamentals of Television. New York: Mancall, 1930.

Hutchinson, R.W. Easy Lessons in Television. London: University Tutorial Press, 1930.

Moseley, S.A., and H.J.B. Chapple. Television To-day and To-morrow. London: Pitman, 1930 (revised editions, 1931, 1933, 1934, 1940).

Philp, C.G. Television for All. London: Percival Marshall, 1930.

Felix, E.H. Television: Its Methods and Uses. New York: McGraw-Hill, 1931.

Collins, A.F. Experimental Television: A Series of Simple Experiments with Television Apparatus. Boston: Lothrop, Lee & Shepard, 1932.

Dinsdale, A. First Principles of Television. London: Chapman & Hall; New York: Wiley, 1932 (reissued by Arno Press in 1971).

Dunlap, O.E. The Outlook for Television. New York: Harper, 1932 (reissued by Arno Press in 1971).

Chapple, H.J.B. Television for the Amateur Constructor. London: Pitman, 1933 (revised 1934).

Hathaway, K.A. Television: A Practical Treatise on the Principles upon which the Development of Television is Based. Chicago: American Technical Society, 1933 (revised 1936).

Tiltman, R.F. Baird of Television: The Life Story of John Logie Baird. London: Seeley Service, 1933 (reissued by Arno Press in 1974).

Camm, F.J. Newnes Television and Short-Wave Handbook. London: Newnes, 1934 (revised 1935, 1937, 1939, 1942).

Reyner, J.H. Television: Theory and Practice. London: Chapman & Hall, 1934 (revised 1937).

Chapple, H.J.B. Popular Television: Up-to-date Principles and Practice Explained in Simple Language. London: Pitman, 1935.

Dowding, G.V. (ed.). Book of Practical Television. London: Amalgamated Press, 1935.

Hutchinson, R.W. Television Up-To-Date. London: University Tutorial Press, 1935 (revised 1937).

Robinson, E.H. Televiewing. London: Selwyn and Blount, 1935 (revised 1936).

Scroggie, M.G. Television. London: Blackie, 1935 (revised 1948).

Ardenne, M. von. (O.S. Puckle, trans.). Television Reception. Construction and Operation of a Cathode Ray Tube Receiver for the Reception of Ultra-Short Wave Television Broadcasting. London: Chapman & Hall; New York: D. Van Nostrand, 1936.

Eckhardt, G.H. Electronic Television. Chicago: Goodheart-Willcox, 1936 (reissued by Arno Press in 1974).

Lewis, E.J.G. Television: Technical Terms and Definitions. London: Pitman, 1936.

Moseley, S.A., and H. McKay. Television: A Guide for the Amateur. London: Oxford University Press, 1936.

Myers, L.M. Television Optics, an Introduction. London: Pitman, 1936.

Television. New York: RCA Institutes Technical Press, 1936-37 (two volumes); and Princeton, N.J.: RCA Review, 1947-50 (four volumes).

The London Television Station, Alexandra Palace. London: BBC, 1937.

Tyers, P.D. Television Reception Technique. London: Pitman, 1937.

Wilson, J.C. Television Engineering. London: Pitman, 1937.

Witts, A.T. Television Cyclopaedia. London: Chapman & Hall; New York: D. Van Nostrand, 1937.

Maloff, I.G., and D.W. Epstein. Electron Optics in Television: With Theory and Application of Television Cathode-ray Tubes. New York: McGraw-Hill, 1938.

Waldrop, F.C., and J. Borkin. Television: A Struggle for Power. New York: Morrow, 1938 (reissued by Arno Press in 1971).

Kerby, P. The Victory of Television. New York: Harper, 1939.

Legg, S., and R. Fairthorne. The Cinema and Television. London: Longmans, 1939.

Cocking, W.T. Television Receiving Equipment. London: Iliffe, 1940.

Fink, D.G. Principles of Television Engineering. New York: McGraw-Hill, 1940 (revised 1952).

Lohr, L.R. Television Broadcasting: Production, Economics, Technique. New York: McGraw-Hill, 1940.

Porterfield, J., and K. Reynolds (eds.). We Present Television. New York: Norton, 1940.

Zworykin, V.K., and G.A. Morton. Television: The Electronics of Image Transmission. New York: Wiley; London: Chapman & Hall, 1940 (revised...in Color and Monochrome, 1954).

Hylander, C.J., and R. Harding. An Introduction to Television. New York: Macmillan, 1941.

Pioneering the Cathode-Ray and Television Arts. Passaic, N.J.: Allen B. Du Mont Laboratories, 1941.

De Forest, L. Television Today and Tomorrow. New York: Dial Press, 1942.

Dunlap, O.E. The Future of Television. New York: Harper, 1942 (revised 1947).

Hubbell, R.W. 4000 Years of Television: The Story of Seeing at a Distance. New York: Putnam, 1942.

Fink, D.G. (ed.). Television Standards and Practice. Selected Papers from the Proceedings of the National Television System Committee and Its Panels. New York: McGraw-Hill, 1943 (selected pages reproduced in this volume).

Lee, R.E. Television: The Revolution. New York: Essential Books, 1944.

Eddy, W.C. Television: The Eyes of Tomorrow. New York: Prentice-Hall, 1945.

Great Britain. Report of the Television Committee. London: H.M.S.O., 1945.

Hutchinson, T.H. Here is Television: Your Window to the World. New York: Hastings House, 1946 (revised 1948, 1950).

Kiver, M.S. Television Simplified. New York: D. Van Nostrand, 1946 (and subsequent editions).

Sarnoff, D. Pioneering in Television: Prophecy and Fulfillment. New York: RCA Technical Institutes Press, 1946 (and subsequent editions).

Tyler, K.S. Telecasting and Color. New York: Harcourt, Brace, 1946.

Hallows, R.W. Television Simply Explained. London: Chapman & Hall, 1947.

Dunlap, O.E. Understanding Television: What It Is and How It Works. New York: Greenberg, 1948.

Kempner, S. Television Encyclopedia. New York: Fairchild, 1948.

Everson, G. The Story of Television: The Life of Philo T. Farnsworth. New York: Norton, 1949 (reissued by Arno Press in 1974).

Gorham, M.A.C. Television: Medium of the Future. London: Percival Marshall, 1949.

Grob, B. Basic Television: Principles and Servicing. New York: McGraw-Hill, 1949, 1954.

Maclaurin, W.R., and R.J. Harmon. Invention and Innovation in the Radio Industry. New York: Macmillan, 1949 (reissued by Arno Press in 1971).

Helt, S. Practical Television Engineering. New York: Murray Hill Books, 1950.

Swift, J. Adventure in Vision: The First Twenty-Five Years of Television. London: John Lehman, 1950.

Dunlap, O.E. Dunlap's Radio and Television Almanac. New York: Harper, 1951.

Horton, D. Television's Story and Challenge. London: Harrap, 1951.

Moseley, S.A. John Baird: The Romance and Tragedy of the Pioneer of Television. London: Odhams, 1952.

Television: A World Survey. Paris: Unesco, 1953 with 1955 Supplement (both reissued by Arno Press in 1972).

Bourton, K. Bibliography of Colour Television. London: Television Society, 1954-56.

Abramson, A. Electronic Motion Pictures: A History of the Television Camera. Berkeley, Los Angeles: University of California Press, 1955 (reissued by Arno Press in 1974).

Fink, D.G. (ed.). Color Television Standards NTSC. Selected Papers and Records of the National Television System Committee. New York: McGraw-Hill, 1955.

Wentworth, J.W. Color Television Engineering. New York: McGraw-Hill, 1955.

McIlwain, K., and C.E. Dean (eds.). Principles of Color Television. By the Hazeltine Laboratories Staff. New York: Wiley, 1956.

Fink, D.G. Television Engineers Handbook. New York: McGraw-Hill, 1957.

Reyner, J.H. The Encyclopedia of Radio and Television. London: Odhams, 1957.

Zworykin, V.K., E.G. Ramberg, and L.E. Flory. Television in Science and Industry. New York: Wiley, 1958.

Fink, D.G. and D.M. Lutyens. The Physics of Television. Garden City, N.Y.: Doubleday, 1960.

BBC Television: A British Engineering Achievement. London: BBC, 1961.

Ross, G. Television Jubilee: The Story of 25 Years of BBC Television. London: Allen, 1961.

Dunlap, O.E. Communications in Space: From Marconi to Man on the Moon. New York: Harper & Row, 1962 (revised 1964, 1970).

Briggs, A. The History of Broadcasting in the United Kingdom. Volume 2, The Golden Age of Wireless. London: Oxford University Press, 1965.

Dizard, W.P. Television: A World View. Syracuse, N.Y.: Syracuse University Press, 1966.

Spottiswoode, R. (ed.). The Focal Encyclopedia of Film and Television Techniques. London: Focal Press, 1969.

Baker, W.J. A History of the Marconi Company. London: Methuen, 1970.

Pawley, E. BBC Engineering: 1922-1972. London: BBC, 1972.

Goldmark, P.C., and L. Edson. Maverick Inventor: My Turbulent Years at CBS. New York: Saturday Review Press, E.P. Dutton, 1973.

Lichty, L.W., and M.C. Topping. American Broadcasting. A Source Book on the History of Radio and Television. New York: Hastings House, 1975.

CONTENTS

Part I: General History

Owen, Clure H.
Television Broadcasting (Reprinted from *Proceedings of the Institute of Radio Engineers,* Volume 50), New York, 1962

Bingley, F[rank] J.
A Half Century of Television Reception (Reprinted from *Proceedings of the Institute of Radio Engineers,* Volume 50), New York, 1962

Part II: Mechanical Systems

Percy, J. D.
John L. Baird: The Founder of British Television, Revised Edition, London, 1952

Ives, Herbert E.
Television (Reprinted from *The Bell System Technical Journal,* Volume VI), New York, 1927

Jenkins, C[harles] Francis
Radio Vision (Reprinted from *Proceedings of the Institute of Radio Engineers,* Volume 15), New York, 1927

Zworykin, V[ladimir K.]
Television with Cathode-Ray Tube for Receiver (Reprinted from *Radio Engineering,* Volume IX), New York, 1929

Ives, Herbert E., Frank Gray and M. W. Baldwin
Two-Way Television (Reprinted from *Transactions of the American Institute of Electrical Engineers,* Volume 49), New York, 1930

Dinsdale, A[lfred]
Television in America To-day (Reprinted from *Journal of the Television Society,* Volume I), London, 1932

Gibas, Hubert
Television in Germany (Reprinted from *Proceedings of the Institute of Radio Engineers,* Volume 24), New York, 1936

Part III: Electronic Television

Campbell Swinton, A. A.
Presidential Address (Reprinted from *The Journal of the Röntgen Society,* Volume VIII, London, 1912)

Farnsworth, Philo Taylor
Television By Electron Image Scanning (Reprinted from *Journal of the Franklin Institute,* Volume 218), Philadelphia, 1934

Zworykin, V[ladimir] K.
Television (Reprinted from *Journal of the Franklin Institute,* Volume 217), Philadelphia, 1934

The Television Committee
Report of the Television Committee, London, 1935

Ashbridge, Noel
Television in Great Britain (Reprinted from *Proceedings of the Institute of Radio Engineers,* Volume 25), New York, 1937

Fink, Donald G., editor
Selections reprinted from *Television Standards and Practice,* New York, 1943

Television Standardization in America—Chapter I

The National Television System Standards—Chapter II

Television Systems—Chapter III

Technical Papers in the N. T. S. C. Proceedings—[Appendix]

Preston, S. J.
The Birth of a High Definition Television System (Reprinted from *Journal of the Television Society,* Volume 7), London, 1953

McGee, J[ames] D.
Distant Electric Vision (Reprinted from *Proceedings of the Institute of Radio Engineers,* Volume 38), New York, 1950

Abramson, Albert
A Short History of Television Recording (Reprinted from *Journal of the SMPTE,* Volume 64), New York, 1955

Abramson, Albert
A Short History of Television Recording: Part II (Reprinted from *Journal of the SMPTE,* Volume 82), New York, 1973

Part IV: Color Television

Ives, Herbert E.
Television in Colors (Reprinted from *Bell Laboratories Record,* Volume VII), New York, 1929

Color Television Demonstrated by CBS Engineers (Reprinted from *Electronics*, Volume 13), New York, 1940

U. S. Senate, Committee on Interstate and Foreign Commerce, Advisory Committee on Color Television
The Present Status of Color Television. Report of the Advisory Committee on Color Television to the Committee on Interstate and Foreign Commerce, United States Senate, 81st Congress, 2nd Session, Document No. 197, Washington, D. C., 1950

PART I: GENERAL HISTORY

EARLY SCHEMES FOR TELEVISION

George Shiers

Early schemes for television

Various proposals for "seeing by electricity" were put forward around 1880. Though only paper plans, they established principles that led to the reality of mechanical television in the 1920s

George Shiers   Santa Barbara, Calif.

More than a dozen schemes for sending visual images by electricity appeared from 1877 to 1884. Some used multiwire lines and mosaic arrays; others used single lines and a scanning method—autographic, spiral, linear. Selenium cells and incandescent filaments were common elements. Basic ideas on scanning speed, repetition frequency, synchronism, picture elements, and beam modulation evolved during these years. Some schemes employed magnetooptic effects, others used polarized light, and an optical equivalent of the cathode-ray tube was proposed for one receiver. Mechanical problems were finally solved by the scanning disk, which, 40 years later, with the aid of electronic techniques, became the foundation for practical mechanical television. Both facsimile and television proposals were covered, partly because they were inseparable during this era and partly to show the continuity of developments.

The possibility of "seeing by electricity" first became of real interest during the late 1870s. Two events of that decade gave impetus to the search for ways of transmitting pictures and views: the discovery of the light sensitivity of selenium in 1873, and the invention of Bell's telephone in 1876. Rising interest in electric lighting elements also promoted interest in "electrical vision." From 1877 a variety of proposals established basic principles that, over seven years, were developed and finally incorporated in the near-practicable invention of Paul Nipkow in 1884. A chronology of this period is given in Table I.

Selenium and light

Willoughby Smith, chief electrician of the Telegraph Construction Company, used selenium rods as high resistances for continuity checks of the Atlantic cable (1866). Selenium proved to be unsuitable, however, because of wide variations in resistance. Later tests to find the cause revealed the fact that the resistance was diminished when a rod was exposed to light. This discovery of the effect of light on selenium was announced by Smith early in 1873.[1,2] Despite the obvious value of this phenomenon, only a few workers studied the physical proper-

ties of selenium during the next decade, and few attempts were made to develop the selenium cell as an electric device.[3] Nevertheless, selenium, though inefficient and capricious, was a key component in all the early schemes and retained its eminent place until efficient phototubes became available after World War I.

Early television,* or "telectroscopy" as it was called, evolved from facsimile telegraphy. This sister art was over 30 years old when the telephone was born. Actually, most of the proposals employing selenium for sending pictures by electricity were directly related to facsimile, with means for instant visual reception thrown in, as it were, as alternatives.

Facsimile, or the copying telegraph

Without going into the history of facsimile,[6,7] a brief reference to the "copying telegraph" is necessary to show what techniques were available during the mid-1870s. The first plan for an "automatic telegraph" was patented[8] in 1843 by Alexander Bain, a Scottish watchmaker, telegraph inventor, and pioneer of electric clocks. To avoid coding and decoding, Bain devised a way to reproduce letters and words of the original message as a series of stains in chemically prepared paper.[9]

Variations of Bain's scheme (based on sequential scanning and synchronized movements) were developed by others, particularly Frederick Bakewell (1847) and Giovanni Casseli (1861). The Casseli system was in service in France during the 1860s. During the next ten years a variety of proposals for other "writing telegraphs" appeared. These schemes, sometimes known as the "telautograph," employed coordinate control whereby the movement of a pen or style is resolved into, and reconstituted by, two separate rectilinear motions. Therefore, the principles of scanning and synchronism—basic to facsimile and television employing sequential signals—were well established when proposals for sending pictures and views by electricity with the aid of selenium

---

* The word "television" was coined in 1900 by a Frenchman named Perskyi.[4] Hugo Gernsback introduced it in 1909 in an article in his magazine.[5]

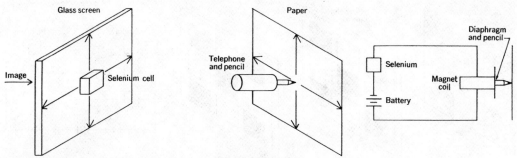

FIGURE 1. Constantin Senlecq's telectroscope, 1877. Facsimile system employing a single selenium cell and a pencil attached to a telephone diaphragm.

entered the literature. (The principle of synchronous control was used by Charles Wheatstone in the early telegraph days; see Refs. 10–13.)

The telectroscope

The idea of using selenium as a pickup element (the counterpart of the microphone) for transmitting graphic materials occurred to several workers around 1877. A French lawyer, Constantin Senlecq, was the first to publish his ideas that year.[14] (In a later account,[15] the claim is made that it "was invented in the early part of 1877.")

His plan for a telectroscope (see Fig. 1) included an early adaptation of Bell's telephone.

Senlecq supposed that his apparatus would reproduce the tonal shades of an image obtained in a camera obscura. He avoided any mention of the mechanical linkages by merely specifying "any system of autographic telegraphic transmission." His plan, although rudimentary, was advanced for the times; and had the merit of combining two new elements in an attempt to extend existing techniques for picture transmission.

A proposal for directly converting electric signals into

I. Chronology of facsimile and television proposals, 1877–1884

Year	Inventor	System	Circuit	Reproducer	Remarks
1877	Senlecq[1]	Fac.	Single	Telephone and pencil	Autographic
1878	De Paiva	—	—	Single incandescent element	Selenium-coated plate
1879	Redmond[1]	TV	Multiwire	Incandescent mosaic	Selenium mosaic
	Redmond[2]	—	Single	Single incandescent element	Single selenium cell
1879	Perosino	Fac.	—	Electrochemical, single-point	Paper record on cylinder
1880	Middleton	Multipurpose	Multiwire	Thermoelectric mosaic	
1880	Ayrton, Perry[1]	TV	—	Apertured mosaic	Electromagnetic shutters
	Ayrton, Perry[2]	—	—	Silvered magnetic mosaic	Plane-polarized light, Kerr effect, analyzing screen
1880	Carey[1]	Fac.	—	Electrochemical, multipoint	Selenium mosaic, wire mosaic
	Carey[2]	TV	—	Incandescent mosaic	Selenium mosaic
	Carey[3]	Fac.	Single	Electrochemical, single-point	Spiral scan, clockwork drive
1880	Sawyer	TV	—	Spark gap	Spiral scan, selenium helix, light pipe
1881	Senlecq[2]	Fac.	—	Electrochemical, multipoint	Selenium mosaic, wire mosaic
	Senlecq[3]	TV	—	Incandescent mosaic	Mechanical selector, rotary switch, synchronizing pulses
1881	Bidwell[1]	Fac.	—	Electrochemical, single-point	Pinhole cylinder
	Bidwell[2]	—	—		Pinhole box, linear scan, flyback
1882	Lucas	TV	—	Optical. Polarized light beam on screen	Beam modulation, beam scanning, continuous scan, receiver only
1884	Nipkow	—	—	Electrooptical. Polarized light and direct viewing	Apertured disks, Faraday effect, automatic flyback, continuous scan

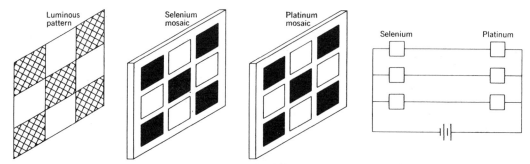

FIGURE 2. Denis Redmond's electric telescope, 1879. Multiwire system with a mosaic of incandescent elements for visual reception.

FIGURE 3. Ayrton and Perry's second proposal for seeing by electricity, 1880. A receiver mosaic of electromagnets with silvered surfaces on the ends of the cores is flooded with plane-polarized light and viewed through an analyzing prism. The plane of polarization of the reflected light is rotated in proportion to the current in the respective circuit.

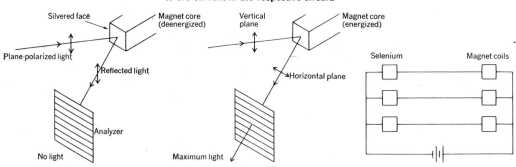

light for reproduction purposes soon appeared. Realizing that an incandescent element could serve as the converse of selenium, a Portuguese physics professor of Oporto, A. de Paiva, outlined his plan for an "electric telescope" early in 1878.* De Paiva suggested projecting an image onto a selenium-coated plate that was scanned by a metal point. Signals from this transmitter were to operate a relay and an incandescent element at the receiver. The motions of the point and the lamp were to be effected by the usual autographic means.

Selenium and platinum mosaics

The relationship between the structure of the ear and its telephonic model and the comparable imitation of the eye in possible apparatus for "electric vision" now began to emerge. An obscure inventor, Denis Redmond† of Dublin, revealed his plans for "transmitting a luminous image by electricity" early in 1879.[17] In his "electric telescope," Redmond employed mosaics patterned after

* De Paiva's proposal was not widely known, apparently, since there are no contemporary accounts in English literature. Another little-known proposal employing selenium for "telephotography" was put forward in 1879 by C. M. Perosino. His plan, employing sequential scanning, single-line transmission, and chemical recording on a revolving drum, is probably the first of its kind.[16]

† Little is known about Redmond. Recent inquiries in Dublin and London have not brought to light any information to supplement his published accounts.

the eye; they consisted of "a number of circuits, each containing selenium and platinum arranged at each end, just as the rods and cones are in the retina."

Redmond's multicircuit plan, shown in Fig. 2, eliminated problems of scanning and synchronism. A luminous pattern was projected onto the selenium mosaic. The bright parts of the image were reproduced by incandescence of the respective platinum elements in the receiver mosaic. There were no moving parts and the optical system was nothing more than a simple camera lens. Although the plan contains only the barest essentials, it is a true television system and appears to be the first of its kind employing selenium.

Unlike other inventors of the period who produced no more than paper plans, Redmond claimed to have done actual experiments. It seems that his efforts were rewarded, because he declared: "I have succeeded in transmitting built-up images of very simple luminous objects." However, he did not describe how the mosaics were made, nor how to prepare selenium or construct the platinum elements.

Redmond, recognizing the practical barrier of a multiwire circuit, also considered a single-wire system. But this introduced the greater problem of synchronizing the motions of the cell and the incandescent element. He therefore proposed to adopt the principle of the copying telegraph, but he did not give any details of this experiment. Like de Paiva, Redmond recognized the importance of persistence of vision and the need for a minimum scanning frequency. He proposed to arrange his system "so that every portion of the image of the lens should act on the circuit ten times in a second, in which case the image would be formed just as a rapidly-whirled stick [brand] forms a circle of fire."

Redmond was also aware of the basic defect of selenium as a light-sensitive element: the sluggishness of response that quickly darkened the inventive horizon. This inertia plagued inventors for many years. Consequently, Redmond's attempt to employ a single circuit "failed through the selenium requiring some time to recover its resistance." (He also stated: "I am at present on the track of a more suitable substance than selenium." His proposal elicited several letters two weeks later[18] and another one in May 1880 asking for further results.[19])

The photophone and its consequences

The potentialities of selenium as a link between electricity and light, as well as its physical properties, were now of great interest. A quite different line of development—the use of a light beam to convey audible tones—was taken up by several workers, including Graham Bell, during the late 1870s. Bell became interested in discussions on the possibility of hearing the sound of light via selenium and the telephone during a visit to England in 1878. Upon returning to the United States in the autumn, he started work on a new telephonic system that employed a beam of voice-modulated light instead of wires between transmitter and receiver.

An extensive series of experiments by Bell and his co-worker Sumner Tainter with selenium cells and "undulatory light" led to the "photophone."[20] Then, before it was practically applied, Bell deposited the related documents with the Smithsonian Institution. This unusual act aroused great curiosity among scientists and stimulated others who suspected that he had developed a system for "seeing by telegraph." The increase in correspondence on this subject and in reports of experiments during 1880—the boom year for incipient television—appears to be a direct result of news and speculations about Bell's new "visual telegraph."

An elegant scheme

Two British professors, William Edward Ayrton and John Perry, referred to Bell's papers when they announced their plan for "seeing by electricity" in Nature, in April 1880.[21] They recognized the practical difficulties, however, and merely advanced suggestions, stating: "It has not been carried out because of its expensive character, nor should we recommend its being carried out in this form." They also declared that "it is well to show that the discovery of the light effect on selenium carries with it the principle of a plan for seeing by electricity." But this pronouncement was rather tardy because they had privately discussed their ideas "some three years" before; besides, Redmond had already published his plan. [Ayrton and Perry wrote, "suggested to us some three years ago more immediately by a picture in Punch." However, their memories were at fault, since the only cartoon published during the period from July 1875 that clearly applies is one in Punch's Almanack for 1879, dated December 9, 1878. This prophetic illustration, showing a wide-view screen, has the imaginary

and somewhat derisive caption, "Edison's Telephonoscope (Transmits Light as Well as Sound)." Their announcement evidently prompted Redmond to call public attention to his plan in a letter to the Times,[22] in which he mentioned "a relay of peculiar construction" and pointed out that he had not patented his apparatus.]

Ayrton and Perry had two multicircuit schemes, both employing transmitter mosaics "made up of very small separate squares of selenium." Their first receiver was an apertured mosaic with light arranged to pass through the openings. A needle, magnetically operated, obscured each opening. An increase of light on a given cell would increase the line current to the respective coil. Thus, the needle would be pulled aside and admit more light through in proportion to the original light intensity.

Their second scheme was boldly based upon a recent advance in physics. In 1845, Michael Faraday passed a beam of plane-polarized light through a block of heavy glass mounted in the pole gap of an electromagnet.[23] He found that the plane of polarization was rotated when the magnet was energized (Faraday effect). In 1877 John Kerr, a Scottish physicist, discovered a related effect when plane-polarized light is reflected from the poleface of a magnet.[24]

The co-inventors felt that Kerr's experiment suggested "a more promising arrangement." Their second receiver plan, shown in Fig. 3, consisted of electromagnets arranged in a mosaic pattern. Silvered soft-iron squares were attached to the ends of the cores. This mosaic was to be flooded with plane-polarized light and viewed through an analyzer. With the transmitting mosaic dark, the magnets would not be energized and the reflected light would be cut off from view. With sufficient light on a given cell, the reflected beam from the respective magnet would pierce the analyzing prism and present that portion of the image with a corresponding intensity.

Although Ayrton and Perry were fully occupied in the new field of electric power, the possibility of seeing by electricity stayed on their minds. A rudimentary model of a somewhat different plan was later used by Perry for lecture demonstrations. This instrument, capable only of reproducing alternate stripes of light and shade, was called a "telephote," or "pherope."[25]

Perry was an early prophet of radio and television. He foresaw communications without wires before the pioneer work (1886–88) of Heinrich Hertz on electromagnetic waves was widely known. Concerning "the people of one hundred years hence," he wrote: "They will probably speak to one another at a distance without any artificial connection between." He also declared: "They will probably be able to see one another's actions at great distances, just as if they were close together."[26]

A thermoelectric plan

In March 1880, a different scheme was announced by Henry Middleton of St. John's College, Cambridge.[27] He spurned selenium, however, and suggested a multipurpose plan employing thermoelectric elements arranged as mosaics and connected by a multiwire circuit.

Middleton, like Redmond, pointed out "a striking analogy between the camera of the instrument and that of the human eye." He also equated "the conducting system" with the "optic nerve." He believed that his plan would be adaptable for all kinds of reproductions. Through the agency of heat, he supposed that "these

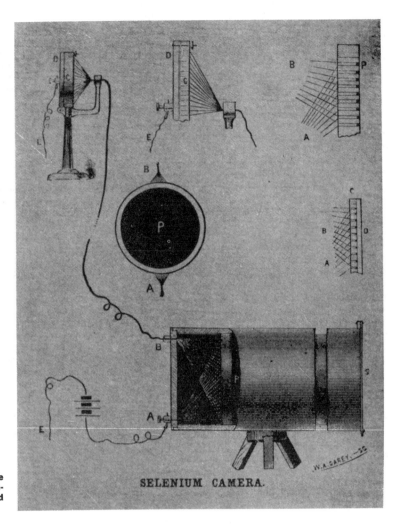

SELENIUM CAMERA.

FIGURE 4. Carey's multiwire scheme for seeing by electricity, 1880. (Reproduced from Scientific American.)

FIGURE 5. Carey's first plan (A) with a receiver mosaic of wire points for electrochemical recording. Alternative proposal (B) with a mosaic of incandescent elements for visual reception.

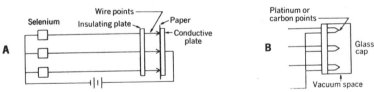

images can be either received directly or by reflected light...and projection on a screen...or by suitable apparatus they can be retained as a photograph, a thermograph, or chemicograph."

Two American plans

Meanwhile, U.S. inventors had not been idle in exploring the new field, nor immune from the enticements of selenium. Comprehensive plans for two schemes appeared in the *Scientific American* early in June 1880.[28] The inventor, who had originally submitted his ideas in the spring of 1879,[29] was George R. Carey, of Boston. The article contained engravings that portrayed a sele-

ium camera, probably the first illustration of its kind; see Fig. 4. Indeed, Carey was the first to publish "constructional" details. The careful delineation of the parts gives the impression that the apparatus was built, although there is no evidence to support this.

In his first scheme, Carey employed a circular mosaic of selenium elements connected by separate wires to a similar mosaic of wire points at the receiver. A sheet of chemically prepared paper was inserted between these points and a metal plate. See Fig. 5.

To furnish a visual image, Carey proposed an incandescent mosaic as a substitute for the wire array at the receiver. Platinum or carbon "points" were mounted

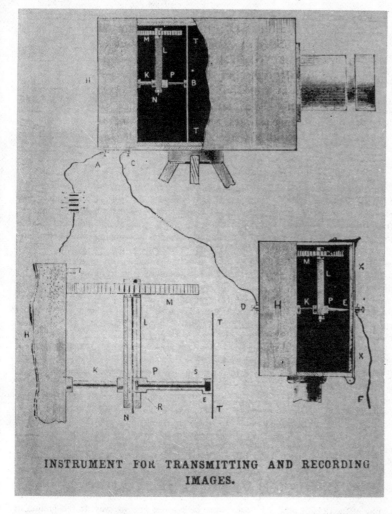

INSTRUMENT FOR TRANSMITTING AND RECORDING
IMAGES.

FIGURE 6. Carey's instruments using spiral scanning for transmitting and recording images along a single-wire circuit. (Reproduced from Scientific American.)

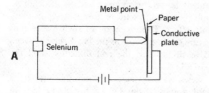

FIGURE 7. Carey's single-circuit plan for electrochemical recording (A). The selenium cell and the metal point, both operated by clockwork, were arranged to travel a spiral path (B).

in an evacuated space "covered with a glass cap." The inventor confidently stated: "These points are rendered incandescent by the passage of the electric current, thereby giving a luminous image instead of printing the same."

Carey, realizing the practical objection to a cable between the camera and the distant station, proposed another method employing a single line. In this second scheme (see Fig. 6), he introduced the novel idea of spiral scanning. As shown in Fig. 7, a single selenium cell in the camera was caused to "describe a spiral line upon the glass, thus passing over every part of the picture" projected on it. The receiver contained a metal

plate carrying a sheet of chemically prepared paper over which a metal point traced a similar spiral path. A clockwork-driven mechanism in each instrument produced the motions of the cell and the point. Since this single-circuit plan was intended only "for transmitting and recording," the inventor did not advance an alternative suggestion for luminous reception. Both proposals refer to a "one-shot" scan.

Except for the alternative suggestion of incandescent elements, both of Carey's schemes relate to facsimile. However, he is often given credit for the first suggestion of a "television" system in 1875. These later accounts[30] describe the mosaic plan with multiwire connections

FIGURE 8. William Sawyer's proposal for transmitting visual images, 1880. The light pipe traverses the image and a flat helix of selenium wire in a spiral path. A spark gap carried by an index describes a similar path in a darkened case at the receiver.

and incandescent elements. They also variously refer to selenium cells or a light-sensitive mosaic of photographic materials, with or without line relays at the receiver. But this disclosure of June 1880 appears to be the first published by Carey related to an electrical system employing selenium.

The rather tenuous link between published reports, private plans, and the growth of technical ideas was strengthened by a quick response to Carey's article published in the *Scientific American* the following week. The issue for June 12, 1880, carried the account[31] of another scheme by William Edward Sawyer, an electrical engineer of New York. This contributor, also a telegraph inventor and pioneer of electric lamps, revealed that his plan dated back to the fall of 1877. At that time he had described to business acquaintances "the principles and even the apparatus for rendering visible objects at a distance through a single telegraphic wire." Interestingly, Sawyer was concerned only with a visual system, although his earlier work was on facsimile. He also proposed spiral scanning, novel optomechanical arrangements in both instruments, and a spark-gap receiver; see Fig. 8.

The transmitter contained a flat helix of fine selenium wire mounted in a darkened case. Light from the image was piped into this case through a fine tube. In an unspecified manner, the tube was arranged to traverse the image and the selenium coil, starting at the outside and finishing at the center. The receiver was vaguely described as a similar darkened case in which an "index" traced a spiral path identical to that of the light pipe, or scanning tube, at the transmitter. The index carried fine platinum points connected to the secondary of a "peculiar induction coil." Tiny sparks between these points were supposed to reproduce the instantaneous level of light of the original image.

Sawyer gave no other details of his plan but he did describe its operation. Like Carey's, this was a one-shot scan, with no mention of repetitive action. Recognizing the need for rapid motion consistent with the persistence of vision, he wrote: "The speed being sufficiently great it is obvious that...an exact image of the object... would be reproduced before the eye of the observer placed at the darkened chamber of the receiver."

Sawyer was aware of the deficiencies of selenium: its sluggishness and inadequate sensitivity. He understood the absolute need for isochronous motion and declared that "isochronism is unattainable." He also recognized that an acceptable reproduction would require a vast number of individual points (picture elements) accurately registered to provide reasonable detail (definition).

The first to state this problem, Sawyer declared (with

reference to about 6.5 cm² of surface): "To convey with any accuracy an image...this surface should be composed of at least 10 000 insulated selenium points." (This figure represents a good-quality halftone screen.) Also, discounting single-line scanning, and with alignment of the picture elements in mind, he added that these points were to be "connected with as many insulated wires leading to the receiving instrument; for the variation of the one-hundredth of an inch either way will 'throw a line out of joint.'"

Sawyer was a severe critic of any scheme, including his own. Pessimistically he stated: "There is no likelihood of any plan of this kind ever being reduced to practice, for some of the difficulties in the way of all of the plans are insuperable." Thus, the problems facing any would-be inventor of a practicable system for transmitting visual images were now emerging from the mists of thought. A fast and sensitive pickup device and an equally reponsive visual reproducer were essential. Practical mosaic arrays would have to contain large numbers of these elements.

The superiority of sequential signals along a single line also was becoming recognized. This approach eliminated cumbersome and costly cables, but introduced the greater problems of constant speeds and synchronous scanning. The related need for repetitive scanning—still a vague idea—at a frequency compatible with persistence of vision had to be satisfied in some way. In recognition of these fundamentals, any proponent had to incorporate adequate optical and mechanical means to put these principles into effect. The invention of such means that would bring any scheme closer to the borderline between fantasy and feasibility was the real difficulty—and the greatest one of all.

Sawyer's views were offset a few months later by a more cheerful opinion from Oliver Joseph Lodge, an English physicist. In December 1880, he lectured on "The Relation Between Electricity and Light" at the London Institution.[32] In concluding his lecture, Lodge said: "I must just allude to what may very likely be the next striking popular discovery...the transmission of light by electricity; I mean the transmission of...views and pictures by means of the electric wire."

Scheme for a fugitive picture

Early in 1881 Senlecq made his second contribution to the inventive pool with a plan[15] that incorporated some notable features already proposed by the American inventors. His new "telectroscope" was a conglomeration of ideas based upon the dial, or step-by-step, telegraph introduced in 1840.[10-13,33] These codeless systems, also known as the "printer" or "ABC" tele-

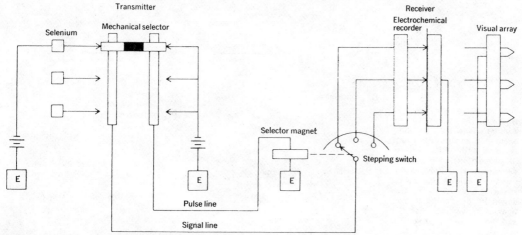

FIGURE 9. Senlecq's double-line facsimile system employing mosaics and contact selectors. An incandescent array was suggested for obtaining a transient visual image.

graph, employed lettered dials. Pulses were sent along the line as the dial was set to a given character.

Senlecq's plan included the first clear proposal for effecting positive synchronism in a picture transmission system. The heart of his apparatus, shown in Fig. 9, was a mechanical selector, or "rectangular transmitter," as he called it; a cumbersome device that permitted only a one-shot scan. The corresponding portion at the receiver connected the signal line to a mosaic array.

The receiver mosaic consisted of wire points and a metal plate with a sheet of paper between them for chemically recording the impression. The receiver elements were individually connected to contacts on a circular disk, or rotary switch. This switch was stepped by pulses along a second line, the pulses being controlled by the duplicate half of the transmitter selector. This selector, mounted vertically, was furnished with a two-circuit slider. Thus, during its motion under gravity, picture elements were connected sequentially to the signal line "intended to conduct the various light and shade vibrations."

Senlecq believed the contacts ought to "insure the perfect isochronism of the transmitter and receiver." However, it seems he ignored the limit to the rate of travel that would be imposed by the inertia of the receiver mechanism. Nevertheless, he firmly stated: "The picture is, therefore, reproduced almost instantaneously."

This plan is primarily a facsimile system. But, to keep up with the modern trend, Senlecq suggested an alternative visual receiver. He adopted Carey's idea of platinum elements and, to allow for contingencies, included Sawyer's induction coil. He described the operation in fanciful terms: "By the incandescence of these wires according to the different degrees of electricity we can obtain a picture, of a fugitive kind, it is true, but yet so vivid that the impression on the retina does not fade during the relatively very brief space of time the slide occupies in travelling over all the contacts." Senlecq was the most persistent of these pioneers. He kept faith

with his dream for sending visual images over wires, and returned with another scheme in 1907.[34]

Interest in the properties of selenium increased during 1880–81, with emphasis on research on obtaining crystal-line selenium in the most photosensitive form and on constructing practical cells. A well-known investigator, Shelford Bidwell, lectured on "Selenium and Its Applications to the Photophone and Telephotography" at the Royal Institution in March 1881.[35] Bidwell described various forms of selenium and their qualities, commented on the results of work done by others, emphasized selenium's capricious behavior, and gave details of his own experiments.

Bidwell was intent on phototelegraphy and made no reference to visual reception. His instrument, simply designed for demonstrations, was essentially the same as Bakewell's of 1847, except that selenium was used in the transmitter instead of an on–off contact. The cell was mounted inside a rotating cylinder with a pinhole midway between the ends. A fixed image about 5 cm square was projected onto the side of this cylinder. The receiver cylinder carried a sheet of chemically prepared paper with a platinum style in contact. The cylinders were coupled mechanically and arranged to move along their axes. As they slowly revolved, the pinhole and point scanned the image and paper, respectively.

This demonstration appears to be the first in public to show successfully the potentialities of selenium for picture transmission. Bidwell had faith in his work, for he made several improvements and devised another version during the next few months. He replaced the transmitter cylinder with a cam-operated box so arranged that the pinhole scanned the image vertically with a flyback between adjacent "lines."[36]

Bidwell was most active during 1881 and retained a keen interest in phototelegraphy for many years. Interestingly, his critical attitude in 1908 toward early forms of television[37] prompted the first bare suggestion for the "employment of two beams of kathode rays" that foreshadowed all-electronic television.[38-41]

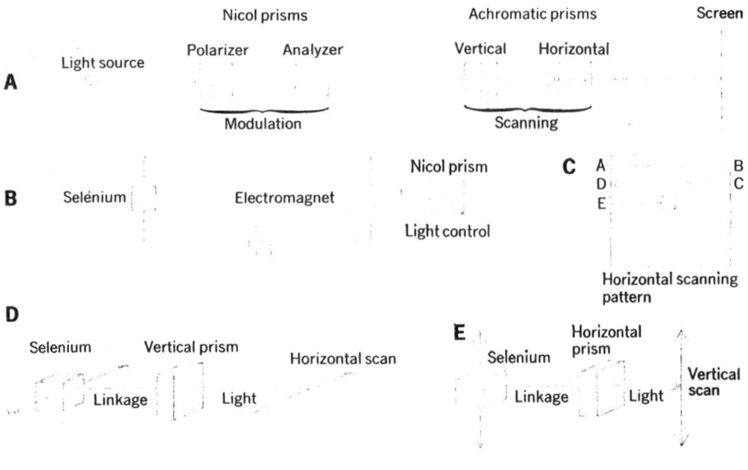

FIGURE 10. A telectroscope receiver (A) proposed by William Lucas, 1882. The light beam is modulated by rotating one of the Nicol prisms (B) proportional to the instantaneous level of light on the selenium cell. A horizontal scanning pattern (C) is produced by the combined motions of the vertical prism (D) and the horizontal prism (E).

An optical receiver

The report of Bidwell's lecture in the *English Mechanic* evidently stimulated one reader, William Lucas,* to pursue his ideas on apparatus for seeing by electricity. In a letter on "The Telectroscope,"[42] published in the same magazine in April 1882, he said he had "lately been giving a good deal of thought to this subject." He clearly saw his objective and asserted that "provided..the practical difficulties can be overcome...an image in light and shade will be formed upon a screen..." However, he was concerned only with a receiver, which he believed could be used with a sequential transmitter, such as Bidwell's cylinder model. Although it was impracticable and only a partial solution to one half of a system, this proposal is the first to incorporate all the features essential for the reception of *continuously moving images.*

Lucas depended wholly upon optical methods. His scheme, shown in Fig. 10, has two important new features: beam modulation and beam deflection by oscillating scanning prisms.[43] Direct lighting from a local source and a moving light spot on a screen are also novel. A strong beam of light from a lantern was projected through a pair of Nicol prisms and then through a pair of ordinary achromatic prisms to illuminate a small spot on a screen.

The Nicol prisms (one the polarizer, the other the analyzer), with one of them rotatable through 90 degrees, comprised the optical modulating system. At extreme settings, light would either pass or be cut off; in between, the intensity of the beam would be proportional to the relative positions. One achromatic prism was placed vertically, the other horizontally, with both arranged to be partially rotatable about their axes. Together they comprised an optical scanning system. When the vertical prism was turned the spot of light would move across the screen; similarly, movement of the horizontal prism would swing the light beam up and down. In considering these parts together, Lucas declared: "Hence, by this arrangement, we can vary the position of the spot of light upon the screen, and augment or diminish its brightness at will."

Lucas suggested an electromagnet for turning one of the Nicol prisms. He gave no mechanical details nor did he suggest ways to operate the scanning prisms. However, he clearly envisioned the total functions and accurately described the operation: "The spot not only moves synchronously with the selenium cell, but its brightness also varies as the brightness of the portion of the image which the cell receives upon it." He described and illustrated the required motions with reference to a back-and-forth linear scan with horizontal traces. He also pointed out that the cell and the spot of light, on completing the last trace at the bottom, would return to the start at the top left-hand corner. This type of scan is the predecessor of today's "raster and flyback."†

A remarkable similarity exists between this optical receiver and the high-vacuum cathode-ray tube. With the electrostatic type, for example (ignoring focusing methods), the essentials are: light source—hot cathode; light beam—electron beam; Nicol prisms—biased grid; rotatable prisms—deflection plates; light screen—fluorescent screen. But the basic cathode-ray oscilloscope was yet to be invented; by Ferdinand Braun in 1897.

There is an interesting sequel to the Lucas proposal. He wondered whether the resistive changes would be sufficiently rapid and of sufficient magnitude to effect control of the Nicol prism. A reply by Llewelyn B. Atkinson appeared in the same magazine two weeks later.[44] Commenting on the scheme, he thought it was "certainly most ingenious, and...in theory is all right." But he also clearly pointed out the main problem: "In practice I am afraid the inertia of the moving parts connected with the Nicol's prism would render it impracticable." These words evidently dismayed Lucas, because he apparently dropped his plans.

Atkinson is said to have devised a rotating drum

---

*Little is known about Lucas. Recent inquiries in London have not brought to light any information to supplement his published accounts.

†Bidwell's pinhole box was the first transmitter employing a linear scan with flyback. Lucas referred to Bidwell's "parallel lines close together." However, since the Lucas scan required a flyback only between frames, his receiver was not compatible with the Bidwell transmitters.

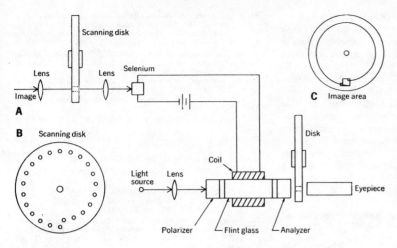

A

B

C

FIGURE 11. Paul Nipkow's plan (A) for an electric telescope, 1884. Two identical apertured disks (B) rotate in synchronism ten times per second. An image is dissected by the transmitter disk to produce sequential signals. The light valve at the receiver is based on Faraday's magnetooptic effect. Light from a local source is modulated in proportion to the line current. The image area (C) is viewed through an eyepiece.

with a series of peripheral mirrors to serve for sequential and repetitive scanning. This was the same year (1882), but he did not publish an account at that time. A similar proposal was made in 1889 by Lazare Weiller, who is generally given credit for this.[45] Late that year a note by Atkinson appeared in a London electrical journal.[46] With reference to "an integrating apparatus for producing the whole image," he observed that the idea "was first, I believe, published in the *English Mechanic* about 1881 or 1882, and was then, as far as publication goes, new."

A faint echo of these distant days returned in June 1936. In a brief note in *Nature*, on the scanning principle, Lucas quoted Atkinson's remarks and pointed out that "The communication thus referred to . . ." was his own.[47]

The master television patent

The process of exploring an image to obtain a sequence of electrical values proportional to the respective light values (and the reverse for reception) was now becoming a basic idea. However, regardless of the scanning means, coordinated motions of a complex order appeared to be essential. A suitable mechanism had to be delicate, precise, constant, and effective at speeds that would provide visual continuity. Although such a mechanism was a vital objective, it was often the least considered; perhaps because it presented the greatest difficulty in the way of a practical solution. Thus, for mechanical reasons, the dream of "seeing by electricity" was still unrealized.

These difficulties were overcome by Paul Nipkow, a student of natural science in Berlin. His solution was exceedingly simple and fundamentally sound—a spinning perforated disk. During Christmas 1883, Nipkow experimented with a disk perforated with a spiral of small holes near the edge. With the disk rotating, each hole revealed the field of view in consecutive strips; or line by line, as in reading. Of course, the process was reversible and identical disks could serve for the transmitter and receiver: one to dissect and the other to reconstitute the image. In a remarkably short time he selected the other elements required and prepared the details for a complete system. He filed his plans for an "electric telescope"

early in January 1884.[48]

Nipkow's scheme is shown in Fig. 11. The transmitter consists of fixed lenses, a scanning disk, and a selenium cell. The composite light values of an image are displayed on the disk. As it rotates, each hole acts like a sliding shutter. Individual beams of light from each aperture are presented to the selenium cell in a continuous series. The "flyback" between each line and between each complete scan (or frame) is automatic. The disks contained 24 holes or lenses and were to rotate synchronously and in phase ten times per second.

While Nipkow thus reduced mechanical scanning in both instruments to its simplest form he also avoided any other moving parts at the receiver. He adopted Faraday's magnetooptic effect as the basis for a "light valve" to modulate the light beam at the receiver. This consisted of a block of flint glass placed inside a magnet coil with a polarizing prism at each end.

Light from a local source, polarized in one plane by the first prism, is blocked by the second prism. Current through the coil is proportional to the conductivity of the selenium. A high level of light at the transmitter therefore increases the strength of the magnetic field and rotates the plane of polarization accordingly. The light passing through the analyzer then has an intensity proportional to the original picture element. The spinning disk distributes these contiguous light values over the image area in the eyepiece with sufficient rapidity to create the illusion of a complete picture.

Nipkow's scheme, simple yet technically elegant, ended a seven-year quest for sending visual images over wires. Mechanical problems related to rapid scanning and synchronism, hitherto "insuperable," were neatly solved by a spinning, apertured disk. The master patent in the television field was issued to Nipkow on January 15, 1885. By all accounts, no apparatus was constructed nor did Nipkow attempt to exploit his invention; the patent lapsed a few years later.

A wide variety of television proposals appeared during the next 40 years for both wire and radio transmission. All of the early methods, and some new ones, were tried—spinning disks, drums, and wheels; rotating mirrors, prisms, and lenses; vibrating mirrors, apertured belts, light banks, and commutators—in attempts to realize "distant electric vision." Nipkow's disk finally prevailed in 1926 when reception of moving images in halftones was first demonstrated by John Logie Baird in London.[49]

During the following seven years, mechanical transmission apparatus was highly developed to its practical limits, particularly in England, Germany, and the United States.[50,51] From 1930, the picture tube replaced all moving parts in receivers. Then, with the introduction of practical camera tubes during the early 1930s, and associated radio techniques, mechanical parts were totally replaced and high-definition television became a reality.

REFERENCES

1. Smith, W., "The action of light on selenium," *J. Soc. Telegraph Engineers*, vol. 2, pp. 31–33, 1873; vol. 5, p. 183, 1876; *Am. J. Sci.*, vol. 5, p. 301, 1873.

2. Garratt, G. R. M., and Mumford, A. H., "The history of television," *Proc. IEE*, vol. 99, pt. IIIA, pp. 25–42, 1952. Most accounts of this discovery are incorrect. The original correspondence given on pages 25–26 describes the circumstances.

3. Barnard, G. P., *The Selenium Cell*. London and New York: Constable, and Richard R. Smith, 1930, pp. 10–17.

4. Wilson, J. C., *Television Engineering*. London: Pitman, 1937, pp. 1, 11.

5. Gernsback, H., "Television and the telephot," *Modern Electrics*, Dec. 1909.

6. Jones, C. R., *Facsimile*. New York: Rinehart, 1949, pp. 1–23.

7. Korn, A., and Glatzel, B., *Handbuck der Phototelegraphie und Telautographie*. Leipzig: Otto Nemnich, 1911.

8. British patent 9745, 1843.

9. Bain, A., "Automatic telegraphy," *Eng. Mech.*, vol. 2, pp. 273–274, 1866. In this account, from a paper read to the Society of Arts on January 17, 1866, the inventor describes his system and subsequent improvements.

10. Prescott, C. B., *History, Theory, and Practice of the Electric Telegraph*. Boston: Ticknor and Fields, 1866, pp. 160–168.

11. King, W. J. "The development of electrical technology in the 19th century: 2. The telegraph and the telephone," Smithsonian Institution, Washington, D.C., 1962, pp. 291–296.

12. Marland, E. A., *Early Electrical Communication*. London: Abelard-Schuman, 1964, pp. 94–98.

13. Walmsley, R. M., *Electricity in the Service of Man*. London: Cassell, 1890, pp. 793–794.

14. "The telectroscope," *Eng. Mech.*, vol. 28, p. 509, 1879; *Sci. Am.*, vol. 41, p. 143, 1879.

15. "The telectroscope," *Eng. Mech.*, vol. 32, pp. 534–535, 1881; *Sci. Am. Supp.*, vol. 11, p. 4382, 1881.

16. Barnard, G. P., *op. cit.*, pp. 264, 288.

17. Redmond, D. D., "An electric telescope," *Eng. Mech.*, vol. 28, p. 340, 1879.

18. *Eng. Mech.*, vol. 28, pp. 585–586, 1879.

19. *Eng. Mech.*, vol. 31, p. 235, 1880.

20. Bell, A. G., "On the production and reproduction of sound by light: the photophone," *Proc. Am. Assoc. Adv. Sci.*, vol. 29, pp. 115–136, 1880; *Engineering*, vol. 30, pp. 253–254, 407–409, 1880, vol. 32, pp. 29–33, 1881; *Nature*, vol. 23, pp. 15–19, 1880; *Proc. IRE*, vol. 50, pp. 1995–1996, 1962; see Ref. 3, pp. 257–258, for numerous other references.

21. Ayrton, W. E., and Perry, J. J., "Seeing by electricity," *Nature*, vol. 21, p. 589, 1880; *Times (London)*, p. 13, Apr. 22, 1880; *Eng. Mech.*, vol. 31, pp. 177–178, 1880.

22. *Times (London)*, p. 10, May 13, 1880.

23. *Phil. Trans.*, 1846; also *Experimental Researches*. London: R. Taylor and W. Francis, 1855, vol. 3, pp. 1–4.

24. *Phil. Mag.*, ser. 5, vol. 3, p. 321, 1877; vol. 5, p. 161, 1878.

25. Walmsley, R. M., *op. cit.*, pp. 782–784.

26. *Ibid.*, p. XXV.

27. *Eng. Mech.*, vol. 31, p. 178, 1880; described at a meeting of the Cambridge Philosophical Society on March 8, 1880.

28. Carey, G. R., "Seeing by electricity," *Sci. Am.*, vol. 42, p. 355, 1880; reported also in *Eng. Mech.*, June 18, 1880, and in *Design and Work*, June 26, 1880.

29. "The telectroscope," *Sci. Am.*, vol. 40, p. 309, May 17, 1879.

According to this short editorial notice, Carey's plans were already fairly complete. "We hope to present to our readers before long the details of these interesting instruments." The reason for the delay (unusually long for those days) is not known.

30. See, for instance, Swinton, A. A. C., "Television, past and future," *Discovery*, vol. 9, p. 337, 1928; Hogan, J. V. L., "The early days of television," *J. SMPTE*, vol. 63, pp. 169–173, 1954; Abramson, A., *Electronic Motion Pictures*. Berkeley and Los Angeles: Univ. Calif. Press, 1955, p. 4; Zworykin, V. K., Ramberg, E. G., and Flory, L. E., *Television in Science and Industry*. New York: Wiley, 1958, p. 4.

31. Sawyer, W. E., "Seeing by electricity," *Sci. Am.*, vol. 42, p. 373, 1880.

32. Lodge, O. J., *Modern Views of Electricity*. London: Macmillan, 1907, p. 358.

33. Wheatstone and Cook, British patent 8345, 1840.

34. "The Senlecq telectroscope," *Sci. Am. Supp.*, vol. 64, pp. 372–373, 1907.

35. *Proc. Roy. Inst.*, vol. 9, pp. 524–535, 1881; also "Selenium and its applications to the photophone and telephotography," *Eng. Mech.* vol. 33, pp. 158–159, 180–181, 1881.

36. Report of the British Assoc. for 1881, p. 777; Lister, W. C., "The development of photo-telegraphy," *Electron. Eng.*, vol. 19, pp. 38–39, 1947; Ref. 2, pp. 27–28. This apparatus is in the Science Museum, London.

37. Bidwell, S., in a letter commenting on optimistic reports that "distant electric vision" would be possible within a year, *Nature*, vol. 78, pp. 105–106, 1908.

38. Swinton, A. A. C., "Distant electric vision," *Nature*, vol. 78, p. 151, 1908.

39. Garratt, G. R. M., and Mumford, A. H., *op. cit.*, p. 31.

40. McGee, J. D., "Electronic generation of television signals," in *Electronics and Their Application in Industry and Research*, B. Lovell, ed. London: Pilot Press, 1947, pp. 138–139.

41. Jensen, A. G., "The evolution of modern television," *J. SMPTE*, vol. 63, pp. 175–176, 1954.

42. Lucas, W., "The telectroscope, or seeing by electricity," *Eng. Mech.*, vol. 35, pp. 151–152, 1882.

43. Maurice Leblanc, French electrical engineer and inventor, is credited with a proposal for mirror scanning. *La Lumière Electrique*, vol. 2, p. 477, 1880; Ref. 2, p. 26.

44. Atkinson, L. B., *Eng. Mech.*, vol. 35, p. 194, 1882.

45. Weiller, L., *La Lumière Electrique*, vol. 34, p. 334, 1889.

46. Atkinson, L. B., *Telegraphic J. Electrical Rev.*, p. 683, Dec. 13, 1889.

47. Lucas, W., "The scanning principle in television," *Nature*, vol. 137, p. 981, 1936.

48. German patent 30 105, Jan. 6, 1884; Ref. 7, pp. 440–444.

49. Demonstrated to members of the Royal Institution on January 27 at 22 Frith Street, Soho. *Times*, p. 19, Jan. 28, 1926; Percy, J. D., *John L. Baird*. London: Television Soc., 1952, p. 5; Moseley, S., *John Baird*. London: Odhams, 1952, pp. 76–77. Original apparatus (1925) is in the Science Museum, London.

50. Wilson, J. C., *op. cit.*, pp. 389–419.

51. Abramson, A., *Electronic Motion Pictures*. Berkeley and Los Angeles: Univ. Calif. Press, 1955, pp. 51–64.

HISTORY OF MODERN TELEVISION

P. K. Gorokhov

HISTORY OF MODERN TELEVISION

P.K. Gorokhov
Member of the Society

The year 1961 is an anniversary in the history of television. In May of this year 50 years had passed since the world's first television transmission was carried out in our country by B. L. Rozing, and 30 years from the beginning of Soviet television broadcasting. In September of the past year, 30 years had gone by since the appearance of the first television transmitting tube with charge storage, now referred to as the iconoscope.[1] These events were important landmarks in the history of television and were highly instrumental in its subsequent development.

1. FROM THE SELENIUM PHOTOCELL TO THE ROZING TUBE

The discovery in 1873 of the photosensitivity of selenium [1] was one of the most significant moments in the history of television. The initial projects, appearing shortly after this event, regarding systems for the transmission of moving images between distant points were based on the application of selenium photocells for photoelectric conversion.

From the initial concept to the practical realization of television,

---

[1] The history of the invention of the iconoscope and the development of television transmitting tubes will be covered in a separate article.

nearly 50 years went by. The initial period in the history of television (1873-1920) was one marked by persistent efforts to uncover methods for solving a most complex problem. It was essential to surmount some tremendous technical difficulties before proceeding to the development of the first practical operational television system. At the then existing level of science and engineering, the problem of television could not be solved completely in the initial stage. For this reason, by means of a number of simplifications (rejecting, for the time being, the transmission of color and three-dimensional form of an object), it was reduced to the transmission of flat black-white images consisting of a given number of elements characterizing the different levels of brightness. The basic efforts were concentrated on the transmission of motion effects.

The first system for transmitting moving images, proposed in 1875 by G.R. Carrey [2], was based on simultaneous transmission of all elements of the image. The transmitting and receiving equipment in this system had to be connected by a number of communication lines corresponding to the number of elements in the image. This type of system was impractical because of its complexity even with low image definition (sharpness).

The majority of inventors at first rejected the idea of a multichannel system and proceeded to develop a single-channel system based on the application of image scanning and utilizing the principle of optical perception. Scanning represents a space-time process of sequential conversion of the image elements into electrical signals proportional to their brightness. In a single-channel transmission system these signals are sent in proper order along a single communication line.

The introduction of the scanning process resulted in significant simplification of the image transmission system, but at the same time it brought in the chief limitation of single-channel systems, which subsequently led to delays in the development of television. During the sequential transmission of the image elements, the period in which the light beam acts on the photoelectric converters decreases as the number of elements of definition increases, and therefore an increase in transmission clarity results in a catastrophic decrease in the image signal level. In contrast to a multichannel system which consisted purely of electrical elements without any mechanical devices, the single-channel systems required the use of more or less complex mechanisms for image scanning. In subsequent systems these mechanisms become more complicated and this resulted in increasingly complex and cumbersome transmitting and receiving apparatus.

The first television systems with image scanning were proposed in 1878-1880, independently, by the Portuguese physicist A. de Paiva [3], the French engineer M. Senlecq [4] and our own countryman, at that time a student, but subsequently a famous physicist and biologist, P.I. Bakhmet'-yev [5].

The telephotograph system designed by P.I. Bakhmet'yev, which laid the foundation for television developments in our country, differed from the systems by de Paiva and Senlecq not only in its improved circuit but also in the painstaking design of all of its parts. In place of a complex multi-element panel of photoconductive cells in the transmitting equipment and a corresponding number of off-on souces of light in the receiver, he used only a single selenium resistor revolving along a spiral path for image scanning during transmission, and a single light source of original design in the receiving equipment. This light source, which possessed a high spot brightness modulated by the image signals, also moved along a spiral path. Bakhmet'yev's telephotograph system contained the principles which characterized the television systems proposed in subsequent years.

An important step toward a practical solution of the problem of television appeared in the invention in 1884 by P. Nipkow (Germany) of a simple optical mechanical system for scanning and reproducing television images [6]. The basic element in the transmitter and receiver of this system was a scanning disc with holes, designated as the Nipkow disc in the history of television. There were well known similarities in the scanning principles employed in the Bakhmet'yev system (with the photocell moving along a spiral path) and Nipkow's system (a disc with spirally-arranged openings rotating relative to a fixed photocell).

Subsequently, a large number of optical mechanical television sets, with scanning elements in the form of a disc with lenses, mirrored wheels, lense type drums, etc., made their appearance, using selenium for the photoelectric converters. Among them we may include the interesting systems proposed in our country by M. Vol'fke in 1898 and by the engineer A.A. Polumordvinov in 1899.

M. Vol'fke [7] obtained a license (patent) on a device for the electrical transmission of images without wires [7]. The image definition during transmission was achieved by means of a Nipkow disc. The image signal taken from a selenium plate excited an inductance coil with the secondary winding containing a vibrator radiating electromagnetic waves. These waves were caught at the receiving point by means of rods connected to a Geissler tube located behind the rotating Nipkow disc. The proposal made by Vol'fke had two interesting features: the transmission of image signals without wires, and the use of an inertialess light-source in the receiver — a gas-discharge tube.

Polumordvinov's system [8] was designed not only for the transmission of black-white but also of color images. During black-white transmission the scan was achieved by means of two discs with slots, rotating at different angular velocities. During color transmission, instead of the discs it was proposed to use two concentrically arranged cylinders with slits covered consecutively by red, green and violet light filters. This, in essence, was the first system of color television with sequential transmission of the colors.

Some of the proposed optical mechanical systems during the initial period in the development of television (P.I. Bakhmat'yev, P. Nipkow, and Ya. Shchepanik) made it possible, in principle, to also carry out the transmission and reception of moving images but none of them was actually constructed and tested in operation. The feasibility of television transmission at the beginning of the 20th century was not yet proved in practice. The principal obstacles in this connection were the inertia of the selenium photoconductive cells and the lack of amplifiers for weak image signals.

The optical mechanical television method which first appeared in the history of television was not the only possible method. Even during the initial stage in the development of mechanical television, far-sighted scientists and researchers were seeking new methods for solving the problem of television, excluding the use of optical mechanical devices.

The first scientist to make a critical evaluation of the feasibility of mechanical television was an instructor in the Petersburg Technological Institute, Boris L'vovich Rozing.

B.L. Rozing was born on April 23, 1869, in Petersburg. In 1887, after completing his education in the gymnasium, he enrolled in the Physics Mathematics Faculty of Petersburg University, which he completed in 1891, and then was retained in a physics chair (professorship) for further training in scientific educational work. In 1893 Rozing defended his dissertation for the Candidate's degree and was transferred to a position with the Petersburg Technological Institute where he began his teaching career. In the Physics Laboratory of the Institute, B.L. Rozing resumed his research in the field of magnetism and began his work in the field of image transmission over distances or, as he called it, the electrical telescope [9].

Over a period of several years he carried out experiments with various mechanical and electromechanical systems for transmission of images. In the primitive optical mechanical devices, he perceived the principal shortcomings of mechanical television and came to the conclusion that it had no future in view of the disparity between the complexity of the problem and the technical means with which it was proposed to solve this problem. He set forth this conclusion in the form of the following basic principle for designing a practical television system:

"The elimination, insofar as possible, of all inertial mechanisms from the electrical telescopes and replacing them by 'inertialess devices, to the full extent of the significance of this word." [10]

It goes without saying that in his search for these inertialess devices B.L. Rozing, as a physicist, turned to the latest scientific discoveries and achievements in the field of physics which typified its development at the end of the 19th and beginning of the 20th century.

The discovery in 1887 of photoelectric emission by H. Hertz and the far-reaching investigations of this phenomena by Prof. A.G. Stoletov of Moscow University, who uncovered its physical essence and established the basic principles for photoelectric emission, led to the creation of inertialess photoelectric devices which played an important role in the development of television.

The invention of radio by the great Russian scientist A.S. Popov in 1895, investigations of the nature of cathode rays, the discovery by Thomson (Lord Kelvin) of the electron in 1897, and the study of its properties, the invention of the cathode (electron-beam) tube by K.F. Brown in 1897, and of the electron tube by J. Fleming and Lee De Forest in 1904-1906 — all of these laid the foundation for the subsequent development of electronics and radio engineering, on the basis of which B.L. Rozing formulated his new principles of television.

Working in laboratories with oscilloscopes, which in those days were used for measuring purposes, and observing how the electron ray formed complex, luminescent curves on the screen, B.L. Rozing decided to use the electron ray for the reproduction of images in an electrical telescope system. Thus, through the ingenuity and far-sightedness of this scientist, an effective means was found for constructing a television system based on the cathode-ray tube (CRT).

"Fortunately — he wrote — science knows of one such ideal mechanism — the cathode ray which appears as the result of the electrical discharge in a Crookes tube. This beam is also suitable for supplying a light signal which will form a bright spot when it falls on a fluorescent screen. This same mechanism, therefore, should be used in the electrical telescope" [11].

Having developed a circuit for a television system based on the new principles, and having tested all of its elements in practice, B.L. Rozing applied for a patent on June 25, 1907 on "A Method of Transmitting Images over a Distance" [12]. The principal difference between his system and those previously proposed was the use in his transmitter of an inertialess photocell with photoelectric emission instead of a selenium photoconductive cell (resistor), and the introduction of a CRT into the receiver.

The use of a photoemissive cell did not change the principle of image definition in the transmitter but it did eliminate a formidable obstacle in the path of practical realization of television as a result of inertialess photoelectric conversion. The use of a CRT signified a basically new direction in the design of television systems — a transition from an optical mechanical to an electronic device.

The CRT, in the simple form in which it was used for laboratory oscilloscope investigations, could not be used in an electrical telescope system, where it had to peform more complex functions. In order to reproduce a moving image with various brightness levels of its elements, it was necessary to vary the intensity of the electron beam in accordance with the value of the electrical signals from the transmitter, i.e., to introduce modulation of the spot brightness on the CRT screen.

B.L. Rozing solved this difficult problem by introducing electron-beam deflection relative to the diaphragm aperture in the CRT through the action of the image signal. The introduction of electron-beam modulation transformed the oscilloscope tube into a television receiver (picture) tube simultaneously fulfilling the functions of a scanning device for the modulated light source and screen, and suitable for reproducing images with half-tones.

Thus, after more than 10 years of persistent research and work, B.L. Rozing found a true and original solution of the problem and uncovered the path along which television research had to proceed. In 1907 B.L. Rozing filed patent claims on his invention in Germany [13] and England [14]. His priority in the discovery of a new method for the reception of images and use of cathode-ray tubes in television was irrefutably established by his Russian and foreign patents.

Late in 1908 B.L. Rozing carried out experiments with actual models of his electrical telescope and attempted to transmit simple images (various drawings on positive slides, moving fingers, hands, etc.). At the same time, he continued to improve his system and on May 9 (22), 1911 he sucessfully carried out the first distant transmission of images. In his writings, which were in the form of a day-by-day account of his investigations, Rozing recorded such events as the following: "On May 9, 1911, a distinct image was seen for the first time, consisting of four luminous bands" [15]. After this, the experiments were repeated in the presence of a group of Petersburg physicists — V.K. Lebedinskiy, V.F. Mitkevich, S.I. Pokrovskiy, and others [10].

This was the first television transmission in the world since none of Rozing's predecessors was able to display his system in operation or to transmit even a simple image. It was an outstanding technical achievement for that period when there were no suitable means for amplifying weak photoelectric currents, and when the CRT was still a somewhat crude instrument with low sensitivity and primitive electron-beam focusing.

B.L. Rozing opened up a basically new path of television development which involved a transition in the problem of television system design from the technical area of rotating mechanical devices to the field of electronic engineering based on the application of the CRT in the receivers and the use of photoemission of image transmission. The main significance of this discovery was not in the proposed concrete system of image transmission in itself, but rather in that it pointed out a new direction for the subsequent development of television.

The patents on B.L. Rozing's inventions granted him in other countries, and their description in many foreign journals made his research widely known and had considerable influence on television research and the development of television techniques throughtout the world.

The idea of using an electron beam to scan the image in the receiver could also have been used as a scanning technique in image transmission, although the inventor, himself, applied the electron-beam principle only to the receiver. Less than a year after B.L. Rozing filed his patent claim with the British Patent Office and published preliminary data [16], the British engineer A.A. Campbell Swinton expressed the concept of using an electron beam to scan the image in the transmitter and the construction of a fully electronic television system [17]. Within three years, in 1911, he presented a complete schematic diagram of an electronic system of image transmission in which the transmitter, in line with the receiver, was designed to use a CRT with photoelectric conversion instead of a luminescent screen [18].

B.L. Rozing had a drect influence on the work of many Soviet and foreign specialists. For example, the famous specialist in the field of electronic television, V.K. Zworykin, who carried out his research in the USA, has written: "When I was a student, one of my teachers was Physics Professor B. Rozing who, as is well-known, was the first to use the CRT for receiving television images. I was very much interested in his work and asked for permission to assist him. We spent much time together discussing the possibilities of television. At that time I was fully aware of the shortcomings of mechanical television and the necessity of using an electrical system" [19].

B.L. Rozing not only formulated a new principle for solving the problem of television but also attained the first practical successes in this direction. Nevertheless, the state of engineering knowledge at that time did not permit the construction of a television system based on these principles. His ideas outstripped engineering developments by several decades, and were not realized in practice until the 1930's.

2. PERIOD OF PREDOMINANCE OF THE MECHANICAL SYSTEMS

Electronic television evolved side-by-side with mechanical television but its path of development was considerably more complex and difficult. The first practical successes in image transmissions over a distance were attained with optical-mechanical equipments.

Research in the field of mechanical television was renewed after World War I in various countries from 1920-1923. The most interesting work on optical mechanical systems was carried out by the following: D. Mikhay (Hungary); J. Baird (England); C. Jenkins, G. Ives, and E. Alexanderson (USA); E. Belen, M. Valency, and A. Deauviller (France); A. Karolus and F. Schroeter (Germany); A.I. Adamian, A.A. Chernyshev, V.A. Gurov, S.N. Kakurin, and others (USSR).

The history of television began a new period characterized by the transition to the development of practical television systems on a scientific basis.

As a result of the progress made in radio engineering and the expansion of radio broadcasting, the necessary technical means for achieving television broadcasting were thereby created. The basic obstacle in the path of practical television was eliminated, namely, the inability in the past to amplify weak image signals.

In the USSR, despite the disruption of the national economy after the imperialist and civil wars, and the absence of an essential engineering foundation, television research was being carried out even in the first years of the reconstruction period (1921-1295). Thanks to the concern of the Party and Government, and of V.I. Lenin, personally, scientific centers were established in 1918 such as the Physicotechnical Institute in Petrograd and the Nizhegorod Radio Laboratory. Other institutions established somewhat later were the Central Radio Laboratory (TsRL) in Leningrad and the State Experimental Electrical Engineering Institute (GEEI) in Moscow. These centers carried out activities on theoretical principles of radio engineering, created new Soviet radio equipment and began research in the field of television. This research was carried out at a time when there was practically no information available regarding the work of other countries in the field of television.

Under the directorship of M.A. Bonch-Bruyevich at the Nizhegorod Radio Laboratory, work was carried out in 1920-1922 on the design and testing of a television transmitter which included a panel of miniature photocells switched by a mechanical switchboard, and receivers with cathode-ray tubes and panels with glow discharge lamps [20]. Both of these projects were brought to the attention of V.I. Lenin who displayed keen interest and issued instructions to provide the necessary assistance for the development of a perfected system and requested that he be advised personally when practical results had been attained [21].

In 1922 Professor (and later Academician) A.A. Chernyshev [22] began his work in the field of television and created a mechanical television system with image scanning achieved by means of a polyhedral mirrored drum, and with light modulation in the receiver based on the use of the Kerr electron-optical effect [23].

In 1924 at the Central Radio Laboratory, V.A. Gurov developed an optical mechanical system of image transmission by means of radio, using an oscillating prism and drum with cylindrical lenses for the resolution (scanning) and synthesis of the images, and a string galvanometer for modulating the light source in the receiver [24].

From 1920, S.N. Kakurin worked on the design of a television system which contained a Nipkow disc in the transmitter and receiver, a photoemissive cell, a radio transmitter for transmission of image and sync signals, and a photocell relay for modulating the light source in the receiver [25].

In 1926, at the Fifth All-Union Convention of Physicists in Moscow, a group of associates of the Leningrad Electrophysics Institute demonstrated the transmission of moving images by means of an optical mechanical television system.

Thus, at a time when television research in the USA, England, and other countries had just been resumed, the Soviet Union already had carried out a number of experiments regarding the practical realization of an optical mechanical television system.

However, the above-indicated work was characteristic only of the initial period in the development of Soviet television. It was carried out in the majority of cases through the initiative of the researchers themselves without a unified plan. On a broader scale, the research and experimental work on television, now unified by an overall plan for the solution of a complete range of problems relating to television research, began in the USSR in 1928-1930. These projects were carried out by the collective efforts of scientists in the large scientific-research institutes and laboratories: Leningrad Electrophysics Institute (in the laboratories of A.A. Chernyshev and Ya. A. Ryftin), the All-Union Electrical Engineering Institute (the laboratory of P.V. Shmakov), the "Comintern" Factory Laboratory (the laboratory of V.A. Gurov), the Central Laboratory of Wire Communications (laboratory of A.F. Shorin), ORPU — The Branch Laboratory for Transmitting Equipment (the laboratory of A.L. Mints), and others. From that

time on, television in our country developed on a foundation of deep theoretical and experimental research.

The First All-Union Conference on Television, held in Leningrad on
December 18-22, 1931, discussed the results of the scientific-research
work and the overall status of television in the USSR [26]. The year 1931
was one of remarkable successes in Soviet television. Besides the solution of general television problems and the development of new television
equipment, during this same year regular experimental television transmissions were begun in the USSR, which was an important, practical
achievement. On April 29 and May 2, 1931, in the Television Laboratory
of VEI (All-Union Electrical Engineering Institute) the first transmissions
of images by means of radio with the short-wave transmitter Model RVEI-1
of the All-Union Electrical Engineering Institute (Moscow), at a wavelength
of 56.6 meters. Live images and photographs will be transmitted. Television amateurs will be able to view those working in the VEI who made
this achievement possible, and to see photographs of the leaders of the
Revolution, and finally, to see the workers — excursionists comprising
1,500 people who will be assembled by the walls of the Institute for an
inspection tour of its laboratories and achievements in building our
national economy.

It is hoped that these trial transmissions will evolve into regular
transmissions by Narkompochtel, will stimulate a new movement of radio
amateurs and, within a short period, will increase the effectiveness of one
of the most important sectors of the cultural front, one which also has
great political significance." The first television transmissions were received in Moscow and Leningrad.

On October 1, 1931, regular television broadcasting began over a
Moscow radio station. In the image transmissions, a transmitter with a
scanning beam was used, developed under the supervision of V.I. Arkhangel'skiy by the engineers N.N. Orlov, N.N. Vasil'yev, and others. The
image was divided into 1,200 elements (30 lines) and 12.5 frames per
second [27]. From the year 1932, the transmissions were carried out by
the Moscow Radio Broadcasting Engineering Terminal from specially designed television equipment; the equipment for the radio terminal was
developed by A.I. Korchmar, A.I. Pilatovskiy and Ya. B. Shapirovskiy,
under the supervision of I. Ye. Goron. These transmissions created great
interest and became a means for propagandizing television. Already by
November-December 1931 the transmissions from Moscow were being
viewed on home made receivers by amateurs in Kiev, Leningrad, Nizhni-
Novgorod, Odessa, Smolensk, Tomsk and many other cities of the USSR.
Thus, the basis for Soviet mass television was established.

Experimental television transmissions were also carried out in 1931
in Leningrad ("Comintern" Factory Laboratory), in Tomsk (Siberian Institute of Physics and Technology), and in Odessa (Communications Institute).
In the "Comintern" Factory a commercial model of a television receiver
with a Nipkow disc was designed for mass production (developed by A. Ya.
Breytbart).

In other countries (England, Germany, USA), at this time, experimental television transmissions were being carried out with a picture
definition of 30-60 lines, but only in the laboratories and by a small number
of amateurs, since none of the countries had as yet organized the mass
production of television receivers. The total number of television receivers
in all countries by 1934 did not exceed 10,000.

The beginning of regular television broadcasts was proof of the fact
that the development of television systems in the laboratories attained a

sufficiently high technical level to warrant their practical application outside the laboratories.

The practicability of arranging television transmissions through existing radio broadcasting stations of large territories, which was particularly important in our country, and the relative simplicity of the television receivers, which could be assembled readily by radio amateurs, paved the way for the expansion of low-definition television.

At the same time, the transmissions showed that the picture quality for the selected definition and frame frequency was not good. In the initial stages, this fact was somewhat concealed by the very importance of achieving actual transmissions, and by the novelty of the matter, but it was clear nevertheless, that much work remained to be done in order to improve picture quality.

Mechanical television, which first received practical application in a system with 30-line definition, attained a high level of development by 1934-1935. In a number of countries (USSR, Germany, USA, England) experimental optical mechanical transmitters with 120-180 line definition were developed for television centers, and with 240-280 line definition for studio transmissions and telefilm transmissions.

With an increase in the number of definition lines, it became increasingly clear that the problem of high-quality television could not be solved on the basis of an optical mechanical system. This was dictated primarily by the basic contradiction of mechanical television, namely, that an increase in the sharpness of definition decreased the sensitivity of the transmitter. The inadequacy of technical means to cope with the problem made it imperative to switch to electronic television, the foundation for which had already been established by the work of B. L. Rozing.

The significance of mechanical television may be attributed to the fact that it permitted the practical realization of the transmission of moving images over great distances and thus proved the practicality of television.

The period in which mechanical television systems dominated experimental television broadcasting continued until 1933, after which there was a sharp transition to electronic television. This transition began with the replacement of optical mechanical receivers with receivers containing cathode-ray (picture) tubes and culminated in a completely electronic television system after the invention of the iconoscope and its practical application.

As a result of the rapid tempo of developments over the last decade, television engineering attained such a high level that even the design of color and three-dimensional television facilities came within the realm of possibility.

Television developed on the foundation established by the work, discoveries and inventions of scientists, engineers and scientific workers of many countries. The outstanding ideas and discoveries which form the basis for modern television may be attributed to representatives of our great nation who in all stages in the history of television found independent, and mostly original and correct solutions of complex problems, and in a number of cases surpassed the corresponding achievements of foreign specialists.

REFERENCES

1. W. Smith. Journal of the Society of telegraph engineers, Vol. 2, 1873.
2. G.R. Carrey. Scientific American, Vol. 40, 1879.

3. A. de Paiva. Instituto d' Oporto. February 20, 1878.
4. M. Senlecq. English Mecanics. V. 28, No. 273, 1877
5. P.I. Bakhmet'yev. The New Telephotograph. "Elektrichestvo", No. 1, 1885.
6. P. Nipkow. German patent No. 30105, January 6, 1884.
7. M. Vol'fke. Device for the Electrical Transmission of Images Without Wires. Patent No. 4498, November 24, 1898.
8. A.A. Polumordvinov. Patent No. 10738, December 23, 1899.
9. P.K. Gorokhov. Boris L'vovich Rozing — Founder of Electronic Television. Gosenergoizdat, 1959.
10. B.L. Rozing. Electrical Telescope (Sight at a Distance). The Approaching Problems and Prospects. Academy, Petrograd, 1923.
11. B.L. Rozing. "Elektrichestvo," No. 20, 1910.
12. B.L. Rozing. A Method for the Electrical Transmission of Images. Patent No. 18076, July 25, 1907.
13. B.L. Rozing. German Patent No. 209320, November 26, 1907.
14. B.L. Rozing. British Patent No. 27570, December 13, 1907.
15. B.L. Rozing. Notebook No. 3, 1911. Archives of the A.S. Popov Central Museum of Communications.
16. B. Rosing. Illustrated Official Journal (Patents), December, 1907, p. 1925, application 27570.
17. A.A. Campbell Swinton. Nature, V. 78, June 16, 1908.
18. A.A. Campbell Swinton. Presidential Address. Journal of Röntgen Society, V. 8, 1912.
19. O.E. Dunlap. The Future of Television. Harper and brothers, New Jork, 1947.
20. Newspaper "Izvestiya," September 24, 1922, No. 215 (1654).
21. Lenin Collection, Vol. XX, p. 317.
22. "EFI, The History of Its Development and the Basic Projects Carried out as of the 15th October Anniversary." "Zhurnal Tekh. Fiziki", Vol. 3, No. 6, 1933.
23. A.A. Chernyshev. "Elektrichestvo," No. 4, 1925.
24. V.A. Gurov. Author's Certificate No. 30723, March 23, 1926.
25. S.N. Kakurin. Soviet Patent No. 144, August 18, 1920.
26. All-Union Conference on Television. December 18-22, 1931, Leningrad. Conference Data.
27. V.I. Arkhangel'skiy. Television. Svyaz'tekhizdat, 1936.

TWENTY FIVE YEARS' CHANGE IN TELEVISION

J[ohn] C. Wilson

Twenty Five Years' Change in Television

by J. C. WILSON (Fellow)

It is not perhaps generally realised that practically the whole of the theoretical side of modern television practice is contained in embryo in disclosures which were made public by various workers prior to the Great War. I have chosen, therefore, as the thesis for the present paper, the enormous strides which have been made in the last twenty-five years in the technique of the art; the sharp contrast, that is, which exists between the amount of thought and knowledge necessary to propound a workable system, and the slow infiltration of practical experience from allied branches of electrotechnology, as well as development of a special technique in television engineering itself.

Electrical transference of images to a distance within the visual retentivity period has not always, since the earliest proposals for effecting it, been called television, for formerly words like telectroscopy and others were used. The British Patent Office, until 1908, employed the term "telescopy," and Ribbe[1] in 1904 used the term "telautography" in a very similar sense. In 1900, however, a continental worker seems to have been the first to coalesce the sections of the present term from the Greek and Latin roots, and so formed the word "television."[2] Apparently without knowledge of the previous existence of this word, an Examiner at the British Patent Office in 1911, in forming a search file devoted to the subject[3] (as a branch of copying telegraphy) re-invented it in its anglicised form, and thenceforward it has been brought into popular use in this country.

There is a popular superstition that the development of the photocell is an affair of the last few years, and that until its advent in commercial form, television necessarily hung fire. This is by no means true.

In 1887 Herz[4] discovered the dependence of an electric discharge upon irradiation of the gap with ultra violet light while he was carrying out his now famous experiments on the propagation of electromagnetic waves. Hallwachs,[5] however, in 1888 was the first to reduce the apparatus necessary for demonstrating the photoelectric effect to its essentials. Following these researches, Elster and Geitel tackled the problem in masterly manner, and from 1889 to 1913 they divided the honours of the development of commercial photocells with Hallwachs; in 1889[6] they discovered an amalgam of sodium which exhibited light-sensitivity in the visible spectrum, and in 1890[7] evolved an enclosed cell which retained its sensitivity almost indefinitely. In 1904 Hallwachs published an account[8] of the preparation of a photocell in which the sensitive coating comprised an oxidised layer on a copper surface. By 1913[9] Elster and Geitel had increased the sensitivity of their cells by two orders of magnitude, using a sensitive coating of potassium hydride; these cells, according to Zworykin and Wilson[10] differ but little from many of those manufactured in 1930.

In 1845, Faraday, that mighty giant of research workers, discovered a relation between light and electricity;[11] the rotation of the plane of polarisation of a light beam passed through silica borate of lead glass under the influence of a magnetic field. Many of us saw a replica of his experiment recently at his Centenary Exhibition. His diary comment is interesting:

" . . . and thus magnetic force and light were proved to have relation to each other. This fact will most likely prove exceedingly fertile. . . ."

Later, in 1875, Professor Kerr discovered the electrostatic birefringence[12] of optical media long sought for by Faraday. The first of these effects, namely, the magnetic rotation was employed by Nipkow in his suggested television receiver, while the use of birefringence developed in carbon-disulphide was suggested by Sutton.

A table has been compiled, showing step by step the building up of knowledge of all the bases—the physical bases, that is—upon which television rests, from the first " light-electric " effect discovered by Becquerel in 1839 to the development of a practical and sensitive form of photocell in 1912-1913 by those splendid co-workers Elster and Geitel; in this period are, of course, included all the various suggestions, based on the state of knowledge from time to time extant, for transmitting images of objects instantaneously to a distance electrically. It will be seen from this table* that all the ingredients of modern television systems had already been proposed before the commencement of the twenty-five years period.

Although many proposals for practical television were made before 1900, notably the notorious " sealed papers " of Graham Bell, the telephone inventor, deposited with the Smithsonian Institute in 1880 and commented upon by Ayrton and Perry in *The English Mechanic* in April of that year,[13] the only ones which have borne fruit in the modern art are those of Henry Sutton[14] and Paul Nipkow,[15] which are both so well-known now that further description is unnecessary.

May's discovery of the photo-resistive effect in selenium in 1873 provided the basis of nearly all the systems of transmission; for example, Senlecq's mosaic screen, which was included also in the suggestions of Ayrton and Perry, Carey and others. A notable exception was Middleton's mosaic of thermo-couple elements at the transmission end. In the years around 1880 there seems to have been a spate of television inventions, engendered no doubt by the commercial importance which electro-technology was rapidly assuming in connection with telegraphy, telephony, railway, signalling and lighting. I think it is obvious that everyone expected a rapid solution of the special problems of this new branch, and that its practical application would be contemporaneous with that of the telephone.

Ribbe in 1904 protected[16] a two-way television system using only one channel of communication, and also a system for transmitting continuously moving message-bands.

In 1905 Bernouchi proposed to transmit electro-optical signals along a light-beam by means of a directly modulated electric arc. This was closely followed in 1906 [17] by a suggestion of Rignoux and Fournier to scan an opaque flat object indirectly (that is, by means of a travelling spot of light) using a selenium cell accompanied by a lens to convert a part of the diffused light into electrical impulses. Previously, it has been proposed to accomplish transmission (apart from cell-mosaic methods) by means of a disc-scanned light sensitive cell, using ordinary floodlighting.

* Included at the end of this paper: Appendix I.

Adamian in 1908 suggested the use of positive-column tubes (or Geissler tubes, as they were then called) to supply the modulated source of light at the receiver. In the following year the Andersons proposed a scheme for transmitting images of objects in their natural colours, and Hoglund in 1912[18] suggested a two-element scanning device comprising slotted discs, following the proposal of Quierno Majorana in 1894.

In addition, in 1907 Rosing, in a British Patent specification and Campbell Swinton, independently in *Nature*, had published suggested systems of television in which a cathode-ray tube was employed to reconstitute a picture at the receiving station.

Ekstrom's independent evolution of the indirect or " light spot " method of transmission for transparencies in 1910 should also be mentioned, although he was certainly anticipated by Rignoux's French patent.

This, therefore, was the state of the television art at about the period when the thermionic valve was commencing to be recognised as the greatest revolutionary agent in the history of the art of the electrical transference of intelligence.

In 1916, Latour[19] expounded the principles of triode amplification and by 1920 [20] the practice of thermionic amplification for speech-frequencies was well-established. A further great advance was made with the introduction of oxide-coated filaments, described by King[21] in 1923.

No element, in the state of physical knowledge of that time, was lacking in the range necessary to the accomplishment of practical television, and yet the whole of the television field was merely littered with unconnected, untried schemes, many involving elements of the greatest importance to-day yet none exhibiting that " combination of best elements " which is essential to success.

In sharp contrast with the haphazard evolution of the bases, over a period of thirty-three years from Senlecq's original proposition in 1877 to the rise of the thermionic valve about 1910, has been the intensive awakening of directed scientific effort inaugurated by the war period and carried on through the last twenty years in unbroken crescendo.

The trend of the art changes from the upspringing of disconnected " bright ideas " to the systematic development of one or two systems; of these, the most pre-eminently satisfactory has been that upon which J. L. Baird in this country,[22] and C. F. Jenkins[23] in America, have independently concentrated, namely that in which the subject at the transmitter is scanned by a rotating optical element strip by strip, each strip being presented in sequence to a light sensitive element, and the resultant image-signal transmitted by a single channel to the receiver, where, by an inverse process, reconstitution is effected using a single modulated light source.

In 1923 both Baird and Jenkins succeeded in demonstrating the transmission of silhouettes, or shadow outlines of objects, within the period of the retentivity of vision, but Baird was undoubtedly the first, in 1925, to show natural pictures in half-tone. A great advance resulted in his invention in 1926 of the inverse method of scanning applied to three-dimensional views, and he followed up his early success by developing intensively the Baird light-spot system in which photo-cells, as far as artistic effect, shadow and contrast are concerned, replace the part played by the individual lamps in the earlier floodlight method. It is to his " good-genius " in recognising the immense possibilities of his system[24] that the rapid advances in technique in this country are entirely due. With greatly increased signal strengths resulting from this inversion without a corresponding increase—with considerable diminution in fact—in the intensity of the illumination to which the subject of the transmission is exposed, Baird was able to simplify his apparatus to bare essentials, and this simplicity has always been the keynote of his success.

At the time of these initial successes, Baird was working with directly-coupled amplifiers to preserve in the reproduced image the mean level of brightness, as well as the variations, of the transmitted scene; those who have worked with a string of nine or ten valves coupled in this way will wonder not that technique took so long to evolve but that to cope with the apparatus at all is within the range of one unassisted man's capabilities. This method of amplification, however, was already obsolescent at the time and has since been completely superseded. A major branch of television technique must here be indicated, and to do so I must again recapitulate a little.

I refer to the phenomenal advance in communication technique in general, opened up by the advent of the three electrode thermionic valve, and to another and perhaps more important branch of knowledge, not then fully or even partly developed : the theory and design of electrical distortion-correcting networks.

With regard to distortion, the underlying electrical theory was originally given by Oliver Heaviside towards the close of last century in his papers in the *Electrician* and is his classic work, " Electro-magnetic Theory." Unfortunately, Heaviside himself was something of a recluse, and owing partly to this and partly to other contributory causes, communication engineering has not even yet experienced the full benefit of his great work. I cannot do better than to quote Louis Cohen, whose book " Heaviside's Electrical Circuit Theory " has done much to lighten the dark places for engineers. In his preface, Mr. Cohen says : —

" The importance of Heaviside's contributions to electrical theory is now generally recognised and appreciated. His teachings, nevertheless, are available to only a comparatively few; to the many engineers and physicists who could profit by it, the work of Heaviside is more or less a sealed book. This may be accounted for largely as due to the novel and original mathematical processes he has introduced and applied with such extraordinary skill in the solution of many problems; and also to some extent to the lack of any attempt to correlate and present his teachings in a systematic manner suitable for one approaching the subject for the first time."

Briefly, therefore, the function of perfect amplification, even given suitable valves and other elements, is by no means easy to obtain; an amplifier is not a box with two terminals for input and two for output, and works inside capable of delivering a faithful replica, on a large scale, of minute varying potential differences applied to it. Indeed, the channel itself, of which the amplifier forms but a part, is almost as deadly an enemy as the lag of selenium cells was originally to the production of anything like a faithful television image; and both are far more lethal to television than they are to sound transmission on account of the predominant importance in television of preserving the phases of the various components of the signal.

In 1925, when Baird first achieved success, and MM. Belin and Holweck, Dauvillier, Denes von Mihaly and others were actively engaged on the subject on the Continent, a considerable part of this additional technique had already been developed.

In 1927 Baird demonstrated his Noctovision apparatus, using infra-red rays, to the British Association meeting at Leeds, and the Bell Laboratories, one of the most efficient and best organised commercial research machines in the world, carried out a television transmission over a circuit between New York and Washington, and placed on record beyond all doubt the practical importance of line-correction. That Frank Gray used Baird's method of indirect or " light spot " scanning[25] is no detraction from the merit of this performance, which probably represents the first piece of co-ordinated commercial television research on record; the full details of the methods and apparatus employed were at once published in what a contributor to *Nature* has called " that remarkably fine and generous periodical," the Bell System Technical Journal.[26]

In the same year Belin and Holweck achieved a measure of success in transmitting outlines and " shadowgraphs " by means of their cathode-ray television system. Thenceforward development

Early mechanical television image (30 lines, 1928)

has been rapid: in 1928 television colours and stereoscopic television on the Baird system were demonstrated to the British Association in Glasgow; early in 1928 Baird carried out a transmission of television pictures across the Atlantic on 45 metres.

The following year, light-spot transmission and cathode-glow lamp with disc reconstitution, also on the Baird system, were demonstrated in engineering form to the British Association in Cape Town. The same year 1929 saw the inauguration of a series of experimental broadcasts on the London B.B.C. transmitter, using the Baird process. At this time the transmission of of wording for instantaneous news broadcasts, telegram transmission in character-languages and other purposes was further developed by Baird[27]

and transmissions of this nature took place in the experimental broadcasts. The first public demonstration of television in a London theatre took place in July, 1930, at the Coliseum, when living artists and cinema films were transmitted from a studio in Long Acre and reproduced on a multi-cellular lamp screen on the stage. The transmission was carried out through the Baird process. In 1930 also M. von Ardenne in Germany commenced his researches on cathode-ray systems for the reconstitution of television images, and was later the first to produce results[28]

Lt.-Comdr. W. W. Jacomb, R.N.
da Vinci of video engineering.

comparable with those of mechanical reconstituting devices. In June, 1931, a demonstration was given by the Gramophone Company at the Exhibition of the Physical and Optical Society of projected television : cinema films were transmitted by a multi-channel process and reproduced by means of a Kerr cell and mirror drum apparatus on a translucent screen. In the same year the Derby was televised by the Baird process, and the directly-modulated arc for big-screen projection was demonstrated by Baird to the British Association meeting in London, in the section devoted to Mechanical Aids to Learning.

Contemporaneously with these advances, P. T. Farnsworth in America had been developing a system of his own in which a special form of photo-electric cathode ray tube transmitter was employed. His system has been described in "Radio Industries," Volume 5, No. 7, 1930, at pp. 386 to 389 *et seq.*[29]

In 1932 two major events took place: the Derby was televised and projected at the time of its occurrence upon the screen of a London cinema using the Baird process; and the B.B.C. installed television transmission equipment designed by Baird Television Limited for regular transmissions from their Studio B.B. at Langham Place.[30] In August, 1932, home television equipment employing, ironically enough, scanning apparatus of the type originally proposed by Weiller in about 1890[31] and a light-modulating device developed by Jacomb directly from Sutton's original suggestion in that year, was commercially available, and the year closed in a flurry of preparations for utilising to the full the increased channel-width which will immediately become available on the development of ultra-short wave radio transmission and reception systems.

A little later in the same year, Marconi's Wireless Telegraph Co. Ltd carried out a demonstration, at the British Association meeting at York, of radio reception of continuously moving printed characters, the transmission taking place from the Company's station at Chelmsford. In March the following year, the same Company published a description of their commercial television apparatus for the transmission of living artists.[32]

This survey would be incomplete without a mention of the extensive development in practical commercial television which has taken place in the United States in the last two or three years. Names of importance in this connection are Sanabria, Hollis Semple Baird (who bears no traceable relation to the British master of the art), Dr. H. E. Ives of the A.T. and T. Associated Companies, Drs. Alexanderson and Goldsmith of the R.C.A. A name worthy of special mention at the moment is that of an indefatigable worker, Vladimir Kosma Zworykin, whose latest form of photo-electric transmitter or "iconoscope" is described in the *Wireless World*, Vol. 33, No. 731, 1933.[33] In Germany, apart from the men already mentioned, Dr. A. Karolus of Telefunken, whose work on the Kerr cell is well known, and Drs. Möller, Schubert, Hudec and Kirschstein should be included.

RECENT DEVELOPMENTS

We are yet too close to the happenings of the past eighteen months to be able to state accurately their relative significance. Briefly, however, we may review what has occurred in order to trace if we can the tendencies in the art at the moment.

It was early realised that television as a mode of intelligence transference makes very large demands upon the frequency-band-width of the channel, radio or otherwise, by which it is to be transmitted, especially when anything approaching the amount of detail in a cinema picture is to be reproduced. This arises from the apparent necessity for conveying the whole of the detail of a scene within the period of retentivity of vision.

Early high-definition cathode-ray image (120 lines, 1933)
Note the zero-error effect due to gas focusing.

For a given available frequency-band it is necessary to effect a compromise between number of pictures per second and number of strips per picture, and in 1928 to 1933 while wavelengths between 200 and 400 metres were the only ones available commercially for television transmissions, it was essential to restrict the channel-width to the conventional 9 kc/sec. However, in England as well as in Germany and the States, during this time experiments were being carried out with a view to making a totally different order of wave-length available, with almost unlimited sideband width: this is the ultra-short wave region between 4 and 8 metres, where modulation-frequencies up to about 1·5 to 2 mc/sec. can be used.

The medium-wave transmissions were therefore confined to low definition pictures with numbers of strips between 30 and 50, but with the coming of ultra-short waves high-definition pictures of 120 to 500 strips have been transmitted. The first ultra-short wave transmission of television in this country was sent out by the Baird Company in 1932, and in 1933 120-line pictures were transmitted experimentally by the B.B.C., using the Baird process.

In March, 1934, shareholders of the Baird Company, at its Annual General Meeting in London, were addressed by the Chairman *via* ultra-short wave radio television from the Crystal Palace, and in 1935 experimental 405- and 240-strip transmissions have been carried out by Electric and Musical Industries and by the Baird Company.

Since the presentation of the report of the Television Commission in January, 1935, in which definite recommendations were made that an ultra-short wave station for high definition television transmission should be set up in London with the least possible delay, there has been a great deal of activity in design of home receiving apparatus. Just as development of ultra-short wave radio transmission for local areas rendered possible the propagation of high definition pictures, the researches of Knoll and Ruska in 1930[34] on electron-optics have paved the way for cathode-ray tube development making possible high-definition reconstitution very much on the lines of the early suggestions of Rosing and Campbell-Swinton in 1908.[35]

In view of advances in the technique of constructing and operating low-voltage cathode-ray tubes of the "sealed-off" type, it has become increasingly probable that one way at least of attacking the home television receiver problem is to produce small cathode-ray tubes in large quantities, at a price not far in excess of that of the standard multiple valve.

Other developments of importance are the "intermediate film" process for outdoor television transmission, or for large screen projection with a short time delay[36] (demonstrated by Fernseh A.G. in Berlin at the Radio Exhibitions of 1933, 1934 and 1935) and the high-speed intercalated system of Baird for instantaneous television transmission or projection.

The quality of television pictures, together with the cost of production of receiving apparatus for small home screens, have taken such a satisfactory turn in the last few months that, slightly misquoting Sir Arthur Edington's remark in another connection,[37] we might say that the art has now reached a state in which an intelligent man might almost take an interest in it.

REFERENCES

(1) Paul Ribbe: Brit. Pat. No. 29428/04.
(2) M. Perskyi: Annexes, Congrès Internationale d'Electricité, 18–25 August, 1900, pp. 54–56.
(3) Patent Office Index, Class 40, 1909–15.
(4) H. Herz: Ann. d. Phys., 31, p. 421 and p. 983, 1887.
(5) Hallwachs: Ann. d. Phys., 33, p. 301, 1888.
(6) Elster and Geitel: Ann. d. Phys., 38, p. 447, 1889.
(7) Elster and Geitel: Ann. d. Phys., 41, p. 161, 1890.
(8) Hallwachs: Phys. Zeit., 5, p. 489, 1904.
(9) Elster and Geitel: Phys. Zeit., 14, p. 741, 1913.
(10) Zworykin and Wilson: ."Photocells and their Application," p. 13, 1st Edition (John Wiley and Sons).
(11) Faraday: Diary, 13th Sept., 1845.
(12) Kerr: Phil. Mag. (4), 50, p. 337, 1875.
(13) Ayrton and Perry: Eng. Mech. and World of Science, 31, No. 788, p. 177, April, 1880.
(14) Henry Sutton: Telegraphic Journal and Electrical Review, Vol. 27, p. 549 (1890) and "La Lumière Electrique," Vol. 38, p. 538 of the same year.
(15) Paul Nipkow: German Patent No. 30105, 1884.
(16) Paul Ribbe: German Patent No. 160813, 1904.
(17) Rignoux: French Patent No. 390435, 1908.
(18) Hoglund: U.S. Patent No. 1030240, 1912.
(19) Latour: Electrician, 78, p. 280, 1916.
(20) Fortescue: J. Inst. El., Eng., 58, No. 287.
(21) King: Bell System Technical Journal, Oct., 1923.
(22) J. L. Baird: Wireless World, May 7, 1924, p. 153; F. H. Robinson, Broadcaster and Wireless Retàiler, April, 1924, p. 47; Amateur Wireless, Vol. IV., No. 101, p. 1; Kinema, April 3rd, 1924.
(23) C. F. Jenkins: Proc. I.R.E., Vol. 15, 11, pp. 958–968.
(24) J. L. Baird: Brit. Pat. No. 269658, Jan., 1926.
(25) Electrical Research Products Inc. (Frank Gray); Brit. Pat. No. 288238 (1929).
(26) B.S.T.J., Vol. VI., No. 4, October, 1927.
(27) J. L. Baird: Brit. Pat. No. 324029.
(28) See photographs in "Fernseh," Nr. 2, April, 1931, pp. 78–80.
(29) And more recently in Journ. Frank. Inst., Sept., 1934, p. 411.
(30) J. C. Wilson: "The Design of Television Transmission Equipment," Journ. Telev. Soc., Series II., Vol. I., No. VIII.
(31) "La Lumière Electrique," 34, p. 334, 1889; Gèn. Civ. 15, p. 570, 1889.
(32) "Television," Vol. 6, No. 61, pp. 108–113, March, 1933. (Reprinted from the "Marconi Review.")
(33) And in Journ. Inst. El. Eng., Vol. 73, No. 442, pp. 437–452, Oct., 1933. Proc. I.R.E., Vol. 22, No. 1, pp. 16–32, Jan., 1934.
(34) Knoll and Ruska: Zeit. für Physik, 78, 318, 1932. British Patent No. 402781.
(35) A. A. Campbell Swinton: Presidential Address to the Rœntgen Society, Nov. 7, 1911; "The Possibilities of Television," R.S.G.B. paper reported in full in "Wireless World," Vol. XIV., p. 51, p. 82 and p. 114. (Read March 26, 1924).
(36) Electrical Research Products: Brit. Pat. No. 297078, 1927.
(37) "Nature," Vol. 129, No. 3250 p. 233.

APPENDIX I: Table of Early Disclosures.

(1) Photo-electric Effects.

Date.	Name.	Reference.	Bearing.
1839	Becquerel	*Vide* Harrison, Proc. Telev. Soc. (Television, Vol. 3, p. 142	Photo-voltaic effect.
1865	Becquerel	La Lumiére 2, p. 121	Barrier-layer cells.
1873	May	Communicated to Soc. Tel. Eng. by Willoughby-Smith.	Sensitivity of selenium.
1887	Hertz	Ann. d. Phys. 31, 421 and 983	Photo-electricity.
1888	Hallwachs ...	Ann. d. Phys. 33, 301	Hallwachs effect.
1889	Elster and Geitel ...	Ann. d. Phys. 38, 497	Sensitivity of sodium and potassium.
1890	Elster and Geitel	Ann. d. Phys. 41, 161	Sodium photocell.
1904	Hallwachs	Phys. Zeit. 5, 489.	Copper oxide cell.
1913	Elster and Geitel ...	Phys. Zeit. 14, 741	Potassium hydride photocell.

(2) Electro-optical Effects.

Date.	Name.	Reference.	Bearing.
1845	Faraday	Diary, 13th Sept., 1845	Rotation of plane of polarisation.
1875	Kerr	Phil. Mag (4), 50, p. 337	Electric double refraction in liquid.
1877	Kerr	Phil. Mag. (5), 3, p. 321; and Ibid. p. 161, 1878	Magneto-optical effects.
1893	Blondin	Lum. Elec. Vol. 43, pp. 259-266	Electro-magnetic mirror modulator.
1905	Bernouchi		Electric arc directly modulated.
1908	Adamian	British Patent No. 7219/08	Modulated gas-discharge tube.
1909	Pictet	British Patent No. 10450/09	Modulated manometric flame.

(3) Thermionics and Amplification.

Date.	Name.	Reference.	Bearing.
1884	Edison		Edison effect: particles emitted from filament.
1889	Fleming	Proc. Roy. Soc., December, 1889	Conductivity from filaments in vacuo.
1904	Fleming	Brit. Pat., No. 24850/04	
1905	Fleming	Proc. Roy. Soc. lxxiv, 488	Two electrode valve.
1907	De Forest	U.S. Pat., No. 879,532	Introduction of grid.
1913	Ges. f. Draht. Tel. ...	Brit. Pat., No. 8821/13	Three electrode valve amplifier
1914	De Forest	Brit. Pat. No. 2059/14	Audion amplifier for line repeater.
1916	Latour	Electrician, lxxviii, 280	Triode amplification.
1923	King	B.S.T.J., Oct. 1923	Properties of oxide-coated filaments.

(4) Networks and Correction (excluding artificial lines and loading).

Date.	Name.	Reference.	Bearing.
1893 to 1912	Heaviside ...	" Electromagnetic Theory "	Theory of electric nets *in extenso.*
1923	Carson	B.S.T.J., Vol. 1, No. 2, p. 43	Phase and amplitude compensating nets.
1923	Zobel	B.S.T.J., Vol. 2, No. 1, p. 1	Extensions to theory.
1924	Zobel	B.S.T.J., Vol. 3, No. 4, p. 567	Same.
1927	Gray	B.S.T.J., Vol. 6, No. 4, pp. 625-8	Application to television.
1928	Zobel	B.S.T.J., Vol. 7, No. 3, p. 438	Constant resistance equivalent networks.

(5) Television References.

Date.	Name.	Reference.	Bearing.
1877	Senlecq	Eng. Mech. Vol. 32, Feb. 1881, p. 534 ... ... Ibid. Vol. 28, No. 723, p. 509	Crude mosaic type of television.
1879	Carey	Sci. Am., Vol. 40, p. 309	Incandescent carbon elements.
1880	Ayrton and Perry ...	Eng. Mech. Vol. 31, 788, p. 177	Mosaic Kerr pole-pieces.
1880	Middleton	Ibid. loc. cit.	Thermopiles and Peltier effect.
1880	Ayrton and Perry ...	" Nature," Vol. 21, p. 589	Kerr pole-pieces again.
1880	Gordon	" Nature," Vol. 21, p. 610	
1880	Ayrton and Perry ...	" Nature," Vol. 22	Reply to Gordon; refers to Faraday effect.
1882	Bidwell	Eng. Mech. Vol. 33, p. 158, 180	Telephotography.
1884	Nipkow	Germant Patent No. 30105	Spirally arranged apertures in disc.

Date.	Name.	Reference.	Bearing.
1890	... ...	Eng. Mech. Vol. 51, Mar. 7	Gas telephone and revolving mirrors.
1890	Sutton	Tel. Jour. and El. Rev. Vol. 27, p. 549; Lum. Elec. Vol. 38, p. 538	Use of Kerr effect in carbon disulphide.
1891	Brillouin	Rev. Gen. des Sci. Jan. 30, pp. 33-38	Two lensed discs at right angles.
1893	Blondin	Lum. Elec. Vol. 43, pp. 259–266	Use of Weiller mirror drum.
1897	Szczepanik	British Patent No. 5031/97	Vibrating mirrors for zig-zag scanning.
1900	Perskyi	Cong. d'Elec. Annexe 1900, p. 54	Word " television."
1901	Woodruffe	Eng. Mech. No. 1900, p. 39	
1904	Ribbe	German Patent No. 160813	Two-way television with one disc.
1904	Belin	British Patent No. 26586/04	Multiple zones.
1904	Ribbe	British Patent No. 29428/04	Scanning band and photo recording.
1906	Rignoux and Fournier	French Patent No. 390435 of 1908; No. 364189 of 1906	Light spot scanning with lens and cell.
1906	Fowler and Thompson	Eng. Mech. No. 2169, Oct. 19, p. 256	Reference to Nisco.
1907	Bidwell	" Nature," Vol. 76, p. 444	Telephotography.
1907	Rosing	British Patents No. 27570/07 and 5486/11 ...	Velocity modulation cathode ray receiver.
1908	Campbell-Swinton ...	" Nature," June 18, 1908, p. 151	Cathode ray system for transmission.
1909	Anderson and Anderson	British Patent No. 30188/09	Colour television.
1910	Ekstrom	Swedish Patent No. 32220	Light spot scanning of transparency.

TELEVISION

AN ACCOUNT OF THE DEVELOPMENT AND GENERAL
PRINCIPLES OF TELEVISION AS ILLUSTRATED BY A
SPECIAL EXHIBITION HELD AT THE SCIENCE MUSEUM—
JUNE–SEPTEMBER, 1937

Edited by G. R. M. GARRATT, M.A.

Assisted by G. PARR

LONDON
PUBLISHED BY HIS MAJESTY'S STATIONERY OFFICE

1937

[*Reproduced by courtesy of the B.B.C.*]

View in the studio at Alexandra Palace during the transmission of a cabaret scene by television.

TELEVISION EXHIBITION

EXECUTIVE COMMITTEE

Col. E. E. B. MACKINTOSH, D.S.O. (*Chairman*).

F. St. A. HARTLEY, A.C.G.I. (*Deputy Chairman*)

H. J. BARTON-CHAPPLE, Wh.Sch., B.Sc., A.M.I.E.E.

N. R. BLIGH, B.Sc. (Eng.)

T. C. MACNAMARA, Esq.

A. E. MOODY, Esq.

G. PARR, Esq.

O. S. PUCKLE, A.M.I.E.E.

G. W. WALTON, Esq.

J. H. A. WHITEHOUSE, Esq.

G. R. M. GARRATT, M.A. (*Secretary*).

Acknowledgement is due for assistance in the preparation of this Handbook to : H. J. Barton-Chapple, B.Sc. ; N. R. Bligh, B.Sc. (Eng.) ; T. C. Macnamara, Esq. ; O. S. Puckle, A.M.I.E.E. ; G. W. Walton, Esq. ; R. McV. Weston, M.A. ; J. H. A. White-house, Esq. Chapters have also been contributed by the Editors.

CONTENTS

INTRODUCTION

There can be no doubt that high definition television is one of the most remarkable technical achievements of our times.

Only a few years ago it seemed to all but a handful of pioneers that the successful solution of the problems involved, though a scientific possibility, was still a long way off. The average member of the public regarded as incredible the possibility of seeing clear reproductions on a screen of events which were taking place simultaneously at a distant place.

Something of this doubt may still persist, for television has made its debut in an advanced state of perfection, before a public almost completely unprepared by the more usual process of steady development over a period of many years.

The technique of television, though related in some respects to that of the transmission and reception of sound, is far more complex. Nevertheless, its general principles are such as can be understood by the layman in scientific matters.

A Television Exhibition has therefore been organised by the Science Museum, in conjunction with the British Broadcasting Corporation and the leading manufacturers, in order to provide the public with an opportunity to appraise the value of this new form of entertainment. It is hoped that the exhibits will help to explain the problems involved and the methods of their solution, and will give some indication of the research and ingenuity which have gone to make this new development possible.

E. E. B. MACKINTOSH,
Director of the Science Museum.

CHAPTER I

THE EARLY HISTORY OF TELEVISION

The possibility of seeing events at places remote from the observer was a dream of humanity for countless centuries, the stock-in-trade of the weaver of stories of romance and magic. The technical advances of recent years, however, have made this fantastic dream first a possibility and to-day an accomplished fact.

The first definite proposals for a system of distant vision by means of electricity were made almost sixty years ago, but only during the last three or four years have advances been made in technique and design which have rendered high definition television a reality.

In tracing the history of television, one must remember that every system of electric vision which has ever been proposed depends fundamentally upon certain physical changes which are produced by light. If the phenomena known as photo-electricity were non-existent all television would be impossible, and the possibility of television, therefore, may be said to date from the original discovery of the electro-chemical effect of light by Becquerel in 1839.[1]

We now know that the chemical changes which Becquerel observed and the electric currents which resulted were produced by ultra-violet light—electro-magnetic oscillations having a wavelength rather shorter than that of visible light. Such electro-chemical changes, or as they are better known, photo-electric effects, are produced in many substances by visible light. An account of the development of photo-electricity will be found in the next chapter.

SELENIUM The chemical changes observed by Becquerel could be put to no practical use, and it was not until 1873 that the first photo-electric effect of practical value was observed by a telegraph operator named May who was using some high resistances composed of the metal selenium. May observed that his instruments behaved erratically whenever the sun shone on the resistances, and the effect was traced to a decrease in the electric resistance of selenium when exposed to light.

The announcement of this discovery led within a short time to speculation on the possibility of transmitting pictures and scenes to a distance by means of electricity, and within a few years a number of schemes were put forward. For various reasons, however, they were all impracticable at the time.

---

[1] Reference numbers throughout the text refer to the bibliographies at the end of each chapter.

AYRTON AND PERRY

One of the earliest proposals of which details were published was that put forward in 1880 by Ayrton and Perry,[2] though it is clear from their letter that the scheme was conceived some years earlier. They proposed to use a transmitter consisting of a large screen made up of small squares of selenium and to project an image of the scene to be transmitted on to this screen by means of a lens. Each square of selenium was to be connected by a separate wire to the corresponding point in the receiver, of which they suggested two types. In their first plan the receiver was to consist of a number of magnetic needles, the position of each being controlled by the electric current from the corresponding selenium cell in the transmitter. Each magnet, by its movement, was to close or open an aperture through which light passed to illuminate the back of a small square of ground glass, and by having a large screen of such cells, the effect of a complete picture might have been obtained.

Ayrton and Perry's second proposal for a receiver consisted in having each little square on the screen made of silvered soft iron forming the end of an electro-magnet energised by the current from the corresponding selenium cell in the transmitter. They proposed to illuminate the screen by a powerful beam of polarised light and to view the screen through an analysing prism. When polarised light is reflected from a magnetic surface, its plane of polarisation is rotated in proportion to the strength of the magnetic field, and Ayrton and Perry proposed to obtain modulation of the light intensity by using this phenomenon.

CAREY'S SCHEME

Another scheme which was put forward at about the same period was that proposed by G. R. Carey of Boston.[3] Carey's scheme, however, related more particularly to picture telegraphy rather than to television because his proposed receiver was capable only of reproducing the image on a piece of chemically sensitised paper. His transmitter was identical in principle with that suggested by Ayrton and Perry; the receiver was to consist of a screen having a large number of contact points each connected by a separate wire to the corresponding selenium cell in the transmitter. Carey proposed to press a sheet of sensitised paper against the receiving screen by means of an earthed metallic plate, and he considered that the currents passing through the paper would reproduce the image. Carey's scheme had the merit of being workable, which is more than can be said for other suggestions of the period, and it was only impracticable on account of the great number of conductors required.

In order to overcome the necessity of having thousands of conductors between the transmitter and the receiver, several inventors proposed to dissect the picture in various ways and transmit each little element in rapid succession. This process is known as "scanning" and necessitates the receiving mechanism being kept in exact synchronism with the transmitter. If the picture is completely scanned sufficiently rapidly and frequently, the eye receives the impression of a complete picture.

Early proposals for scanning systems were made by Sawyer, de Païva, Senlacq, M. Leblanc and others. The best known, however, was that suggested by Paul Nipkow in 1884.[4]

Leblanc's proposal is of interest as there have been numerous attempts to utilise the same principles. He proposed to scan the image by means of a small mirror arranged to oscillate about two axes simultaneously, but with widely different frequencies, and so to project each little area of the image in turn on to a selenium cell. Although this method of scanning is very efficient optically, the mechanical difficulties of such a system are considerable, and it would have been difficult to maintain accurate synchronism.

NIPKOW'S PATENT, 1884　　Nipkow's patent of 1884 has received from later experimenters by far the greatest attention of all the early proposals. This can only be attributed to the fact that it was the most simple device proposed, and it could be used with stationary optical systems readily

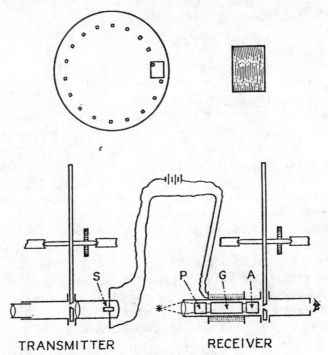

FIG.. 1—Nipkow's Patent of 1884.
　　S.　Selenium cell.
　　P.　Polarising prism.
　　G.　Flint glass.
　　A.　Analysing prism.

obtainable. Nipkow's method of scanning employed a disc of large diameter near the periphery of which was a series of small holes arranged in the form of a spiral. The area of image which could be scanned by this arrangement was small in comparison with that of the disc. The apertured disc is very inefficient as a receiving scanner, though it is still used to-day for one system of film transmission.

Nipkow's transmitter employed a selenium cell, the only practical form of photo-element known at the time. His receiver, however, employed an ingenious system of light modulation which depended on the property possessed by flint glass of rotating the plane of polarisation of light when situated in a magnetic field. With the small currents passed by a selenium cell and the absence of any means of amplification this system was too insensitive to give practical results, but a similar system was adopted by Baird in which specially designed Kerr cells were used in place of flint glass.

MIRROR DRUM Another system of scanning which was given much attention by later experimenters was that proposed by L. Weiller in 1889. In place of the disc with a series of holes arranged on a spiral, Weiller proposed to use a drum fitted with a number of tangential mirrors, each successive mirror being orientated through a small angle so that, as the drum rotated, the area of the image was scanned in a series of lines and projected on to a selenium cell. Such an arrangement was first used by Ll. B. Atkinson in 1882, and his apparatus is now preserved in the Science Museum, but no descriptions were published at the time. Weiller, therefore, is generally recognised as the inventor of the mirror drum.

One of the chief difficulties which faced the early experimenters was the time-lag of selenium and its consequent insensitivity to rapid changes of light intensity. An ingenious attempt to overcome this sluggishness was suggested by Szczepanik in 1897.[5] He proposed to construct the selenium cell in the form of a ring which was to rotate steadily. By means of a pair of oscillating mirrors the image was to be scanned and projected through an aperture on to a part of the selenium ring. As the ring rotated, it continually exposed a fresh surface to the aperture and thus became capable of responding to higher frequencies. There is no evidence, however, that Szczepanik ever attempted to construct his apparatus which would have been quite unworkable as described in his patent, as its main principles are founded on technical fallacies.

Although there had been almost a glut of schemes for " seeing by telegraphy " during the period 1877–1885 it was gradually recognised that the characteristics of selenium were unsuitable, and interest in the possibility of distant vision appears to have waned.

ROSING'S PATENT The gradual development of photo-electric elements, however, and, more particularly, the introduction of the Braun tube—a primitive cathode ray oscillograph—aroused interest once more, and in 1907 a patent was granted to Boris Rosing for a system of electrical transmission of images in which the Braun tube was used as the receiver.[6] Rosing's transmitting arrangements were similar to many others, and used two mirror drums revolving at right angles to each other at widely different speeds to scan the image. The varying current impulses from the photo-cell were transmitted to the receiver where they were caused to charge two condenser plates in the cathode ray tube. The fluctuating charges on these plates caused the beam of electrons projected from the cathode to be deflected away from an aperture, and the amount of the beam which passed through the aperture was proportional to the potential of the plates and thus to the degree of light and shade in the original scene. Having passed through the aperture the electron beam was caused to scan the surface of a fluorescent screen placed at the far end of the tube and so to reproduce the original image.

Successful results were never obtained with this scheme, partly on account of the crude forms of photo-cells and cathode ray tubes then available and partly on account of the lack of any means of amplification.

In 1911 Rosing also evolved a system of reproduction by a different method, in which the variations in the picture brightness were produced by varying the speed of travel of the light beam instead of its intensity. This scheme was later applied to cathode ray tube reproduction by Thun in 1931 and von Ardenne in 1932, and is generally known as "velocity modulation." The brightness of the trace on the fluorescent screen of the tube is governed by the rate at which the beam sweeps across it and a retardation in speed will thus give a brighter line. A complete system based on this principle was independently invented and described by Bedford and Puckle in 1934.[7]

CAMPBELL SWIN-
TON'S SCHEME
The idea of using cathode ray tubes for television occurred almost simultaneously to an Englishman, A. A. Campbell Swinton, and in a letter to *Nature*, dated 18th June, 1908, he made a suggestion that the problem might be solved by using two cathode ray tubes of an appropriate design, one at the transmitter and one at the receiver.[8] Campbell Swinton's proposal was made entirely independently of Rosing, whose patent had not been published at the date of Campbell Swinton's communication. Furthermore, it possessed the fundamental difference of using a totally different form of cathode ray tube as the transmitting element. In 1908, Campbell Swinton's scheme was only a brief and vague suggestion, but in November, 1911, in the course of a presidential address to the Röntgen Society, he greatly amplified his proposal and gave comprehensive details of the scheme.[9]

He conceived the idea of a mosaic screen of photo-electric elements which were to form a part of a special type of cathode ray tube. The image of the scene to be transmitted was to be projected on to the mosaic screen by means of a lens, and the back of the screen was to be scanned by a beam of cathode rays controlled magnetically by the currents from two alternating current generators. The cathode ray beam in the receiver was synchronised with that in the transmitter by means of deflecting coils connected by wires to the same generators as the transmitter, and a separate conductor carried the photo-electric currents for modulating the receiver beam.

Campbell Swinton never attempted to construct a working demonstration of his proposals, and he realised the difficulties which would have to be overcome before such a system could be made to work. With characteristic sincerity, he said, " It is an idea only and the apparatus has never been constructed. Furthermore, I do not for a moment suppose that it could be got to work without a great deal of experiment and probably much modification.''

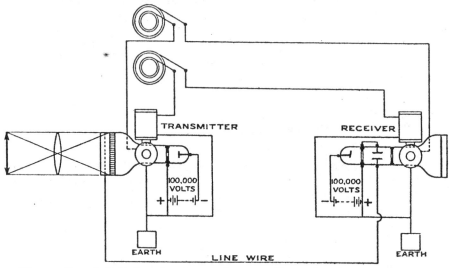

Fig. 2—Campbell Swinton's scheme of 1911. (Reproduced from the *Journal of the Röntgen Society*.)

Although it must be admitted that modern television owes its achievement to the researches of many different and distinguished workers, it is interesting to realise that the system of television now in use by the British Broadcasting Corporation has been developed on the fundamental lines first suggested by Campbell Swinton.

In 1911 the three-electrode amplifying valve was practically unknown. During the next ten years, however, thermionic valves and multi-stage amplifiers were developed, and fresh interest was taken from 1921 onwards in the possibility of television, notably by Mihaly of Budapest, Jenkins in America, and Baird in England.

Mihaly's initial experiments were confined to systems employing oscillating mirrors and were of no practical utility, but he subsequently developed a mechanical system of television which employed a small double-sided mirror rotating on the axis of a stationary drum fitted with a number of small mirrors on its interior surface.

The pioneer of television in America was C. F. Jenkins, who devised a scanning system which employed a pair of bevelled-edged glass discs, the angle of bevel changing continuously around the disc in one revolution. The bevelled edges formed prisms which deflected the beam of light as the discs rotated, and by rotating one disc many times faster than the other the entire surface of the image was scanned. Jenkins gave his first demonstration in June, 1925, transmitting by wireless the shadow-graph image of a slowly revolving model windmill.

J. L. BAIRD The name best known among the pioneers of television in this country is unquestionably that of John Logie Baird. Although in his early experimental work he used methods and apparatus proposed by others it was his dogged persistence which brought results which had not previously been obtained. His faith in the ultimate success of television was instilled into others, and by means of frequent demonstrations and wide publicity he aroused public interest and stimulated research. Baird commenced his experiments in 1923, and in April, 1925, he gave a public demonstration at which crude images were transmitted between two machines.[10] Nine months later, in January, 1926, a demonstration was given to members of the Royal Institution, at which the images of moving human faces were shown, not as plain black and white outlines, but having tone gradations of light and shade. The results were crude in the extreme, but it was just possible to recognise individual faces. Improvements were made gradually, and further demonstrations were given ; for example, the transmission of low definition images by telephone wires between London and Glasgow in May, 1927, and by radio in February, 1928, between London and New York. Prior to June, 1928, Baird's experiments had been performed by artificial light, but in that month he gave a demonstration of daylight television, and in August, 1928, he demonstrated a method by which a crude picture could be built up in colours.[11]

In September, 1929, an experimental service of television broadcasts was commenced by the Baird Company and the British Broadcasting Corporation, and in August, 1932, the B.B.C. took over the transmissions entirely, the vision signals being radiated on 261 metres.

Prior to 1932 the television image as seen on the disc model receivers measured only about 4 in. $\times$ 2 in. The introduction of the " mirror drum " type of receiver in June, 1932, provided a much brighter image on a screen measuring nine inches high and four inches wide, and this form of receiver became the standard type until the termination of the low definition broadcast service in September, 1935.

Hitherto the television images had been built up of thirty " lines." Such a coarse texture of picture rendered the transmission of small detail impossible, and the programme could have but little entertainment value.

Research has therefore been directed in numerous laboratories towards the development of a higher standard of definition. For television to be acceptable, it was necessary to develop the highest definition commercially practicable, and in view of the many problems involved, it became a task for the large engineering organisations with comprehensive research laboratories, rather than for the independent investigator.

During the past ten years concerted attacks on the problems involved have been made by highly trained engineers, each a specialist in his own field. The technical progress made, for example, in photo-electric cells, cathode ray tubes, valves, amplifier design and scanning oscillators is recorded in the technical journals and the Patent Offices of America, England, France and Germany.

In this brief summary it is impossible to mention by name the many workers whose united contributions have made modern high definition television a reality. Simultaneous development has taken place in many different laboratories, and there have been many whose work has been responsible for the development of the rudimentary scheme proposed by Campbell Swinton into the practical television system of to-day.

BIBLIOGRAPHY

1. Becquerel. *Comptes Rendus*, Vol. 9, p. 145. 1839.
2. Ayrton and Perry. *Nature*, Vol. XXI, p. 589. 1879–80.
3. Carey. *Design and Work*, June 26th, 1880.
4. P. Nipkow. German Patent No. 30,105. 1884.
5. Szczepanik. British Patent No. 5031. 1897.
6. B. Rosing. British Patent No. 27,570. 1907.
7. Bedford and Puckle. *Jour. I.E.E.*, Vol. 75, p. 63. 1934.
8. Campbell Swinton. *Nature*, Vol. LXXVIII, p. 151. 1908.
9. Campbell Swinton. *Journ. Röntgen Soc.*, Vol. VIII, pp. 1–13. 1912.
10. Baird's experiments. *Nature*, Vol. CXV, p. 505. 1925.
11. Baird's experiments. *Nature*, Vol. CXXII, pp. 233–4. 1928.

CHAPTER II

PHOTO-ELECTRICITY

Since any system of television must depend fundamentally on some form of electric effect produced by light, it is clear that the discoveries and developments in the field of photo-electricity are of prime importance in any account of the development of television. In this chapter the principal characteristics and types of photo-electric cell will be briefly described.

The term " photo-electric cell " is used as a general term to include all devices which are capable of producing an electrical effect when acted upon by visible, infra-red or ultra-violet light. For convenience and to avoid ambiguity it is usual to classify cells under the following heads :—

1. *Photo-conductive*, in which the internal resistance of the cell is influenced by incident light.
2. *Photo-emissive*, in which the operation of the cell depends upon the emission of free electrons from a metallic surface.
3. *Photo-voltaic*, in which a potential difference is actually generated by the cell. Cells of this type are usually sub-divided into two groups, namely :—

   (a) *Electrolytic* ; containing a liquid electrolyte.
   (b) *Electronic* ; dry cells, containing no liquid.

PHOTO-CONDUC-TIVE CELLS The most important cell of this type is the well-known selenium cell. It was discovered in 1873 by May,[1] the assistant of Willoughby Smith, that the resistance of selenium in its " metallic " form is reduced when light falls upon it.

Selenium cells are usually made by forming, when hot, a very thin layer of the material over two interlaced metallic structures which form the electrodes, the whole being supported on an insulating and heat-resisting base such as mica, glass or porcelain. The vitreous selenium layer so obtained is cooled and reheated to convert it to a grey crystalline light-sensitive form. The characteristics of any individual cell depend to a great extent upon this heat treatment during manufacture.

The sensitivity of selenium is greatest in the region of the red and infra-red portions of the spectrum, but the cells are irregular in their action and suffer from an appreciable time lag. When the cell is suddenly illuminated the current through it continues to increase during its exposure to light, and if the illumination is extinguished the current continues to fall for an appreciable time.

This behaviour makes selenium cells unsuitable for photometry and relatively insensitive to very rapidly varying illumination. Special cells have been designed for use up to a frequency of illumination of 10,000 per second, such, for example, as that designed by Thirring,[2] and such cells were used in early talking picture reproducers, but the very high frequency response required of photoelectric cells used for television transmission altogether precludes the use of selenium for the purpose.

PHOTO-EMISSIVE CELLS The discovery of this type of photo-electric effect is due to Hertz, who observed, in 1887, that the potential required to produce a spark across a small gap was reduced when the spark gap was illuminated by means of ultra-violet light from another spark.[3] This discovery was immediately investigated by Hallwachs, who showed that an insulated zinc electrode would lose a negative charge, but would not lose a positive charge when illuminated by the ultra-violet light of burning magnesium or of an electric arc.[4] If the zinc electrode was originally uncharged, it was found to acquire a positive charge on illumination. This effect is now known to be due to the emission of free electrons by the zinc electrode, the phenomenon being instantaneous with the incident light.

Elster and Geitel subsequently showed that sodium and potassium and their alloys exhibit this type of photo-electric effect when subjected to daylight and other sources of visible radiation.[5] These metals are so active chemically that they must be enclosed in evacuated glass cells furnished with wire connections to the light sensitive layer (the cathode) and to the electrode to which the electrons are drawn (or anode). The sodium or potassium is usually introduced by distillation *in vacuo*. Elster and Geitel also showed in 1912 that a cell consisting of a layer of potassium in an atmosphere of rarified hydrogen could be made more sensitive by passing an electric discharge through the cell, and cells of this type were widely used until 1928.[6]

Rubidium and cæsium behave similarly to sodium and potassium, but their sensitivity extends more into the red end of the spectrum.

Each photo-electric metal has a spectral sensitivity differing from the others, and this can be seen from the curves shown in Fig. 3. Sodium is most sensitive to the blue, violet and ultra-violet, and for this reason some sodium cells are fitted with quartz windows to admit ultra-violet radiation, which would be cut off by ordinary glass.

Cells of the photo-emissive class can be of two types : in one the vacuum is made as perfect as possible, whilst in the other a trace of inert gas at reduced pressure is present. Gas-filled cells give a larger output current than vacuum cells, as the residual gas enables a supply of ions to be produced by the electrons emitted from the cathode. The electrons on their passage to the anode collide with some of the gas molecules liberating further electrons and also positive ions. This process is sometimes referred to as " gas magnification."

The degree of ionisation produced is dependent upon a number of factors including anode potential and the intensity of illumination. Gas magnification is attended by a time-lag, which limits the frequency to which the cell can respond, and gas-filled cells therefore cannot be used for the very high frequencies of television. For such purposes vacuum cells must be employed.

All the cells so far considered have a thick layer of metal as the emitting surface. Work on another type of cathode, known as the thin film cathode, was started in 1924 by Ives.[7] In cells of this type the emitting layer is so thin as to be of atomic dimensions, and in certain cases a very great increase in sensitivity is obtained by

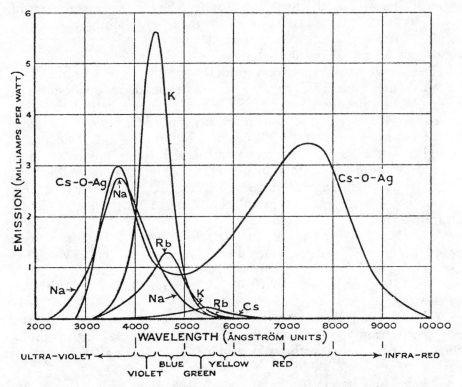

FIG. 3.—Spectral sensitivity of various photo-sensitive surfaces.

depositing this thin film on a specially treated metal surface. The standard type of cell for general purposes nowadays is of this kind, the alkali metal used being cæsium and the base oxidised silver. The method of production, due originally to Koller,[8] results in a complex formation of silver and cæsium oxide in which cæsium is embedded, and on which is a layer of cæsium probably only one atom thick. The spectral sensitivity curves * (Fig. 3) show that this type of cathode (referred to as Cs–O–Ag) is sensitive over a much wider range of wave-lengths than are the plain metal cathodes, the range extending, in fact, into the near infra-red. Since the radiation of

* The spectral sensitivities of different cells of the same type may in practice differ quite widely, especially in the case of modern cells by different makers. Hence the curves given should be regarded as approximate averages.

filament lamps is much richer in red and infra-red rays than in rays of shorter wavelengths, the superiority of these surfaces over the older type, whose sensitivity is confined to the shorter rays, is in practice much more marked than the curves indicate.

PHOTO-VOLTAIC CELLS (a) *Electrolytic.*—A cell of this type consists of two electrodes immersed in a suitable electrolyte in a similar manner to an ordinary galvanic cell. The fact that such a cell will develop an electromotive force when illuminated was first observed by Becquerel in 1839.[9] These cells are now of merely historical importance.

(b) *Electronic.*—The electronic type of photo-voltaic cell is variously known as the Rectifier, Barrier Layer, or Self-generating cell. Its action depends on the fact that light falling on certain solid " semi-conductors " causes electrons to leave the semi-conductor for an adjoining metal.

The fundamental discovery underlying their use was made by Grondahl and Geiger in 1927.[10] It was also made independently by Lange,[11] who produced, in 1930, the first practical cell of this type consisting of a layer of cuprous oxide on copper (similar to that used in the cuprous oxide rectifier), covered by an electrode transparent to light. Later, a similar effect was found in selenium between an iron plate and a very thin transparent upper electrode. This type (which must be distinguished from the photo-conductive or resistance selenium cell mentioned on p. 17) is now more widely used than the cuprous oxide cell.

The characteristic feature of these cells is that they yield a current when illuminated without the aid of an external battery. They are durable and give an output considerably better than that of the best gas-filled photo-emissive cell. Their maximum sensitivity is in the green region of the spectrum.

On the other hand they have the disadvantage of low internal resistance (from 300 to 2,000 ohms) which makes it difficult to employ them in conjunction with valve amplifiers, and their large capacity makes them unsuitable for high frequencies.

Their chief use is to operate a sensitive moving coil measuring instrument in photometers in which simplicity and portability are more important than accuracy, and for such purposes they are ideal.

THE ELECTRON MULTIPLIER The foregoing brief discussion shows that vacuum photo-emissive cells are alone suitable for the high frequencies of television, but their small output must be amplified millions of times. Amplification by means of thermionic valves always involves the introduction of unwanted disturbances or noise, and a definite limit is set to the permissible degree of amplification by the increase of the ratio of noise to signal.

Recently an old idea has been applied to remove this limitation. When an electron impinges on a surface with sufficient speed, it causes the emission of several secondary electrons from the target. By causing these secondary electrons in turn to impinge on a further

20

target so that each one liberates several more secondaries, and by repeating this process sufficiently often, an enormous multiplication of electrons (or amplification of the current) can be obtained. Like the emission of primary electrons with incident light, the liberation of secondary electrons is instantaneous, and secondary emission multipliers can therefore be used at television frequencies. Some noise is introduced in this multiplication, but the signal-to-noise ratio is considerably greater than that obtained by using thermionic valves.

Two developments have brought this idea to the fore. The first is the discovery of surfaces which give about nine secondary electrons per incident primary, where ordinary metals only give two or three. These surfaces are the same as the cæsium-treated oxidised silver cathodes mentioned above.

The second is the devising of methods for guiding the electrons from one surface to another. In one method proposed originally by

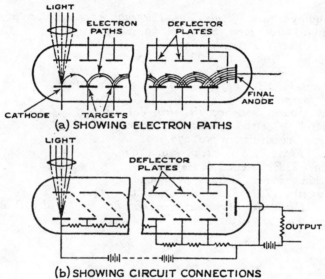

(a) SHOWING ELECTRON PATHS

(b) SHOWING CIRCUIT CONNECTIONS

FIG. 4.—Principles of the Zworykin electron multiplier.

Slepian [12] in 1919, but improved by Zworykin [13] in 1935, the electrons leaving a primary cathode under the action of light are attracted to the first potential electrode by means of a positive potential applied to the latter (see Fig. 4); but a magnetic field perpendicular to the plane of the drawing forces them to curve round so as to fall on the first secondary cathode. The secondary electrons from this electrode are then made to fall on a second cathode and so on, the electrons from the last secondary cathode being collected by the final anode from which the output current is carried away.

Another type, due to Farnsworth,[14] comprises two secondary emitting cathodes to which are applied a high frequency alternating potential which accelerates the electrons to and fro between the surfaces. This is carried out many times until the amplification is built up and the electrons collected by an annular anode.

In the Weiss type multiplier,[15] the electrons liberated from a sensitised grid are drawn through the meshes of the grid and accelerated towards further grids at each of which secondary multiplication takes place.

Electron multipliers can be made with many stages of amplification, and a ten-stage multiplier, giving an overall amplification of about 5×10^6, has been developed by Zworykin. Such multipliers will have a wide application for talking picture reproduction and in certain forms of television cameras.

TELEVISION CAMERAS Any form of television camera must employ a photo-electric element as the light-sensitive component. The two chief types of television camera are described in Chapter V.

BIBLIOGRAPHY

1. Willoughby Smith. *Journ. Soc. Telegraph Engrs.*, Vol. 2, p. 31. 1873.
2. Thirring. *Proc. Phys. Soc.*, Vol. 39, p. 97. 1926.
3. Hertz. *Ann. d. Phys.*, Vol. 31, p. 983. 1887.
4. Hallwachs. *Ann. d. Phys.*, Vol. 33, p. 301. 1888.
5. Elster and Geitel. *Ann. d. Phys.*, Vol. 41, p. 161. 1890.
6. Elster and Geitel. *Phys. Z.*, Vol. 14, p. 741. 1913.
7. Ives. *Journ. Astrophys.*, Vol. 60, p. 209. 1924.
8. Koller. *Journ. Opt. Soc. America*, Vol. 19, p. 135. 1929.
9. Becquerel. *Comptes Rendus.*, Vol. 9, p. 145. 1839.
10. Geiger. British Patent No. 277,610.
11. Lange. *Phys. Z.*, Vol. 31, p. 139. 1930.
12. Slepian. U.S. Patent No. 1,450,265.
13. Zworykin. *Proc. I.R.E.*, March, 1936.
14. Farnsworth. *Jour. Franklin Inst.*, Oct., 1934.
15. Weiss. *Funktech. Monatshefte.—Fernsehen u. Tonfilm.*, Vol. 7, No. 6, pp. 41–47. 1936.

CHAPTER III

PICTURE DISSECTION AND SYNTHESIS BY SCANNING

No picture or scene is properly intelligible to the human eye unless it can be perceived instantaneously as a complete whole. Unfortunately no practical electrical communication system is capable of handling more than one element of information at any instant. The inability of electrical communication systems to transmit a picture as a whole makes it necessary to dissect the picture into a large number of small elemental areas, to transmit them one by one, and to reassemble them in their appropriate positions at the receiver, in order that the observer may view the scene as a whole. If this process of dissection and reconstitution or synthesis is performed a sufficient number of times per second, the eye receives the impression of a complete picture, due to the phenomenon known as "persistence of vision." This dissection of the picture into small elemental areas is known as scanning.

Although scanning may be performed in several ways, it is usual to scan the picture in lines, from left to right, and to proceed line by line from top to bottom, in much the same way as one's eye travels when reading the pages of a book. This system, with a modification known as interlacing which will be described later, has been adopted by the B.B.C.

NIPKOW DISC   The earliest practical means of scanning was that invented by Nipkow in 1884. The method makes use of a rotating disc having a number of holes, regularly spaced on a spiral path, pierced in it. A photo-cell is mounted behind the disc and, as the latter rotates behind an aperture, the scene is scanned in a series of slightly curved lines, the rotation of the disc showing each individual portion of the scene to the photo-cell in rapid succession.

A similar result is obtained if the source of illumination and the photo-electric cell are interchanged. That is to say that instead of floodlighting the whole subject and examining each element with the photo-cell, it is practicable to illuminate the scene by a spot of light projected through the scanning disc while the whole scene is exposed directly to the photo-cell. The light falling on the sensitive surface of the photo-electric cells is converted into electrical energy, as described in Chapter II and, after suitable amplification, is radiated by the transmitter.

The resulting radio signal is picked up by the receiver and is caused to modulate a source of light therein. Besides modulating the light source, however, it is necessary to move the light beam rapidly over the surface of a screen in the same sequence and in exact synchronism with the scanning spot at the transmitter in order to ensure that the picture elements are reconstituted in the correct order. This may be done by means of a second Nipkow disc or any other scanning system.

MIRROR DRUM Besides the Nipkow disc, which is now practically obsolete, a number of different devices have been proposed for picture dissection and synthesis. One type is known as a mirror drum and consists of a drum, the periphery of which carries a number of mirrors, each successive mirror being set at a slightly different angle to the drum axis. If a narrow beam of light is projected on to a mirror through an appropriate lens, and if the drum is rotated so that the beam falls upon each mirror in turn, the reflected light will scan the surface of a suitably placed screen in a series of lines. Such a series of parallel lines is known as a " scanning field."

MODERN METHODS Although picture dissection and synthesis may be performed by both electrical and mechanical methods, only the latter system can be used for large screen pictures. Electrical dissection of the picture is performed by means of a cathode ray tube and a photocell or electron multiplier, or by means of an electron camera, while electrical synthesis is performed by means of the cathode ray tube. These devices are described in detail in Chapter V.

For small pictures, the cathode ray tube has several advantages over mechanical methods of synthesis. It combines the functions of light source, light modulator and picture screen in one unit, it is of convenient size and has no moving parts other than the electron beam, and, due to the negligible inertia of the electron beam, it can function at the very high frequencies necessary for high definition television. On the other hand, mechanical methods have the advantage of a flat screen, a linear scanning characteristic and a picture size which is not limited by the size of the apparatus.

In order to reproduce the picture on the screen of the cathode ray tube, it is necessary to deflect the electron beam to form the scanning field, and this must be done in exact synchronism with the scanning spot at the transmitter. Deflection of the beam may be performed by electromagnetic or electrostatic means.

In the case of electromagnetic deflection, two coils are placed on either side of the neck of the tube, and the deflection is proportional to the current flowing through these coils. If two deflecting systems are arranged so that they function at right angles to each other, the cathode ray beam may be deflected horizontally and vertically, and may therefore be caused to form a scanning field on the screen if the relative frequencies of the deflections are suitably chosen.

A beam of electrons may also be deflected by means of an electrostatic field. By applying suitable potentials to a pair of deflector plates arranged one on each side of the beam, the latter may be caused to swing across the screen and, by using two pairs of deflector plates at right angles to one another, it is possible to obtain a complete scanning field.

TIME-BASE GENERATORS The apparatus for providing the potentials for forming the scanning field is known as a time-base oscillation generator. A time-base generator in its simplest form, consists of a condenser which is charged from a direct current source through a high resistance and discharged by short-circuiting. The potential across the condenser is applied to the deflector plates, and the deflection of the beam is proportional at each instant to the potential across the condenser. It is necessary to ensure that this potential increases linearly with time if the received image is to be undistorted.

The discharge of the condenser is automatically accomplished by means of a gas-filled relay or a high vacuum valve which is connected across the condenser. The valve is normally in a non-conducting condition, but is " triggered " by the synchronising signal from the transmitter. A number of circuit diagrams of time-base generators are shown in the exhibition and the method of action of the various circuits is described. The bibliography at the end of this chapter also gives references to some important developments in these circuits.

It is usual to apply the time-base deflecting potentials to the deflector plates through a form of push-pull amplifier since a balanced deflection is necessary if the cathode ray spot is to remain in focus. A somewhat similar circuit is used when the beam is deflected electromagnetically, but the inductance of the deflecting coils must be taken into consideration in the design of the circuit.

INTERLACED SCANNING Unless the picture is scanned at a high speed an objectionable flicker is observed, particularly at high picture brilliancies. This is because the rate of scanning, when less than about 35 to 40 traversals per second, is not sufficient to prevent the scanning movement from being appreciated by the eye. A method of overcoming this defect is to scan the picture at a more rapid rate, say 50 pictures per second, but this would increase considerably the band of frequencies required for transmission. The method of overcoming flicker which has been adopted for the B.B.C. transmissions employs the principle known as " interlacing." The scene to be transmitted is scanned in two movements, the first scan covering the " odd " lines 1, 3, 5, etc., a space being left between each line corresponding to the width of a line. On the completion of the first scan, *i.e.* when half the total number of lines in the picture have been scanned, the spot scans the even lines. The whole picture is thus scanned in two portions separated by the width of one line and the *apparent* frequency so far as flicker is concerned is 50 per second, whereas the complete picture is actually scanned at the rate of 25 pictures per second.

BIBLIOGRAPHY

D. Prinz and W. Wehnert. "A Time-Base in which the Condenser is discharged by an Oscillating Valve." U.S.A. Patent No. 2,036,719.

Report of the Radio Research Board for the period ending March 31st, 1929. "A Linear Time-Base for the Cathode Ray Oscillograph."

A. H. Kobayashi. "A Time-Base in which the Condenser is discharged by an Oscillating Valve." U.S.A. Patent No. 1,913,449.

R. V. L. Hartley. Transmission of Information. *Bell System Tech. Journ.*, July, 1928.

R. S. Holmes, W. L. Carson and W. A. Tolson. "An Experimental Television System (The Receiver)," *Proc. I.R.E.*, Vol. 21, No. 12. December, 1933.

D. M. Johnstone. "The Design of a Line Scanning Transformer," *Journ. Television Soc.*, June, 1936. Vol. 2, Part 5.

O. S. Puckle. "A Time-Base employing Hard Valves," *Journ. Television Soc.*, June, 1936. Vol. 2, Part 5.

CHAPTER IV

LIGHT CONTROL

Every television receiver requires some method of light control by means of which the electrical signals received from a transmitter can be made to vary the intensity of a source of light at the receiving end. These variations are then distributed over a surface by the scanning device to reproduce the picture. Various types of light controls are fully described in the Exhibition, and the following explanation is intended as a general guide to the principles involved.

DISCHARGE LAMPS Discharge lamps containing neon gas were used by Baird in his early experiments in conjunction with a Nipkow disc. When the gas pressure has been suitably chosen, the variation of current through the lamp is linear with respect to the applied voltage over a wide range. The television signal variations produced by the changing light values at the transmitting end are applied to the neon tube and the overall brightness change in the gas glow spread over the negative electrode is observed through the apertures of the synchronised scanning disc. In this way the picture is built up in its correct form as a coarse miniature replica of the subject at the transmitting end.

A combination of " positive column " discharge tubes in which one tube contained substantially pure neon, and the other a mixture of mercury and helium was used for the first demonstration of " colour television " by Baird in 1928. These tubes were employed in conjunction with a triple spiral apertured disc having red, blue and green filters.

In addition to the flat plate neon lamp used in conjunction with an apertured disc scanner, neon and other lamps of the point cathode type have also been employed with varying degrees of success when used with picture reconstituting devices of the oscillating mirror or rotating mirror drum type, but the limited frequency response and the restricting effect of inter-electrode capacities make this class of light control unsuitable for high definition television.

LIGHT CONTROL CELLS A second class of light control uses a device which is independent of the light source, the luminous intensity remaining unaltered, and the modulation being accomplished by means somewhat analogous to the action of a shutter or diaphragm. Numerous types have been proposed : mechanical, optical-mechanical and electro-optical. Again because of the very high frequency response required, only the electro-optical method need be considered, and of these only two, the Kerr cell and the supersonic light control, are of importance.

The history of electro-optical methods of light control may be said to date from 1845, when Michael Faraday discovered the rotation of the plane of polarisation of polarised light when passed through a magnetically stressed section of heavy lead glass.

Since that date many other media which exhibit the Faraday effect have been discovered, and it was this scheme which was proposed by Nipkow in 1884 in his German patent for modulating a beam of light in conjunction with a scanning disc.

KERR CELL In 1875, however, Dr. John Kerr discovered an entirely different effect which is now known by his name. He found that when a liquid medium such as carbon disulphide is subjected to electrostatic stress between a pair of plates immersed in the fluid, it becomes " bi-refringent," *i.e.* doubly refracting. As is well known, a beam of light consists of transverse vibrations in all directions at right angles to the direction of propagation. Certain arrangements of natural crystals have the property that only light vibrations in one particular plane are transmitted through them, and such crystals are said to be bi-refringent. Crystals of Iceland Spar are an excellent example of this media, and the action of an arrangement of this type (*e.g.* a Nicol prism) is to select the component of all light vibrations lying in a given plane fixed with respect to the prism.

If a second prism is set to pass all components in a plane at right angles to that of the first, then the net result is that no light will pass through the double prism combination. A Kerr cell inserted between the two sets of prisms has the effect of distorting the plane of vibration of the light which has passed through the first prism into an ellipse of eccentricity which progressively changes as the voltage across the cell plates is increased, passing through a circle and eventually becoming a line of vibration at right angles to the initial direction. Accordingly, a progressively increasing component of the light is available for passage through the second prism. In practice, it is found that a simple two-plate Kerr cell requires extremely high voltages to make it operate successfully, and to overcome this difficulty a multi-plate cell was devised. The cell consisted principally of a set of very thin interleaved electrodes sealed and immersed in chemically pure nitro-benzine. This cell has negligible inertia and the variations of the light beam passing through the combination are practically proportional over a definite range to the corresponding voltage variations due to the applied television signal.

DOUBLE-IMAGE KERR CELL An improved type of Kerr cell uses double image prisms which separate the incident light into two images, light in one image being polarised differently from that in the other. Once separated, the Kerr cell controls both components simultaneously, and a double image analysing prism brings about re-combination, or still further separation, of the lights of the two images, according to the method used. The light emerging from the arrangement is preferably adjusted so that three images are formed, light being transferred by the signal from the centre image to the outside images and *vice versa*. A stop is provided to eliminate the side images, the centre image only being utilised. Alternatively, by stopping out the central image, the side images can be utilised. The double-image Kerr cell passes at least twice as much light as the normal Kerr cell.

THE SUPERSONIC LIGHT CONTROL The supersonic light control used by Scophony Ltd. does not use polarised light, but employs the principles of diffraction to vary the intensity of the emerging beam. It consists of a transparent medium, generally a liquid contained in a suitable cell having transparent sides. A piezo-electric crystal is energised by a high-frequency oscillator and produces a series of supersonic waves which are applied to the medium transversely to the optical axis. These waves have a very short wave-length and consist of alternate half-waves of compression and rarefaction, so that the refractive indices of the two half-waves are different. Light passing through the layer of waves is retarded by the compressive and accelerated by the rarified half-waves. This brings about interference of the emergent light, and produces diffraction spectra. The system of supersonic waves produced by the crystal thus acts as a diffraction grating, and the emergent light is brought to a focus in the same way as when using such a grating. At this focus appear the diffraction spectral images and a central image. Models in the Exhibition show working examples of a double image Kerr cell and a supersonic light control.

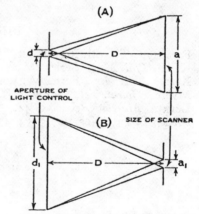

FIG. 5.—Diagram to show the reduction in size of scanner when using a number of elemental areas simultaneously active.

If the supersonic light control were used to illuminate a single scanning spot on the screen, like other light controls at present known, it would show no advantage over the Kerr cell.

The advance which has been made possible with the new device is due almost entirely to the illumination of a large number of the picture elements *simultaneously*, the number used being in some cases equal to a whole line of picture elements (*i.e.* about 500 for a 405-line picture).

For a given quantity of light, it is optically more simple to project on to the screen the long image of a long aperture such as that of the supersonic light control, than to project an image of element size from the tiny aperture which is all that can be made available in a Kerr cell suitable for high definition.

This can be appreciated from Fig. 5. In Fig. 5A an aperture of element size d such as would be used with a Kerr cell passes light to fill a scanner surface a at a distance D, the quantity of light being thus proportional to $\dfrac{da}{D}$. In Fig. 5B is an arrangement such as is used with the supersonic light control, the aperture d_1 having a size corresponding to many elements. In order to handle the same quantity of light, the scanner surface a_1 at the distance D need only have a size $\dfrac{da}{d_1}$. From this it will be seen that for the same amount of light, the supersonic light control greatly reduces the scanner surface necessary; in fact it reduces it n times, n being the number of elements simultaneously active.

LIGHT CONTROL IN CATHODE RAY TUBE In the case of the cathode ray tube, the modulation of the light is accomplished by varying the potential of the grid which surrounds the cathode. The action is similar in many respects with that of the thermionic valve, the flow of electrons being reduced by increasing the negative grid potential, and full brilliance of the spot occurring when the grid bias is reduced to a low value. The cathode ray tube is described further in Chapter V.

BIBLIOGRAPHY

Faraday. " Exptl. Researches," Vol. III. 1845.
Kerr. *Phil. Mag.*, Series IV., Vol. 50, p. 337. 1875.
Kerr. *Phil. Mag.*, Series V., Vol. 9, p. 157. 1880.
Myers. *Journ. Tel. Soc.*, Vol. I., Pt. 9. 1933.
Levin. *Marconi Rev.*, No. 44. 1933.
British Patents Nos. 407,385 ; 439,236 ; 328,286 and 451,132 ; 433,945.

CHAPTER V

CATHODE RAY TUBES AND ELECTRON CAMERAS

The modern cathode ray tube used in television reproduction is a development of the Braun-Wehnelt tube which was in laboratory use at the beginning of the century. The effects produced by the discharge of electricity through a vacuum had been investigated since 1859, and the name " cathode ray " was given by Plücker to the discharge of electricity from the cathode of a vacuum tube when a high potential was applied to an electrode (the anode) at the opposite end of the tube. It was later shown that this discharge produced fluorescence on the glass walls of the tube, due to bombardment by particles of electricity to which the name " electron " was given by Johnstone Stoney in 1890.

BRAUN TUBE   In the Braun tube (1897) the electron stream emitted from the cathode when a high potential was applied to the anode A (Fig. 6) was directed up the tube, and after passing through an aperture D impinged on a mica screen coated with fluorescent material. The point of impact of the

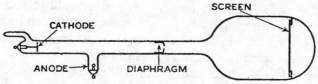

FIG. 6.—The Braun cathode ray tube of 1897.

beam was thus made visible, and by applying an external magnetic field the beam could be moved so as to fall on any point on the fluorescent screen. It was this tube which Campbell Swinton had in mind when forecasting his system of television.

In 1905 Wehnelt added an important improvement to the Braun tube by using a hot cathode—a strip of platinum coated with oxides and heated electrically to red heat. The increased electron emission obtained with this cathode enabled a lower potential to be applied to the anode of the tube and gave a fluorescent spot of greater brilliance, although the life of the emitting surface was short.

Up to this time no attempt had been made to reduce the spot on the screen to the smallest dimensions, although the advantage of such a refinement was understood. The early tubes had metal diaphragms which limited the section of the electron beam, but in 1902 Ryan showed that a magnetic coil surrounding the neck of the tube had a focusing action on the electron beam, and by varying both the position of the coil and the value of current through it an exceedingly sharp spot could be obtained.

Another improvement was the insertion of a cylinder or shield surrounding the cathode (the " Wehnelt cylinder "), which, on applying a suitable negative potential, acted as a pre-concentrator of the electron stream and directed it up the tube in the form of a narrow jet.

GAS FOCUSING In 1921 Van der Bijl and Johnson showed that a trace of gas left in the tube after exhaustion acted as a focusing medium and enabled sharp focus to be obtained at low voltages (300 v.). The action of the gas-focused tube can be briefly described as follows. The electron beam on its way up the tube collides with molecules of the residual gas and forms positive ions by collision. These ions, being relatively slow-moving on account of their greater mass, remain in the path of the beam and act as a central positive core which attracts electrons from the outer edge of the beam towards the middle. The effect is thus to constrict the section of the beam as it travels towards the screen and to produce a small sharply defined spot, the size of which is governed by the beam current, *i.e.* the number of electrons in the beam. Until a few years ago the gas-focused tube was widely used in research laboratories on account of its convenience and the low accelerating voltage required for the production of the spot, but one or two inherent defects have precluded its use in television reproduction.

ELECTROSTATIC FOCUSING About 1932 the theory and practice of focusing the electron stream was extended by the introduction of " electrostatic focusing " or the converging of the stream by means of one or more electric fields. The researches of Brüche, Knoll and Ruska showed that the electron beam could be focused in a similar manner to a beam of light by the action of electric fields arranged to form " electron lenses."

These electron lenses take the form of electrostatic fields between discs or cylinders (mounted co-axially along the path of the beam) to which various potentials are applied. By suitably adjusting the potentials between the electrodes the beam can be caused to converge to a small point on the screen. The science of electrostatic focusing has been developed under the name " electron-optics," and plays a very important part in the various electronic devices used in modern television.

A diagram of a modern electrostatically focused cathode ray tube is shown in Fig. 7. The electrons emitted by the cathode C are constricted by the negative potential applied to the Wehnelt cylinder G, now usually termed the " grid " following the nomenclature of thermionic valves. The first anode A_1 has a positive potential of a few hundred volts applied to it, which serves to accelerate the electrons towards the screen end of the tube. After emerging from the aperture in the anode A_1 the electron stream tends to diverge, due to the mutual repulsion which exists between individual electrons. A second anode A_2, having a potential of 1,000 volts or more, is mounted axially above the first, and the field between the two anodes converges the beam as shown by the dotted outline. It is preferable to focus the beam on the screen in two stages to avoid the equivalent of " spherical aberration " and other defects, and hence a third anode, A_3, is mounted above the second, having a higher value of positive potential (3,000–6,000 volts). The field between the second and third anodes constitutes a second lens which finally determines the point at which the beam is focused.

In practice, the potential of the final anode is fixed and that of the second anode varied to produce the sharply focused spot. The

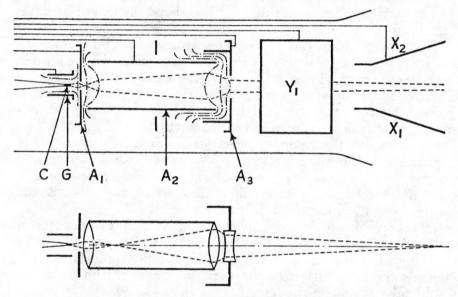

FIG. 7.—Diagram showing typical electrode arrangement and electrostatic focusing fields in a modern cathode ray tube.

action of the grid is important in television reproduction, as its potential controls the number of electrons arriving at the screen and hence the brightness of the spot. The vision signal is applied to this electrode to modulate the light intensity on the screen, and hence in the cathode ray tube we have a source of luminosity which can be controlled in intensity and at the same time caused to move in any direction over the surface of the reproducing screen.

MAGNETIC FOCUSING

Owing to its relatively greater simplicity the use of a magnetic field for focusing the beam has been revived in a large number of modern cathode ray tubes whose internal structure is thereby simplified. The grid and first anode are retained for the purpose of modulating and accelerating the beam, but the remaining anodes are replaced by a coil external to the tube and co-axial with it. The action of the magnetic focusing coil can be briefly explained as follows : An electron leaving the axis of the tube at an angle will be travelling across the magnetic field and will experience a deflecting force proportional to the strength of the field and to the velocity with which it is moving. Electrons travelling up the tube parallel to the lines of force of the coil will experience no deflecting force and will not be affected by the field. The actual path which an electron will be constrained to take will be a combination of that due to its initial velocity along the tube and its radial velocity under the action of the field.

Diverging electrons will move up the tube in helical paths of differing pitch and conical angle, but all eventually arrive on the axis at the same point. This point is caused to coincide with the surface of the screen. Examples of magnetically focused tubes are shown in the exhibition, and a special exhibit has been constructed to show the effect of varying the current through the coil and the length of travel of the beam.

The focused beam can be deflected in any direction across the fluorescent screen by magnetic or electrostatic means. In the former, which was the original method of producing a trace on the screen, an electromagnet is mounted externally to the tube and the beam is deflected by the magnetic field in the same way that any current-carrying conductor is influenced by a magnetic field.

For deflecting the beam by electrostatic fields, a pair of plates is mounted in the tube above the final anode, the beam passing mid-way between them (X_1 and X_2 of Fig. 7). The application of a potential between the plates will deflect the beam towards the positive one, the movement being proportional to the deflecting potential applied. A similar pair of plates, Y_1 and Y_2, mounted at right angles to the first will deflect the beam in a plane at right angles to the original movement due to X_1 and X_2, and by applying suitable potentials simultaneously to both pairs, the beam can be caused to take up any position on the screen.

When used for television reproduction, the output potentials of the two time-base generators (one oscillating at 10,125 cycles per second, and the other at 50 cycles per second for the present B.B.C. transmissions) are applied respectively to the two pairs of deflector plates. The cathode ray beam is thus deflected in both the vertical and horizontal directions simultaneously and a scanning field formed as described in Chapter III. Modulation of the instantaneous intensity of the beam produces the picture on the screen.

BIBLIOGRAPHY

Knoll and Ruska. *Ann. d. Phys.*, Vol. 12, pp. 583, 656. 1932.
Johannson and Scherzer. *Zeit. f. Phys.*, Vol. 80, p. 464. 1933.
Taylor, Headrick and Orth. *Electronics*, December, 1933.
Maloff and Epstein. *Proc. I.R.E.*, Vol. 22, p. 1386. 1934.
McArthur. *Electronics*, June, 1932.
Busch. *Arch. f. Elek.*, Vol. 18, p. 583. 1927.

ELECTRON CAMERAS

The function of the microphone in sound broadcasting has its counterpart in television in a special form of camera of which there are two principal types, (1) the storage type first suggested by Campbell Swinton, and (2) the image dissector tube first patented by Dieckmann and Hell in Germany in April, 1925. As these are both electronic in action and are closely related to the cathode ray tube already described, they will be dealt with in the present chapter.

The function of a television camera is to convert the variations of light and shade in an image into corresponding electrical variations which can then be amplified and transmitted by wire or radio.

THE EMITRON TELEVISION CAMERA   The Emitron camera shown in Fig. 8 consists of a highly evacuated glass envelope A which has a flat plate glass window B and a neck C in which an electron-producing system is carried. This system consists of a cathode D,

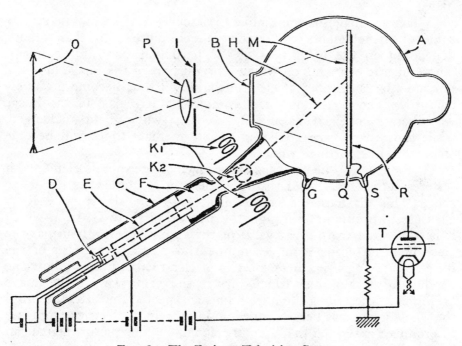

FIG. 8.—The Emitron Television Camera.

the electrons from which are accelerated and focused into a small spot on the mosaic screen M by the electrodes E and F in a manner similar to that in an ordinary cathode ray tube as already described. The electrode F consists of a silver coating on the glass walls of the tube to which a contact is made at G. As the electron beam H passes through the constricted portion of the neck it is subjected to the influence of two mutually perpendicular magnetic fields produced by the two pairs of coils K_1 and K_2. These coils are supplied with alternating currents of special wave-form and frequency from the pulse generator described in Chapter VI, and cause the electron beam to be deflected and to scan the surface of the mosaic screen in a series of parallel lines which form a scanning field as described in Chapter III.

A lens P forms an image of the scene to be transmitted on the mosaic screen M. The screen consists of a thin sheet of insulating material Q, such as mica, covered on one side with a conductive metal coating R, known as the " signal plate," to which an electrical contact is made by means of a wire sealed through the glass wall at S. On the side of the mica sheet Q, which faces the window B and lens P, a mosaic M is formed which consists of a very large number of minute globules of silver deposited in such a way that each globule is separate and electrically insulated from its neighbours. The surface is then treated by a special process which deposits a photo-sensitive coating on the silver globules. When light falls upon the surface of the mosaic, each globule liberates electrons in proportion to the quantity of light falling upon it. Further, each element of the mosaic forms a minute condenser with the signal plate.

When an image of some object is focused on the mosaic screen, some of the elements will receive much more light than others, depending on the brightness of the particular part of the image. Those mosaic elements on which a bright part of the image falls will liberate a large number of electrons and will thus become positively charged. Those elements in a completely dark part of the image will liberate no electrons and their charge will remain unchanged, while those parts corresponding to intermediate tones will become charged to intermediate values.

As the electron beam scans the surface of the mosaic elements, it restores sufficient electrons to each element to reduce its charge to zero. This discharge is communicated to the signal plate R by electrostatic induction through the insulating sheet Q, from which it is conducted to the first valve of the amplifier T. Hence, as the electron beam scans along each line and passes over parts of the image of varying brightness, a series of electrical impulses follow one another to the amplifier which vary in the same way as the image brightness varies along the line. This is repeated line after line until the whole image has been scanned and the train of electrical impulses so produced is termed the " picture signal." This picture signal can be amplified and, after suitable treatment, radiated by the transmitter.

It is important to notice that while the electron beam passes over the mosaic twenty-five times each second, it only remains on an element for less than one millionth of a second. The light of the image, however, remains on each element continually, and consequently the building up of the electrical charge on the elements goes on continually, until, as the beam returns to each element, it discharges it. .

The ability of the device to store up the effect of the light in this manner increases its efficiency greatly in comparison with transmitting devices which depend on the action of the light only during the time of scanning each picture point.

Another interesting and useful feature of the Emitron camera is what may be termed its " electric memory." If an image is formed on the mosaic and the electron beam is not allowed to scan it, no picture signal will be produced. If the light is cut off after it has acted on the mosaic for a few seconds it is found that the electrical charges produced by the light will remain on the mosaic for some minutes, and if during this time the electron beam is caused to scan over the mosaic, a picture signal is produced which corresponds to the original image. It may thus be said that the device has the faculty of " remembering " an image of an object that it has " seen " and of reproducing it some time later. This feature is made use of in one method of film transmission as it enables the image to be impressed on the mosaic and subsequently scanned while the film is taking up a new position.

BIBLIOGRAPHY

1. Campbell Swinton, *Journ. Röntgen Soc.*, Vol. VIII., pp. 1–13, 1912.
2. Zworykin. *Jour. Franklin Inst.*, Vol. 215, p. 535. 1933.
3. Zworykin. *Proc. I.R.E.*, March, 1936.

THE BAIRD ELECTRON CAMERA

In the Baird Electron camera, a development of the image dissector tube,[1] an optical image of the scene to be scanned is focused upon a large uniform photo-electric cathode of high sensitivity. Electrons are liberated from the cathode at any particular point in direct proportion to the degree of illumination at that point. This produces an " electron image " corresponding to the optical image, at or very near the surface of the cathode. In a normal photo-cell the emission of electrons becomes diffuse inside the tube, but in this type of electron camera the electron image is brought to a sharp focus in a plane parallel to the cathode, but at some distance removed from it, by a combination of magnetic and electrostatic fields. This electron image is naturally invisible and is composed of variations in electron density corresponding to variations of illumination on the cathode. The focusing of the electron image is similar in principle to the cathode ray tube operation dealt with earlier in this chapter, and although all the electrons travel in different paths towards the plane of the scanning aperture, on arrival at that plane, they will all bear the same relative positions and velocities as they had at the instant of emission.

For television purposes, scanning is accomplished by displacing the focused electron image by two auxiliary magnetic fields perpendicular to the focusing field and thus sweeping the electron image across a fixed scanning aperture placed before a collecting anode. At any instant the electrons forming the image at the particular point coincident with the scanning aperture enter the aperture and are attracted to the collecting anode. These electrons are then amplified by means of a secondary emission multiplier and the current from this multiplier constitutes the output current from the camera.

This secondary emission multiplier amplifies the electron current entering the scanning aperture about 2,000 times without frequency

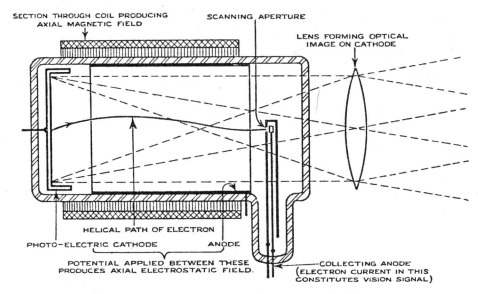

FIG. 9.—Electron Camera as used by Baird Television, Ltd.

distortion within the range required for a good high definition picture, and without the introduction of amplifier noise.

The remainder of the circuits associated with the camera consist of vision signal amplifiers, generators for the scanning currents which cause the traversal of the image over the aperture, a master frequency generator for synchronising these scanning generators and pulse generators for injecting the synchronising pulses into the vision signal.

BIBLIOGRAPHY.

1. Dieckmann and Hell. German Patent No. 450,187. 1925.
2. Roberts. British Patent No. 318,331. 1928.
3. Farnsworth. British Patents No. 368,309 and No. 368,721. 1930.
4. Farnsworth. *Journ. Franklin Inst.*, October, 1934.

CHAPTER VI

THE VISION TRANSMITTER AT ALEXANDRA PALACE

The problems involved in the transmission of high definition vision signals are considerably more complicated than the analogous problems associated with a sound transmitter. This is principally due to the far higher modulation frequencies involved and partly, also, to the necessity for incorporating synchronising pulses in the transmission.

As has been briefly described in Chapter III, it is impossible to transmit simultaneously all the thousands of small areas of light and shade which go to make up even a simple picture, but the picture or scene has to be dissected into very small elements, and the tone values of the individual elements transmitted in rapid succession. This means that the circuits have to be capable of responding to all frequencies between zero and about 2·5 megacycles, *i.e.* 2½ million cycles per second.

CHOICE OF WAVELENGTH When a high frequency carrier wave is modulated by waves of a lower frequency, such, for example, as sound frequencies, additional waves or " sidebands " are produced having frequencies both higher and lower than that of the carrier and extending over a total band of twice the modulation frequency. For example, consider a broadcast transmitter on a wavelength of 300 metres. This has a carrier frequency of one megacycle, and if this wave is modulated by sound waves having frequencies up to 10,000 per second, the sideband frequencies will extend from 990,000 cycles to 1,010,000 cycles, corresponding to wavelengths between 297 and 303 metres. If interference is to be avoided therefore, the wavelengths of stations must be so chosen as not to encroach on the waveband occupied by any other. There is thus a definite limit to the number of stations which can occupy any given band of wavelengths.

The modulation of a high frequency carrier wave by a band of frequencies ranging as high as 2·5 megacycles produces sideband frequencies extending over a total band-width of five megacycles—2·5 megacycles on either side of the fundamental frequency. Such frequencies are higher than the fundamental carrier frequency of the medium-wave broadcast transmitters, and for this reason it is necessary to employ a shorter wavelength for television. Considerations connected with the propagation characteristics of the available wavelengths led finally to a choice of 6·67 metres for the vision transmitter. An important characteristic of the ultra-short wave-lengths is that their range tends to become limited to the areas within visual range. At wavelengths of the order of 6–8 metres, the range is rather greater than the visual range, and refraction of the waves tends to prevent the occurrence of undue " blind-spots " behind hills. Even so, the effective area in which reliable reception can be obtained is limited.

TELEVISION CAMERAS Returning to a consideration of the transmitter, the first and most distinctive component of a vision transmitter is the camera. Externally, its appearance is similar to a conventional motion picture camera. It is fitted to a similar stand, it is capable of similar movements, and it is fitted with an ordinary lens and focusing system. Internally, however, it is very different, but as the two chief types of television camera have been described in Chapter V they need not be considered here.

It is necessary to employ more than one camera for general work, and at the London transmitting station at Alexandra Palace, at which the Marconi–E.M.I. system is now employed, no less than six cameras are in use, four being used in the studios and two for the transmission of films.

The signals generated by the cameras are extremely feeble and are first amplified in a unit built into the camera itself. After amplification in the four-stage " head amplifier " they pass down a special multi-core screened cable to the main amplifying equipment. This cable not only carries the vision signals from the camera to the main amplifiers, but it also carries the high and low tension supplies to the camera and its associated amplifier as well as the scanning pulses from the time-base generators to the deflecting coils of the camera.

" A " AMPLIFIER After passing through the multi-core connecting cable, the vision signals enter the " A " amplifier, one of which is provided for each camera. Each " A " amplifier incorporates an " illumination corrector unit," the purpose of which is to apply certain correcting impulses to the picture signals. The signals generated by the cameras tend to fluctuate about a mean level, and this results in the production of a light or shadow effect over part of the picture. It is the function of the illumination corrector units to adjust for this effect, but as the fluctuations vary continually with changing scenes or illumination, the correcting controls require corresponding manual adjustment.

After correction for evenness of illumination, the signals enter the " phase-reversing unit," the purpose of which is to enable either positive or negative films to be used for transmission. When a negative film is used this unit, which employs a single phase-reversing valve without amplification, is brought into use, and the picture is transmitted as a positive.

Up to this point, the equipment described is appropriate to each camera, there being an " A " amplifier and an illumination corrector unit for each one, but at this stage the camera signals are fed into a "mixer unit" which is operated by electrical remote control from the producer's control desk. This unit enables the signals from any particular camera to be used at will and controls are incorporated for fading or superimposing any of the camera outputs to enable artistic or dramatic effects to be obtained.

Electrically, the mixer unit employs two groups of six valves, each group having a common anode circuit which is connected to a separate "B" amplifier. Control of the picture signals from the individual cameras is obtained by adjustment of the appropriate camera valve and, by connecting the anode circuits of each group together, it is possible to select and pass on the signals from a single camera, or, if necessary, a combination of cameras may be selected in any desired order.

There are two independent output channels from the mixer unit, the two channels being identical and interchangeable. This provision not only permits the observation on a monitor tube of the picture actually being transmitted, but also enables the producer to watch the picture on the camera which is to come into action next.

" B " AMPLIFIER The duplicate " B " amplifiers each comprise a four-stage amplifier with appropriate gain controls. The " B " amplifiers also contain a unit for the partial removal of certain spurious transient signals which arise in the camera output during the intervals between successive lines and frames. These transients arise when the steady conditions which obtain in the camera during the scanning period are interrupted by the cutting off of the electron beam during the " fly-back " period at the end of each line and frame.

From the " B " amplifiers the signals are fed into duplicate " C " amplifiers which comprise a further three stages of resistance-capacity amplification and continue the process of removing the spurious transients referred to above.

After passing through the " C " amplifiers the picture signals are fed into the " suppression mixer unit," which finally suppresses the transients and, in addition, part of the vision signals at the ends of the lines and frames in order to provide the requisite space for the insertion of the synchronising signals.

The signals now enter the " gamma panel," a unit which is chiefly used during the transmission of films for the purpose of adjusting the picture contrasts. The word " gamma " is derived from the photographic term used to denote contrast ratios.

The next stage is the " synchronising signal mixer unit " in which the synchronising pulses are applied to the vision signals in the intervals between the lines and frames prepared by the suppression mixer unit. With the addition of the synchronising pulses, the signal becomes the complete modulating signal and, after suitable amplification, it is now ready to control the radio frequency transmitter. Before amplification, however, it passes into the " distribution unit," which provides five separate output circuits, one being

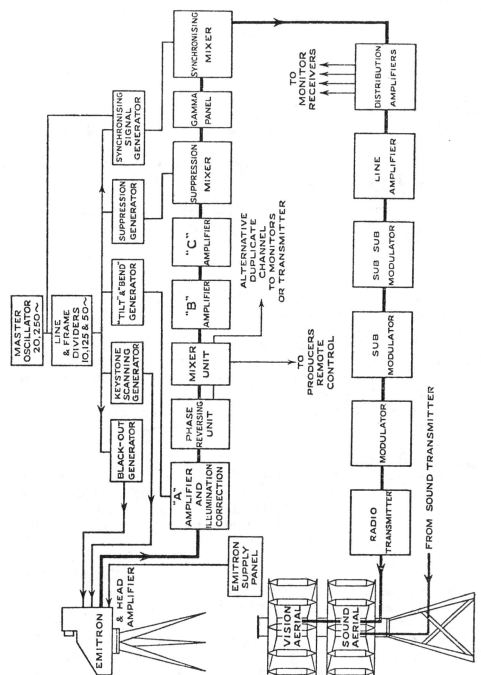

FIG. 10.—Arrangement of Marconi-E.M.I. television transmitter at Alexandra Palace.

used to feed the modulating amplifiers and four being available for the operation of monitor receivers.

IMPULSE GENERATORS There is a further series of units which have not been considered hitherto and which are concerned with the generation of the numerous different forms of impulse which are required at various points throughout the system. These units are controlled by a master oscillator which is electrically locked with the 50-cycle supply mains and which generates a rectangular wave-form having a frequency of 20,250 cycles. This frequency is supplied to two units, known as the line and frame dividers, which select the other two frequencies required in the system, namely 10,125 and 50 cycles respectively, these being the line and frame repetition frequencies.

The three frequencies, 20,250, 10,125 and 50 cycles, are supplied to the " synchronising signal generator " which generates the necessary impulses for incorporation with the vision signals in the synchronising signal mixer unit referred to earlier, and the 10,125 and 50-cycle pulses are also supplied to the " black-out generator," the " keystone generator," the " tilt and bend " generator, and the " suppression generator."

The black-out generator produces a rectangular wave-form for application to the Emitron cameras in order to suppress the return stroke or " fly-back " between line and frame scanning periods. The " keystone " or scanning generator provides the " saw-tooth " impulses for deflecting the scanning beam in the cameras in the line and frame directions. Owing to the assymetrical position of the electron gun in relation to the photo-electric mosaic of the camera, the scanning impulses require to have a special wave-form to counteract the geometric distortion which would otherwise occur. The " tilt and bend " generator produces specially shaped pulses for feeding to the " A " amplifier illumination correcting controls which, if mixed in the correct proportions, can neutralise any illumination distortion which is likely to arise. The suppression generator produces rectangular-shaped pulses which are supplied to the suppression mixer for the purpose of removing the spurious transients to which reference was made previously.

" D.C." MODULATION The system of modulation employed in the vision transmitter is known as a " D.C." system on account of the introduction of a direct current component by means of a d.c. restoring circuit. It is well known that the electrical output from a photo-electric device is a unidirectional current. Under the influence of light of varying intensity the output may be resolved into an alternating current plus a unidirectional component, the latter being a measure of overall picture brightness. This picture brightness component is transmitted as an amplitude modulation of the transmitter so that a definite carrier wave value is associated with a definite brightness. This has been called direct current working, and results in there being no fixed value of average carrier wave, since this quantity varies with picture brightness.

The advantages of D.C. working are manifold, but three of the more outstanding may be cited. First, by transmitting the D.C. component all gradations of light and shade in the picture are given their proper tonal values by the receiver. Secondly, there is no tendency for the signals to float and distribute themselves about an arbitrary centre line, as in the case of A.C. working ; so that the extremes of modulation of the transmitter are known and fixed. In consequence much more efficient use may be made of the available transmitter characteristic, since excursions to the extreme limits of permissible amplitude may be made in the full confidence that the signals will not float up in mean potential and so overload the transmitter. Thirdly, definite amplitude levels may be fixed to correspond to extremes of black and white in the picture.

The actual wave-form transmitted is shown diagrammatically in Fig. 11, and it will be observed that while a signal of maximum amplitude corresponds to a white picture, a black signal is represented by a carrier having an amplitude of 30 per cent. of the peak value,

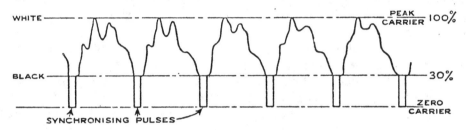

FIG. 11.—Typical line signals as radiated by the vision transmitter showing the synchronising pulses between each scanning line.

while the synchronising pulses are constituted by a complete momentary cessation of the carrier signal. The percentages of carrier amplitude allocated to the picture and to the synchronising pulses are maintained at the predetermined value whatever type of scene is being transmitted and irrespective of how much the scene changes in overall illumination.

MODULATION AMPLIFIERS

Returning to a consideration of the passage of the signal through the equipment, the picture signals with the synchronising pulses are passed from the distribution unit to the modulation amplifiers, where they pass through six successive stages of amplification before finally being applied to the radio frequency circuits of the transmitter. The modulation amplifier circuits are designed to give linear amplification over a range of frequencies extending from zero to four megacycles.

The two final stages in the modulation amplifier employ high power water-cooled valves, and these are arranged to modulate the final stage of the radio frequency transmitter by means of a grid-control circuit.

The radio frequency circuits of the vision transmitter at Alexandra

Palace consist essentially of a Franklin temperature-compensated master oscillator operating at half carrier frequency, followed by a frequency-doubler and five stages of radio frequency amplification at the carrier frequency of 45 megacycles. Water-cooled valves are used in the last two stages of the transmitter, the peak output of which reaches 17 kW.

THE SOUND TRANSMITTER A second ultra-short wave transmitter is required for the purpose of broadcasting the sound which accompanies the vision signals. The radio frequency circuits of the sound transmitter, which radiates on a wavelength of 7·23 metres, are of similar design to those of the vision transmitter, modulation being arranged at the final stage by means of a conventional choke control circuit. The sound transmitter, which has an output of 3 kW., was designed to have a flat frequency response between 30 and 10,000 cycles per second, and the distortion factor is less than 2 per cent. at 90 per cent. modulation. As a result, the quality of reproduction reaches a high standard.

CHAPTER VII

TELEVISION RECEIVERS

I. The Radio Receiver

In the autumn of 1936 several types of high definition television receivers were made available to the public. These all employed the cathode ray tube to produce the picture, though receivers using mechanical-optical systems have also been developed. Both types of receiver are, however, similar up to that section in the receiver where the vision frequency signals are reproduced in a form suitable for controlling either the cathode ray tube or the mechanical-optical projection system, and they can be considered together up to that point.

It has been explained in the previous chapter that for a complete television service two carrier waves are normally radiated, one modulated with " sound " frequencies up to about ten thousand cycles per second, and the other with " vision " frequencies up to some two million cycles per second. Owing to the high values of the vision frequencies, a carrier wave of unusually high frequency is required, and the " vision carrier frequency " will be in the neighbourhood of at least forty megacycles. Transmissions at this frequency cover only a limited range and the sound transmitter which is designed to have a similar range works on a neighbouring frequency.

The reception of signals radiated at these frequencies is preferably carried out by means of special aerial systems as described in Chapter VIII. The simplest aerial consists of a straight conductor, slightly less than half a wavelength long. Such a conductor, if placed in the plane of the electric field of the radiation, will have the maximum energy developed in it. It is usual to have the same aerial for both sound and vision carrier waves, and to choose the aerial length to resonate between these two frequencies.

In the receiver itself, the signal path may be most clearly followed in conjunction with Fig. 12.

The incoming signal has first to be amplified and this is done in one of two ways. It may be amplified at the incoming signal frequency, or the carrier wave frequency can be reduced to enable the amplification to be carried out more conveniently at a lower frequency.

Taking the former case first, the incoming signals, vision and sound, may be applied simultaneously to a valve input circuit so as to produce an amplified output across its anode load. This output voltage is passed on to the next valve, where it is again amplified. After passing through seven or eight valves the signal magnitude is sufficient to operate the detector and output stages. After the first few stages of combined amplification, the sound carrier is filtered off and amplified separately.

AMPLIFIER DESIGN Owing to the unusually high carrier frequency and the total frequency range of the signals to be dealt with, the amplifier is of unusual design. To deal with a total frequency range of three or four megacycles, the couplings between the aerial and the first valve and between the subsequent valves must each be of the band pass type and must be capable of giving the same performance at all the sideband frequencies. Alternatively, each stage of the amplifier is coupled by circuits having slightly different frequency

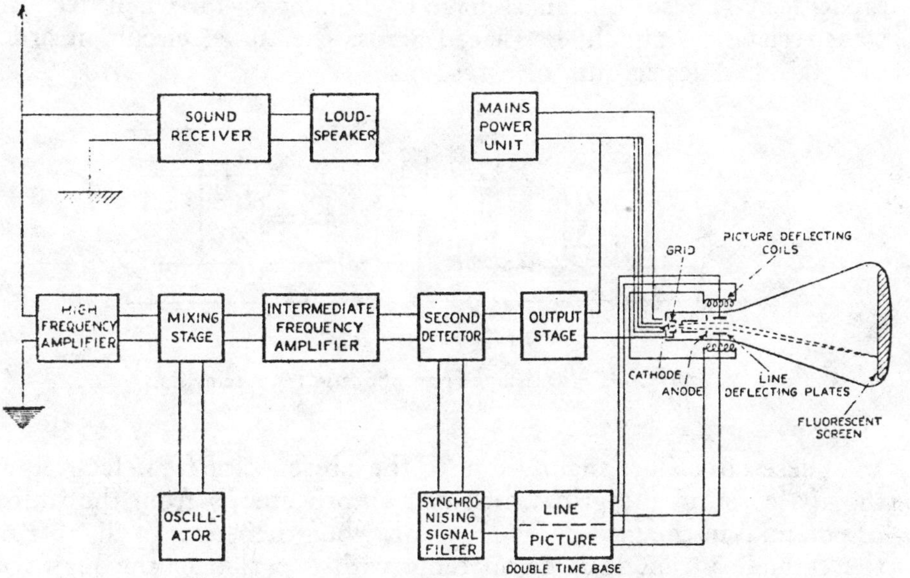

FIG. 12.—Typical lay-out of a television receiver.

characteristics, so chosen that the overall amplification obtained from their combined characteristics is sufficiently uniform.

In either case a common difficulty in obtaining amplification arises. The amplification in any section of the amplifier is largely decided by the product of the figure for the mutual conductance of the valve and the magnitude of the load in its anode circuit. For a given range of frequencies which are to be amplified substantially uniformly, the impedances obtainable are inversely proportional to the capacities in the tuned circuits. These capacities are therefore kept as low as possible and they are limited to the inevitable stray capacities of the coils, the valves and the wiring. A typical stage of amplification would therefore be as shown in Fig. 13. The coupling between the coils and the resistances joined across them are chosen to give the required uniformity of amplification over the desired frequency range. Even with such a circuit, the impedances are

47

unusually low. The impedance is also inversely proportional to the frequency band to be amplified, and the circuits are therefore arranged to pass only the band of frequencies necessary, a range of some four megacycles for the Alexandra Palace 405 line transmission. To obtain appreciable gain it is generally necessary to use special valves having a higher mutual conductance than normal.

A feature encountered in amplifiers working at the carrier frequencies is the special behaviour of valves at these frequencies. When normal valves are used at the television carrier frequencies, their performance differs in two ways from their performance at lower frequencies. At the lower frequencies the internal impedances between the grid and the other electrodes and between the anode and the other electrodes are simple capacities whose effects can easily be allowed for, but at the television carrier frequencies the valves behave as though relatively low resistances were connected across these capacities. These resistances have to be allowed for when deciding what resistances should be placed across the tuned circuit in order to obtain the desired uniform response.

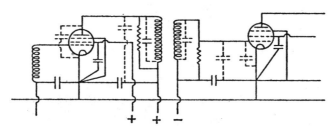

FIG. 13.—Typical stage of amplification in a television receiver.

These effects are mainly due to the presence of free electrons in the space round the grid which can absorb energy from the source of potential in the grid-cathode circuit when their time of flight from the cathode to anode is comparable with a period of the high frequency oscillation.

Another effect which occurs at these frequencies is the modification of the frequency response of the circuit connected between the grid and cathode of a valve by energy from the amplified signal in its anode circuit, fed back through the anode-grid capacity. At low frequencies the internal screening of the valve reduces this capacity to such a value that little energy can be passed back. At the television carrier frequencies, however, the impedances in the valve leads, even inside the valve itself, are such as to make it difficult to maintain the auxiliary electrodes, such as the screen grid and suppressor electrodes, at the cathode potential. In these circumstances, the effective grid-anode capacity is increased many times, and the potentials developed on all the electrodes of the valve modify the signals produced on the grid to a marked extent, and render the amplification very variable with frequency. For this reason special precautions have to be taken with the by-pass condensers which are used to anchor these auxiliary electrodes to cathode potential and

48

they are joined up by the shortest possible leads. These difficulties have led to an alternative form of amplifier where the circuits in the grid and anode of each valve are tuned to slightly different frequencies.

Special miniature valves have been made which overcome the defects mentioned above, but they have not yet come into general use.

SUPER-HETERO-DYNE RECEIVERS Before considering how the modulated carrier wave is rectified, the alternative method of amplification will be considered —namely, the super-heterodyne method.

Again the receiver may start off with one or more stages of carrier frequency amplification through which the modulated vision and sound channel signals are passed.

These signals are then applied to a frequency-changing valve. In this valve, all the signals are combined with a locally generated oscillation. Assuming, for example, that the local oscillator has a frequency of 42 megacycles per second, and that the vision signals have a frequency of 45 megacycles per second with upper and lower sidebands due to picture modulation of 46·5 and 43·5 megacycles respectively, the output of the frequency-changing valve will consist, among other currents, of signals having frequencies extending from 1·5 to 4·5 megacycles per second.

Such currents can be passed into a circuit having a uniform high impedance and uniform transmission properties over the band of approximately 1·5 to 4·5 megacycles and the resultant signal will have all the characteristics of the applied signal except that its mean, or carrier, frequency has been reduced to a frequency where amplification can be more readily carried out. A similar reduction can be made in the sound carrier frequency.

The amplifiers working at these new frequencies are known as the " Intermediate Frequency " amplifiers, and in the output circuit of the frequency-changer the filters used can readily separate the intermediate frequency sound channel signals from the intermediate frequency vision channel signals and lead them to their own respective amplifiers. The modulated sound carrier is rectified in the usual way to obtain signals of audio-frequency and these signals, after amplification, are reproduced by the loudspeaker.

The vision channel intermediate frequency amplifier is very similar to the carrier frequency amplifier previously described, but is free from the secondary troubles peculiar to very high frequency amplification. Thus it will generally consist of a series of about five valves coupled together by band-pass filters.

At the present time, it is doubtful what order of intermediate frequency is to be preferred. If the frequency is low, care must be taken to remove the carrier frequency from the rectified output signal, or it will show as a pattern of fine dots on the final picture, and precautions must be taken to see that the local broadcasting of comparable frequencies is not picked up on the amplifier. If the frequency is too high, most of the advantages of the superheterodyne receiver will disappear.

In most cases, the circuit is so arranged that both the sound and vision intermediate frequency carriers are produced simultaneously by a single variable local oscillator, and as the sound channel may be relatively narrow, the design can be such that the user is forced to tune the set to a frequency giving the best possible value of the intermediate vision carrier frequency.

The detector used for the vision signals is similar for either the superheterodyne or the vision carrier frequency amplifier. It usually consists of a diode valve, but because of the low impedances which must be used in its load circuit if the wide band of frequencies is to be accurately reproduced, a special diode itself having a low impedance is employed. As already stated, the carrier frequency must be removed from the rectified output, and for low carrier frequencies, this is most easily done by balancing it out in a two-diode rectifier circuit in which the products of rectification add, while the carrier frequencies cancel. At higher carrier frequencies, a simple filter can be used, but is not always necessary if the carrier frequency oscillations are too rapid to be perceptible on the screen of the cathode ray tube.

Beyond the point which has now been reached, the cathode ray and mechanical-optical systems differ considerably.

II. Cathode Ray Tube Reproduction

As described in the previous section, the output of the rectifier contains the picture signals and also the synchronising impulses to lock the scanning circuits of the receiver with those of the transmitter. If the diode output is sufficient, these two sets of signals can be separated by filter circuits and the respective components used to modulate the light beam, as described in Chapter V, and also to control the scanning circuits.

It is quite usual, however, to pass the total rectified output signal through an amplifier before this separation is carried out. This amplification must take place at the lowest frequencies required by the synchronising and the picture signals, and this means that if the signal is to show the very slowly changing mean brightness of the scenes transmitted, it must be amplified uniformly down to zero frequency, *i.e.* the direct current component of the rectified signal must be passed.

In either case, with or without amplification, the synchronising signals appear as part of the combined signal available. The voltage to be applied to the cathode ray tube must contain these signals as they suppress the cathode ray beam during the fly-back period. The combined signal requires no more modification before it is applied directly to the modulator electrode of the cathode ray tube as described in Chapter V.

The time-base generator circuits, on the other hand, must not have any voltage which is characteristic of the picture modulation applied to them. The synchronising signals are therefore separated by means of a biased valve which only passes current when the signals fall below a predetermined level, and the resultant current is therefore not affected by the picture modulation signals above that level.

This current, as has been said, contains impulses of the frequencies with which the " lines " and " frames " recur, and, if it is passed through suitable filters, separate signals at the " line " frequency and at the " frame " frequency will be available.

These signals are now available to lock the time-base generators exactly in step with the scanning signals at the transmitter, and in practice both gas-filled relay circuits or back-coupled vacuum valve circuits may be used to produce either potential variations for electrostatic scanning or current variations for electromagnetic scanning.

III. Optical Mechanical Reproduction

EARLY SYSTEMS Although early mechanical systems of television such as the Nipkow disc and the Weiller mirror drum were satisfactory for picture definitions up to 60 lines, they became difficult for higher definitions and practically impossible at definitions above 90 lines. It is necessary briefly to consider these systems in order to appreciate the problems which had to be overcome in modern mechanical high-definition television reception.

The best known mechanical system prior to the introduction of high definition television made use of a Nipkow disc, and the original commercial receiver of the Baird Co. employed a disc of this type having a diameter of 20 in. with 30 apertures. The disc was driven at 750 r.p.m. by a small universal motor with a flat-plate neon lamp as the light source. Synchronising of the disc was effected by a 30-tooth cogged wheel mounted on the shaft of the motor and rotating between the poles of an electromagnet through which the synchronising pulses incorporated in the vision signal were passed.

The next important mechanical system was that using the Weiller mirror drum described in Chapter III. This form of scanner was used widely for low definition television, but at definitions higher than 30 or 40 lines its size becomes prohibitive.

Another form of scanner employing mirrors is the " Mirror Screw " developed by Okolicsanyi and Tekade. The mirror screw consists of a series of long flat strips bolted on a central spindle. The assembled strips are then twisted about the spindle so that they form a spiral, the edges being silvered, to act as mirrors. An important feature of the mirror screw is that its physical dimensions determine the dimensions of the reproduced picture. The mirror screw is an excellent scanning device up to a definition of 120 lines, and by means of special optical systems reproduction up to 240 lines is just possible.

LIGHT EFFICIENCY The most important problem in the development of mechanical optical systems has been that of the light efficiency of the optical arrangement, as only a small fraction of the total illumination available can be projected on to the screen at any instant.

In order to understand the improvements which have been made it is necessary to consider briefly the fundamental problems. In Fig. 14 an illuminated aperture A projects light on to a screen S at a distance D. If the illuminated area on the screen is RY by Y it can be shown that the light falling on it is equal to $\dfrac{ARY^2BE}{D^2}$, where B is the brightness of the light source, and E is the efficiency of the system. This assumes that the area is illuminated all over simultaneously, but in practice only a small fraction equal to the size of the picture element is illuminated at one time. If the number of scanning lines is N then the area of the whole picture is RN² greater than that of a single element.

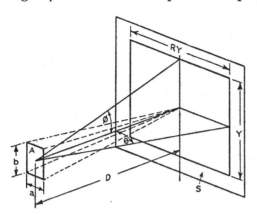

FIG. 14.—Diagram to illustrate the problems of large screen projection.

The area RY² can be expressed in terms of the angles ϕ and θ subtended at the centre of the aperture A, giving

$$RY^2 = 4D^2 \tan\theta . \tan\phi$$

If the area of the aperture A is $a.b$, and the number of elements of the picture simultaneously illuminated is n, then it can be shown that the total light falling on the screen is

$$\frac{4BEn\,(a\tan\theta)\,(b\tan\phi)}{RN^2}\ \text{lumens.}$$

It will thus be seen that the screen illumination decreases with the square of the number of scanning lines. The real situation is worse than this, however, because in a picture of N lines the scanning rays must sweep through the angle 2θ, N times per picture. For a mirror drum this means that N mirrors must come into operation during the time taken to scan the complete picture, *i.e.* the product of the number of mirrors on the drum and the drum speed (revs. per picture) must equal N. The angle of scan (2θ) is proportional to the angle between mirror faces on the drum and becomes very small as the number of faces is increased. This also applies to the scan in the ϕ direction. It is clear therefore that as the definition is increased not only does the illumination on the screen decrease in proportion to N², but in addition decreases as θ and ϕ become smaller. For this reason practically all modern mechanical systems employ methods of increasing ϕ and θ.

MIHALY'S ARRANGEMENT One method of doing this is to make the scanning surface act twice, thus in effect doubling the scanning angle. This was used by Mihaly in his " stationary mirror ring," which was a modification of the Weiller drum. The mirrors were arranged inside the periphery of a stationary drum, the modulated light source being directed upon a single rotating mirror arranged at the centre. The light was thrown from the central mirror to the drum, back to the central mirror and thence to the screen. The method can be used satisfactorily up to a definition of about 100 lines.

A further development of this system due to Traub consists of the use of a " line multiplier " drum in place of the single rotating mirror. In this case the circular mirror ring is reduced to a small arc of inwardly facing stationary mirrors. Thus the definition is determined by the product of the number of mirrors used on the stationary arc and on the rotating " line multiplier " drum. This form of receiver is useful up to about 200 lines.

By increasing the speed of the central polygon, or " line multiplier drum " and decreasing the number of its faces, the system is further improved, because the angles θ and ϕ are increased. This is the basis of the modern Mihaly-Traub system, which for the London Television standard of 405 line uses a small arc of five stationary inwardly-facing mirrors in conjunction with a " line multiplier " drum having nine faces and rotating at nine times the frame speed.

SCOPHONY SYSTEM
The split focus principle, as used by Scophony, Ltd. improves the light efficiency by increasing " b " in the formula given previously, where " b " is the size of the optical aperture at right angles to the direction of scan. Instead of using only normal spherical lenses the required optical power is obtained by crossed cylindrical lenses, the light beam being brought to a focus at different stages in the process in the two transverse directions.

The advantage to be obtained from separating the two focal planes may be shown as follows : It is required to get as much light as possible reflected from a small scanner mirror, and this might be attempted by focussing the beam on to the mirror, using a second lens to form an image of the illuminated mirror on the screen. Unfortunately, this action of focusing means that the light coming from the mirror surface comes to a focus at the same point on the screen irrespective of the angle at which it leaves that surface. Thus an angular movement of the mirror produces no scanning action on the screen under these conditions. However, no interference in the scanning results if the beam is focused on to the scanner surface only in the plane at right angles to the direction of scanning. In the scanning plane the beam comes to a focus on the screen. Hence the focus is said to be " split " or separated in the two perpendicular planes, and this is accomplished by cylindrical lenses. In this way N times more light may conveniently be handled for a given surface, or alternatively one dimension of the mirror can be reduced N times.

A further improvement in optical efficiency is made by use of the " beam converter," a device which, in effect, performs an optical transformation. This has for its object an increase in the dimension *a* and makes it correspond to the axial length of the scanning surface at *right angles* to the direction of scan instead of

53

being the dimension in the scanning direction. In the beam converter the beam from the scanning surface is split into a number of sections in a direction at right angles to the direction of scan and each of these individual sections is rotated through 90° so that the emergent beam is moving in a direction parallel to the axis of the scanner.

The effect is similar to the action of an inverting prism in which objects moving horizontally will appear to move vertically if the prism is rotated through 45° about the optic axis. An appreciable gain in light efficiency is possible by this system, as much as 20 times being obtained.

The remaining factors in the expression for the screen illumination are E, the transmission efficiency of the optical system, the brightness of the light source B, and n, the number of elements simultaneously illuminated. E can be maintained at a reasonable value by keeping the number of refracting and reflecting surfaces to a minimum. To obtain a high value of B low voltage projector lamps of the incandescent type may be used for home receivers, while for large screens the brightest sources available are the high intensity arc lamp and the high pressure mercury lamp. The latter appears to have great promise, and is likely to be much used in mechanical television systems when its development is completed.

There remains the factor n which has received but little attention as yet, because it appeared that in order to have n elements of the picture active simultaneously it would be necessary to employ n separate channels for the transmission of the picture. It is possible, however, to work with values of n as high as 500, when a definition of 405 lines is being used with only a single transmission channel by making use of the "storage" of information occurring in the super-sonic light control described in Chapter IV.

By utilising fully the methods above described, good pictures can be obtained to-day with high definition. It must be remembered, however, that a television receiver, either of the cathode ray tube type or of the optical-mechanical type employs a radio receiver of very similar design and the distinction between the two types occurs only in the final stages of actual picture reproduction.

BIBLIOGRAPHY

G. L. Beers. "Description of Experimental Television Receivers." *Proc. I.R.E.*, Vol. 21, p. 1152. 1933.

V. K. Zworykin. "Description of an Experimental Television System and the Kinescope." *Proc. I.R.E.*, Vol. 21, p. 1655. 1933.

L. H. Bedford and O. S. Puckle. "A Velocity Modulation Television System," *Journ. I.E.E.*, Vol. 75, p. 63. 1934.

Manfred von Ardenne. "An Experimental Television Receiver using Cathode Ray Tubes," *Proc. I.R.E.*, Vol. 24, p. 409. 1936.

W. T. Cocking. "The Television Receiver." *Wireless World*, Vol. 40, pp. 151, 233, 250. 1937.

D. C. Espley and G. W. Edwards. "Television Receivers." *The G.E.C. Journal*, Vol. 18, No. 2. 1937.

R. R. Beal. Equipment used in Current R.C.A. Television Field Tests, *The R.C.A. Review*, Vol. 1, No. 3, p. 37, 1937.

CHAPTER VIII

AERIALS AND FEEDERS FOR TELEVISION RECEPTION

For broadcast reception the aerial used is generally short in length compared with the wavelength employed. In the case of television reception, however, where the wavelength employed is of the order of 7 metres it is practicable to employ aerials whose length is of the same order, or even longer than, a wavelength. The aerial most generally used is approximately one half a wavelength long; were it erected in free space it would be exactly one half a wavelength long. Since, in practice, it is impossible to erect an aerial in free space, it becomes necessary to reduce the length somewhat, in order to compensate for the extra capacity due to the proximity of the ground. The reduction factor involved varies from about 0·85 to 0·95 depending upon the thickness of the wire, the factor becoming larger the thinner the wire.

A half-wave aerial of this type exhibits a resonance effect, *i.e.* it acts in a selective manner over a narrow range of frequencies, the band-width being of the order of ±5 per cent. of the carrier frequency. This factor also depends on the thickness of the wire, the band-width increasing as the diameter of the wire is increased. A half-wave aerial is always tuned, no matter what the diameter may be, and the radiation resistance is approximately 75 ohms.

FEEDERS In the case of a broadcast aerial, the lead-in wire is, in effect, a portion of the aerial, but for television reception the connection between the aerial and the receiving set is not normally allowed to form part of the aerial and, instead of using a single wire, a double connection is used. This lead-in arrangement is known as a feeder and may consist of two parallel wires or of a concentric cable. The characteristic impedance of a concentric cable is obtained from the formula

$$Z = 138 \log_{10} \frac{R_2}{R_1}$$

where R_2 is the inside radius of the outer conductor,
and R_1 is the outside radius of the inner conductor.

It is essential that a feeder shall be terminated with an impedance equal to its characteristic impedance if reflections and loss of energy are to be avoided; hence, the impedance of the feeder is chosen to equal that of the aerial and the input impedance of the receiver is adjusted to be of the same value. Concentric feeders can be made to have characteristic impedances from about 10 to 150 ohms.

The characteristic impedance of an open wire (2 wires) feeder is given by

$$Z = 276 \log_{10} \frac{R_2}{R_1}$$

where R_2 is the distance between the wires,
and R_1 is the radius of the wire.

Open wire feeders can be made to have characteristic impedances from about 75 to 1,000 ohms.

The aerial is a tuned circuit and the impedance between its ends is high. A lower impedance exists, however, between any other two points, and there may be an impedance of the order of 100 ohms between two points close to the centre. If the aerial is split at the centre the impedance measured across the gap is about 75 ohms, and it is therefore convenient to arrange that the feeder be connected to the centre of the aerial on account of the ease of obtaining a low impedance feeder. The aerial is split at the centre for this purpose. However, it is possible to make use of a transformer if it is desired to connect a low impedance feeder to points on the aerial between which a high impedance exists.

Such a transformer consists of two wires each one quarter wavelength long, adjusted to a specific distance apart such that

$$\sqrt{Z_1 Z_2} = 276 \log_{10} \frac{R_2}{R_1}$$

where Z_1 is the impedance at the end of the aerial,
               Z_2 is the impedance of the feeder,
               R_2 is the distance between the wires of which the transformer is made, and
               R_1 is the radius of the wire of which the transformer is made.

The attenuation introduced by the feeders commonly in use is of the order of ten per cent. per hundred feet.

REFLECTORS   At places remote from the transmitter where the field strength is low or where there is an excessive interference field, the signal strength may be increased by the addition of a reflector. A reflector may consist of a sheet of metal arranged in the form of a parabola with the aerial at its focal point. In practice, however, such a reflector is cumbersome and difficult to erect while its efficiency is generally greater than is necessary. For practical purposes it is generally sufficient to use another rod or wire similar to, although very slightly longer than, the aerial and mounted at a distance of one quarter wave-length behind it. Such a reflector may be considered as being a small portion of a parabolic reflector. Approximately twice the voltage is obtained by using such a reflector.

The waves radiated by a television transmitting station are, in general, strongly polarised, and it is therefore essential that the receiving aerial be mounted in such a direction as to be affected to the greatest extent by the transmitted waves. For example, if the transmitting aerial is mounted vertically, the receiving aerial should also be mounted vertically, unless the polarisation of the wave concerned has become distorted during transmission.

A large number of more complicated aerial structures, all of which are designed either to increase the received signal or to make the aerial directionally selective, are available for use, and reference to a number of works dealing with the subject are given in the bibliography at the end of this chapter.

BIBLIOGRAPHY

Mead. " Wave Propagation over Parallel Tubular Conductors," *Bell System Tech. Journ.*, Vol. 4, pp. 327–338. 1925.

Sterba and Feldman. " Transmission Lines for Short-wave Radio Systems." *Bell System Tech. Journ.*, Vol. 11, pp. 411–450. 1932.

Sterba and Feldman. " The Uniform Short-wave Aerial," *Marconi Review*, No. 16. 1930.

Bird. " Determination of the Terminal Impedance of Short-wave Feeders," *Marconi Review*, No. 17. 1930.

Green. " The Energy Magnification of Broadside Aerial Arrays used for Reception," *Marconi Review*, No. 31. 1931.

Ladner. " A Graphical Synthesis of Aerial Arrays," *Marconi Review*, No. 33. 1931.

CHAPTER IX

THE LONDON TELEVISION STATION

ALEXANDRA PALACE The studios and transmitters of the television service now established in the London area are situated at Alexandra Palace in North London. The station was built by the B.B.C. to give effect to the Television Committee's recommendation that a high-definition television service should be inaugurated in this country. On the recommendation of this Committee, the Postmaster-General set up an Advisory Committee to guide the B.B.C. in the planning and early development of the service.

Alexandra Palace is a well-known North London landmark lying between Muswell Hill and Wood Green, some six miles north of Charing Cross, and is situated about 300 ft. above sea level, one of the highest points in London. The height of the ground was a deciding factor in the choice of this site for the station, as the height of the aerial above surrounding territory is important in governing the service area of any ultra-short-wave transmitter.

Before the site was finally selected, surveys of the topographical detail of the surrounding country were carried out by the Advisory Committee to make sure that serious " blind spots," where poor reception might be experienced, would not be likely to occur, as the result of " shadow " effects of other high ground in the environs of London.

The B.B.C. acquired a lease of part of Alexandra Palace, including the South-East Tower and the adjoining ground and first floor areas, totalling some 30,000 sq. ft., and also the North-East Tower, together with the Alexandra Palace theatre which has a further superficial area of about 25,000 sq. ft.

The theatre has so far been used only for experimental purposes, whilst the station premises proper are established in the portion of the building adjacent to the South-East Tower. The tower itself has been modified to form several floors of offices to house the members of the technical, executive and production staffs, and the ground floor forms the main entrance hall.

THE AERIALS The aerial mast is erected on the S.E. tower, from which the original cupola top was removed. A certain amount of structural modification and reinforcement was necessary to permit the extra weight and windage stresses of the mast to be carried with safety, and, as an additional precaution, the lower members of the mast were carried down the tower through several floors of offices to distribute the stress more evenly.

The mast has a tapering square cross-section to a height of 200 ft. above ground level, while for the remaining 100 ft. it has a parallel-sided octagonal section. On this upper portion, which is more than 600 ft. above sea level, are mounted four sets of wooden radial arms, eight to each set, supporting the aerial arrays one above

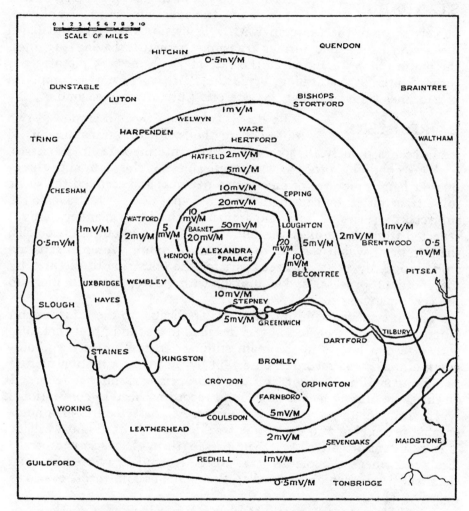

Fig. 15.—Map showing approximate field strength of the Alexandra Palace transmitter at a height of 10 to 15 metres above ground level.

the other. The upper array is the aerial for the vision transmitter, while the lower one is used for the sound transmission. The actual arrangement is as depicted on the cover of this handbook.

The aerials have been designed to give the maximum useful radiation, special precaution being taken to reduce vertical radiation so that the risk of undesirable reflections from the ionised layers of the upper atmosphere is reduced to a minimum. In addition to the aerial arrays, the radial arms carry reflectors which prevent loss

59

of power in the mast and also ensure that the radiation is uniform in all directions.

The aerials are coupled to the respective sound and vision transmitters through the medium of large diameter copper tube concentric feeders of low-loss type which extend from the output stages of the transmitters right up the mast to the aerial termination points.

STANDARDS

At the inception of the service two standards of vision transmission were employed, namely those of the Marconi-E.M.I. Television Co. Ltd., and Baird Television Ltd. The former transmits a picture having 405 lines, the picture being scanned at the rate of 50 frames, 25 pictures per second, interlaced scanning, while the Baird picture was composed of 240 lines transmitted at the rate of 25 pictures per second.

BAIRD SYSTEM

The Baird Television System employed two different methods for the transmission of studio scenes, namely, the spotlight, and the intermediate film processes.

The spotlight process was applicable to close-up and three-quarter length pictures of one or two persons, and scanned the scene to be transmitted by causing a projected spot of light to traverse the subject in a series of lines and frames in the usual manner. Light reflected from the subject was allowed to fall on photo-electric cells, the output of which was fed to the control room. A small separate studio 10 ft. by 15 ft. was set apart for the television of pictures by the spotlight process, and equipped with microphones to pick up the accompanying sound.

The intermediate film process was applicable to the transmission of more extensive scenes, and was installed in a sound proof enclosure with glazed windows in the main studio. In this process a photographic image was recorded on a film by means of a motion picture camera of modified design, and the appropriate sound was recorded on the same film by means of a sound recording head of conventional type. Immediately the film was exposed it passed through photographic processing tanks where it was developed, fixed and washed. The resultant negative was then passed through a scanner of the disc type where the picture was dissected and converted into an electrical image, which was fed to the control room through a low-capacity cable.

In addition to the spotlight and intermediate film methods of transmission, experiments were carried out using an electron camera for the transmission of scenes. This camera was entirely electronic in action, and contained no moving parts.

Standard cinematograph films were televised by means of cinema projectors associated with scanners of the disc type.

MARCONI-E.M.I. SYSTEM

After a period of trial, the Television Advisory Committee decided that future transmissions should take place on one standard only, having 405 lines and 50 frames per second with interlaced scanning. This standard therefore is now exclusively used by the B.B.C. for transmission of Television from the London Station. The transmitter now in use is that of the Marconi-E.M.I. system, which was described in Chapter VI.

Turning to the layout of the station, the disposition of the apparatus may be described. The sound and vision transmission apparatus with the associated plant and switchgear is installed on the ground floor. The ground floor also accommodates a film viewing room which is equipped for the projection of sound films for the purpose of selecting and timing excerpts for inclusion in the television programme.

The distribution switchgear controlling the supply of electricity to the various sections of the equipment is also situated on the ground floor. The incoming supply to the sub-station is taken from a ring main at a pressure of 11,000 volts and is stepped down to 415 volts by means of two 500 K.V.A. outdoor transformers arranged to operate in parallel. The low tension output from the transformers is taken to the main circuit-breaker of the distribution switchgear from whence it is fed to various parts of the building for transmitter supplies, ventilating, heating, special studio lighting, general lighting, and other services.

The first floor is entirely devoted to the business of production, and accommodates the studios and artists' dressing rooms. In addition a band rehearsal room is provided and a large space for the storage of scenery in current use.

STUDIOS Two main studios are provided, each 70 feet long by 30 feet wide and 23 feet high. Considerable attention was paid to the acoustic characteristics of these studios in order to ensure that the sound picked up by the microphone is of suitable quality to accompany the televised action. It has been found that the acoustic properties desirable in a studio intended for television differ to some extent from those appropriate in a studio used exclusively for sound broadcasting.

In the case of sound broadcasting, the ear is the only criterion of the reproduced performance, and the effect produced can be materially enhanced by the introduction of a certain degree of reverberation or echo. Such effects, however, require rather careful arrangement of the performers before the microphone so that a pleasing balance of sound is obtained. Moreover, the degree of reverberation which is acceptable varies widely with the type of performance—thus a studio suitable for a variety performance would not be suitable for a symphony orchestra, and in general, different studios would be used for different types of programmes. Above all, the placing of the performers from an appearance point of view is not important in sound broadcasting, so long as a correct sound balance is maintained.

In the case of television, however, the proper location of performers from the point of view of appearance is of paramount importance in the interests of artistic production, so that the sound requirements must, of necessity, be subordinated to this consideration. Added to this, it is not at present economically possible to provide a series of studios of divergent acoustic properties, each fully equipped with the manifold requirements of television.

In order that the sound accompanying scenes of widely divergent character should be of uniformly good quality, the studios themselves should be heavily damped and thus caused to possess an extremely short reverberation time, relying upon local echoes from scenery and properties to give " life " to the sound picked up by microphones placed as close to the source as possible. To bring about this effect, all four walls of the studios were lined with sound absorbent material about 1 in. in thickness, having an asbestos base, and in addition, the ceiling was covered with building board to give further absorption of sound so that the reverberation time of the studio was made very small.

The studio requirements necessitate the use of several different types of microphone and suspended microphones are used in some cases. The telescopic microphone boom as used in film studios has proved itself of considerable value. By means of this device an operator is able to follow the movements of an actor keeping the microphone at all times within a few feet of him. Considerable dexterity is required for this process as it is essential that the microphone is kept clear of the field of camera vision, neither must it be permitted to cast an undesirable shadow.

CAMERAS The Emitron cameras are not dissimilar in outward appearance to the conventional type of film camera, and are manipulated in much the same way, being provided with optical systems of varying focal length, including telephoto lenses for distant work. The cameras are mounted on tripods and stands of various types possessing tilting and panoramic movements and vertical racking adjustments which allow the flexibility of movement necessary to follow the action being televised.

A tripod mounted on a pneumatic-tyred truck is employed to effect " dolly " shots, as, for instance, where an artist first appears in a full length shot, after which the camera is gradually advanced into a close-up position where the face of the artist occupies the whole picture. During this process, as well as in the course of other action, it is necessary to keep the subject in accurate focus, a procedure which is carried out by means of a focusing and view-finding device which also indicates the extent and composition of the scene falling within the field of vision of the camera. Each camera thus requires the service of a camera-man, upon whose skill largely depends the excellence of the televised picture. In the case of " dolly " shots, the camera-man is perched on an adjustable seat mounted on the camera truck. The camera-men receive telephonic instructions from the producer at the control position *via* a pair of headphones, and are thus kept informed of the periods when the actual camera under their control is in action.

The producer is situated at a control position adjacent to the studio and is able to command a view of the scene being enacted through sound-proof windows of double plate glass. The control position is provided with desks carrying a series of knobs, by means of which it is possible to fade from one camera to another to change the viewpoint of the televised picture. Superimposition of two or more camera outputs may be effected and almost unlimited illusions and artistic effects can be achieved.

At the same time the accompanying sound may be selected from any microphone or combination of microphones by means of the appropriate controls. The producer and technicians are able to judge the effect of the transmitted programme through the medium of suitably located picture producers and loud speakers. The picture and sound signals are fed to the main control room apparatus where they are amplified and corrected, and the appropriate synchronising impulses added to the picture signals. They are then passed to the vision and sound transmitters on the floor below, where they are again amplified, converted to radio frequency, and emitted from the aerial.

LIGHTING The provision of effective studio lighting involves many and complex problems, in the attempt to find a solution for which a great deal of the technique of stage and film studio production lighting has been invoked.

It has been found in practice, however, that the demands of television production lighting have not entirely been met by application of these principles, and it has become clear that the development of a new and specialised technique is required. Great strides have been made in this direction since the inception of the service, as a result of which it will, it is hoped, be possible in the future to embark on more ambitious productions.

Each studio is provided with a lighting control switchboard fitted with banks of dimmers, with individual and group control, by means of which it is possible to pre-select any desired lighting sequence. Each switchboard has a capacity of approximately 60 Kw. at 110 volts, and is connected to a distribution network embracing the whole studio so that any type of lighting may be available at any position as demanded by production requirements.

OUTSIDE BROADCASTS In addition to productions actually in the studios, a number of successful outside broadcasts have been televised from the grounds of Alexandra Palace and from within the confines of the Palace buildings. These have included demonstrations of golf and horsemanship in the grounds, and boxing contests and exhibits from the North London Exhibition, within the Palace, to mention only a few.

The most important of these events which has taken place up to the present time was the broadcast by television of the Coronation Procession of Their Majesties the King and Queen on May 12th, 1937.

INDEX

Printed under the Authority of HIS MAJESTY'S STATIONERY OFFICE
by William Clowes & Sons, Ltd., London and Beccles.

(315) Wt. 200—4078. 10M. 6/37. W. C. & S., Ltd. Gp. 389.

S.O. Code No. 29-1051

CREATING AN INDUSTRY

Robert C. Bitting, Jr.

Creating an Industry

By ROBERT C. BITTING, JR.

Note: "Creating an Industry: A Case Study in the Management of Television Innovation" is a master's thesis submitted by Robert C. Bitting, Jr., at the Massachusetts Institute of Technology in 1963. The author chose the Radio Corporation of America as the organization most suitable for special study. Much of his thesis is concerned with problems of management, but of more general interest is his anecdotal approach to the history of the Radio Corporation of America and of the development of television. The following article is comprised of Chapters II and III from the thesis, in somewhat abbreviated form.

Masters and doctoral theses are always filed in the archives of universities after they have been accepted by the authorities and the degree granted to the graduate student. A relatively small number of these theses get published in such form that the information is readily available as a reference.

The subject of Mr. Bitting's thesis is related closely to the interest of many members of this Society and others. In view of the tremendous growth of the television industry over the past quarter century and its

great potential for further expansion, it seems very appropriate that much of this useful survey should be made available for convenient reference in the pages of our Journal.

Our committee wishes to express sincere thanks to Dr. E. W. Engstrom, R. C. Bitting, Jr., and other members of the R.C.A. organization for their helpful cooperation in making this information available and revising it for publication.

Other papers on the history of television that have been published recently by the Society are: "Historical sketch of television's progress," L. R. Lankes, Jour. SMPTE 51: 223–229, Sept. 1948; "The early days of television," J. V. L. Hogan, ibid. 63: 169–173, Nov. 1954; "The evolution of modern television," A. G. Jensen, ibid. 63: 174–178, Nov. 1954; "A short history of television recording," A. Abramson, ibid. 64: 72, Feb. 1955.

GLENN E. MATTHEWS, *Chairman*
SMPTE Historical and Museum Committee

Part I: Formation and Growth of RCA

Created in 1919 through an unusual set of circumstances, RCA has grown and prospered through the years until, in 1961, it was the twenty-sixth[1] largest corporation in the United States. This article will trace the major events in the history of the company, with particular emphasis on those which were of significance to the development of monochrome television.

The Formation of RCA: 1919

During World War I, the United States learned the importance of international wireless communication to the conduct of its affairs, both military and public. The only nonmilitary source of this service within the United States was the Marconi Wireless Telegraph Co. of America, an organization which was owned and controlled by foreign (British) interests. The complete dependence upon such foreign-controlled and sensitive facilities was a source of considerable concern to the Navy Department, at that time probably the leading user and proponent of long-distance wireless communication within the nation's military organization. This concern became acute in early 1919, when the General Electric Co., holder of certain basic and extremely vital patents covering the Alexanderson alternator, revealed[2] that it was negotiating exclusive licensing rights thereof with the Marconi company. Such an arrangement would have made foreign control of international wireless communication even tighter (at least, until some new invention appeared on the scene).

With President Wilson's knowledge

This contribution to the *Journal* was solicited by the Society's Historical and Museum Committee. Permission to publish these extracts has been granted by Dr. Howard W. Johnson, Dean, Alfred P. Sloan School of Management, Massachusetts Institute of Technology. This paper was submitted in final form on Sept. 23, 1965, by Robert C. Bitting, Jr., Manager, Financial Planning, Radio Corp. of America, 30 Rockefeller Plaza, New York, N.Y. 10020.

and approval, Admiral Bullard, Director of Naval Communications, informed GE's Owen D. Young of the government's desire for an American-controlled organization. The government proposed that an American-controlled company be established to provide international wireless communication for the United States, and that it be provided with the necessary GE licenses to use the Alexanderson alternator. GE not only agreed to such a licensing arrangement, it seized upon what may well have appeared to be a handsome opportunity to expand into a government-blessed business area. Whatever the pressures and motivation, GE did in fact create such a company and, on Oct. 17, 1919, the Radio Corporation of America was incorporated. Almost immediately (on Nov. 20, 1919), RCA acquired a controlling interest in the American Marconi company and the government's aim of an American wireless company was achieved.

Ownership and Organization at Time of Formation

Ownership of the new company was preponderantly by General Electric. Young was made Chairman of the Board (a position he held until his resignation in 1933), Edward J. Nally was installed as President and David Sarnoff was designated as Commercial Manager. Nally and Sarnoff had previously been associated with the Marconi enterprise which RCA acquired in 1919. Nally, who continued as President until 1923, remained as a director of the company until his retirement in 1950. David Sarnoff eventually moved to the chairmanship of the corporation, a post which he holds to this day (1963).

The Beginning Years: 1920 to 1930

Initially, RCA devoted its efforts to wireless point-to-point communication service. In February, 1920, stations which

the government had taken over from the Marconi company during the war were turned over to their new owners for commercial use. International service with England, Germany, France, Norway, Japan and Hawaii was thus inaugurated by RCA. Expansion of facilities was undertaken and construction of high-power Alexanderson alternator stations in California, Massachusetts and Hawaii was started. This work, however, led into patent problems.

Patents were proliferating in the burgeoning wireless field. Many of the more important ones had been obtained by GE, Westinghouse Electric and Manufacturing Co. and American Telephone and Telegraph Co. Commercial operation of wireless stations demanded ownership of or license rights to many different patents. No one person or firm controlled even a substantial fraction of them, and thus the acquisition of needed coverage was quite difficult and expensive. RCA, immediately aware of this very practical operating problem, took steps to establish extensive cross-licensing agreements among a group including GE, Westinghouse, AT&T and RCA. Some of these were "suggested" by the government in "the best interests of the country," a circumstance which had bearing on later antitrust action.

These cross-licensing arrangements thus brought Westinghouse and AT&T into the so-called "radio group." These companies received RCA stock in exchange for use of their patents. By June 30, 1921, when an agreement was consummated with Westinghouse, RCA had rights to more than 2,000 patents in the radio field, thereby clearing its way to engage not only in wireless communication, its raison d'être, but in many other radio pursuits as well.

In 1916, David Sarnoff, then Assistant Traffic Manager of the Marconi company, had sent its General Manager,

Edward J. Nally, a memorandum outlining a rather radical concept of public broadcasting. In it Sarnoff said:

"I have in mind a plan of development which would make radio a household utility in the same sense as a piano or phonograph. The idea is to bring music into the home by wireless . . .

"Should the plan materialize, it would seem reasonable to expect sales of 1,000,000 'radio music boxes' within a period of three years. Roughly estimating the selling price at $75 per set, $75,000,000 can be expected."

The concept was a valid one, and the prediction amazingly accurate; for although the Marconi company did not follow Sarnoff's memo, RCA did, realizing sales of $83 million during the first three years of its home instrument radio line.

A. N. Goldsmith, a prolific inventor in his own right, has related (in an interview, Jan. 11, 1963) an interesting anecdote of the early "radio music box" days:

Anecdote Recounted by A. N. Goldsmith

"In 1920, I had built what, to my knowledge, was the first uni-control radio broadcasting receiving set. It had one knob for tuning and one for volume. It was self contained in a small wooden cabinet which included the necessary dry batteries and a loudspeaker. I had installed it in my home and used it frequently to listen to the stations of the day.

"One evening Mr. Sarnoff came down to my house with a certain high commercial official of his organization who was an enthusiastic amateur. I tuned the set to the Wanamaker station—it was at Ninth and Broadway in Manhattan—and music emerged from the box without any need for earphones. The commercial man—and this is very interesting psychologically—said, 'Well, what a useless thing this is! I'm astonished that you waste your time on it. After all, everybody knows that the great joy of radio is to pick up very distant stations, to have lots of knobs to tune quite accurately, and to wear your headsets so that you don't disturb other people . . .' He was rambling on thusly when Sarnoff—and this is a measure of the man—said to him, 'Please be quiet, I'm thinking . . . You're completely wrong!' He turned to me and said, 'This is the radio music box of which I've dreamed . . .'

"The General Electric and Westinghouse companies were the sole manufacturers at that time for RCA. Neither saw much promise in this uni-control set, sharing the view of our commercial man. And indeed, it was at our own risk that the thing was going to be built—if we insisted on it being built! It *was* produced and sold amazingly well at a rather high unit price. It is considered to be one of the factors that started the broadcasting ball rolling . . ."

Network Broadcasting

While busily and profitably engaged in the sale of radios to the public, RCA was also aggressively developing the "other end of the business," i.e., broadcasting itself. Two stations were opened in 1923, one in New York and the other in Washington, while 1925 saw the debut of WJZ at Bound Brook, N.J. Broadcasting flourished and competition was soon intense. At this juncture RCA, GE, Westinghouse and others ran into a stone wall in the shape of AT&T, for, despite the earlier cross-licensing agreements, the telephone company chose not to provide wire-line distribution (network service) to radio stations which were competing with AT&T stations. Finally, however, the telephone interests withdrew from broadcasting.[3] RCA acquired the key AT&T station, WEAF, for $1 million, and, together with its other stations as a base, formed the National Broadcasting Co. (NBC) in 1926. RCA owned 50% of NBC, GE 30% and Westinghouse 20%. The first full year of operation, 1927, saw a total of 48 stations, nationwide, do a gross business of nearly $4 million. This gross increased to almost $10 million in 1928.

NBC, founded as a separate company, has retained this organizational relationship to RCA throughout its life span, i.e., that of a wholly-owned subsidiary as opposed to a division. It played an important role in the subsequent television development program, and has provided a significant portion of RCA's income, being a good diversification buffer. With the formation of NBC, then, RCA had established two of what in retrospect have been its three major businesses—radio communication and radio broadcasting. The third, manufacturing, was soon to come.

The Van Cortlandt Park Activity

As the merchandising arm for two large radio manufacturers, GE and Westinghouse, RCA had found it necessary to establish some kind of engineering activity of its own to test and evaluate existing products and to plan new ones. In 1924, therefore, the Technical and Test Department was established under A. N. Goldsmith's direction. It was located on the south side of Van Cortlandt Park in uptown New York City. Goldsmith had earlier provided a similar service to RCA under a contractual arrangement involving the electrical engineering laboratory of the City College of the City of New York where he was in charge.

Kilbon, in his unpublished *Pioneering in Electronics*,[4] says the group consisted of about seventy engineers, technicians, mechanics, carpenters, administrative and service personnel. In addition to product evaluation and planning work, there was a considerable amount of rather fundamental development activity. For example, the velocity micro-

phone—perhaps the single most ubiquitous adjunct to modern sound broadcasting—was invented by Harry Olson while at Van Cortlandt Park. The earliest strictly RCA (as opposed to GE and Westinghouse) efforts on television were undertaken by this group.

The Third Arm of the Business: Manufacturing

By 1929, many problems had become quite evident in the merchandising relationship between RCA and its manufacturing associates, GE and Westinghouse. Standardization of product was one such problem, production scheduling and control another, and competitive pressures on profitability a third. Both radio sets and tubes were involved, the latter quite importantly since early practice was to market tubes separately from the radio itself.

Product standardization difficulties arose largely from the organizational fact that RCA had only what amounted to a customer's relationship to the design and production functions. Further compounding the structural difficulty was the autonomy of the two quite separate manufacturers. Standardization and design committees were used to provide coordination and some measure of control but never achieved the same results that an integrated one-company counterpart could be expected to produce.

The same comments pertain to production scheduling and control, with problems further compounded by diversity of plants even within the separate companies. Thus, General Electric operated at Schenectady, N.Y., and Harrison, N.J., while Westinghouse had facilities in East Pittsburgh, Pa., Indianapolis, Ind., and Bloomfield, N.J. The arrangement was ponderous.[5] Production runs of sets already obsolete because of overnight advances made by some small and highly flexible manufacturer would continue too long and inventories all too frequently became glutted with largely unsalable stock.

Lacking any effective management control over costs of product and incurring an additional markup in the distribution process posed significantly increasing profit problems on "merchandising-only" RCA. The pressure, of course, was intensified by the economic events of 1929 and the ensuing years.

Thus in February, 1929, RCA acquired the assets—including a large manufacturing facility in Camden, N.J.—of the Victor Talking Machine Co. In the fall of that year, agreement in principle was reached among RCA, GE and Westinghouse that RCA would become a unified, highly self-sufficient organization. It would bring together ". . . research, engineering, manufacturing and selling activities of the three companies in connection with radio receiving sets and accessories, phonographs and vacuum tubes."[6]

This agreement in principle was achieved in practice the following year, 1930. General Electric turned over its Harrison Tube Plant to RCA and Westinghouse its Lamp Works at Indianapolis. RCA acquired all stock owned by both companies in NBC, RCA Victor Co., Inc., RCA Photophone, Inc., and the General Motors Radio Corp.* With the exception only of the last-named (51% of which remained with General Motors), RCA thus gained full ownership of all its subsidiary organizations, bringing them under a common management structure and completing the essential skeleton of the communications, broadcasting and manufacturing corporate entity as it is known today.

Status at 1930:
A Recapitulation and Summary

The company described itself to its stockholders in its 1930 Annual Report in the following manner:[7]

"For convenience in administration of a business of such diversified character, Radio Corporation of America has become largely a holding company . . . In the radio sales and manufacturing fields [the] Corporation is represented by RCA Victor Company, Inc., and RCA Radiotron Company, Inc.; in the field of wireless telegraphic communication by RCA Communications, Inc., and Radiomarine Corporation of America; in broadcasting by the National Broadcasting Company, Inc.; and in sound recording and reproduction for talking motion pictures by RCA Photophone, Inc.; RCA Institute, Inc., trains students for radio work; E. T. Cunningham, Inc., distributes Cunningham radio tubes; and the Radio Real Estate Corporation of America has charge of real estate holdings."

RCA in 1930 was (1) organized and equipped for business in its three major areas of interest, viz., communications, broadcasting and radio manufacturing; (2) managed by an aggressive, knowledgeable and already highly successful entrepreneur, David Sarnoff; (3) owned and controlled by the two largest electric companies in the country, General Electric and Westinghouse; (4) already devoting serious effort and thought to future products—it was becoming increasingly research and development oriented; and (5) staffed by 22,000 employees.

It was also entering the doldrums of the great depression. Gross income in 1930 was $137 million, down from 1929's high of $182 million. (This high was not equalled again until 1942.) Earnings were $5.5 million in 1930, as compared to 1929's $15.9 million. The balance sheet showed, at the end of 1930, total assets of $169 million and total liabilities of $40

* This company had been formed jointly by RCA, General Motors, GE and Westinghouse in 1929 to enter into the automotive receiver field.

million, of which about $6 million was long-term debt. No common stock dividend had ever been paid.

The Years 1930 to 1940: Solidification

The national economic environment was not, of course, conducive to growth in the thirties; in fact, RCA fared considerably better than many U.S. businesses during the period,† perhaps in no small measure because of its diversified operations. The manufacturing acquisitions at Camden, Harrison and Indianapolis were integrated into the whole, with an ever-increasing array of new products being introduced. In addition, two particularly significant series of events took place early in the decade: a major ownership change and the initiation of a unified research program.

The ownership change resulted from a consent decree entered into by RCA, GE and Westinghouse. The action increased the spread of ownership and control of RCA from 85,000 stockholders (in 1930) to 300,000 by the end of 1932 and removed GE and Westinghouse from the RCA picture. AT&T, one of the original owners of RCA, had earlier divested itself of ownership. Thus, early in the decade, RCA was for the first time fully in control of its own destiny.

The unification of research, a step taken in 1934, met a growing need for centralized direction of research and development activities within the growing complex of subsidiaries that comprised RCA. It was implemented through the corporate funding of research projects as opposed to the subsidiaries bearing such costs alone. A policy was also established whereby a large part of the total research costs of the corporation would be covered by the royalty payments of licensees. This arrangement was an important step toward creating a research-oriented environment in which lower levels of management could make considerably more objective research expenditure decisions.

Status at 1940

Gross income in 1940 stood at $128 million; earnings at $9 million. The employment level had reached a new high of 25,000, and dividends were being paid on both preferred and common stock.‡ There were 242,000 stockholders, all equity interests of GE and Westinghouse having been transferred to the public. The balance sheet showed total assets of $104 million as against total liabilities of $38 million, only $5 million of which represented long-term indebtedness. RCA was healthy and growing.

† For example, the number of employees dropped relatively modestly, from 22,000 in 1930 to about 18,000 (the low) in 1934. Moreover, only two loss years were experienced, 1932 and 1933.

‡ RCA's first common stock dividend was paid in 1938 at the rate of $0.20 per share.

Research and development had produced, among many other notable accomplishments, a commercially practical television system which was introduced to the public at the New York World's Fair in 1939. Delayed by World War II, it emerged as a nationwide service and industry shortly thereafter.

The Years 1940 to 1950: Giant Steps

Between 1940 and 1945, RCA, like the rest of U.S. industry, turned its whole attention to wartime products and service. Existing plants at Camden, Harrison, Indianapolis, Bloomington (Indiana), New York City, Long Island and Hollywood were augmented by new facilities, the principal example being the Lancaster, Pa., cathode-ray and power tube plant. Employment grew from 25,000 in 1940 to 35,000 in 1945–1946 and jumped to almost 55,000 by 1950.

RCA's gross income in 1950 was $586 million, and its earnings $46 million, an increase of more than 400% over the 1940 figures. Television, more than any other single factor, appears to have been responsible for this phenomenal growth. Some organizational consolidation had occurred, notably in the creation of an RCA Victor Division from the previous RCA Victor Manufacturing Co.—a subsidiary. This change, made effective on Dec. 31, 1942, represented a major step away from the earlier "holding company" concept.[8]

An International Division was created during the forties to specialize in the marketing and, where appropriate, the manufacture of products abroad. Although RCA had for many years owned and/or operated certain relatively small foreign enterprises, no overall coordination and control had previously been established.

As of the end of 1950 RCA's balance sheet showed total assets of $312 million and total liabilities of $139 million, $60 million of which was long-term debt. Common stock dividends had been thrice increased during the ten years—from $0.20 per share to $0.30 (1947), then to $0.50 (in 1948), and finally to $1.00 (in 1950).

The Period 1950 to 1962

RCA's gross income grew to about $1.7 billion in 1962, its earnings to more than $50 million, and its employment to the order of 85,000 people.

There had been a decade and a half of commercially successful monochrome television and RCA was pioneering in color television. Man was reaching into interplanetary space—and RCA was technologically involved with such projects as SCORE, TIROS, RELAY and others. A whole new range of products had come on the scene—electronic computers and, more especially, their application to commercial data processing.

The capital investment required merely to get into the Electronic Data Processing business came to a staggering $100 million, rivaled only by RCA's earlier investments in monochrome and color television.

About a third of the company's gross sales was from defense contracts. Broadcasting was a stronger-than-ever contributor, *Fortune Magazine*[9] estimating that about 40% of RCA's earnings had come from NBC. Communications, still

very healthy, was a relatively small part of the business. Thus, forty-three years after its inception as an American-owned wireless communication firm—largely at government behest—RCA had reached fame and fortune of very major proportions.

Part II: Television Development Within RCA

At the turn of the century, the imagination and interest of technical and scientific minds the world over were being stimulated by prospects of being able to see at a distance using electrical means of communication. In a dozen countries, an awakening realization of electrical communication stimulated experimentation and investigation of ways of extending that medium to include vision as well as sound. Hubbell[10] records that an American, G. R. Carey of Boston, set forth the principles of a rudimentary television system as early as 1875.

In 1884, Paul Nipkow, a Russian inventor, patented a much more practical television approach, one which was to form the basis of most of the serious experimentation in the field for the next thirty or forty years. (His basic contribution was that of *scanning* an image in some regular fashion so that the communication channel would be required to handle only one bit of information at a time.) The art was primarily device-limited at the beginning, and its early growth therefore was closely related to inventions such as the photocell, the vacuum tube and the cathode-ray tube.

Shortly following the turn of the century, enough tools were within the early investigators' reach to allow crude but operable embodiments to be achieved. One of the very first of these was conceived and carried out at the St. Petersburg Technological Institute in Russia by Boris A. Rosing, Professor of Physics. Assisting him (in 1911) was a young man named Vladimir K. Zworykin who went on to achieve what was ultimately called "fully electronic television."

Through the 1920's and 1930's a number of distinguished experimenters were active: A. A. Campbell-Swinton and John Baird in England; Philo T. Farnsworth, C. F. Jenkins, E. F. W. Alexanderson, A. B. DuMont and many others in the United States. And as the industrial research team came upon the stage, large business organizations such as General Electric, American Telephone and Telegraph, Westinghouse and RCA began to play increasingly significant roles.

Mechanical vs. Electronic Television

Nipkow had shown that the way to transmit visual intelligence over an electrical communication channel was to sample the brightness of the scene, element by element, in some regular and recurrent sequence. An electrical analog

of the resulting stream of brightness values would then be produced by some appropriate transducer (e.g., a photocell) and transmitted. The process would be reversed at the receiving end where varying brightness values would be reproduced by a transducer (e.g., a neon tube), under the influence of the incoming electrical signal, and reassembled by another device into a reproduction or image of the original scene. This process of scene analysis and image synthesis is called "scanning," the key to all practical television systems.

Mechanical Means

The first practical embodiments of scanning apparatus took the form of rotating discs. A series of small holes therein was arranged so as to pass over the area representing a scene, one by one, thereby sampling but one picture element at any instant. The scanning disc, as it was called, appeared in virtually identical form at both transmitting and receiving points, the former employing a photocell to generate the signal and the latter using a neon or glow tube to reproduce it. The discs whirled merrily away at from several hundred to several thousand revolutions per minute and produced an image having a height dimension which was perhaps a tenth of the disc diameter. Typical images were $1\frac{1}{4}$ by 2 in.

Many ingenious variations of the basic scanning disc subsequently were invented: multifaced drums, rotating mirrors and prisms, and a host of variations on each. All are classified as mechanical, being characterized by the gross—or at least readily discernible—physical motion of man-made objects. In general, the method is limited by mechanical considerations of speed, precision and inertia, which impose severe upper limits on further (required) improvements. This conclusion was far from obvious during the 1920's, and serious broadcasting work was done in England using this method even during the 1930's.

Electronic Means

Electronic television, though still based on fundamental scanning principles of scene analysis and image synthesis, achieves the scanning process by nonmechanical means. Electron beams are caused to move back and forth over scene and image in synchronization, replacing the optical-steering functions of the disc. Virtually inertialess and capable of being controlled in intensity at extremely high

rates, these electron beams provided a potential system capability which far transcended that of their mechanical counterparts. I say "potential" advisedly, since although the general concept of electronic scanning was reasonably well understood by many early experimenters, its practical embodiment was extremely difficult to achieve, and thus only the dedicated few were actually working with it. This was true through the 1920's and the choice of an electronic versus a mechanical system was not made in the RCA organization until the early 30's.

Westinghouse and Zworykin

Vladimir K. Zworykin, the Russian-born scientist mentioned above, emigrated to the United States in 1919. He had an intense personal dedication to the creation not only of a practical television system, but one which would be completely "electronic." This concept was indeed an advanced one for the era, since "electronics" was largely limited to the rather simple technology of the amplifying vacuum tube. But Zworykin had dreamed of it since his early days with Professor Rosing at the St. Petersburg Technological Institute and in 1923 he applied for a patent on what was subsequently to become the historic keystone of electronic television, the iconoscope or image pickup tube.

Zworykin's Story

Zworykin (according to an interview on Jan. 17, 1963) was an undergraduate in electrical engineering when he first encountered Rosing. Rosing, who was actively experimenting with television, developed an interest in Zworykin and invited him to assist with the television work. In 1907, Rosing had patented a receiver concept which used a cathode-ray tube as the image-synthesizing means; scene analysis was accomplished by mechanical means. The implementation of this concept constituted the work which Zworykin was invited to join. Virtually everything had to be made by the experimenters themselves. This included the very basic components such as the photocells and the cathode-ray tube. The latter, patterned after the Braun Tube (disclosed in 1897), was not of the sealed-off variety; i.e., continuous high-vacuum pumping was required. No amplifiers were available. Results were poor, at best, but the young Zworykin developed an interest he never lost.

Graduating with honors from the

Institute in 1912 Zworykin went to the College de France in Paris for graduate work in theoretical physics and x-rays. Returning to Russia, he spent several years working on vacuum tubes and aviation equipment for the Signal Corps (of the Russian Army). With the coming of the revolution, he left Russia and came to the United States where, in 1920, he was hired by Westinghouse at its Pittsburgh Research Laboratory.

After a year with Westinghouse, during which he had filed a number of patent applications on behalf of the company, Zworykin spent a brief period with a small firm in Kansas City, the C and C Development Co. Westinghouse asked him to return, the invitation stemming largely from the ideas he had disclosed earlier. Zworykin insisted on an employment contract which guaranteed him an opportunity to conduct experiments in television, and in late 1922 or early 1923, Zworykin started his "serious work in this country on television."

By November 1923, working mostly alone, he had assembled a working version of the iconoscope and had applied it in the world's first all-electronic pickup and display television system. Employing more sophisticated cathode-ray tube technology than had been available ten years earlier in Russia, the system was capable of transmitting simple geometric figures, albeit at low brightness and contrast, and with a 60-line scanning structure.

Zworykin spent the next three or four years in related areas of research: facsimile, sound motion-picture recording and basic photocell investigations. It was during this period that he developed a highly sensitive cesium-magnesium photocell, the chemistry of which was to be applicable to the first television electronic pickup tube which he designed.

On April 5, 1923, in a report to the RCA Board of Directors, David Sarnoff had said:

"I believe that television, which is the technical name for seeing as well as hearing by radio, will come to pass in due course . . . It may be that every broadcast receiver for home use in the future will also be equipped with a television adjunct . . . which . . . will make it possible for those at home to see as well as hear what is going on at the broadcast station."

In 1927 or 1928, Sarnoff met Zworykin and made arrangements with Westinghouse for Zworykin to proceed with his television research and development. The ensuing work at Westinghouse resulted in an operating television system which, although of considerably greater sophistication and performance than Zworykin had demonstrated in 1923, was still a mechanical/electronic hybrid. It employed vibrating mirrors for scanning at the transmitter end of the circuit and a cathode-ray tube for scanning and display at the receiving end.

Other Early Television Work

E. F. W. Alexanderson, a GE consulting engineer and inventor of the wireless telegraph alternator that bears his name, took an early interest in television. As early as Dec. 16, 1926, he had announced an all-mechanical system, based on the use of revolving mirrors for scanning, for producing "moving pictures by radio." On Jan. 13, 1928, the first public demonstration of GE television was given in Schenectady, using an improved version of the equipment; a rotating disc was used to provide the scanning.

Alexanderson and his group went on through 1930 or so to refine and demonstrate television apparatus. Experimental broadcasts were made in 1928 by WGY, Schenectady; half-hour transmissions, three afternoons a week, were broadcast over the regular station facilities. This was possible because the low picture quality required only a narrow bandwidth, one reasonably compatible with the medium frequency (regular AM) broadcast band. One of the more challenging series of demonstrations of this era was that at Proctor's Theater in Schenectady; a huge seven-foot square television picture was produced by means of an arc lamp, a Kerr cell light modulator or valve and a mechanical scanning disc. Ray D. Kell, who worked with Alexanderson on this project, and who shortly thereafter was transferred to RCA, recalled (in an interview on Jan. 17, 1963) the hectic quality of those particular shows. They were interspersed with the then popular afternoon vaudeville acts at the theater and thus required rather critical scheduling. This involved considerable coordination with the television transmitting studio some miles away, no easy task in view of the highly primitive nature of the equipment. I rather gained the impression from Kell that the operating staff was not overly grieved upon the termination of this particular series of demonstrations. Such demonstrations, however, served to stimulate public interest—or at least, curiosity—about the new medium. This, in turn, led more than one manufacturer to petition the Federal Radio Commission for a commercial go-ahead even with the crude systems then at hand. Such commercial standardization was withheld, in retrospect wisely so, until considerably more system sophistication had been achieved around 1940.

Kell and others of Alexanderson's television group were transferred to RCA's Camden group during the 1929-1930 changeover. There, they became part of the newly forming television cadre.

Van Cortlandt Park's Television Work

Throughout most of the 1920's RCA maintained a laboratory group in New York City (at Van Cortlandt Park) called the Technical and Test Department. Experimental investigations of mechanically scanned television systems were undertaken there during the period 1927 to 1930; this work was similar in nature and approach to Alexanderson's efforts at GE in Schenectady.

In April 1928, this group applied for the first RCA television station construction permit. Issued as W2XBS and initially put on the air from the Van Cortlandt Park location, this station (under a variety of call letters) was to play a key and continuing role throughout the whole development effort. Its descendant is currently operating, thirty-five years later, as WNBT, NBC's prime commercial television outlet in New York City.§ NBC assumed management of the original W2XBS in July 1930, about the time that the Technical and Test Department was disbanded.

An excerpt from a report[11] made by T. J. Buzalski of WNBT, describing the history, equipment and results obtained at W2XBS, serves to give some of the flavor of the era:

". . . The program material usually consisted of (a) posters with black images upon a white background, such as W2XBS–New York–USA . . . (b) photographs, (c) moving objects such as Felix the Cat . . . revolving on a phono turntable . . . and (d) human talent. Communications from observers indicate that the service area of the station is quite large and under favorable conditions good television reception has been reported at distances of several hundred miles, the farthest point being Kansas. There are letters on file from some 200 observers and numerous telephone calls have been received from time to time reporting local reception. Many interesting tests were conducted between W2-XBS at Times Square and the RCA Laboratory at 711 Fifth Avenue and also with Mr. O. B. Hanson at his residence at Sunnyside, N.Y."

The original image size, I estimate, was about $1\frac{1}{2}$ by 2 in. This image was rendered in a reddish-orange color, of low, flickering brightness, 60 lines/picture, 20 pictures/sec; it is not difficult to understand the somewhat less-than-universal public appeal which was stimulated in 1930.

By mid-1930, the Van Cortlandt Park activity had largely been dispersed. Of those who joined the newly forming Camden organization, the only television researcher that I could identify was T. A. Smith, now RCA Executive Vice-President for Corporate Planning. He had participated in some of the earliest attempts[12] to establish television standards requirements on a systematic and rational basis. Smith went on to play a significant role in television engineering

§ Other call letters under which the station operated through the years were W2XK, WNBC, W2XF and WRCA-TV, according to T. J. Buzalski, now Engineer in Charge at WNBT (interview of Jan. 28, 1963).

and marketing, particularly in the studio and transmitter terminal equipment end of the business.

Thus, all basic television activities were now concentrated in Camden. Groups from Westinghouse, GE and RCA's own Technical and Test Department had been brought together as part of a larger corporate plan to coordinate and strengthen the nascent manufacturing and development activity at Camden. It was from this conjunction of diverse technical talent—and from this point in time—that RCA television development can truly be said to have had its serious start.

Getting Started in Camden

Two major groups of development engineers were established initially at the Camden Plant under L. W. Chubb who, in turn, reported to W. R. Baker, Chief Engineer for RCA Victor.[13] One, designated as the General Research Group, was under the direction of the former GE engineer, E. W. Engstrom, and the other was called (prophetically) the Electronic Research Group under Zworykin, formerly of Westinghouse. Totaling some 45 technical specialists initially, these groups proceeded almost immediately to tackle television research and development problems.

Although there were other areas of technology under investigation (e.g., sound broadcast receivers, phonographs, photophone), apparently major emphasis was placed on television—not only initially but throughout most of the 1930's. This emphasis is interestingly demonstrated in other ways, too; for the first time since its founding, the corporation commented on television in its Annual Report for 1930 (p. 26):

"While television during the past two years has been repeatedly demonstrated by wire and by wireless on a laboratory basis, it has remained the conviction of your own Corporation that further research and development must precede the manufacture and sale of television sets on a commercial basis. In order that the American public might not be misled by purely experimental equipment and that a service comparable to sound broadcasting should be available in support of the new art, your Corporation has devoted its efforts to intensive research into these problems, to the preparation of plant facilities and to the planning of studio arrangements whereby sight transmission could be installed as a separate service of nationwide broadcasting."

A Systems Approach

Today, the concept of systems engineering is well understood and widely practiced. But not so in 1930 or 1940, even given such notable examples as the Bell System and the nation's interconnected power companies. Thus the work that went into RCA's television development represents a very interesting beginning in the realm of major systems engineering.

One of the characteristics of the systems approach is a concern with formulating and understanding the fundamental requirements to be met by that complex of physical things which will subsequently comprise "the system." In the case of television, it being a visual medium for the transmission of intelligence, these fundamentals included the nature of the image as finally viewed. Just what *were* the requirements of a "commercially useful" television image with regard to resolution, brightness, size, flicker, color and a host of related characteristics? A beginning had been made at Van Cortlandt Park in 1929 by Weinberger, Smith and Rodwin when they considered "Standards for Commercial Television."[14]

During 1931 and 1932, Engstrom and his group carried out an extensive series of experimental investigations aimed at quantitative specificity in television imagery. This work[15] constituted a solid foundation for many subsequent decisions.

Another important characteristic of the systems approach is "interconnectedness" —that quality whereby all elements of the complex fit together as an integrated whole, and do so under conditions that typify the environment in which the system must perform its intended function. The Camden group evidently recognized the importance of this requirement from the very beginning, since the many field tests which they conducted started in 1931, barely a year after the group was organized. Major, distinctly separate, field tests were conducted in 1931–1932, 1933, 1934 and 1936–1939, and involved virtually all segments of the corporation.

One more systems engineering problem was posed by television for commercial broadcast: that of the technical standards which must be imposed. Unlike radio broadcasting, television transmission carries with it special synchronizing signals and the image itself is, in effect, encoded in ways which a receiver must be specifically designed to accommodate. (The analogy of a "lock and key" relationship between receiver and transmitter is often used.) These more stringent standards obviously required not only rather exhaustive technical investigation but also —because broadcasting is a public medium—government regulatory blessing and industry acceptance.

Development and Field Test: A Recurrent Pattern

As mentioned earlier, the 1930–1940 period was devoted to developing, field testing and refining four or five generations of prenatal television systems. For completeness and perspective we will touch upon each, although briefly; the literature coverage of these tests is excellent, containing many interesting descriptions and photographs, to which reference will be made as we proceed.

The 1931–32 Field Test

The first-generation mechanical/electronic hybrid system developed by the Camden Group (1930–1931) was given a practical test in the New York metropolitan area in late 1931 and the first half of 1932.[16]

The system utilized 120-line scanning and a picture repetition rate of 24/sec. Pickup means were mechanical and included both studio and motion-picture-film scanners. The studio scanner was of the "flying-spot" type in which the only source of illumination for the televised subject came from the scanner itself. An intense, but tiny, beam of light swept repetitively across all televised areas of the scene. Reflected light therefrom excited a bank of photocells, the output of which constituted the television video signal. Ray Kell recalled for me (in the interview on Jan. 17, 1963) the very interesting effect this rather bizarre illumination had during a performance given by a harpist. Essentially acting as a stroboscope, the scanner illumination made the vibrating harp strings appear to be undulating as if blown by some magical, invisible wind—an effect which, as Kell recalls, "virtually drove the harpist crazy . . ."

An important first was achieved by this initial field test: transmission was accomplished in the 40- to 80-mc band. Previous television broadcasting (all experimental) by RCA and other organizations had been at lower frequencies, mostly in the 2- to 3-mc region. (W2XBS had, for example, first operated on 2.1 to 2.2 mc.) This move to what is now termed the "very high frequency" (VHF) region of the spectrum was a result of earlier propagation investigations, both theoretical and experimental, by RCA researchers such as Beverage, Peterson, Hansell, Jones and others. Characterized by so-called line-of-sight limitations of range, the VHF region afforded the only feasible room for subsequent commercial operation of television stations and was thus an important part of any serious television system investigation. The line-of-sight characteristic posed many unknown performance dimensions: shadowing by and echoes from nearby structures, antenna sizes and locations, man-made and natural static or interference, and a host of others. Merely to design receivers and transmitters for this region was a significant challenge in the early thirties, the principal limitation being established by available vacuum-tube technology. This problem was, of course, the special province of the Radiotron Division of RCA.

Receivers for the tests used cathode-ray tubes as the image display device; such tubes came to be called "kinescopes," a term coined by Zworykin[17] to designate that class of cathode-ray tubes specially designed for television imaging. A small number of receivers were placed in and around New York City, many of them in the homes of the technical personnel, in order to collect subjective viewer reaction data as well as objective engineering measurements.

Engstrom's own summary[18] of the results of this first RCA field test provides us with a good feel for the 1932 television situation:

"... Some of the major conclusions and indications are of general interest. The frequency range of 40 to 80 megacycles was found well suited for the transmission of television programs. The greatest source of interference was from ignition systems of automobiles and airplanes, electrical commutators and contractors, etc. It was sometimes necessary to locate the receiving antenna in a favorable location as regards signal and sources of interference. For an image of 120 lines the motion picture scanner gave satisfactory performance. The studio scanner was adequate for only small areas of coverage. In general the studio scanner was the item which most seriously limited the program material. Study indicated that an image of 120 lines was not adequate unless the material from film and certainly from studio was carefully prepared and limited in accordance with the image resolution and pickup performance of the system. To be satisfactory, a television system should provide an image of more than 120 lines... The superiority of the cathode ray tube for image reproduction was definitely indicated. With the levels of useful illumination possible through the use of the cathode ray tube, the image flicker was considered objectionable with a repetition frequency of 24 per second. The receiver performance and operating characteristics were in keeping with the design objectives."

Less than a year later, an improved system was assembled and a second round of field testing began.

1933, A Breakthrough Year

Zworykin announced his iconoscope or television "electric eye" during 1933, thereby freeing the system from mechanical imaging limitations. This device was the practical embodiment of Zworykin's 1923 patent disclosure and working model and has since earned him worldwide recognition. In essence the iconoscope was a cathode-ray tube which could scan an optical image of the scene being televised and produce—in and of itself—a video signal. The tube was able to store the electrical charge corresponding to any given scene element throughout the scanning period, thereby accumulating a much higher signal than otherwise possible. This greatly enhanced sensitivity represented a significant advance.

Philo T. Farnsworth, a young experimenter backed first by private interests and later by the Philco Corporation, invented another type of electronic television pickup tube about 1930. This tube, the "image dissector," was the only other new device to appear in television development prior to the 1940's. But the image dissector suffered very seriously in sensitivity, requiring extremely high light levels to develop useful output signals. Because of this, it achieved but brief experimental use.

The 1933 Field Test

Improved imaging standards, a virtually all-electronic system‖ and the radio-relaying of television signals were the major improvements tested in the field during the first several months of 1933.[19] The iconoscope was used for studio and film pickup; at Camden, where the tests were centered, this allowed 240-line scanning to be employed, the picture repetition or frame rate being maintained, as before, at 24/sec. A remote studio and relay link transmitter were located on the eighty-fifth floor of the newly erected Empire State Building in New York City, serving as a source of test programs.

In Camden, too, there were additional sophistications. The main program transmitters and antennas—picture was on 49 mc, sound on 50 mc—were located in downtown Camden (Building 2). The main studio and terminal equipment, which was about 1,500 cable feet away from the transmitters (Buildings 5, 6 and 7), also received program feed from two other remote sources. One of these was the Empire State Building installation (via a relay at Arney's Mount); the other, a temporary setup at Building 53, about one mile away. A "typical receiver" was located some four miles distant from the main Camden transmitter, at Collingswood, N.J. It employed a vertically mounted kinescope (having a screen about 9 in. in diameter) with a 45° mirror to allow viewing from a position in front of the set.

These tests were of particular value since they utilized essentially all the elements needed for an actual commercial system. This was another "first." Engstrom[20] summarized the results as much in terms of the problems that remained to be tackled as in accomplishments. The problems had to do with the need for greater power for the picture transmission (to provide more reliable coverage of a given area), still greater resolution and elimination of image flicker. Results with the inconoscope had been most gratifying and a great deal of practical and very

‖ The only remaining mechanical device was, interestingly enough, the generator of synchronizing signals. More as a matter of convenience than of necessity, the device employed an optical technique (rotating disc with serrations corresponding to the desired timing signals, which interrupted a light beam) to generate synchronizing impulses for the system.

valuable operating data and experience had been obtained. Thus armed, the laboratory work proceeded.

The 1934 Field Test

Three principal changes appeared in the 1934 version of the field-test system. These were:

1. An increase in the number of scanning lines to 343.
2. Use of an all-electronic synchronizing generator.
3. Introduction of interlaced scanning which provided an effective picture rate of 60/sec, thereby virtually eliminating flicker.

Testing was confined to the Camden area with relatively low transmitted power.

In passing, note should be made of the fundamental importance of the interlaced scanning principle. The idea, basically simple and practical to implement, involves "fooling" the eye in much the same manner as does a motion-picture projector. There, a mechanical light chopper is used to increase the apparent flicker frequency, whereas the television counterpart chops one scene into two, geometrically interlacing them. A member of the Camden group, Randall C. Ballard, holds the basic patent on the method, having disclosed the idea somewhat before the need became crucial in the field tests. Interestingly, another member of the group, Alda V. Bedford, proceeding under the acute stimulus of the field-test flicker problem, also conceived the idea and reduced it to practice. Thus even in a small organization (perhaps fifty people at that time), intergroup communication could be a problem. Ballard and Bedford were in different groups, Bedford being more directly concerned with the flicker problem per se.

Now fully implemented with all-electronic apparatus, the television system which RCA was developing had come a long way in the four years since the group responsible for it had come into being. Most of the troublesome pieces of the jigsaw puzzle had been found; polishing and refining became the principal tasks for the next four years.

The 1936–1939 Field Tests

Results of the initial New York and the two subsequent Camden field tests, together with the system development throughout the period, convinced the corporation that another large-scale New York test was in order.

The tests[21] involved a system based on 343-line, 30 frames/sec, 2:1 interlaced scanning standards. Transmission, from the Empire State Building, was at 49.75 mc for picture and 52 mc for sound, the video bandwidth being 2 mc. Transmitted power had been increased to 8 kw and an antenna having an effective power gain of 2:1 was put into service.

Approximately a hundred receivers

were built on a short-run production line basis (thereby beginning the development of production "know-how") for placement in the homes of RCA personnel. These receivers were equipped with 9-in. diameter kinescopes which produced an image approximately $5\frac{1}{2}$ by $7\frac{1}{4}$ in. They employed a total of 33 tubes and were of the upright-console type in which viewing was via a 45° mirror in a lift-lid arrangement.

The tests ran on intermittently through 1938. Results were so encouraging that a major corporate decision was reached in that year to undertake "commercialization" of the medium. Such a course involved not only the actual design and production of receivers but also the much more difficult job of obtaining the approval of the Federal Communications Commission and winning the support of the broadcasting and manufacturing industry regarding television standards.

Inauguration of a Public Service

On April 30, 1939, with the opening of the New York World's Fair, RCA presented the results of ten years' intensive research and development in television for public broadcasting.

RCA petitioned the FCC for full commercial status of its television broadcasting in 1938 but received no approval until the following year. At that time a limited form of program sponsorship was allowed on a temporary basis pending public hearings scheduled for Januray 1940. The findings of this hearing by the FCC were interpreted by RCA as allowing full-go-ahead. Thus it undertook an active merchandising campaign to sell television receivers to the public and stepped up its programing activities. In April of the same year, however, the FCC rescinded its go-ahead on the basis that it felt RCA's actions would tend to "freeze" television standards prematurely. Further consideration, investigation, experimentation and consultation by the industry itself were urged by the Commission and a period fraught with wrangling, bitterness and frustration ensued.

On July 1, 1941, fully commercial television was authorized in the United States. When the country went to war the next December, commercial television had been born but left in a state of suspended animation. Although there was limited television broadcasting by RCA in New York City during the war (largely for civil defense purposes) commercial growth awaited a postwar environment.

A Wartime Interruption

Most of the wartime television work was conducted at the Princeton Laboratories which were opened in 1942. Television received very little attention by the military early in the war. Several classified military television projects were undertaken later, however. These were primarily aimed at developing small television apparatus which could be used in various types of airborne vehicles for reconnaissance or control purposes. Going under exotic project code names such as Block, Bat, Roc, and Glomb, the work resulted in a number of useful advances in the state of the art, particularly as regards miniaturization.

Work was pressed on development of more sensitive pickup or camera tubes. For, as great an advance as the iconoscope represented, it still required scene illuminations which for most military applications were out of the question. Other promising approaches had been under investigation during the mid-30's and shortly before the war the "orthicon" became available. It was approximately five times more sensitive than the iconoscope and, of equal importance, was free from many of the operational idiosyncrasies of the latter. During wartime research the orthicon became the amazingly sensitive, complex-to-build, "image orthicon." This tube provided a sensitivity gain of about one hundred times that of the iconoscope, thus greatly enhancing the flexibility of camera operation. This tube formed the backbone of postwar camera technology; it is still the principal tube in use.

High-power transmitting tubes, capable of producing many kilowatts of radiofrequency energy at hundreds of megacycles also came out of wartime research and development activities. Such tubes became an indispensable part of postwar television broadcasting technology. So, too, did cathode-ray and other special tubes used during the war for radar applications. Advances in these arts greatly aided the rapid maturation at war's end of a much more sophisticated version of the 1941 television system which had finally been blessed by the FCC. It was a case of the same girl in a new gown, with all new accessories; to those who loved her, she was even more attractive than before.

The Postwar Era

After the war, it took RCA the better part of two years to develop a home instrument for the market. This new product was the "630-TS" 10-in. table model television receiver, the first practical, widely marketed device of its kind. It was offered in the fall of 1946 at a retail price of $375.

The 630-TS television receiver was designed to use a 10-in. kinescope, the 10BP4, which had been developed specifically for mass production and utilization in home instruments. The Radiotron Tube Division and the Victor Division, both still under the RCA Manufacturing Co. organizational structure, coordinated their plans closely and effectively. Mass production of cathode-ray tubes was a very new technology, having grown up during the war. RCA had operated a new plant in Lancaster, Pa., built by RCA for the Navy specifically for producing cathode-ray tubes for military applications. RCA bought the plant from the Navy immediately after the war, largely on the gamble that television would be successful. Television *did* catch the public's fancy so much so that by 1948 a kinescope shortage actually developed within the industry. Lancaster had a considerable inventory of 10BP4's by 1948, and this stockpile proved the appropriateness of the earlier decision to produce substantial quantities.

We have mentioned that RCA's introduction of the 630-TS met with immediate and vigorous public acceptance. In large measure this was a result not only of the novelty and amusement attraction of the product but also of the pent-up demand created in the buying public by wartime consumer goods scarcity. The combination appeared to RCA to be insatiable for any one manufacturer, however "big and ready" he might be. Thus a policy was adopted, *prior* to the actual realization of demand in the market place, whereby all RCA licensees were thoroughly briefed on the sets that RCA would shortly offer to the public. So-called Licensee Symposia were held in Camden throughout mid-'46 in which the in-house manufacturing know-how which RCA had acquired in producing the 630-TS was made available to any interested licensee. This act very probably did more to accelerate the phenomenal volume of television set manufacture by American industry over the next two years than any other taken by RCA. Frank M. Folsom, at the time Executive Vice President in charge of the Victor Division and later President of the Corporation, is credited with this policy decision.

Television Broadcasting

Until a substantial number of receivers were in the hands of the public, television broadcasting revenues, derived solely from advertising fees, were virtually nonexistent. This, in turn, was not conducive to the expansion of television broadcasting and programing which would attract more viewers and thus more advertising revenue. The age-old "chicken or egg" paradox became painfully evident soon after the initial novelty appeal began to fade.

David Sarnoff appealed to the NBC broadcasting affiliates not to buck television as some silent motion-picture makers had bucked "talkies," cable companies had balked at wireless, and phonograph companies had bucked radio. C. B. Jolliffe, then Executive Vice President of the RCA Laboratories

Division, recalled (in an interview on Jan. 29, 1963) that many NBC broadcasters—simply on the strength of Sarnoff's fervor and sincerity—made immediate plans to "go television." Figure 1 shows that the rate of TV station growth increased significantly from 1947 through 1949; station growth exceeded receiver growth during this period, a condition vitally needed to overcome the earlier-described "chicken and egg" hiatus. Certainly other factors must have been at work to encourage this trend too, but equally certain is the conclusion that David Sarnoff's personal counsel and appeal was an important factor in moving the broadcasting engine off dead center. Those who heeded his advice certainly had no economic regrets, particularly since television station licenses—then easy to obtain—later became an extremely scarce and thus valuable asset.

Color Television

This treatment of television's history within RCA has concerned itself solely with the black-and-white or monochrome system. Although this thesis does not encompass it, there is an equally interesting history from the very recent past regarding the developing of color television. We should note, however, that research and development on color television was under way at the Princeton Laboratories in the late 1940's. By 1950, other industry groups, notably CBS, were pressing for adoption of a national color television system of a noncompatible type; i.e., color programs could not be received (even in monochrome) on the existing television receivers, of which there were between ten and eleven million in the hands of the public. RCA violently disagreed with this concept and as rapidly as possible, pushed to completion a so-called "compatible" color system. Again there was a long, frequently bitter wrangle involving diverse industry groups and the FCC. Finally, in December, 1953, a system of standards was adopted by the FCC based upon the compatibility principle. It took eight years for the "chicken and egg" problem to run its course this time; but run it, it did— and today (spring of 1963) color tele-

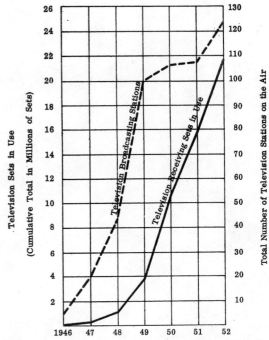

Fig. 1. Growth of television stations and receiving sets in the beginning years. Data from *Electronic Industries,* **January 1963, p. 101.**

vision broadcasting is about at the point that black and white was in 1947-1948. Like its cousin, it promises big things for the years ahead.

References

1. *Fortune Magazine,* 171–190, July 1962.
2. Letter from Owen D. Young of GE to Franklin D. Roosevelt, Acting Secretary of the Navy, in the early part of 1919. See W. Rupert Maclaurin, *Invention and innovation in the radio industry,* The Macmillan Company, New York, 1949, p. 100.
3. Maclaurin, *op. cit.,* pp. 114–117.
4. R. Kenyon Kilbon, "Pioneering in electronics," unpublished two-volume manuscript, Jan. 1960, Princeton, N.J., Vol. I, p. 26.
5. "Blue Chip," *Fortune Magazine,* 46, May 1932.
6. RCA Annual Report for 1929, p. 4.
7. RCA Annual Report for 1930, pp. 4–5.
8. *Ibid.,* p. 4.
9. Walter Guzzardi, Jr., "RCA: The General never got butterflies," *Fortune Magazine,* 102, Oct. 1962.
10. Richard W. Hubbell, *4000 years of television,* G. P. Putnam's Sons, New York, 1942, p. 59.
11. T. J. Buzalski, *Experimental television station W2XBS,* Development Group Engineering Report No. 95, National Broadcasting Co., Department of Technical Operation and Engineering, New York, Mar. 15, 1933.
12. Julius Weinberger, T. A. Smith and George Rodwin, "Standards for commercial television," *Proc. IRE, 17,* no. 9, 1584–1594, Sept. 1929.
13. Kilbon, *op. cit.,* Vol. I, p. 35.
14. Weinberger, Smith and Rodwin, *op. cit.*
15. E. W. Engstrom, "A study of television image characteristics," *Proc. IRE, 21,* no. 12, 1631–1651, Dec. 1933.
16. E. W. Engstrom, "An experimental television system," *Proc. IRE, 21,* no. 12, 1652–1654, Dec. 1933. (Introduction to a group of papers by associated RCA authors.)
17. V. K. Zworykin, "Description of an experimental television system and the kinescope," *Proc. IRE, 21,* no. 12, 1655–1673, Dec. 1933.
18. See Ref. 16.
19. E. W. Engstrom, "An experimental television system," *Proc. IRE, 22,* no. 11, 1241–1245, Nov. 1934.
20. *Ibid.*
21. R. R. Beal, "Equipment used in the current RCA television field tests," *RCA Rev.,* Jan. 1937.

TELEVISION BROADCASTING PRACTICE IN AMERICA

1927 TO 1944

Donald G. Fink

TELEVISION BROADCASTING PRACTICE IN AMERICA—1927 TO 1944*

By DONALD G. FINK.†

(The paper was received 5th July, 1944. It was read before the RADIO SECTION 24th January, the NORTH-WESTERN RADIO GROUP 23rd February, and the HAMPSHIRE SUB-CENTRE 2nd June, 1945.)

SUMMARY

This paper reviews the history of television broadcasting in America from 1927 to the present, with particular emphasis on current practice. Section 1, the historical survey, traces the evolution of standards of transmission, frequency allocations, and broadcasting practice. Noteworthy programmes are recalled. Section 2, on present practice, gives a detailed account of the standards of transmission governing public broadcasting under the current regulations of the Federal Communications Commission. The stations currently operating are listed. Typical equipment used in these stations is described in four categories: studio equipment, transmitters, radiators, and mobile pick-up equipment. The design of current (i.e. immediately pre-war) receiving equipment is described. The paper concludes with a digest of post-war prospects.

Definitions of American Television Terms

Antenna Field Gain—The ratio of the effective free-space field intensity produced at 1 mile in the horizontal plane from the antenna, expressed in millivolts per metre for 1 kW antenna input power, to $137 \cdot 6$.

Aspect Ratio—The numerical ratio of the frame width to frame height, as transmitted.

Black Level—The amplitude of the modulating signal corresponding to the scanning of a black area in the transmitted picture.

Field Frequency—The number of times per second the frame area is fractionally scanned in interlaced scanning.

Frame—One complete picture.

Frame Frequency—The number of times per second the picture area is completely scanned.

Frequency Modulation—A system of modulation of a radio signal in which the frequency of the carrier wave is varied in accordance with the signal to be transmitted, while the amplitude of the carrier remains constant.

Interlaced Scanning—A scanning process in which successively scanned lines are spaced an integral number of line widths, and in which the adjacent lines are scanned during successive cycles of the field-frequency scanning.

Negative Transmission—Applied to a system in which a decrease in initial light intensity causes an increase in the transmitted power.

Scanning—The process of analysing successively, according to a predetermined method, the light values of picture elements constituting the total picture area.

Scanning Line—A single, continuous, narrow strip, containing highlights, shadows, and half-tones, which is determined by the process of scanning.

Vestigial Sideband Transmission—A system of transmission wherein one of the generated sidebands is partially attenuated at the transmitter and radiated only in part.

* Radio Section paper.
† Office of the U.S. Secretary for War (on leave of absence from the staff of *Electronics*).

(1) HISTORICAL SURVEY

Television broadcasting in America began in 1927, when the Federal Radio Commission issued the first television licence to Mr. Charles F. Jenkins, authorizing broadcast transmissions from a station in the suburbs of Washington, D.C. Prior to that time, development of television techniques was not open to public participation. V. K. Zworykin applied for a patent on his iconoscope (emitron), which may fairly be called the cornerstone of modern television, in 1925. In 1923, Jenkins in America and John Baird in England had demonstrated the transmission of crude images over wires. Early in 1927 the Bell Telephone Laboratories demonstrated a low-definition picture over wire circuits, between New York and Washington. But the concept of providing broadcast emissions, available to experimenters not otherwise connected with the transmitting organization, did not gain wide currency in America until 1929. In that year some 22 stations were authorized by the Federal Radio Commission to broadcast visual images. In the 15 intervening years, construction permits and licences have been granted to no fewer than 104 stations. Of these, 21 never got beyond the construction permit stage, 48 were licensed but later ceased transmission, and 35 remain in operation to-day. Of the latter, 9 stations are licensed for regular transmissions to the public, and may accept programmes from commercial sponsors. The remainder are experimental stations, many of which render occasional public service. Fig. 1 shows the trend of activity, as

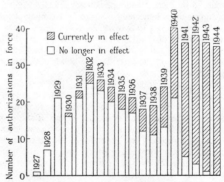

Fig. 1.—The trend of American television activity from 1927 to the present, as indicated by the number of broadcasting station authorizations (licences and construction permits) in force each year.

judged by the number of stations authorized each year from 1927 to the present.

During this period there has occurred a definite evolution of standards of transmission and frequency assignments. The earliest stations had wide latitude in choice of frequency, almost any frequency above 1 500 kc/s being permitted if no interference was caused to other services. But this latitude was soon withdrawn, as the short-wave region became crowded with other, more vital services. In 1929, emissions were limited to a bandwidth of 100 kc/s, within the regions 2 000–2 100, 2 100–2 200, 2 200–2 300, 2 750–2 850, and 2 850–2 950 kc/s. The powers employed varied from 100 W to 20 kW, the majority of stations operating at 5 kW.

The quality of the early images was primitive, judged by any standard. The pictures were commonly transmitted at a rate of 20 per second. At this rate the number of picture elements capable of being transmitted by double sidebands in a 100-kc/s band is limited to 5 000. This number of elements provides sufficient resolution to distinguish the features of a man's face, not much more. Equal resolution in vertical and horizontal dimensions was achieved within the band limits by employing a square image of about 70 lines, but the preferred figure was 60 lines.

Many different types of scanning were employed during this period. The most popular was the flying-spot method. In this system (see Fig. 2, Plate 1) the subject remained in partial darkness before a bank of eight (or more) stationary photocells, while a narrow, intense beam of light scanned the scene.

The typical experimental receiver, which persisted in use up to 1932, consisted of a scanning disc, from 2 to 4 ft' in diameter, through the perforations of which was viewed the surface of a flat electrode enclosed in a gas-filled lamp. Synchronization, when achieved, was by virtue of the power-line connection. Three or more stages of resistance-capacitance-coupled amplification were customarily used to produce the signal to actuate the gas-filled lamp.

Many of the major American television stations of the present day can trace their origin to this early period. The National Broadcasting Company's station in New York was first licensed as W2XBS in July, 1928, and has since evolved from the 2 000–2 100 kc/s band to the 50–56 Mc/s band, from 60-line pictures to 525-line pictures. In 1942 the call letters W2XBS were withdrawn in favour of the "commercial" call letters WNBT. Similarly, the Columbia Broadcasting System station in New York, now WCBW, started in July, 1931, as W2XAX. This station operated with 60-line pictures, 20 per sec, for a total of 2 500 hours in the period ending February, 1933. The General Electric station in Schenectady operated on similar standards, with 20 kW power from 1929 to 1932.

The programmes of the day were suited to the capabilities of the system. A toy doll in the form of "Felix the Cat," was a familiar test subject in the early days of the N.B.C. transmitter (see Fig. 3, Plate 1).

One of the early stations in Boston, W1XAV, was a favourite of engineering students at the Massachusetts Institute of Technology, several of whom possessed scanning-disc receivers. The subject-matter often consisted of a small group of students who went to the studio to be scanned by the flying spot, while another group, viewing a receiver nearby, endeavoured to recognize the first. Facial recognition was not always possible, although such simple acts as exhaling the smoke from a cigarette were usually evident enough. The elementary nature of the technique may be judged from the synchronization method employed by the author, while a student during this period. The scanning-disc motor, of the non-synchronous variety, was driven above synchronous speed. A rubber eraser, held against the edge of the disc, provided sufficient friction to bring the disc into synchronism.

In 1931 it was evident that progress could not be made on the restricted channels of the 2-Mc/s band, and the trend toward higher frequencies began. One of the earliest to apply for permission to use frequencies above 40 Mc/s was the Don Lee Broadcasting System in Los Angeles, California. In December, 1931, the licence of station W6XAO was granted to this organization, authorizing the use of the bands 43–46, 48·5–50·3 and 60–80 Mc/s. In 1941 this station became a commercial station with the call letters KTSL, operating on 50–56 Mc/s with a regular public programme service. The Don Lee station has, throughout its 13 years of operation, offered a regular and consistent high-definition programme service, a fact which makes it pre-eminent among American television stations.

Permission to use the 43–80-Mc/s bands was granted to several other stations, including N.B.C.'s W2XF and W2XBT in New York, Jenkins's W3XC in Wheaton, Maryland, W1XG in Boston, and W8XF in Pontiac, Michigan. In 1933 and 1934 several additional v.h.f. stations were licensed to use the bands 42–56 and 60–86 Mc/s. In 1936, all activity in the 2-Mc/s band ceased and all v.h.f. stations were placed in the bands 42–56 and 60–86 Mc/s.

In 1937, the frequency allocation was set up for the first time on the basis of channels 6 Mc/s wide. Nineteen such channels were set up between 44 and 294 Mc/s. The channel from 44 to 56 Mc/s was transferred in 1940 to frequency-modulation (sound-broadcasting) stations and replaced by an additional channel from 60 to 66 Mc/s, and the channel from 156 to 162 Mc/s abolished. The present allocation (Table 2) comprises 18 channels, each 6 Mc/s wide, from 50 to 294 Mc/s.

(1.1) Evolution of High-Definition Standards

Shortly after permission to operate in the v.h.f. bands was given, attention was focused on purely electronic methods of scanning and cathode-ray tubes for reproduction. Dr. Zworykin had demonstrated a cathode-ray receiver before the Institute of Radio Engineers in Rochester, New York, in November, 1929. In 1932, an "all-electronic" system was demonstrated by RCA, transmitting 240-line images from New York to Camden, near Philadelphia, over an air-line distance of about 80 miles, with one intermediate relay point at Arney's Mount, New Jersey. This was one of the earliest demonstrations of cathode-ray equipment, but the term "all-electronic" was something of a misnomer, since no satisfactory electronic synchronizing circuits had been developed, and the synchronizing pulses were derived by passing light to a photocell through apertures in a whirling disc.

In 1934, the march toward higher definition got properly under way. Each transmitter was free to employ any scanning method, but between 1932 and 1934 agreement was reached that interlaced transmission, based on the "odd-line" principle, was the simplest and most satisfactory method of avoiding flicker in the reproduced images. The first "odd-line" value chosen was 343, an odd number composed of odd factors ($343 = 7 \times 7 \times 7$). From this root sprang many other choices, all tending toward greater definition in the pictures, all odd numbers, composed of odd factors. The list of scanning specifications, in approximately chronological order, is given in Table 1. It will be seen that the progress toward higher numbers of lines was not steady, indicating that there were differences of opinion. The majority opinion was that no more than 441 lines could be accommodated in a picture sent by double-sideband methods within the limits of a 6-Mc/s channel. The dissenting opinion was that it would be better to err on the high side in the number of lines, with consequent excessive definition in the vertical dimension, in the hope that better utilization of band-width would be possible as time went on. This dissenting opinion was in fact justified in 1939, when vestigial-sideband transmission was proved feasible and adopted as standard. The 441-line figure then proved too small for a 6-Mc/s channel, and the standard was eventually changed (in 1941) to 525 lines, the present value specified in the F.C.C. regulations.

The decision concerning the rate at which the pictures were to

Table 1
SCANNING SPECIFICATIONS

Date	Organization	Number of lines	Frame rate, per second
1934	R.C.A.	343 (7 × 7 × 7)	24 and 30
1937	Philco	441 (3 × 7 × 3 × 7)	24
1938	Philco	525 (3 × 5 × 5 × 7)	30
1939	Philco	605 (5 × 11 × 11)	24
1939	R.M.A.	441 (3 × 7 × 7 × 3)	30
1940	R.C.A.	507 (3 × 13 × 13)	30
1941	N.T.S.C.	525 (3 × 5 × 5 × 7)	30

be transmitted centred about the two values indicated in Table 1, 24 frames per sec and 30 frames per sec. The figure of 24 frames per sec was highly attractive from two points of view. In the first place it coincides with the previously established standard frame rate for motion pictures, which constitute a more-than-appreciable fraction of the television programme material. In the second place, the 24-per-sec rate admitted 25% more detail in the picture than did the 30-per-sec rate, all other factors considered equal. But when an attempt was made to operate at 24 frames per sec with a 60-c/s source of alternating current (mains power), economic factors began to obtrude because the 24-per-sec rate is not simply commensurate with 60-per-sec. Very complete shielding of the cathode-ray tube and expensive filtering of the d.c. supply, especially in the scanning circuits, was found necessary to avoid deterioration of the scanning pattern caused by the resulting vertical and horizontal shifts of the scanning lines from their proper positions.

Argument proceeded for some time on the question whether the added cost incident to reducing this effect was justified by the increase in detail possible with the 24-per-sec rate. Finally the economic factors won the day. The value of 30 frames per sec was standardized, in 1936, and has not been seriously questioned since that time.

This left unsettled the question of operating motion-picture film, recorded at 24 frames per sec, on a television system operating at 30 frames per sec. Running the film at 30 frames per sec would involve 25% excess speed of motion, and corresponding increase in the pitch of the associated sound reproduction. It was soon found that these effects could not be tolerated, so it was necessary to find means of preserving an *average* frame rate of 24 per sec, while scanning each frame at an "instantaneous" rate of 30 per sec. The means was found in the fortunate fact that the fraction 1/24 is the average of two other fractions, 2/60 and 3/60. The television system scans the frame twice in two interlaced fields in 1/30 sec, i.e. the scanning of a single interlaced field occupies 1/60 sec. Thus, if one frame is held stationary in the projector for 2/60 sec, and the next frame is held stationary for 3/60 sec, synchronism with the field scanning rate is maintained, while the average rate of passing the frames through the projector is maintained at 24 per sec. This system works surprisingly well. It is necessary only to devise an intermittent pull-down mechanism which allows the film to "dwell" for unequal lengths of time (2/60 and 3/60 sec) on alternate frames. To ensure equal exposures of each frame, it is customary to expose the film during the vertical retrace time, between field scansions, and to count on the storage property of the iconoscope mosaic to preserve the charge configuration throughout the subsequent field scansion. This procedure also provides plenty of time in which to move the film between frames. Continuous-motion projectors have also been devised in which a moving lens system counteracts the motion of the film, and a shutter system divides the exposures into the required unequal intervals.

It is safe to say that if this problem had not been so satisfactorily solved, the cost of television receivers to the American public would be substantially increased to permit operation at 24 frames per sec. Alternatively, with 30 frames per sec adopted as the standard, it would be necessary to produce motion-picture film especially for television. Our British cousins are indeed fortunate that the mains frequency in the British Isles is 50 c/s, permitting a frame rate of 25 per sec, which is so close to the motion-picture standard that no bizarre effects are produced.

A frame rate of 15 per sec was seriously urged by the DuMont organization in 1940 as a means of obtaining a two-fold increase in picture detail. The problem of flicker was to be solved by employing cathode-ray tube phosphors which retained the image for substantially the full frame interval and then suddenly reverted to darkness just in time for the next image. Unfortunately, the search for such phosphors was not successful, since the characteristic exponential decay of light could not be circumvented. Flickerless images could indeed be demonstrated at low light levels without evident flicker at 15 frames per sec, but motion in the image was accompanied by a very evident and annoying "smear" of light, resulting from light carried over from one frame to the next. The project was abandoned in 1941 after a thorough investigation by the National Television System Committee.

Another revolutionary proposal made by DuMont was that favouring, not single values of number of lines and frames per second, but a flexible standard, the lower limit of frame rate at 15 per sec and the upper limit of number of scanning lines in the neighbourhood of 800. It was urged that such a flexible standard would permit the broadcasters and the public to adopt the best compromise at each stage of the art. This proposal was also investigated at length by the National Television System Committee, who concluded that provision for such flexibility would so increase the cost of receivers, and provide so little advantage in picture quality (which is more fundamentally limited by the band-width than by the scanning standards), that it was not justified.

The foregoing paragraphs may indicate that the adoption of standards in the United States was accompanied by not a little dissension in the ranks. This must be admitted. In fact, the five-year period from 1936, when official sanction for public programmes was given in the United Kingdom, to 1941, when similar sanction was granted in America, was characterized by a very vigorous debate on standards. Moreover, the debate was not confined to technical meetings and committees, where it rightly belonged. Rather, the issues were debated in the Press and at hearings before the Federal Communications Commission. During this period the Commissioners took the attitude that public service must wait until the engineers could agree.

The debate finally came to an end in the meetings of the nine panels of the National Television System Committee (N.T.S.C.) This group of 168 television specialists, in the period from August, 1940, to March, 1941, devoted 4 000 man-hours to meetings, witnessed 25 demonstrations of the comparative merits of different proposals, and finally left behind them a record of reports and minutes some 600 000 words in length. Out of this monumental effort came virtually complete agreement on a set of 22 standards which were presented to the Federal Communications Commission for approval and adoption. This approval was granted, and commercial operation of television broadcast stations was authorized, to be effective the 1st July, 1941. The stage was set for a rapid advance. On the 7th December, 1941, the United States entered the war and the state of the art was frozen by lack of man-power and materials. Since then, commercial broadcasting has continued, but at a "bare-subsistence" level. The F.C.C. required a minimum of 15 hours per week of public programmes from each station before the war; after Pearl Harbour this was reduced to four hours per week. No television receivers have been produced for public sale since 1941.

In recent months, a thorough review of the standards for post-war use has been conducted by the Television Panel of the Radio Technical Planning Board (R.T.P.B.). The findings of this group are described at the conclusion of this paper.

(1.2) The Evolution of Equipment

The evolution of standards is but a reflection of a much more fundamental achievement, the evolution of equipment. This latter evolution represents the substance of television development during the past 10 years.   Space is available to discuss only a few of the more basic types of equipment—camera tubes, picture tubes (including projection types), and relay equipment—all of which have received particular attention in American laboratories.

Camera tubes fall naturally into two categories: storage-type tubes, like the iconoscope, which employ the radiant energy of the scene continuously, and non-storage tubes, like the image dissector, which employ the radiant energy only during the instant the scanning agent passes over each picture element.

The non-storage image dissector was developed by Philo T. Farnsworth. In the image dissector the electron image, formed at the surface of a flat photo-cathode, is drawn bodily down the length of the tube by the action of a uniform attracting field. As it moves, the electron image is moved vertically and horizontally by magnetic forces imposed by scanning coils, into which are fed saw-tooth waves of current at the vertical and horizontal scanning rates. The electron image is thus caused to move laterally past an aperture in the tube, opposite the photo-cathode. The aperture acts as the scanning agent, stationary to be sure, but capable of exploring the image by virtue of the motion of the image past it. The aperture leads to a secondary-emission electron multiplier, which multiplies the electron current by a factor of some 10 000. At the last (collecting) anode of the multiplier structure the video signal appears, ready for application to conventional video line amplifiers.

The image dissector has one advantage over storage-type tubes in that it is a direct-coupled device which provides in the video signal a d.c. component representative of the average brightness of the scene, and a second advantage in that it is free from the shading difficulties caused by redistribution of secondary electrons in the storage-type tubes. But the fact that the dissector makes such inefficient use of the available light has put it at a disadvantage for all direct pick-up work and has relegated the tube principally to motion-picture transmission, where plenty of light can be made available.

The storage-type camera tubes, which enjoy almost universal use in studio and outdoor pick-ups, were evolved from the basic iconoscope, the development of which, as previously mentioned, actually pre-dates the pioneer work of Baird and Jenkins.

The principle of the iconoscope is now so well known that little description is required. A mica plate covered with millions of separate, sensitized silver particles is exposed to the optical image, which causes the photo-electric emission of electrons proportionate to the lights and shadows of the picture. The charge configuration thereby produced is scanned by a narrow beam of high-velocity (1 000-volt) electrons, which excites secondary emission from each silver particle, the amount of which is governed by the previous emission of the photo-electrons. A collection electrode collects this secondary emission, the variations of which constitute the video signal.

The insulation of the silver particles from each other permits the charge configuration to maintain its form for appreciable lengths of time, so the charge is effectively stored from one scansion to the next. This accounts for the great sensitivity of the device. The insulation, however, also prohibits any charge leaving the mosaic except as it is replaced. The average potential of the mosaic must therefore remain unchanged, and it is thus impossible for the iconoscope to evaluate the average brightness of the scene, which must be transmitted by some separate means. Moreover, the collection of the secondary electrons is incomplete. Those electrons not collected fall back on the mosaic, causing a spurious signal which introduces unwanted variations in background shading. These variations must be compensated by circuits which introduce a waveform of inverse shape to that produced by the spurious signal.

A considerable amount of effort has been expended on the iconoscope on both sides of the Atlantic. The technique of "silver-sensitization" of the mosaic surface (depositing silver over the caesiated mosaic) has increased its sensitivity, while at the same time matching the spectral response more closely to that of the human eye. The employment of bias lighting (illumination of the tube behind the mosaic) has made still further sensitivity practically possible and reduced troubles from electrons trapped on the walls of the tube.

By far the most impressive improvement from a practical standpoint is the development of the orthicon (short for "orthi-conoscope") principle. The orthicon tube, first publicly described by Iams and Rose in 1939, has a sensitivity five to ten times that of the iconoscope, has no shading troubles due to random redistribution of secondary electrons, and has a linear relationship between light input and voltage output.

The secret of this remarkable device, which has been consistently used in broadcasting since 1940, lies in the fact that the scanning electrons themselves are collected as the video signal, rather than secondary electrons as in the iconoscope. This is made possible by maintaining the mosaic and its stored electron image at an average potential equal to that of the cathode, so that the electrons are turned back toward the cathode upon reaching the mosaic. The variations which constitute the video signal are caused in the returning current by variations in the potential of successively scanned picture elements. The sensitivity of the orthicon makes it a serious competitor of the fastest motion-picture film. In fact, orthicon cameras have continued to televise football games in the late autumn afternoons (with recognizable, if imperfect, images) after the newsreel cameramen (with admittedly higher professional standards) have given it up as a bad job.

Another method of improving the sensitivity of the iconoscope, which has seen wider use in England, is the secondary-emission multiplier mosaic ("image-iconoscope" in America, "super-emitron" in Britain). In this tube an electron image is drawn from a transparent photo-cathode and moved bodily, as in the image dissector, down the length of the tube to a "mosaic" surface opposite, which is a uniform insulating surface capable of high secondary-emission ratio. The electron image there finds itself multiplied, by the excess of lost electrons, and ready to be scanned by a high-velocity scanning beam in the conventional iconoscopic manner.

It must be admitted that the limit of sensitivity of storage-type television cameras is not yet in sight. In fact Iams and Rose have published a theoretical analysis which indicates that the storage television camera may some day be the most sensitive continuously registering optical device in existence, save only the eye of a cat.

The opposite end of the television system, the cathode-ray picture tube, has been improved in many small ways which in the aggregate have greatly advanced image quality. "Settled" phosphor screens, with high contrast range, were on the point of wider availability at the outbreak of the war. The previously available sprayed screens were, in general, inferior in contrast to the tubes available in England and on the Continent. Tube sizes have increased to 20-in diameter, although the 12-in tube seems the practical upper limit for home use. Ion traps have been developed to remove the source of the ion spot so noticeable in magnetically-deflected tubes.

The most significant recent advance in picture tubes is the projection-type tube. Tubes of small diameter, with high accelerating voltage and correspondingly small-diameter beam, and with high-efficiency phosphors have been developed. These tubes are capable of producing images of such intense brilliance as to be dangerous to the eye when viewed directly. An image of 4 by 3 in, formed on such a tube with 60 000 volts acceleration has been magnified to 15 by 20 ft through the medium of a Schmitt optical system, with sufficient brilliance to hold the attention of a critical audience (about 20% of the normal brightness of motion-picture projection). Such a system (see Fig. 4, Plate 2) is considered to have great promise for theatre television.

More recently a smaller version of this tube, intended for use in domestic receivers, has been demonstrated using a Schmitt system with a cast plastic correction plate. Pictures of 24 by 18 in, bright enough to be viewed satisfactorily against normal incandescant room illumination, and possessing all the detail inherent in the 525-line picture, were demonstrated with this equipment in 1943. It is expected that this type of receiver (see Fig. 5, Plate 2) will be offered to the public, in the upper price range, following the war.

Television relaying, a matter of paramount importance in a country with widely separated centres of population, has received intensive study and development in America. Two competitive systems have appeared: the coaxial line and the radio repeater station. Following the epic relay at Arney's Mount in 1934, the radio repeater has sought higher frequencies, reaching 500 Mc/s just before the war. Highly directive (parabolic) antenna structures have been used to obtain a high signal/noise ratio over the optical relay paths, and to avoid image reflections from objects off the direct path. The last fact has permitted the successful use of frequency modulation in relaying the picture signal (wave-interference effects are serious when frequency modulation is employed, since they may result in cancellation or phase reversal of the synchronizing pulses and in shifting beat-note patterns over the image which have no evident relation to the basic image).

The coaxial cable has been the special province of the Telephone Group (Bell Laboratories and American Telephone and Telegraph Co.) who have developed it to the point of readiness for nation-wide use, subject only to economic justification. The first link in the nation-wide chain was laid down in 1937 between New York and Philadelphia. At first limited to a band-width of 1 Mc/s, it was later revised for 2-Mc/s operation. This permits transmission of half the pictorial detail possible in the 525-line picture, but is adequate for many types of programmes. The band-width can be expanded, by decreasing the spacing between repeater stations, to cover the full requirements of the present standards. The repeater stations are unattended and derive their mains power from alternating current sent along the cable itself. To avoid crosstalk at the low-frequency end of the video spectrum, the wide-band signal is modulated upward several hundred kc/s.

An extension of the New York–Philadelphia coaxial link is already in place as far as Washington, D.C., but construction of the repeater facilities is not yet completed. As it has been found that the coaxial cable is an economical carrier of telephone traffic, compared with multi-conductor cable and open wire, coaxial links will in all likelihood be installed to meet normal increases in telephone long-distance service. Plans for its use in a post-war television network are described later in this paper.

(1.3) Evolution of Programme Service

Prior to 1936, the American public was an incidental partner in the television enterprise. But in that year, broadcasting of programmes especially designed for public consumption began, although no regular source of receivers was available and no official sanction had been given for other than experimental transmissions. The occasion was one of considerable international competition between England and the United States. The first move came from the British Broadcasting Corporation. Plans for the opening of the London station in Alexandra Palace were announced early in 1936. This was the cue for the R.C.A.-N.B.C. transmitter to "get busy." On the 29th June, 1936, R.C.A. started a field test of its television system with 100 of its engineers in and around New York as the observers. The images were transmitted by double sideband, at 343 lines, 30 frames per sec. Film was used copiously in the early stages. But by the 6th November of that year, the New York *Times* announced "the first complete programme of entertainment over the N.B.C. system," viewed enthusiastically by the Press. Four days before, on the 2nd November, the Alexandra Palace station had begun a regular public service. The American service was not public in the same sense, and did not become so for five years. But programmes designed to entertain R.C.A. engineers were transmitted regularly from this point, and more or less regularly since.

By 1938 the R.C.A. field test had progressed to the point where its directors were ready to take a step in the direction of inviting the public to participate actively. It was announced in October, 1938, that, coincidentally with the opening of the New York World's Fair in May, 1939, regular public service would be offered.

In the meantime, vestigial-sideband transmission had been adopted as standard by the Television Committee of the Radio Manufacturers' Association and had been incorporated in the N.B.C. transmitter. The scanning pattern had been increased to 441 lines, and the effective video band to 4 to 4·5 Mc/s. In February, 1940, the Federal Communications Commission adopted rules permitting "limited" commercial operation of stations.

The stage seemed set. The N.B.C. transmitter was offering 10 to 15 hours of programme a week, including elaborate dramatic presentations from the studios, regular outside rebroadcasts of sporting events of every description, educational programmes and films. But in April, 1940, the F.C.C. retracted the offer of limited commercialization, following an announcement by R.C.A. that receivers would be offered to the public in greater volume and at reduced prices. The F.C.C. stated that this action by R.C.A. tended to freeze the then accepted standards, those formulated by the Radio Manufacturers' Association, without official sanction from the Government. This action effectively held up further progress until the standards had been studied and essentially reaffirmed by the National Television System Committee. In July, 1941, the impasse was cleared and television broadcasting for the public officially began.

From 1936 to 1941 the programme departments of the N.B.C. and C.B.S. were busy televising practically everything which reflected the necessary amount of radiant energy. The programme listings show that the following sports were offered to the viewing audience: baseball, basketball, football, hard-court and grass-court tennis, boxing and prize-fighting, horse racing, track and field events, wrestling (of the highly entertaining American "professional" type), ice hockey, swimming, figure skating, bicycle racing, aeroplane racing, and rodeo contests.

In the political arena, the two major political conventions were televised in 1940, the Republican directly from Philadelphia (Fig. 19, Plate 5) via the coaxial cable, and the Democratic indirectly by films flown from Chicago and shown the following day. The 1944 political conventions will be televised by the second method.

From 1939 to 1942, the N.B.C. consistently polled members of its New York audience to determine the most popular type of programme material. Dramatic productions, produced in the studio, were most popular. Sports came next, followed by films. The variety show was variously received, depending on the quality of the talent.

During 1942 and 1943, consistent network linking of three stations has occurred without benefit of coaxial cable or radio repeaters. The stations involved are WNBT in New York, where the programmes originated, and WPTZ in Philadelphia and WRGB near Albany, New York, which rebroadcast them to local audiences. The air-line separations between these stations are roughly 80 and 130 miles, respectively. Despite the fact that the horizon intercepts both transmissions, regular service has been possible through the use of directive receiving arrays, mounted on the highest eminence available.

Programmes of particular interest to British viewers were broadcast in April and May, 1941, by the N.B.C. transmitter. On these occasions the programme was a transatlantic telephone conversation between British evacuee children in America, and their parents in England. The children were televised in the studio, and the varied expressions of delight, apprehension and nonchalance were shown to the New York audience. Fig. 6 (Plate 3) is an image of one of these children, photographed from the screen of a domestic receiver in New Jersey. The happy concentration on a familiar, far-away voice is clearly evident.

More recently, in 1942 and 1943, a considerable portion of the programme activity has consisted of instruction courses for air-raid wardens and similar A.R.P. personnel (Fig. 7, Plate 3). Television receivers have been distributed to precinct police stations, where they are viewed regularly by the A.R.P. workers of the local district. It has been stated that the efficiency of this method of instruction is high and the interest in it great, although it has less appeal to the general audience.

(2) PRESENT PRACTICE

Before examining the American television standards in detail, it is necessary to point out a difference between the American and British terminologies. The British terms given in B.S. 204—1943 (Section 5) agree closely with the American definitions. The difference, which may lead to confusion, is in the use of the term "frame frequency." The American "frame frequency" coincides with the British "picture frequency," that is, the number of complete images transmitted per second. The British "frame frequency," is equivalent to the American "field frequency," that is, the number of partial scannings per second in interlaced scanning. The term "field" relates to the alternate lines in scanning, and two fields comprise a complete picture or frame. These and other American definitions are listed on page 145.

(2.1) The Standard Television Channel

Fig. 8 shows the configuration of the standard 6-Mc/s channel assigned to television broadcast stations in the United States. The channel is intended for vestigial-sideband transmission. The frequency scale (abscissae) starts, at the left, at the lower frequency limit of the channel. At this point the emission is required to be substantially zero, actually no more than 0·05% of the picture-carrier amplitude. At a point 0·5 Mc/s higher in

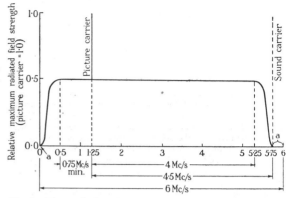

Fig. 8.—The standard vestigial-sideband channel for television broadcasting in the U.S.A. (from the F.C.C. "Standards of Good Engineering Practice").

a—Relative field strength of picture sideband not to exceed 0·0005. Drawing not to scale.

frequency, the sideband emission reaches its maximum (equal to one-half the carrier amplitude). The picture carrier itself is located 1·25 Mc/s above the lower channel edge, and occupies an asymmetrical position with respect to the channel limits. Thus only a portion of the lower sideband is transmitted, hence the term "vestigial-sideband transmission." The upper sideband is transmitted fully over a region 4 Mc/s wide, i.e. it maintains maximum amplitude to a point 5·25 Mc/s above the lower channel edge. At this point the sideband energy is attenuated with increasing frequency until, at a point 5·75 Mc/s above the lower channel edge, it is attenuated to 0·05% of the carrier amplitude. The carrier of the associated sound transmission is placed at this point. The remaining 0·25 Mc/s of the channel is reserved as a guard band.

This arrangement of carriers and sidebands was originally devised in 1938 and has persisted substantially without change through the deliberations of the National Television System Committee and those of the more recent Radio Technical Planning Board. The vestigial-sideband principle permits a maximum unattenuated vision frequency of 4 Mc/s, and permits attenuated transmission of vision signals up to a maximum of 4·5 Mc/s. In comparison, the double-sideband system permits a maximum vision frequency of 2·5 Mc/s. The vestigial-sideband transmission thus offers an increase in pictorial detail of 80%, with substantial improvement·in picture quality.

The wide spacing between carriers (4·5 Mc/s) was chosen in preference to the alternative narrow spacing (1·25 Mc/s) which would be possible if the sound carrier were transferred from the right-hand edge of the diagram to the left-hand edge. In the presence of cross-modulation the wide spacing produces a high-frequency beat note which is outside the limits of the vision frequency band and hence has no visible effect. The narrower spacing would produce a beat note within the vision band resulting in a visible pattern on the picture.

The vestigial-sideband system is intended to operate with a receiver whose response characteristic is that shown in Fig. 9. It will be noted that from 0·5 Mc/s above the lower edge of the channel to a point approximately 1·5 Mc/s higher, the response increases linearly with frequency. This "slope" region corresponds to the portion of the transmitter spectrum where both sidebands are transmitted. By virtue of the sloping receiver response the picture carrier is attenuated to one-half, and the sum of the two sideband voltages is constant throughout the

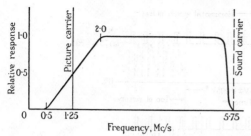

Fig. 9.—Receiver response characteristic for use with the emission characteristic shown in Fig. 8.

region, and equal to the value of sideband voltage at higher frequencies outside the "slope" region. Thus the vision-frequency voltage developed at the output of the demodulator is the same for all sideband frequencies. One theoretical disadvantage of this system is the greater amount of phase distortion arising from mistuning, relative to that introduced in conventional double-sideband transmission. But in practice the effects of such additional distortion have been found to be negligibly small.

The ratio of sound-carrier power (radiated by the sound radiator) to picture-carrier power (radiated by the picture radiator) has been set within the limits 0·5 and 1·0. The tolerance has been set up to allow for inevitable variations in radiated power. The general range has been chosen to provide approximately equal areas of coverage of the sound and picture signals. The sound power is lower than the picture power because the sound transmission employs frequency modulation, with an inherent signal/noise ratio superior to that of amplitude modulation. Account has been taken, in setting up this ratio, of the fact that interference (particularly impulsive noise) in the sound channel is usually more objectionable than interference arising from the same noise voltage in the picture channel.

The F.C.C. standards also allow the use of frequency modulation for the picture carrier, although use is not made of this type of modulation in stations currently operating. At the time the standard was drawn up, it was not realized that frequency modulation produces very annoying image patterns in the presence of multi-path transmission (as outlined earlier), and is thus unsuitable for broadcast emissions. More recently the Television Panel of the R.T.P.B. has recommended that the standards be changed to permit only the use of amplitude modulation for picture transmissions.

(2.2) The Scanning Specifications

The standard American television picture is scanned in 525 lines from the beginning of one frame to the beginning of the next. Each frame is broken up into two fields of 262·5 lines each. The half-line portion at the end of a field causes the lines of one field to fall between the lines of the previous field (the usual odd-line interlacing sequence). Hence an odd number of lines was chosen to give this half-line relationship between fields. The value 525 consists of odd integral factors (as shown in Table 1). This permits multivibrators or counting circuits in the synchronization generator (used to divide from the line frequency of 15 750 c/s to the frame frequency of 30 c/s) to operate in the most stable condition.

The frame frequency (British "picture frequency") is 30 per sec interlaced two-to-one for reasons given earlier in this paper. The aspect ratio of the scanning pattern (ratio of width to height) is 4/3, to agree with the ratio previously adopted as standard for motion-picture projection. The F.C.C. standards specify that the active scanning of the picture shall occur at uniform velocity from left to right horizontally and top to bottom vertically.

The choice of 525 lines was made from among several proposed values, including 441, 495 and 507 lines. Assuming 4·25 Mc/s as the maximum usable video frequency, equal vertical and horizontal resolution at 30 frames per sec is obtained with a scanning pattern of about 500 lines, which would indicate that all the proposed values are equally suitable. The number 507 has the disadvantage of two large integral factors which require two of the frequency-dividing circuits in the synchronizing signal generator to count by a factor of 13. The choice between 495 and 525 was finally made on the basis of fineness of line structure, which indicates a slight preference for the higher number of lines. The value of 441 lines was discarded as it did not make full use of the maximum available video frequency.

(2.3) The Video (Vision-Frequency) Signal

The basic form of the video signal is shown in Fig. 10, with a permitted alternative form of the synchronization pulses in Fig. 11. Both diagrams represent the carrier envelope of the amplitude-modulated picture signal. Referring to Fig. 10, it will be seen that the basic level of the video signal is the so-called black level, at 75% (plus or minus 2·5%) of the maximum carrier-voltage amplitude. Negative polarity of transmission is specified, so that carrier-voltage amplitudes higher than the black level drive the signal into the intra-black region. This region is occupied by the synchronizing signals. Voltage levels lower than the black level are occupied by the picture information. The maximum brightness capable of being depicted by the transmission corresponds to zero carrier, but since few transmitters can modulate to this low level it is specified that the maximum white level shall occur at a level not greater than 15% of the maximum carrier-voltage level. This assures that at least 60% of the carrier-voltage range is reserved for the contrast range of the picture.

The black level is a fixed reference level in the carrier envelope, i.e. a level which does not change with picture content. This level acts as a d.c. reference level against which the background brightness of the picture is set. This technique of d.c. transmission is now unanimously approved as providing the maximum possible carrier-amplitude range for variations in picture brightness, while at the same time offering a fixed and simply ascertained level for operating the automatic sensitivity control of the receiver.

The synchronizing signals, above the black level, are of two types, the horizontal (line) pulses and the vertical (field) pulses. The line pulses are simple closely-rectangular 5-μs pulses occurring 15 750 times per sec, and occupying 25% of the maximum carrier amplitude. At the receiver these pulses may be differentiated after demodulation to provide the driving signal for the line-scanning generator.

The vertical pulses are of more complex shape. In Fig. 10 the so-called "serrated" vertical pulse is shown. It consists of six 27-μs pulses spaced 5 μs apart. This basic group is integrated in the receiver to produce the driving signal for the vertical scanning generator. The serrations between pulses are introduced to preserve the continuity of the horizontal synchronizing pulses. Immediately preceding and following the serrated vertical pulse are groups of six "equalizing pulses" which are half the width of the horizontal pulses and occur at intervals one-half as great. These equalizing pulses are such that, when they are added to the basic vertical serrated pulse, the integrated effect is the same on each successive field. If the equalizing pulses were not present, intervals of unequal length would obtain on successive fields between the last horizontal pulse and the serrated vertical pulse. Long experience with this type of synchronizing signal has shown that it will provide an adequate lock of the reproduced picture if the signal strength is sufficient to provide

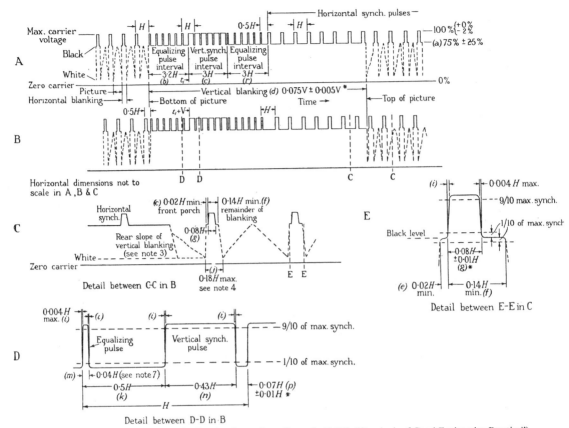

Fig. 10.—The standard synchronization signal waveform (from the F.C.C. "Standards of Good Engineering Practice").

1. H is time from start of one line to start of next line.
2. V is time from start of one field to start of next field.
3. Leading and trailing edges of vertical blanking should be complete in less than $0·1H$.
4. Leading and trailing slopes of horizontal blanking must be steep enough to preserve min. and max. values $(e + f)$ and (j) under all conditions of picture content.
5. Dimensions marked with an asterisk indicate that tolerances given are permitted only for long-time variations, and not for successive cycles.
6. For receiver design, vertical retrace should be complete in $0·07V$.
7. Equalizing pulse area shall be between $0·45$ and $0·5$ of the area of a horizontal synchronizing pulse.

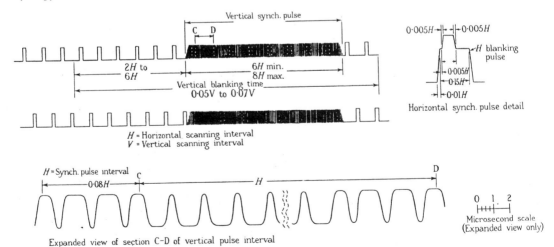

Fig. 11.—Alternative "500-kc/s" form of vertical synchronization signal (from the F.C.C. "Standards of Good Engineering Practice").

Fig. 2.—Early low-definition installation of the National Broadcasting Company. Notice the figure of "Felix the Cat" suspended before the flying-spot scanner and the photocells.

A   B   C

Fig. 3.—The progress of definition, 1929 to the present: A, 60-line image; B, 120-line image; C, 525-line image.

(*Facing page* 152).

Plate 2

Fig. 4.—Theatre television: An image of Miss Lucy Monroe singing in the N.B.C. television studios and reproduced on a 15 ft by 20 ft screen in a nearby theatre. The projector employs a Schmitt optical system, 60 ft from the screen.

Fig. 5.—A 1941 model of the R.C.A. projection receiver for home use. The image is $13\frac{1}{2}$ in by 18 in, projected on a translucent, retractable screen.

Fig. 14.—Typical scene in the N.B.C. studios.   Note three iconoscope
cameras in use simultaneously.

ig. 6.—A British evacuee child in America, listening to his
parents in England via transatlantic telephone while
being televised in the N.B.C. studios. Photographed
direct from the screen of a domestic receiver in New
Jersey.

ig. 7.—Typical war-time programme material: air-raid instruction for the
benefit of A.R.P. workers viewing the images at precinct police stations.
The camera is of the orthicon variety.

Fig. 23.—The transmitting radiators of the N.B.C. transmitter
in New York, on top of the Empire State Building (elevation 1 250 ft).

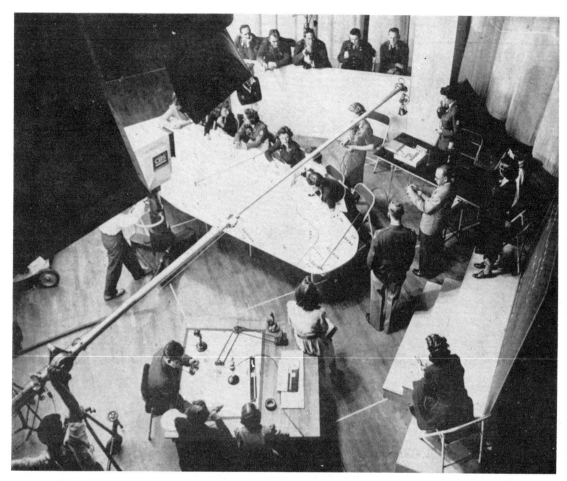

Fig. 15.—Typical scene in the Columbia Broadcasting System studios. The use of the plotting board to follow aircraft is televised for the instruction of A.R.P. workers.

Fig. 16.—Typical high-fidelity image photographed from the monitor screen of the N.B.C. transmitter.

Fig. 18.—The N.B.C. mobile television vans. The microphone is mounted in parabolic reflector to pick up the sound at a distance, while the camera fitted with a telescopic lens. The transmitter van is at the rear.

19.—The orthicon camera in use televising the Republican National Political Convention in 1940. Note the large lenses used to secure sufficient light on the long-distance shots.

Fig. 21.—An elaborate television repeater station, constructed by R.C.A. Communications to relay programmes from Long Island to New York. The picture signal is sent by frequency modulation in this equipment.

20.—A television relay station in a tent, used by the Don Lee station in California to carry programmes back to the transmitter for rebroadcast to the public.

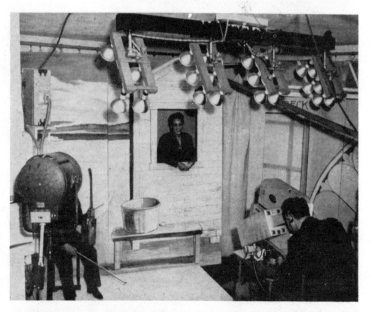

Fig. 17.—A scene in the DuMont studio, illustrating the use of the "electronic view finder." The camera operator views a true television image in operating the camera, rather than an optical image formed by an auxiliary lens system.

Plate 6

Fig. 22.—The transmitting radiators of the C.B.S. transmitter in New York, on top of the Chrysler Building (elevation 1 046 ft).

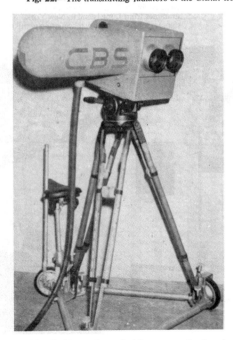

Fig. 24.—Orthicon colour-television camera developed by P. C. Goldmark, of C.B.S. The housing contains a rotating optical filter system to produce successive images in the three primary colours.

Fig. 25.—Internal view of a typical colour-television receiver used in the C.B.S. tests.

adequate picture content. With recently-developed "flywheel" scanning circuits this type of signal has been shown to be capable of holding synchronization even when the signal level is so low that the picture is beyond recognition.

The alternative type of vertical pulse, permitted by the F.C.C. standards, is shown in Fig. 11. This pulse has been used by several transmitters, but the recently expressed opinion of the R.T.P.B. panel is that this type of pulse is not generally so satisfactory as that shown in Fig. 10, and their recommendation is that only the latter pulse be permitted in the future. Briefly, the waveform in Fig. 11 consists of a "burst" of 500-kc/s sine-wave signal, lasting from six to eight lines (380 to 510 μs) and superimposed on the horizontal pulses. Limiter circuits cut off the top half sine-waves during the horizontal pulses, and the bottom half sine-waves during the intervals between horizontal pulses. The primary advantage of such a pulse is that it permits simple 500-kc/s resonant circuits to be used to distinguish between vertical and horizontal synchronizing information.

Two F.C.C. standards set up tolerances on the intervals between horizontal pulses (not more than $0 \cdot 5\%$ variation from the average) and on the rate of change of the horizontal pulse frequency (not greater than $0 \cdot 15\%$ per sec). These tolerances were intended to tighten the performance of the synchronizing generators so that they would match the inertia characteristics of such mechanical methods of scanning as might be employed at the receiver.

(2.4) Sound-Signal Standards

A major difference between British and American television practice lies in the method of modulation employed for the sound transmission. The American standard specifies frequency modulation with a maximum frequency deviation, corresponding to the maximum audio level, of 75 kc/s either side of the unmodulated carrier frequency. Frequency modulation, employing a spectrum considerably wider than that required for the corresponding amplitude-modulated signal, has been shown to offer a substantially higher signal/noise ratio than that offered by amplitude modulation. This advantage obtains over all types of noise, provided only that the peak signal voltage is at least twice the peak noise voltage. Since natural atmospherics are rarely present in the v.h.f. spectrum, the principal advantage of frequency-modulated transmission is the mitigation of impulse noise such as is generated by automobile ignition systems, and noise generated in valves and circuit elements of the receiver. Tube and circuit "hiss" is important when reproduction of the full audible spectrum is contemplated, but ignition noise is important under any condition, since it usually arises from cars on a nearby highway and hence has every opportunity to maintain a r.f. voltage level comparable with that of the desired signal. The reduction of ignition noise is the principal justification for frequency-modulated transmission of the sound signal. The disadvantages are a somewhat more complex receiver, and the necessity of considerably more precise tuning to obtain maximum benefit, both relative to the corresponding amplitude-modulation system.

The maximum deviation of $\pm$ 75 kc/s was chosen to conform with the previously established standard for frequency-modulated sound broadcasting in the 44–50 Mc/s band. This deviation, taken in conjunction with a maximum audio frequency of 15 000 c/s, provides a signal/noise ratio 27 db greater than that offered by an amplitude-modulated transmitter of the same peak power. This substantial improvement allows the use of less power in the frequency-modulated sound transmitter than is required in the picture transmitter.

To assist in the reproduction of the upper register of the audible spectrum, it has been customary in frequency-modulated sound transmissions to introduce audio pre-emphasis at the transmitter. The standard pre-emphasis characteristic is that of a series resistive-inductive impedance whose time-constant (resistance times inductance) is 100 μs. The converse de-emphasis is inserted in the receiver (usually by a resistive-capacitive impedance of the same time-constant). The advantage of such pre-emphasis is well understood: the sound power associated with the higher register is generally lower than that of the lower register and hence is less inefficiently transmitted with respect to the noise level. Artificial emphasis and de-emphasis thus add to the overall signal/noise ratio, by reducing the noise level in the upper register.

The remaining standard is the direction of polarization of the electric vector of the radiated wave. This was chosen as horizontal as early as 1938, and although polarization has been the subject of intensive investigation by the N.T.S.C. and the R.T.P.B. the advantage of the horizontal direction has been consistently upheld. Here, again, we find a departure from British practice, which favours vertical polarization. The N.T.S.C. Panel charged with investigating this matter found that horizontal polarization was superior in respect of the effects of multipath transmission, and slightly superior with respect to inherent noise, whereas vertical polarization was preferable in respect of freedom from fading and miscellaneous signal variations, and of the directive properties of receiving antennas. No difference was found in the propagation of signal intensity with transmitting and receiving antennas at the heights normally employed. Faced with this balance of small differences, the Panel found a slight preponderance of evidence in favour of horizontal polarization, and so the standard was set.

(2.5) Rules governing Allocation of Television Broadcasting Facilities

As the demand for broadcasting facilities has consistently exceeded the supply, it has been necessary for the F.C.C. to set up equitable rules whereby the available portions of the spectrum may be allocated to serve the public interest.

The basic allocation, shown in Table 2, consists of 18 channels,

Table 2

AMERICAN TELEVISION CHANNELS

Channel No.	Frequency	Channel No.	Frequency
	Mc/s		Mc/s
1	50–56	10	186–192
2	60–66	11	204–210
3	66–72	12	210–216
4	78–84	13	230–236
5	84–90	14	236–242
6	96–102	15	258–264
7	102–108	16	264–270
8	162–168	17	282–288
9	180–186	18	288–294

each 6 Mc/s wide, from 50–56 Mc to 288–294 Mc/s. Channels as high as 78–84 Mc/s have been assigned thus far to commercial stations, but little experience has been gained on higher frequencies, except in relay service.

A conflict in allocation arises, by definition, when interference occurs among stations. The interference is defined in terms of (1) the signal level required to give satisfactory service in an area from the station serving that area, and (2) the level of signal which creates interference in that area, arising from another station on the same channel in an adjacent area. The problem of interference between stations in the same area but assigned to adjacent channels must also be considered.

The basic level of service is defined as a field strength which must be equalled or exceeded over 50% of the distance along a radial line from the transmitter. The field strength thus specified for built-up city areas and business districts is 5 mV/m. For residential and rural areas the specified field strength is one-tenth as great, or 500 μV/m. These figures properly surpass the figure of 100 μV/m commonly regarded by engineers as the lower limit for "marginal service," in the absence of man-made sources of noise.

The applicant for a television construction permit or licence must show that his proposed transmitter will offer service in accordance with the above rules, and must estimate the population lying within the 5-mV/m and 0·5-mV/m contours.

The coverage of the transmitter is specified in terms of a field-strength contour map. It is desirable to have, in addition, a simple figure representative of transmitter performance for purposes of comparison. The power of the output stage of the transmitter, commonly used in medium-wave broadcasting, is an inadequate index in the v.h.f. region. In its place a quantity known as "effective signal radiated" (e.s.r.) has been set up by the F.C.C. to take account of the antenna gain and antenna height as well as the output power of the transmitter. It is defined as the product of the square root of the peak power input (kW) to the antenna, the antenna field gain, and the antenna height in feet above ground level. Thus a transmitter delivering 5 kW peak power to an antenna of field gain 2·0 at a height of 500 ft has an e.s.r. of $\sqrt{5} \times 2·0 \times 500 = 2\,230$. The e.s.r. figure has admitted limitations in indicating the comparative performance of transmitters, but it is superior to a simple statement of output power.

The interference ratio (ratio of desired signal to undesired signal) which must be equalled or exceeded within the service area has been set at 100 : 1 for stations on the same channel and 2 : 1 for stations on adjacent channels. This is in keeping with the commonly-held engineering opinion that an interfering signal must be at least 40 db below the desired signal if it is to have negligible effect. For adjacent-channel interference, the selectivity

of the receiver circuits will introduce sufficient additional rejection if the interfering signal voltage does not exceed one-half the desired signal voltage. It is the practice of the F.C.C. to avoid assigning adjacent channels in the same metropolitan area; otherwise the 2 : 1 ratio could be met in but a small portion of the normal service area.

The problem of finding sufficient facilities for television without interference is most critical in the highly populated areas along the eastern seaboard. In general an allocation plan to provide sufficient service to the cities of Boston, Providence, Hartford, New York, Philadelphia, Baltimore and Washington (combined metropolitan population in 1940 about 20 150 000), will take care of all other population centres in North America. A suggested allocation proposed by the R.T.P.B. Television Panel is described at the conclusion of this paper and illustrated in Fig. 28.

Other F.C.C. regulations which govern the operation of commercial television stations have to do with the minimum hours of programme service and the question of multiple ownership of stations. In the immediate pre-war period, each commercial station was required to offer at least 15 hours of programme per week. Currently this requirement has been reduced to 4 hours per week, as a war measure, to take account of man-power and equipment, particularly transmitting valve, shortages. No single organization can own, operate or control more than three commercial television stations.

(2.6) Stations Currently Licensed

Tables 3, 4 and 5 list American television transmitters at present authorized. The commercial transmitters listed in Table 3 are required to offer regular programme service to the public. The geographical distribution of these stations is shown in Fig. 12. The experimental stations are listed as "relay stations" (used in conjunction with other broadcast transmitters) and "other than relay stations" (operated for research and the development of the art). The geographical distribution of the experimental stations is shown in Fig. 13.

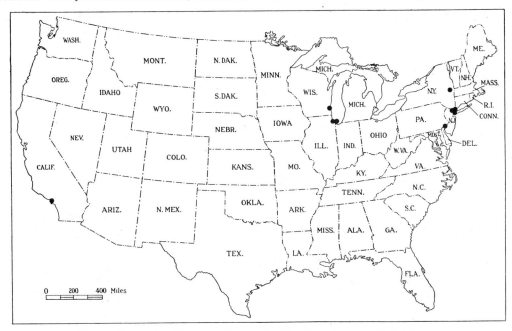

Fig. 12.—Geographical distribution of commercial television stations currently in operation.

Table 3

COMMERCIAL TELEVISION BROADCAST STATIONS IN THE U.S.A.

(As of the 5th May, 1944)

Location	Call letters	Licensee	Frequency	Effective signal radiated
			Mc/s	
New York, N.Y., Empire State Building	WNBT	National Broadcasting Company	50–56	1 800
New York, N.Y., Chrysler Building ..	WCBW	Columbia Broadcasting System	60–66	1 000
New York, N.Y...	WABD	Allen B. DuMont Laboratories	78–84	—
Schenectady, N.Y., New Scotland	WRGB	General Electric Company	66–72	3 100
Philadelphia, Pa., Springfield Township	WPTZ	Philco Radio and Television Corp.	66–72	500
Chicago, Illinois	WTZR	Zenith Radio Corporation	50–56	1 270*
Chicago, Illinois	WBKB	Balaban and Katz Corporation	60–66	550
Milwaukee, Wisconsin	WMJT	*The Milwaukee Journal*	66–72	1 200*
Hollywood, Calif., Mount Lee	KTSL	Don Lee Broadcasting System	50–56	5 600*

* Construction permit only. Licence to cover permit not yet issued.

Table 4

EXPERIMENTAL TELEVISION RELAY BROADCAST STATIONS IN THE U.S.A.

(As of the 5th May, 1944)

Location	Call letters	Licensee	Frequency	Power Visual	Power Aural	Associated station
			Mc/s	W	W	
Area of New York ..	W2XBU	National Broadcasting Company	282–294	15	—	WNBT
Area of New York ..	W2XCB	Columbia Broadcasting System	346–358	25*	—	WCBW
Area of New York ..	W2XBT	National Broadcasting Company	162–168	400	—	WNBT
Area of New York ..	W10XKT	Allen B. DuMont Laboratories	258–270	50	—	WABD W2XVT
Schenectady, New York	W2XGE	General Electric Company	162–168	60	50	WRGB
New Scotland, New York	W2XI	General Electric Company	162–168	50	—	WRGB
Area of Philadelphia ..	W3XP	Philco Radio and Television Corp.	230–242	15	—	W3XE
Area of Philadelphia ..	W3XPA	Philco Radio and Television Corp.	230–242	15	—	WPTZ W3XE
Area of Philadelphia ..	W3XPC	Philco Radio and Television Corp.	230–242	15	—	WPTZ W3XE
Area of Philadelphia ..	W3XPR	Philco Radio and Television Corp.	230–242	60 (peak)	—	WPTZ
Area of Chicago	W9XBT	Balaban and Katz Corporation	204–216	40	—	W9XBK
Area of Chicago	W9XBB	Balaban and Katz Corporation	384–396	10	—	W9XBK
Area of Los Angeles ..	W6XDU	Don Lee Broadcasting System	318–330	6·5	50	W6XAO
Area of Los Angeles ..	W6XLA	Television Productions Inc.	204–216	100* (peak)	—	W6XYZ

* Construction permit only. Licence to cover permit not yet issued.

Table 5

EXPERIMENTAL TELEVISION BROADCAST STATIONS (OTHER THAN RELAY) IN THE U.S.A.

(As of the 5th May, 1944)

Location	Call letters	Licensee	Frequency	Power Visual	Power Aural
			Mc/s	W	W
New York, N.Y.	W2XMT	Metropolitan Television Inc.	162–168	50 (peak)	50
Passaic, N.J.	W2XVT	Allen B. DuMont Laboratories	78–84	50	50
Philadelphia	W3XE	Philco Radio and Television Corp.	66–72	10 000 (peak)	11 000
Washington	W3XWT	Allen B. DuMont Laboratories	50–56	1 000	1 000*
Cincinnati, Ohio	W8XCT	The Crosley Corporation	50–56	1 000	1 000*
West Lafayette, Indiana ..	W9XG	Purdue University	66–72	750	750*
Chicago, Illinois	W9XBK	Balaban and Katz Corporation	60–66	4 000	2 000
Chicago, Illinois	W9XPR	Balaban and Katz Corporation	384–396	10	—
Chicago, Illinois	W9XZV	Zenith Radio Corporation	50–56	1 000	1 000
Iowa City, Iowa	W9XUI	University of Iowa	50–56 210–216	100	—
Los Angeles, California ..	W6XAO	Don Lee Broadcasting System	50–56	1 000	150
Los Angeles, California ..	W6XYZ	Television Productions Inc.	78–84	4 000	1 000

* Construction permit only. Licence to cover permit not yet issued.

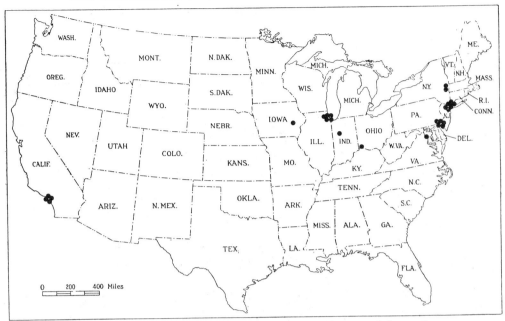

Fig. 13.—Geographical distribution of experimental television stations currently in operation.

(2.7) Studio Methods and Equipment

Typical scenes taken in the N.B.C. and C.B.S. studios are shown in Figs. 14 (Plate 3) and 15 (Plate 4) respectively. Fig. 14 shows three cameras in use on separate portions of the set: they operate simultaneously and independently. The output of the desired camera is switched to the line circuit by a control operator in an adjacent booth. Green lights on the front of the camera housing warn the performers that the camera is "on the air."

The cameras are predominantly of the iconoscope variety for studio work, and are provided with bias lighting. Several stages of the camera pre-amplifier, blanking and horizontal deflection circuits are included in the housing. The cable between camera and control booth contains 32 conductors, including the video coaxial line. The microphone is manipulated by an operator on a boom which was first developed for sound motion-pictures.

The lighting consists of groups of six 500-W "bird's-eye" reflector lamps. About 75 kW of lighting is available, the heat being conducted from the studio by an extensive, but practically noiseless, air-conditioning system. The direction of each group of lamps is controlled by a rope-and-pulley mechanism manipulated by a lighting engineer at the side of the studio. Large banks of fluorescent lamps are employed in the C.B.S. studios, one of which is visible at the upper left in Fig. 15. A typical high-fidelity image is shown in Fig. 16 (Plate 4).

Associated with the studio is the control element of the system, the synchronizing signal generator. The action of this device begins with a 31 500-c/s sine-wave oscillator, which is frequency-controlled by an automatic-frequency-control discriminator to maintain a rigid relationship with the 60-c/s mains frequency. The output of the oscillator is divided into four counter-circuit stages by factors of 7, 5, 5 and 3 (total division by a factor of 525). The output frequency of the last counter stage is accordingly $31\ 500/525 = 60$ c/s. This frequency is compared with the mains frequency in the discriminator previously mentioned. A further and separate division by two of the output of the oscillator provides pulses at 15 750 c/s. The two frequencies, 15 750 c/s and 60 c/s, are respectively the line scanning frequency and the field (British "frame") scanning frequencies which control the synchronizing pulse-forming circuits. These pulse-forming circuits produce groups of horizontal pulses and vertical pulses of the proper shape and frequency, which are intermixed in accordance with the standard shown in Fig. 10. The intermixing is accomplished by so-called keying pulses, which block the horizontal pulse during the vertical pulse but allow the equalizing and serrated pulses to pass in this interval. The generator is a highly complex electronic maze judged by pre-war standards, although quite simple when compared with more recent developments in the radar field.

(2.8) Mobile Pick-Up and Relay Equipment

The outstanding example of mobile pick-up equipment, and virtually the only one of its type in use in America, is the N.B.C. mobile unit shown in Fig. 18 (Plate 4). The unit consists of two 10-ton vans. One houses the video equipment, cameras, pre-amplifiers, synchronizing generator, mixers, etc. The other, housing a 400-W transmitter operating on the relay carrier channel of 162–168 Mc/s, is surmounted by a multi-element stacked antenna array. Power for the unit is obtained from mains supply, a circumstance which prohibits the use of the equipment while the vans are in motion. Two complete camera "chains" are available to permit quick switching from one scene to another. The cameras are equipped with 250 ft of multi-conductor cable to enable them to be set up at a distance from the vans. It is common practice to use the orthicon camera for maximum sensitivity on one chain and the iconoscope for maximum resolution on the other. A typical use of the orthicon camera, during the Republican National Convention in Philadelphia in June, 1940, is shown in Fig. 19 (Plate 5). The large long-focus lenses shown were necessary to obtain telephoto shots without resorting to special illumination. The vans were driven to Philadelphia on this occasion, and the video signal was relayed to New York by coaxial cable, as previously described. The N.B.C. mobile unit has originated some of the most important programmes thus far broadcast in America.

Another relay installation, of much simpler design, is shown in Fig. 20 (Plate 5). This is a transportable relay transmitter employed by the Don Lee System of Los Angeles. It is housed in a tent and is set up in advance at the point of origin of a programme.

A typical point-to-point relay receiver-transmitter, developed before the war and installed at Hauppage, Long Island, is shown in Fig. 21 (Plate 5). Parabolic receiving and transmitting reflectors are enclosed in the cylindrical wooden structure at the top of the tower. The relay equipment and monitoring facilities are housed at the base.

(2.9) Transmitting and Radiating Equipment

Three different approaches to the transmitter problem are represented in the N.B.C., C.B.S. and General Electric stations. The N.B.C. transmitter generates the carrier by crystal control at a sub-harmonic frequency, followed by doubler stages to produce the carrier frequency of 51·25 Mc/s. The final r.f. stage is grid-modulated by a high-level video amplifier stage, conductively coupled to the grids of the final r.f. power amplifier. A peak r.f. power of 7·5 kW is achieved.

The C.B.S. transmitter is somewhat similar, but employs a transmission-line-controlled oscillator in place of the crystal control, followed by a buffer stage, an intermediate stage and a final r.f. amplifier, grid modulated as in the N.B.C. transmitter. The peak picture-signal power is somewhat less than that of the N.B.C., owing to the higher carrier frequency (61·25 Mc/s). Coaxial filters are employed in these transmitters to suppress a portion of the lower sideband, in accordance with the standard channel shown in Fig. 8.

The modulation of high-level stages, illustrated in these examples, imposes a practical upper limit on the peak power output by virtue of the difficulty of obtaining high voltage levels at the output of the video modulating amplifier. High voltage output may be obtained over the band of vision frequencies only if the modulating valve can supply heavy peak currents to compensate for the necessarily low value of coupling impedance. But such valves are large and inherently possess large capacitance between elements and to earth, which requires still lower values of coupling impedance and still larger currents. A practical upper limit is quickly reached in this series of requirements. With pre-war valves, it was considered impossible to obtain much more than 10 W of r.f. peak power by high-level methods of modulation.

An alternative approach to the high-power case is the low-level modulation scheme adopted by the General Electric station, WRGB. In this transmitter the vestigial-sideband characteristic is introduced in the studio transmitter, which relays the programme to the transmitter proper on the 162–168 Mc/s channel. At the transmitter the relay carrier signal is heterodyned to the final carrier frequency of 67·25 Mc/s in a low-level (10-W) stage, and thereafter is amplified in nine successive stages of class-B amplification at carrier frequency. An extreme degree of linearity is required of these class-B stages to prevent re-insertion of the attenuated sideband, but this requirement has been satisfactorily met. The peak power of the final power amplifier is about 40 kW, or roughly five times that of comparable high-level transmitters.

The radiators used for picture and sound transmission take various forms, of which two examples are shown. The first (Fig. 22, Plate 6) is the crossed dipole arrangement, employed by the C.B.S. transmitter in the Chrysler Building, New York. The upper elements radiate the picture signal, the lower the sound

signal. An approximately circular radiation pattern is obtained by proper phasing of the four dipoles in each array. The second antenna (Fig. 23, Plate 3) is that of the N.B.C. transmitter at the top of the Empire State Building, New York. The upper structure consists of four folded dipoles bent into circular shape for radiating the sound signal. A circular radiation pattern is thus obtained and no mutual coupling is experienced with the picture radiator below. The picture radiator is essentially two dipole radiators, arranged at right angles in "turnstile" fashion. The dipoles are fed in quadrature to produce a circular pattern of radiation. The unusual bulbous shape is used to effect a wide-band radiation characteristic. The impedance of this radiator is essentially constant over the assigned range from 50 to 55·75 Mc/s, over which it must radiate without discrimination.

(2.10) Television Receivers

Present receiver practice is typified by pre-war equipment, no receivers having been manufactured since 1941. In the pre-war allocation, channels as high as 102–108 Mc/s were contemplated but no specific allocations were made above 78–84 Mc/s. None of the pre-war receivers provided channels higher than 84–90 Mc/s, and many of the inexpensive sets were limited to two or three channels. Few of the pre-war sets were built with frequency-modulation detectors, but most of them could be adjusted to receive frequency-modulated transmissions by tuning the sound carrier to one slope of the i.f. band-pass characteristic. Maximum sensitivity to the picture signal was seldom better than 100 μV, and no r.f. amplifier stage was employed prior to the frequency-convertor stage. The most commonly used intermediate frequencies were 12·75 Mc/s for the picture and 8·25 Mc/s for the associated sound carrier. In sets employing small picture tubes (5-in diameter) the i.f. picture channel was usually limited to a maximum band-width of about 2·5 Mc/s, since this figure corresponded to the spot size of the picture tube, relative to the picture size. Amplifications of 10 to 15 per stage were possible with this band-width, using high-slope valves ($G_m = 9\,000\ \mu$A/V). In the cheaper receivers three stages of i.f. amplification usually sufficed. The more expensive receivers, with 12-in cathode-ray tubes, used an i.f. band-width of about 3·5 to 4 Mc/s, with four or five stages at an amplification of about 8 per stage. Electric deflection was the rule for picture tubes of diameter less than 7 in, and magnetic deflection for larger tubes. Second-anode voltages varied from 2 000 for small electrically-deflected tubes to 7 000 for large magnetically-deflected tubes. No projection-tube receivers were offered to the public.

(2.11) Colour Television

Intensive development of colour television has been undertaken by the Columbia Broadcasting System under the direction of Dr. Peter Goldmark. Experimental broadcasts from W2XAB began in August, 1940, and were continued over WCBW throughout 1941. The N.T.S.C. and R.T.P.B. standardization committees, after considering colour television at length, decided that it was not possible to set standards so early in the development. Presumably, therefore, colour television will not be a factor in the immediate post-war activity, although its eventual importance cannot be doubted.

The C.B.S. colour television system is a synchronized colour-disc system. The initial scanning standards employed were 120 fields per sec, 343 lines. When the black-and-white standard was raised from 441 lines to 525 lines in 1941, the colour standard was raised a similar amount to 375 lines. The colour disc at the transmitter causes successive fields to be scanned in successive primary colours in the sequence red, blue, green. At the end of three fields (1/40 sec) all three colours have been presented to the eye. Consequently the basic colour frame is reproduced at a

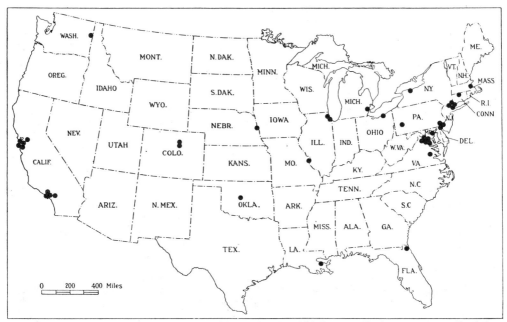

Fig. 26.—Geographical distribution of post-war television stations for which licence applications are now pending before the F.C.C. (as of the 5th May, 1944).

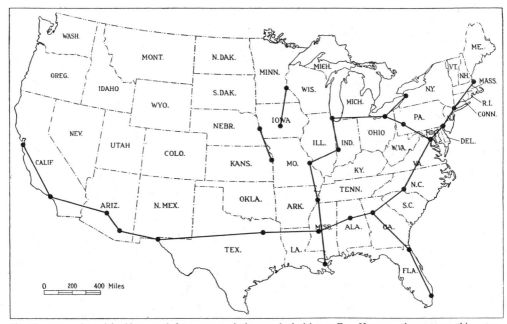

Fig. 27.—Proposed coaxial-cable network for post-war telephone and television traffic. If present plans mature, this system will be completed by the American Telephone and Telegraph Co. by 1950.

rate of 40 per sec. This rate is high enough to avoid visible flicker, provided the picture brightness is not too high. A given element of the picture is passed over by the three colours, each in 1/20 sec. To avoid pairing of the interlaced lines very careful shielding of the cathode-ray tube is required.

Linear camera tubes are used to simplify the problem of colour balance. An image dissector is used to televise colour slides and film. An orthicon of special design (employing thicker mica to reduce the mosaic capacitance and hence allow more complete discharge of the mosaic during a single scansion) is used for direct pick-up. The colour disc at the transmitter uses Wratten filters numbered 47, 58 and 25, whereas the disc in the receiver uses filters numbered 47, 58 and 26. Additional overall filtering is needed at the transmitter to remove excessive infra-red rays from the light source.

Several very impressive demonstrations of this system were viewed by technical people. The majority opinion was that the presence of colour did much to off-set the lack of detail in the images, but that the lack of detail was evident. It was also evident that the colour pictures offered satisfaction only when the colours were in proper balance. Separate gamma controls were provided for each colour and switched synchronously with the colour sequence. This "colour mixing" circuit proved essential to the system, but it was not always successful in coping with wide ranges of brightness. Simple disc-synchronizing mechanisms (a magnetic brake controlled by the phase difference betwen the horizontal scanning frequency and the disc rotation frequency) were developed, and the discs were made to run without appreciable noise. Figs. 24 and 25 (Plate 6) show an external view of the orthicon direct pick-up colour camera and an internal view of a typical colour-disc receiver.

Present technical opinion in America holds that a channel wider than 6 Mc/s is required for a colour television system if it is to be competitive with the present black-and-white images. Moreover, many problems remain to be solved in providing proper balance of colour under the wide variety of spectral distributions and brightness encountered in light sources. A non-mechanical method of introducing the colour sequence is also highly desirable. These problems will receive attention as soon as facilities for research are again available.

(2.12) Post-War Prospects

The long period of inactivity in television since America's entry into the war has obscured the future course of events, but there are indices which point to a post-war period of intense activity. One such index is the list of applications for commercial television station licences, now pending before the Federal Communications Commission. There are now 39 such applications, over four times the number of commercial licences at present in force. The geographical distribution of the proposed stations is shown in Fig. 26.

Another index of coming activity is the plan tentatively announced by the American Telephone and Telegraph Co. for extending the coaxial cable. The schedule, contingent on the progress of the war, calls for extensions of the present New York–Philadelphia link to Washington in 1945; to Boston and Charlotte, North Carolina, Chicago to St. Louis, and Los Angeles to Phoenix, Arizona, in 1946; Chicago to Buffalo and the beginning of a southern transcontinental route in 1947. By 1950, at the completion of the present programme, the coaxial system will extend as shown in Fig. 27. This gigantic system of coaxial cable is not necessarily predicated upon television network connections; rather it is planned as normal extension of long-distance telephone circuits. But sufficient reserve facility is planned to accommodate television demands where they may appear.

A definite index of post-war plans is the review of standards recently made by the Television Panel of the Radio Technical Planning Board. This group of 30 experts voted unanimously in June, 1944, to recommend standards substantially the same as the pre-war standards, stating that "the proposed standards are the best on which to resume television activity, based on all presently available information."

The changes in the standards proposed by the R.T.P.B. Panel are simple. The alternative standards permitted by the F.C.C. and previously discussed in this paper are restricted to permit

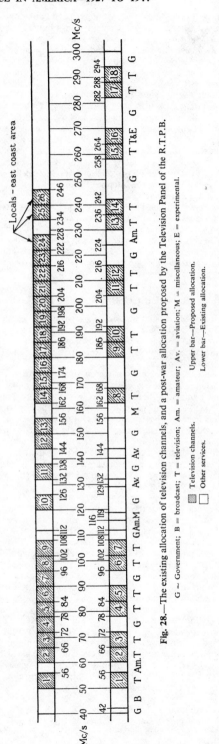

Fig. 28.—The existing allocation of television channels, and a post-war allocation proposed by the Television Panel of the R.T.P.B.

G = Government; B = broadcast; T = television; Am. = amateur; Av. = aviation; M = miscellaneous; E = experimental.

Television channels.   Other services.

Upper bar—Proposed allocation.
Lower bar—Existing allocation.

but one method in each case. Frequency modulation is no longer proposed for alternative use in transmitting the picture signal, and the alternative waveform shown in Fig. 11 is no longer recommended. Definite changes in the standards relate, strangely enough, only to the sound channel. Frequency modulation is retained, but the maximum frequency deviation is reduced from $\pm$ 75 kc/s to $\pm$ 25 kc/s, to reduce the difficulties associated with local oscillator drift on the channels at 200 Mc/s and above, which are now proposed for active use. To compensate partially for the reduction in signal level brought about by the reduction in frequency deviation, it is proposed to increase the radiated power of the sound transmitter to the range from $1 \cdot 0$ to $1 \cdot 5$ times the radiated picture power, a maximum increase of three times over the prevailing standard. Finally, the time-constant of the audio-emphasis characteristic has been reduced from 100 μs to 50 μs, in the interest of more efficient operation of the sound transmitter.

The recommendation of the R.T.P.B. Panel which points most clearly to the expected level of post-war activity is a proposed allocation (Fig. 28) which calls for twenty-six 6-Mc/s channels between 50 and 246 Mc/s. Whether these facilities will actually be assigned remains to be seen. But the need is clearly apparent for a truly nation-wide service, comparable with the present standard broadcasting system.

(3) CONCLUSION

In conclusion, a brief comparison between present American standards and the pre-war British standards is appropriate. Since the British standards are eight years older than the most recent revision of the American standards, the British standards might profitably be reviewed to take account of recent practice.

If this were done, comparison with the American standards would suggest that vestigial-sideband transmission be adopted in preference to the double-sideband system, and that the number of lines be increased from 405 to some value between 500 and 600. The 25-per-sec frame rate should of course be retained; this allows approximately 10% more lines than the 30-per-sec frame rate. A suitable figure corresponding to the American value of 525 lines would then be 567 lines ($= 3 \times 3 \times 3 \times 3 \times 7$). In all other respects, except polarity of transmission and equalizing pulses, the American and British systems are closely similar. The method of modulation for the sound channel is a question on which much international discussion is still required, but the issue is not a major one in any event.

The close co-operation between British and American television engineers which has come about through solving mutual problems during the war, will, it is hoped, continue after the war so that the television systems of both countries will benefit from a continued interchange of information and equipment.

DISCUSSION BEFORE THE RADIO SECTION, 24TH JANUARY, 1945

Mr. H. Bishop: It is an unusual and pleasurable occasion for the Radio Section to have a paper from an American engineer, and if this is an indication of still closer collaboration between American engineers and ourselves I feel sure that everyone in this country will welcome it.

The paper is of value as a record of American television development. The subject is one which is at present under active discussion in this country, for over a year ago the Lord President of the Council set up a committee under Lord Hankey to advise the Government on what should be the future of television. The report of this committee will appear shortly, and it is therefore a particularly appropriate moment for us to have such a clear statement on what has been done in America. It is interesting to read of the vicissitudes through which American television has passed in the last 17 years, because progress in this country has taken a very similar course.

The first part of the paper deals with the evolution of standards, and the author says that the adoption of the present standards in the United States was accompanied by not a little dissension in the ranks. The same happened here. The B.B.C. started low-definition television in the medium-wave band in 1929. The results were poor, the service, which was experimental, did not appeal to the public, and hardly any receivers were sold. But these early experiments supplied the necessary impetus for the development of high-definition television, and as a result of the recommendations of a Government committee—a way we have of doing things in this country—a high-definition service started from Alexandra Palace in November, 1936. At first two systems were radiated, one using 405 lines and the other 240 lines, but after a few months it was found that the higher-definition picture was so much better that it was decided to continue with this exclusively. Thus, the B.B.C. was able to give a high-definition service on 405 lines, 50 frames interlaced, for a period of nearly three years up to the beginning of the war, when the service was shut down.

In the evolution of American standards, I should like to pay tribute to the painstaking work done by the Radio Technical Planning Board. Their final recommendation to retain pre-war standards on which to resume television activity is of great in-

terest. I did not know that this decision was unanimous, as I thought that the Columbia Broadcasting System in particular was in favour of starting up again with a much higher definition system. In fact, I believe that it has already ordered equipment for colour television on about 800 lines.

We are not yet aware of the recommendations to be made by Lord Hankey's committee. Some engineers are in favour of re-starting on pre-war standards so that receivers already in the hands of the public may be used without alteration; this, it is claimed, will revive interest in television and allow time for the development of a higher definition system. The experts says that this development will take some years to complete, and if television does not restart on the old system there will inevitably be a long period before a start can be made with any improved system. Others say that to re-introduce the pre-war system would be a retrograde step, because the public would have to be satisfied with a relatively inferior picture which would be a poor advertisement for television. Further, the higher definition system will inevitably arrive and not only existing receivers but new receivers bought for the old standards will become useless.

I notice that the Americans have adopted vestigial-sideband transmission as standard. This is an interesting development which, of course, has been tried for sound broadcasting. I am bound to say that my own somewhat limited experience of it has been disappointing, because quality always seems less good than with double-sideband transmission. According to the author, there is, however, no distortion of picture quality with this system. I notice also that frequency modulation has been adopted as standard for the sound channel. The theoretical advantages of this system for the suppression of noise are, of course, well known, but so far we have not had much practical experience. It must be remembered that the receiving end of the broadcasting service is in the hands of non-technical people, and we must therefore be satisfied that any system of transmission, such as frequency modulation, is capable of being received on apparatus which is cheap to buy and easy to tune.

The author says that with the latest type of projection-type tube, the brilliance is still only 20% of the normal brightness of motion-picture projection. I am rather disappointed in this figure, and would like to know what hope there is of an increase up to, say, 50% or 60% without the equipment becoming too bulky or expensive.

The ratio of sound-carrier power to picture-carrier power has been set within the limits of 1·0 to 1·5. The Alexandra Palace sound transmitter was 3 kW, and the picture transmitter had a peak white power of 17 kW. In practice this appeared to be about the right ratio to provide equal coverage of sound and picture signals. In other words, interference from motor-cars, etc., on the fringe of the service area was, I think, about equally annoying to both.

I see that the Americans have come to the conclusion that there is very little to choose between horizontal and vertical polarization. I am inclined to agree. In this country we adopted vertical polarization before the war, and in due course a decision will have to be made whether this is to be continued or whether we shall follow the American lead and change to horizontal. The important points to consider are, of course, the signal/noise ratio with a given radiated power and the conditions of reception in towns where tall buildings are likely to cause shadows.

The author has given us the F.C.C. frequency-allocation recommendations, and I note that the band around 50 Mc/s is being retained. Our pre-war service was also in this band. The allocation of these ultra-high frequencies is certain to be the subject of close study to accommodate all the new services developed during the war, and whether these allocations shall be national or international has yet to be decided.

Dr. D. C. Espley: There can be no doubt that this panorama of American television leaves everyone very much impressed by the scale and courage of the total effort.

In this country we find difficulties in interpreting some of the American proposals. For example, Fig. 27, which shows the proposed coaxial-cable network for post-war telephone and television traffic, should give television engineers some anxiety. The main circuit is about 4 000 miles in length and is likely to have a transmission loss greater than 10 000 db after equalization. It is surprising that the literature has given no clue as to the technique for maintaining the quality of a television signal through, say, 500 stages of amplification. I think that the word "television" should not be associated with such projects, until a sufficient explanation has been given of the technical position. Alternative techniques may be retarded by the implied promise of a satisfactory cable system, and so far we have not been justified in our faith in the cable possibilities.

We join the author in the hope for future co-operation between American and British television engineers, but it is difficult to see what can be gained by our adoption of some of the main points of the American standards. The number of lines has been changed many times since 1934, and I hold the view that the 525-line standard may very easily be changed in the next few years. We shall probably jump from 405 lines to a very much higher definition. This process of evolution is very difficult to reconcile with the capital cost of inflexible cable networks.

The vestigial-sideband concept is economically very tempting, but I have never been completely satisfied about the residual signal distortion, which is very difficult to remove in simple circuit arrangements. Apart from the signal overswing effects there is an apparent gamma modification which is dependent on the signal level. It will be very interesting to have the author's views on this most important difference from British practice.

Dr. E. L. C. White: Since the American standards were fixed 4 or 5 years after the British, it behoves us to consider the differences carefully, and see whether any changes are desirable in the light of more recent experience. As I was associated with the development of purely electronic pulse-generators for odd-line interlaced scanning, my interest is mainly in the standard wave-

form, and I shall confine my remarks to one or two points under this head.

The two standard waveforms show general similarity, but there are certain important differences. Regarding frame-synchronizing pulse waveform, we assumed that, if interlacing was to be successful, the synchronizing of frame scanning must be very accurate. We considered that separation of the frame-synchronizing pulses from the line-synchronizing pulses by purely linear networks, such as low-pass filters, would produce too slow a rising edge to the frame pulse to give the required accuracy.

We concluded, therefore, that separation should be made with non-linear devices, such as diodes or other valves, and should be so designed that it took no account of previous synchronizing pulses but delivered a kick (or not) to the frame-scanning circuit, depending merely on whether the particular synchronizing pulse arriving was a wide (or a narrow) one. On this basis, the American "equalizing" pulses preceding the wide pulses are unnecessary. Pursued to its logical conclusion, our thesis indicated that only one wide pulse was required, but eight were actually included, partly in case interference should blot out the first one, and partly to cater for the cheapest possible receivers, which might use a simple low-pass filter. Accurate interlacing would probably be unnecessary in this class, owing to the size of the scanning spot.

That our waveform, which is simpler than the American standard, does not unduly restrict receiver design is shown by the fact that at least two different methods of separation have been produced by different firms, which yet hinge essentially on this feature of observing the length of the particular pulse arriving, without dependence on previous ones.

Secondly, there is the question of the sign of modulation, which turns almost entirely on the effects of interference: whether interference with either the synchronization or bright white spots on the picture is regarded as the more objectionable. From tests carried out about ten years ago, we concluded that interference with synchronization was the more irritating, and this was very largely removed by the use of zero carrier level for synchronization. A simple receiver is bound to experience one form of interference, and for such receivers—doubtless the majority—I still believe positive modulation to have been the right choice. Since then, methods of reducing either type of interference have been developed. For positive modulation there is the method known colloquially as "black-spotting," and for negative modulation there is the "flywheel" scanning mentioned in Section 2.3.

Thus the more expensive receivers on either side of the Atlantic can have greatly improved immunity from interference by suitable additions, and there is not much to choose in the cost.

Dr. R. C. G. Williams: In Section 1.2, the author discusses transmitting tubes with their development towards extreme sensitivity, and later refers to projection tubes for large-screen reception. I should like to ask the author what degree of resolution is sacrificed in the transmitting tube as sensitivity is increased, and also if he has any information on the expected life of projection tubes running at incandescent brilliance. The technical comparison between radio and coaxial linkage of television relay stations is interesting, but some figures of comparative cost would be helpful.

The review of American television receiver design practice in Section 2.10 merits comparison with our own pre-war experience. It is surprising that it is common to have no r.f. amplifier stage, as in this country we have found it important for improving the signal/noise ratio. The other characteristics of receiver design are very similar, including an intermediate frequency of the same order. The reference to picture-tube sizes, however, does indicate a different trend. In this country, sizes appeared to be settling down to a 9-in tube for the less expensive sets and a 12-in tube for the more expensive ones, and such few models as were produced using 5-in tubes proved not very acceptable to the public. This small size of picture is apparently regarded as primarily for solitary viewing, while a television programme is normally seen by a family or larger group.

The reference to colour television in Section 2.11 agrees very closely with our own conclusions, and provides a refreshing comparison with American advertising in the cautious view which is taken of the introduction of colour. Even when the technical problems are overcome, there will still remain the programme problem, and it is possible that only special transmissions with a planned high-colour content, such as stage shows, ballet, etc., will be suitable for colour transmission. American radio engineers seem, like ourselves, to dislike mechanical solutions in their search for a non-mechanical method for introducing the colour sequence.

In his conclusion, the author suggests that the British post-war television standards might profitably be based on the American standard, with due allowance for the difference in mains frequency. There is no doubt that a proper review of this matter is of vital importance, if only to avoid the development of conflicting "spheres of technical influence" in television export. On the other hand, many engineers in this country feel that the difficulties of changing from the pre-war system to one with slightly higher definition are on balance outweighed by the importance of being able to restart the television system quickly and hence promote a rapid rehabilitation of the industry. Ultimately, there is no doubt that a radically higher definition system will be introduced both here and in America, and it will no doubt be the vehicle for the introduction of colour; but it is certainly arguable whether, as an interim stage, a relatively small advance is desirable, in view of the inevitable delays which would result. It would be interesting to have the author's comments on this point of view.

Mr. L. C. Jesty: We have heard of the development of American television to its present standard of 525 lines. I was recently privileged to see some of the New York service, and, making due allowance for war-time conditions in America and the inaccuracy of my recollections of our own pre-war pictures, I feel that the difference between the 525-line and 405-line standards is insignificant from the entertainment point of view. The difference between one receiver-transmission combination and the next, even the difference between successive items in a transmission, was much more noticeable—as noticeable as it used to be with our own system.

In view of the above, I should like to ask American engineers what they have in mind for the future. It is suggested in Section 3 that 567 lines would be a suitable standard for us. Is this the last word in definition? Is colour the next step? We are told, in Section 2.11, that a 375-line colour picture does not compare in definition with a 525-line monochrome picture. Will a 525-line colour picture compare with a 750-line monochrome? The experimental channels proposed in the United States for future development are 16 Mc/s wide, which would be consistent with either of the latter alternatives.

An aspect of the more immediate future which is of interest, is the question of the suitability of projection pictures for the home, at the present standard of definition. With regard to the direct viewing cathode-ray tube, is there any move in America towards standardization, particularly of picture size? In a recent issue of *Electronics* it was suggested that the bulb manufacturers proposed to force standardization on a 10-in diameter tube. An interesting example of the way in which further improvement in standards affects the design of cathode-ray tubes and associated receiver components is given by a rough calculation based on the alternatives mentioned above. The change to colour or to a higher definition would require approximately twice the accelerator voltage on the cathode-ray tube for the same brightness of picture, which in terms of present designs will mean about 10 kV on a 12-in tube, or about 40 kV on a home-projection tube.

Mr. D. A. Bell: I fully agree with the author's suggestions for British television. In particular, I assume that the Americans have practical evidence to support the view, which I hold on theoretical grounds only, that horizontal polarization is preferable to vertical polarization from the point of view of multipath transmission. The greatest technical problem facing television, especially as carrier frequencies rise in order to accommodate more channels, is that of overcoming multipath troubles; and if horizontal polarization will help, that should outweigh any alternative advantages of vertical polarization, such as simplification of aerials.

This question of aerial design illustrates the effect of economic factors on technical problems. A high-definition system using a carrier frequency of hundreds of megacycles and relying on very directional receiving aerials to eliminate multiple transmission paths has been proposed here on an earlier occasion; since in Britain we are convinced that economic limitations will restrict us to a single station in each area, this will be quite feasible; but the Americans are already committed to two or three stations in each of the major cities, and therefore have to design in terms of receiving aerials which are multi-directional and cover a wide frequency band.

I must refer to two points on frequency modulation. First, the requirement of a 2 : 1 field-strength ratio in order to obtain the signal/noise gain on *all types of noise* is not inconsistent with the advantage of f.m. in the specific case of ignition interference at a level higher than the carrier. As I have endeavoured to explain on many occasions, there is a difference between "fluctuation noise" and "impulse noise" in f.m. systems; a positive signal/noise ratio is required for the former, but on the latter the f.m. system serves as a perfect limiter, cutting at the instantaneous modulation level, whatever the amplitude of the impulse. The advantages of f.m. are underlined by the fact that it has been chosen even for the vision signal in relaying. This allows intermediate relaying stations to be of the simplest design, consisting of a frequency-changer and Class C amplifiers; there is no need for rectifying and re-modulating, as will probably be required in a.m. relaying.

Mr. W. A. Beatty: One of the major problems in television broadcasting in a large city such as New York is that of reflections from various buildings. I do not think that the proposed American post-war standards and frequency allocations offer a solution to the reflection problem, more especially when different programme transmissions are radiated from different localities. From a television angle New York can be compared to a building with very poor acoustics, where good sound reproduction can be obtained by using a large number of small loudspeakers properly located. Similarly, there may be advantages in having a large number of small transmitters in New York rather than one large transmitter for each programme. This would allow each transmitter location to be shared by the competing programme services, and a single optimum alignment could be given to receiving antennae, at any one receiving location, for the various television services.

The paper recommends that the post-war sound transmission should be frequency modulated, with a frequency deviation of 25 kc/s rather than 75 kc/s as on the present transmissions. The reason given for this proposed reduction in frequency deviation is that it is difficult to maintain receiver oscillators stable enough to allow correct operation with the higher deviation of 75 kc/s. I do not think this can be the only reason for this decision. If instability of receiver oscillators is likely to be a source of trouble, an i.f. band-width in the receiver greater than that which would be required with a stable oscillator must be proposed. Such an increase in i.f. band-width immediately in front of the limiter would largely nullify the small signal/noise improvement likely to be obtained with a frequency deviation of 25 kc/s, assuming a transmitted radio frequency of 15 kc/s. The increase in band-width necessary to allow for an unstable oscillator is independent of frequency deviation, and will add a smaller proportion of overall noise in the case of a large deviation frequency, while the larger deviation gives an improved signal/noise ratio as compared to the smaller deviation frequency. For these reasons, I feel that something other than oscillator stability lies behind the recommendation that the frequency deviation be reduced.

The reduced frequency, moreover, renders the sound transmission more liable to interference from its own reflected signals, since the improvement likely to be obtained against such reflections is that obtainable for random noise without the use of pre-emphasis. However, if the problem of reflections is solved generally for the vision signal, reflections on the nearby sound channel should not give trouble, since the vision transmission, being amplitude modulated, requires a discrimination of 40 db between the main signals and any unwanted reflections, and if this discrimination against reflections is obtained in vision, sound transmission will automatically be free from this trouble.

Major R. H. Tanner: I am surprised both at the complication of the American system for televising films, and at the fact the previous speakers have suggested that film was transmitted from Alexandra Palace at 25 film-frames a second. Actually, the film could be run at the standard rate of 24 or at any other speed from zero to about 30 frames per second, while being transmitted at 25 television-frames per second. This apparent impossibility was achieved by means of a very clever projector, unfortunately of German origin. Through this, the film moved continuously past a gate rather more than two frames long. Compensation for the motion of the film was obtained by means of mirrors tilted by cams in such a way that the image on the screen remained stationary. At the same time, the mirrors moved sideways across the light field, and this movement resulted in a fade from each frame to the next. In this manner the screen was kept continuously illuminated, irrespective of the speed at which the film was moving; it could thus be scanned at the standard rate of 25 frames per second. This immensely simplified the use of film in television programmes, as it avoided all the complications of synchronization and also facilitated the use of very short lengths of film for routine captions.

Mr. H. L. Kirke: The main point I wish to make in connection with this paper is that a note of warning should be sounded in respect of the tendency to strive for high definition mainly in terms of the number of lines. For a given band-width and a given picture repetition frequency, there will be an optimum number of lines if the definitions in the horizontal and vertical planes are to be approximately equal. The overall result, however, is dependent upon the proper use of the full specified band-width. If for any reason the band-width is limited, such as, for example, by the aerial, the i.f. circuits or even the video circuits, then the horizontal definition will suffer, and the result may be a worse picture than will be given by fewer lines. This did occur in a number of cases in this country before the war, for in many receivers the equivalent band-width corresponded to probably 1 or $1\frac{1}{2}$ Mc/s and sometimes less, instead of the full band-width of at least $2\frac{1}{2}$ Mc/s.

In the conclusions to the paper, it is recommended that for this country a vestigial-sideband system should be adopted, and that the number of lines should be increased to some value between 500 and 600. The question whether vestigial sideband is worth while has not been proven as far as this country is concerned, and most engineers appear to favour the double-sideband system. Only comparative tests will provide the necessary data

for a decision to be made. The problem of distortion is not quite the same as for sound transmission, and in any case the distortion is very considerably reduced in sound by the use of vestigial sideband rather than the single-sideband system. Were we to adopt this system, there would be no doubt that we could increase the number of lines. I suggest, however, that the value of 567 lines recommended will be above the optimum and will put a premium on receiver band-width in regard to overall picture definition. In the majority of cases, a number of lines around 500 or even less might be better in practice.

Mr. E. C. Cherry (*communicated*): I am interested in that part of the paper dealing with the vestigial-sideband characteristics of the transmitter and receiver, but feel that the figures quoted can be somewhat misleading. For instance, it is stated in Section 2.1 that "the vestigial-sideband transmission thus offers an increase in pictorial detail of 80% with substantial improvement in picture quality," after giving comparative band-widths of 4·5 Mc/s (one sideband) and ± 2·5 Mc/s (double-sideband). It is true that an improvement in picture detail will result, but only to this degree provided that the depth of modulation is kept to an extremely low value, as it is well known that the distortion effects of vestigial-sideband systems increase with the percentage modulation. One effect is the deterioration of the rate of rise of a step-envelope (sudden increase in carrier level) which is most pronounced at the zero carrier level, but negligible at maximum carrier level.

From this point of view, the American practice of "negative transmission," in which the picture modulation is from nearly zero carrier to 75% full carrier, would seem to be somewhat less suitable for the vestigial-sideband operation than would the British system with picture modulation between 30% and 100% full carrier. In the British system, it would be the synchronizing impulses that would be affected more than the picture definition, but in practical circuits this could be only of secondary importance.

No mention seems to have been made of the possible improvement of 3 db in signal/noise ratio that may be obtained by the attenuation of one sideband at both the transmitting and receiving ends, provided that the radiated transmitter power remains unchanged.

Mr. O. S. Puckle (*communicated*): The author seems to wonder why vertical polarization has been used in this country for television transmission. I make the following suggestions, which appear to me to be reasonable. Vertical polarization provides the easiest method of obtaining omni-directional transmission, although noise and multipath interference are stated to be increased. I assume that it was considered more important to avoid blind spots and so to serve everybody, rather than to improve the quality at the expense of reduced availability.

I believe it is correct to state that the use of ordinary telephone lines for television transmission over short distances was first suggested by Mr. H. B. Rantzen of the B.B.C. Lines Department. Cable lengths of the order of half a mile with special correcting networks were employed, and these very greatly increased the available number of programme sources which could be fed into a length of coaxial cable.

It will be interesting to know when colour television is expected to become available for a public service in the United States. I do not expect to see it in this country much before 1952, but certain public statements unfortunately tend to lead the public to believe that colour television and even stereoscopy may be available almost immediately after the war, with better quality and at a much lower price.

It will also be of interest to know whether the proposed coaxial cable network shown in Fig. 27 will be used for a television-telephone service, i.e. for private vision and speech communication.

Mr. E. O. Willoughby (*communicated*): I should like to ask the author whether the main reason for the American choice of horizontally polarized propagation, instead of vertically polarized, is not due to superiority from the viewpoint of spurious re-radiation from power and telephone lines.

It would appear that u.h.f. vertically polarized waves could excite an overhead line as a type of Beverage aerial; this might be very effective owing to the large ground losses and the fact that the exciting propagation was directed substantially parallel to the earth. Moreover, as the transmitter is usually located in a city which is a centre of a radial road system, long lengths of overhead line almost co-phased with the wavefront are very probable, with considerable signal pick-up and re-radiation at discontinuities, whereas for horizontally polarized waves appreciable coupling to an overhead line would seem to be practically impossible.

THE AUTHOR'S REPLY TO THE ABOVE DISCUSSION

Mr. D. G. Fink (*in reply*): The greater part of the discussion relates to the comparative merits of British and American standards. The most important difference between them is the maximum vision frequency. The records of the N.T.S.C. and the R.T.P.B. show substantially unanimous opinion among American television specialists that 2·5 Mc/s is too low a value for satisfactory public service when the inevitable degradations of band-width, encountered in practice, are considered. Mr. Kirke mentions instances of the use of a band-width of 1 Mc/s "or less" when the standard was set at 2·5 Mc/s. In America it has been deemed wise to set 4 Mc/s (or somewhat more) as the mark to aim at, in the hope that at least 3 Mc/s will be utilized even in the cheaper receivers.

Here an important difference between the British and American situations is evident. America is committed to a policy of private ownership of television broadcasting stations, with competition as a regulating agent. This implies several stations in major cities, and the need for a multiplicity of channels, particularly along the eastern seaboard. The width of the channels must be set in advance, to avoid later encroachment on neighbouring channels. Hence it is sound policy to establish channels sufficiently wide to permit improvement within the existing standards. This will encourage technological advances without reducing the initial utility of the equipment in public hands.

In Great Britain, as Mr. Bell points out, a smaller number of stations is in prospect, and it may therefore be possible to expand the sidebands of a station, as the state of the art may dictate, without danger of encroachment. Such expansion would have precisely the same effect as reserving the space in advance, namely it would encourage improvement of receiver design without reducing the utility of existing receivers.

If provision for such expansion is made, consideration must also be given to the number of lines standardized. As Mr. Kirke states, the number of lines is but a partial index to picture quality. M. W. Baldwin, jun., of the Bell Telephone Laboratories, has shown experimentally that the subjective sharpness of a television image decreases very little with departure from the condition of equal vertical and horizontal resolution, provided the band-width is fixed. But if eventual expansion of the transmitted sideband from 2·5 Mc/s to 4 Mc/s is contemplated, it would appear to be sound policy to set the number of lines at a figure which gives equal vertical and horizontal resolution when the maximum vision frequency is near the upper limit of this range. This would indicate a number of lines greater than 405.

Respecting vestigial-sideband transmission, the choice is basically one of receiver design. It must be pointed out that a receiver possessing the vestigial characteristic of Fig. 9 gives the same performance, whether the signal is double-sideband or vestigial-sideband (of the type illustrated in Fig. 8). Moreover, such a receiver is appreciably cheaper to construct relative to one which accepts both sidebands. This fact leads to the possibility that, if double-sideband transmissions are resumed in Great Britain, designers of receivers may adopt the vestigial characteristic of Fig. 8 as a means of reducing the number of high-gain valves required.

The relative merits of the vestigial- and double-sideband systems rest in the matter of waveform distortion. This distortion is theoretically apparent but not readily identified in practice. Other distortions (those introduced prior to modulation and subsequent to demodulation) are usually of greater severity. When these distortions are removed, the photometric deficiencies (brightness, contrast, tonal gradation) take the centre of the stage. When these are remedied, a very tolerable picture results using vestigial-sideband transmission, and no noticeable improvement appears when double-sideband transmission is introduced. Nor is this good performance limited to low percentages of modulation. In fact, as Mr. Cherry points out, all American pictures of substantial brightness must be highly modulated. In short, few if any of the theoretical difficulties have been experienced, and the vote in favour of vestigial-sideband transmission in all polls since 1940 has been unanimous.

Respecting frequency modulation for the sound channel, another difference between British and American prospects appears. America is to have an extensive frequency-modulated sound-broadcasting service, on frequencies adjacent to the television channels. A considerable economy is secured in using the same circuits in a receiver intended for both services. Mr. Beatty suggests that the narrow deviation, 25 kc/s, does not introduce much noise advantage especially when a wide (actually, 200 kc/s) i.f. channel is employed. This fact is recognized, but it is hoped that improvements in oscillator stability will permit the deviation figure to be raised at a later date.

Dr. Espley's fears concerning the transmission of television images over 4 000 miles of coaxial cable are understandable. The repeater equipment developed for this project has been designed for a 2·75-Mc/s band. This is admittedly below the 4-Mc/s standard for local transmission (as indeed the 5-kc/s standard of transcontinental sound-programme circuits is below the 10-kc/s standard for local circuits), but it is nevertheless acceptable for long-distance service. I am informed by Mr. M. E. Strieby, of the American Telegraph and Telephone Company, that highly acceptable television signals have been transmitted over a circuit of 800 miles' length, and that no insuperable difficulty is anticipated in longer circuits unless wider-band-width or higher-contrast-range requirements are set up. For wider bands, of course, the radio-relay method of transmission offers many practical and theoretical advantages. No decision has been reached, and both methods are being developed by active groups.

Regarding polarization, and in specific reference to Mr. Puckle's communication, it is the American view that an omnidirectional receiving antenna (simply achieved when vertical polarization is used) is a definite disadvantage when reflected signals abound, as they do in large cities. Such reflections are a most serious problem; they may make it virtually impossible to receive the optimum signal from more than one station with a single aerial system.

I wish to acknowledge with gratitude the generous assistance of Dr. David B. Langmuir, who presented the paper before The Institution in London.

THE HISTORY OF TELEVISION

G[erald] R. M. Garratt and A. H. Mumford

Paper No. 1320
RADIO SECTION

621.397(091)

THE HISTORY OF TELEVISION

By G. R. M. GARRATT, M.A., and A. H. MUMFORD, O.B.E., B.Sc., Members.

(The paper was first received 24th January, and in revised form 21st February, 1952. It was presented at the TELEVISION CONVENTION *28th April, 1952.)*

DEFINITION

Although the word "television" has come to be associated in the public mind with the broadcasting of entertainment by means of radio, such a limitation seems unacceptable and illogical in dealing with the history of the subject. The authors have therefore interpreted their task as one involving a broad historical review of the various methods whereby visual phenomena may be reproduced at a distance but without regard to the method of transmission or to the time occupied in transmission. In consequence of this interpretation, the earlier part of the paper deals with a number of proposals which aimed at providing a reproduction on paper of the original image in a form which, to-day, we should associate more with the practice of photo-telegraphy than with modern television. Since all practical systems of either photo-telegraphy or television involve some form of scanning, it will be realized that both arts have a common ancestry.

----

(1) EARLY SCHEMES

(1.1) The Electrochemical Effect of Light

In tracing the history of television, one may recall that every system of electric vision which has ever been proposed makes use of certain physical changes which are produced by light. If the phenomenon known as photo-electricity were non-existent all television would be impossible, and the real history of television may therefore be said to date from the discovery of the electro-chemical effects of light by Edmond Becquerel in 1839.

Alexandre Edmond Becquerel belonged to a French family, several of whose members became distinguished scientists. Born in 1820, he was only nineteen years of age at the time of his observation of the electrochemical effect of light, but it seems likely that he was working under the guidance of his father, Antoine César Becquerel, who probably assisted him in presenting the account of his experiments to L'Academie des Sciences, in July, 1839, which was subsequently published in *Comptes Rendus*.[1] Edmond Becquerel must not be confused with his son, Antoine Henri, who became world-famous in 1896 on account of his discovery of radioactivity.

Briefly, Becquerel observed that when two electrodes are immersed in a suitable electrolyte and illuminated by a beam of light, an e.m.f. is generated between the electrodes. This type of electrolytic photocell is now of historic interest only, but it represents the first stage in the linking of light phenomena with electricity and may thus be regarded as the earliest step in the development of television.

(1.2) Selenium

Chronologically, the next stage in the development of photo-electricity which had any application to television occurred in 1873, with the discovery of the effect of light upon the resistance of selenium. The discovery is important historically, not so much for any practical value which selenium might have had for the purpose, but for the glut of schemes and proposals which were made for television systems in the years which followed it.

Mr. Garratt is at the Science Museum, South Kensington.
Mr. Mumford is at the General Post Office.

The details of the discovery have so often been misrepresented that it may be of value to record again the circumstances in which it was first reported and with which a former President of The Institution, the late Mr. Willoughby Smith, was intimately concerned.

Selenium had been discovered by Berzelius in 1817 and it had been found to possess a very high electrical resistance. It was this property which led to its use in certain experiments by Willoughby Smith and to the discovery which he reported in a letter to Latimer Clark, then Vice-President of the Society of Telegraph Engineers:

Wharf Road,
4th February, 1873.

My dear Latimer Clark,

Being desirous of obtaining a more suitable high resistance for use at the shore station in connection with my system of testing and signalling during the submersion of long submarine cables, I was induced to experiment with bars of selenium—a known metal of very high resistance. I obtained several bars, varying in length from 5–10 centimetres, and of a diameter from 1 to 1½ millimetres. Each bar was hermetically sealed in a glass tube, and a platinum wire projected from each end for the purpose of connection.

The early experiments did not place the selenium in a very favourable light for the purpose required, for although the resistance was all that could be desired—some of the bars giving 1 400 megohms absolute—yet there was a great discrepancy in the tests, and seldom did different operators obtain the same result. While investigating the cause of such great differences in the resistances of the bars, it was found that the resistance altered materially according to the intensity of light to which they were subjected. When the bars were fixed in a box with a sliding cover, so as to exclude all light, their resistance was at its highest, and remained very constant, fulfilling all the conditions necessary to my requirements; but immediately the cover of the box was removed, the conductivity increased from 15 to 100 per cent, according to the intensity of the light falling on the bar. Merely intercepting the light by passing the hand before an ordinary gas-burner, placed several feet from the bar, increased the resistance from 15 to 20 per cent. If the light be intercepted by glass of various colours, the resistance varies according to the amount of light passing through the glass.

To ensure that temperature was in no way affecting the experiments, one of the bars was placed in a trough of water so that there was about an inch of water for the light to pass through, but the results were the same; and when a strong light from the ignition of a narrow band of magnesium was held about nine inches above the water the resistance immediately fell more than two-thirds, returning to its normal condition immediately the light was extinguished.

I am sorry that I shall not be able to attend the meeting of the Society of Telegraph Engineers to-morrow evening. If, however, you think this communication of sufficient interest, perhaps you will bring it before the meeting. I hope before the close of the session that I shall have an opportunity of bringing the subject more fully before the Society in the shape of a paper, when I shall be better able to give them full particulars of the results of the experiments which we have made during the last nine months.

I remain, yours faithfully,
Willoughby Smith.

Latimer Clark, Esq., C.E.

This letter was published in the *Journal of the Society of Telegraph Engineers*[2] and gave rise to much interest and speculation in the scientific world at the time. Inaccurate statements and claims were soon being made, however, and these prompted Willoughby Smith to make a further communication to **the**

Society.[3] Extracts from his letter dated the 3rd March, 1876, are as follows:

Gutta Percha Works,
18, Wharf Road,
City Road.
3rd March, 1876.

To the Secretary of the
    Society of Telegraph Engineers.
Sir,
    On February 4, 1873, I brought before the notice of the scientific world, through the Society of Telegraph Engineers, the effect which light has on the electrical conductivity of selenium. . . . I think it would be well if the Society . . . will allow me to place on record a few particulars of the actual discovery. While in charge of the electrical department of the laying of the cable from Valentia to Heart's Content in 1866* I introduced a new system by which ship and shore could communicate freely with each other during the laying of the cable without interfering with the necessary electrical tests. To work this system it was necessary that a resistance of about one hundred megohms should be attached to the shore end of the conductor of the cable. . . . While searching for a suitable material the high resistance of selenium was brought to my notice but at the same time I was informed that it was doubtful whether it would answer my purpose as it was not constant in its resistance. I obtained several specimens of selenium and instructed Mr. May, my chief assistant at our works at Greenwich, to fit up the system we adopt on shore during the laying of cables. . . . It was while these experiments were going on that it was noticed that the deflections varied according to the intensity of light falling on the selenium. . . .

Yours very truly,
Willoughby Smith.

The discovery of the effect of light on the resistance of selenium had no immediate sequel, but within a few years the exciting era of progress which followed Bell's invention of the telephone, in 1875, provided the background for numerous schemes for "seeing by electricity." It is perhaps not surprising that many ideas and inventions were put forward which were impracticable at the time, either because a great deal of development would have been required to perfect them or, more often than not, because other discoveries and developments essential to the working of the main idea had not yet been made.

The fact that an idea is impracticable at the time of its original suggestion does not necessarily detract from its historic interest. Indeed, for more than half a century the history of television consists almost exclusively of plans and suggestions, all of them incapable of immediate practical embodiment and all of them waiting for other inventions and parallel developments which would enable them to be incorporated in a practical system. By surveying these plans, impracticable though they may have been initially, we see how the broad front of science advances, a step or two forward here and there until, in the lapse of time, the whole state of the art is seen to have made considerable progress.

(1.3) M. Senlacq

One of the earliest of the schemes put forward for distant vision by means of electricity was that proposed by M. Senlacq of Ardres, in 1878. From the brief account of his "Telectroscope," which was published in the *English Mechanic*,[4] we find that it was proposed to employ the properties of selenium "to reproduce telegraphically at a distance the images obtained in the camera obscura."

An interesting feature of Senlacq's proposal was the implication of the need for a process of scanning. Senlacq's idea was to use "an ordinary camera obscura containing at the focus an unpolished glass (screen) and any system of autographic telegraphic transmission; the tracing point of the transmitter intended to traverse the surface of the unpolished glass will be formed of a small piece of selenium held by two springs acting as pincers,

* This was the first really successful Atlantic submarine cable to be laid.

insulated and connected, one with a pile and the other with the line."

It must be remembered that the earliest proposals for electric vision did not attempt to provide for the almost instantaneous viewing of distant events such as we know to-day. Instead, they usually aimed, as in Senlacq's scheme, at providing a reproduction on paper in the form which we now associate with the practice of photo-telegraphy. Senlacq's proposed receiver was almost amusingly naïve; he conceived it as consisting merely of

a tracing point of blacklead or pencil for drawing very finely, fitted to a very thin plate of soft iron held almost as in the Bell telephone and vibrating before an electro-magnet governed by the irregular currents in the line.

The pencil, in fact, was expected to reproduce on a sheet of paper the graduations of light and shade in the original image in response to the varying pressure exerted on the paper in consequence of the movements of the telephone diaphragm.

Should the selenium tracing point at the transmitter run over a light surface, the current will increase in intensity, the electro-magnet at the receiver will attract the plate to it with greater force and the pencil will exert less pressure on the paper. The line thus formed will be scarcely, if at all, visible; the contrary will be the case if the original image be dark, for, the resistance of the current increasing, the attraction of the magnet will diminish and the pencil, pressing more on the paper, will leave upon it a darker line.

Senlacq put forward his original proposals in 1878, but he seems subsequently to have realized some of the practical difficulties to be surmounted, as, in 1880, he published details of a new scheme in which the transmitter screen consisted of a large number of very small selenium cells. Each cell was connected individually to a form of distributor through which contact could be made with the single-line wire by means of a falling slider, the arrangement being such that each cell was scanned once only.

The receiver consisted of a multi-cellular mosaic screen of fine platinum wires, each connected to contacts on a distributor plate which was associated with a clockwork mechanism intended to ensure synchronism with the falling slider at the transmitter. Each platinum wire was thereby to be rendered momentarily luminous in proportion to the light falling on the corresponding selenium cell at the transmitter. The description of the scheme which appeared in the *English Mechanic* is somewhat vague in regard to many of the details, and although it is stated that

The picture is, therefore, reproduced almost instantaneously; . . . we can obtain a picture, of a fugitive nature, it is true, but yet so vivid that the impression on the retina does not fade. . . .

one is, in fact, left with some doubt as to whether M. Senlacq's imagination was not more highly developed than his Telectroscope.

Whether or not M. Senlacq's scheme could have worked quite as well as his words implied, there is no doubt that the publicity which was given to his ideas in Continental and American journals

. . . everywhere occupied the attention of prominent electricians who have striven to improve upon them. Among these we may mention Messrs. Ayrton and Perry, Sawyer (of New York), Sargent (of Philadelphia), Brown (of London), Carey (of Boston), Tighe (of Pittsburgh) and Graham Bell himself. (*English Mechanic*, 1881.)

(1.4) Ayrton and Perry

It will suffice to refer in detail to only two of these further schemes, namely that of Ayrton and Perry and that of George Carey of Boston. Ayrton and Perry's proposals were contained in a letter to *Nature* dated 22nd April, 1880, but their letter[6] seems to have been written more with the idea of diminishing any claim which might have been put forward on behalf of Graham

Bell, who was reported to have described, in a sealed document, an invention for the same purpose, than with the intention of putting forward a definite invention of their own. Nevertheless, their letter is of interest from several aspects and is reproduced below:

"Seeing by Electricity."

We hear that a sealed account of an invention for seeing by telegraphy has been deposited by the inventor of the telephone. Whilst we are still quite in ignorance of the nature of this invention, it may be well to intimate that complete means for seeing by telegraphy have been known for some time by scientific men.

The following plan has often been discussed by us with our friends, and, no doubt, has suggested itself to others acquainted with the physical discoveries of the last four years. It has not been carried out because of its elaborate nature, and on account of its expensive character, nor should we recommend its being carried out in this form. But if the new American invention, to which reference has been made, should turn out to be some plan of this kind, then this letter may do good in preventing monopoly in an invention which really is the joint property of Willoughby Smith, Sabine, and other scientific men, rather than of a particular man who has had sufficient money and leisure to carry out the idea.

The plan, which was suggested to us some three years ago more immediately by a picture in *Punch*, and governed by Willoughby Smith's experiments, was this: Our transmitter at A consisted of a large surface made up of very small separate squares of selenium. One end of each piece was connected by an insulated wire with the distant place, B, and the other end of each piece connected with the ground, in accordance with the plan commonly employed with telegraph instruments. The object whose image was to be sent by telegraph was illuminated very strongly, and, by means of a lens, a very large image thrown on the surface of the transmitter. Now it is well known that if each little piece of selenium forms part of a circuit in which there is a constant electromotive force, say of a Voltaic battery, the current passing through each piece will depend on its illumination. Hence the strength of electric current in each telegraph line would depend on the illumination of its extremity. Our *receiver* at the distant place, B, was, in our original plan, a collection of magnetic needles, the position of each of which (as in the ordinary telegraph) was controlled by the electric current passing through the particular telegraph wire with which it was connected. Each magnet, by its movement, closed or opened an aperture through which light passed to illuminate squares at B as of selenium squares at A, and it is quite evident that since the illumination of each square depends on the strength of the current in its circuit, and this current depends on the illumination of the selenium at the other end of the wire, the image of a distant object would in this way be transmitted as a mosaic by electricity.

A more promising arrangement, suggested by Professor Kerr's experiments, consisted in having each little square at B made of silvered soft iron, and forming the end of the core round which the corresponding current passed. The surface formed by these squares at B was to be illuminated by a great beam of light polarized by reflection from glass, and received again by an analyser.

It is evident that since the intensity of the analysed light depends on the rotation of the plane of polarization by each little square of iron, and since this depends on the strength of the current, and that again on the illumination of the selenium, we have another method of receiving at B the illumination of the little square at A. It is probable that Professor Graham Bell's description may relate to some plan of a much simpler kind than either of ours; but in any case it is well to show that the discovery of the light effect on selenium carries with it the principle of a plan for seeing by electricity.

Scientific Club, April 21.
John Perry.
W. E. Ayrton.

(1.5) Carey's Schemes

The scheme put forward by George Carey was described in an article in *Design and Work*[7] and is of special interest on account of the detail with which the proposals were described and of the drawings which accompanied the article. Once again, however, we find that, although the article describes the inventor's "wonderful instruments," there is an unfortunate lack of any report of their having worked, and one is left with the impression that Carey's schemes were never subjected to the test of actual construction.

Carey's schemes—there were two distinct proposals—for seeing by electricity both envisaged the reception of the image on chemically prepared paper, and so, like Senlacq's scheme, they were more akin to what we know as photo-telegraphy than to television. His first scheme comprised a multi-cellular mosaic of selenium cells, each cell being separately connected by a line wire to a point in a corresponding mosaic at the receiver, against which was pressed a sheet of prepared paper.

Carey's second scheme is perhaps of greater interest as it incorporated the elements of a practical mechanism for scanning the image, and thus it required only the use of a single line-wire for transmitting the signal currents. He seems, however, to have overlooked the need for any means of synchronizing the clockwork scanning mechanism which he proposed to use at transmitter and receiver, a further indication that the scheme was never put into a practical form. Figs. 1 and 2 illustrate both of Carey's schemes.

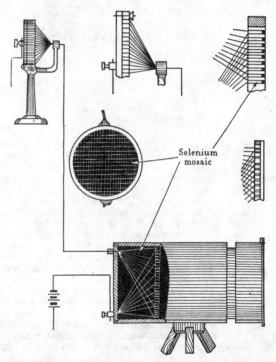

Selenium mosaic

Fig. 1.—Carey's first scheme, 1880. It was proposed to project an image on to a multi-cellular mosaic of selenium cells, each of which was to be separately connected by a wire to corresponding points at the receiver where the image was to be reproduced on paper by electrochemical decomposition. (Reproduced from *Design and Work*.)

(1.6) Shelford Bidwell

If there is much doubt as to whether any of the proposals made by Senlacq, Carey, Ayrton, Perry and certain other contemporary inventors were ever carried into practical effect, there is certainly none in regard to those made by Shelford Bidwell in 1881, for not only did he lecture on and demonstrate his apparatus[8, 9, 10] to members of the Royal Institution, the Physical Society and the British Association for the Advancement of Science, but his apparatus has actually been preserved in the Science Museum for many years. In the interests of accuracy,

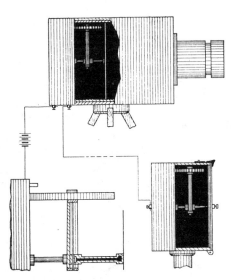

Fig. 2.—Carey's second scheme, 1880. Put forward at the same time as the first scheme, this proposal incorporated an elementary form of scanning whereby the image was to be explored by a selenium cell which traversed a spiral path. No provision was made for synchronization. (Reproduced from *Design and Work*.)

Fig. 3.—Shelford Bidwell's transmitter, 1881. A shadowgraph image was projected on to the front of the box with a pin-hole aperture containing a selenium cell. The image was scanned by the rise-and-fall motion of the box and its slow lateral traverse.

it must be recorded that Shelford Bidwell's apparatus was acquired by the Science Museum in 1908, and while, unfortunately, no documentary evidence exists to prove that it was made in 1881, it conforms so exactly with the description given in the Report of the British Association for that year that there can be little doubt on the matter. The point is of some importance since the apparatus does not conform to the description published in *Nature*, but since the Report of the British Association for the same year refers to the apparatus exhibited to the members as being "a development of that described in *Nature* . . ." it is safe to assume that Bidwell made two machines within a few months and that it is the second of these which has been preserved. Figs. 3 and 4 show details of the transmitter and receiver.

Bidwell's machine is of special interest, not only because it worked effectively in transmitting silhouettes, but also because of the ingenuity of the scanning motion. In operation, an image of the picture to be transmitted was projected by means of a lens onto the front of a small box containing a selenium cell and having a small pin-hole in the centre of the front face. By means of a cam mechanism, the box could be made to rise vertically to scan the image, a quick flyback was caused at the end of each vertical line, whilst progression to the next line was caused by mounting the cell-box on a horizontal shaft, one part of which had a fine screw-thread cut upon it.

The receiver consisted of a platinum-covered brass cylinder mounted horizontally on a spindle similar to that of the transmitter. A platinum point attached to a flexible brass arm pressed on the surface of the cylinder around which was wrapped a piece of paper soaked in a solution of potassium iodide. As the cylinder rotated in synchronism with the shaft at the transmitter, the platinum point traced out brown lines upon the paper, the intensity of which varied in accordance with the current through the selenium cell.

So far as is known to the authors, no practical application or commercial use was ever made of Bidwell's invention. Nevertheless, this tangible evidence of the state of the art in 1881 is a

salutary reminder to those of us who, unthinkingly, regard photo-telegraphy and television as quite recent innovations.

As has already been mentioned, the need for a system of scanning was foreseen by many of the early workers in this field. Among others, Senlacq, Sawyer, de Paiva, Leblanc, Carey and Bidwell all made proposals, some of which have already been noted in the paper. The proposal put forward by Leblanc is of some interest since there have been numerous attempts to utilize the same principles. He proposed to scan the image by means of a small mirror arranged to oscillate about two axes simultaneously, but with widely different frequencies, and so to project each little area of the image in turn on to a selenium cell. Although this method of scanning is efficient optically, the mechanical and synchronizing difficulties proved considerable, and no real success ever attended the attempts to use this principle.

(1.7) Nipkow's Patent

Of all the early scanning systems, that suggested by Paul Nipkow[11] in 1884 is the best known, a fact which can be attributed to its simplicity and, in more recent times, to its adoption by Baird in many of his early experiments. Nipkow's method of scanning employed a disc of large diameter near the periphery of which was a series of small holes arranged in the form of a spiral. The area of image which could be scanned by this arrangement was small in comparison with that of the disc, but it could be used with a stationary optical system, and, although very inefficient as a receiving scanner, it was used until recent times in one system of film transmission.

Nipkow's transmitter employed a selenium cell—the only practical form of photo-element known at the time. His receiver, however, employed an ingenious system of light modulation which depended on the property, possessed by flint glass, of rotating the plane of polarization of light when situated in a

Fig. 4.—Shelford Bidwell's receiver. The receiving drum was mechanically coupled to the transmitter for demonstration purposes. The drum carried paper soaked in potassium iodide, the image being reproduced in a series of closely-spaced brown lines.

magnetic field. With the small currents passed by a selenium cell and the absence of any means of amplification, this system was too insensitive to give practical results, but a similar system was used successfully by Baird in which specially designed Kerr cells were used in the place of flint glass.

In common with so many of the early schemes, Nipkow seems to have made no attempt to put his ideas into practice. Perhaps he tried and failed, but so far as is known, the only record of his work is the German Patent No. 30105, which he was granted in 1884 and from which Fig. 5 has been taken.

(1.8) Mirror Drum

Another system of scanning which was given much attention by later experimenters was that proposed in 1889 by Lazare Weiller.[12] In place of the disc with a series of holes on a spiral track, Weiller proposed to use a drum or disc fitted with a number of tangential mirrors around its periphery, each successive mirror being orientated through a small angle so that, as the drum rotated, the area of the image was scanned in a series of lines and projected onto a selenium cell. Such an arrangement was in fact first used by Ll. B. Atkinson, in 1882, but he did not publish a description at the time, and Weiller is therefore generally recognized as the inventor of the mirror drum.

One of the difficulties which faced the early experimenters was the time-lag of selenium and its consequent insensitivity to rapid changes of light intensity. It was perhaps the existence of this time-lag, more than any other factor, which gradually brought about a realization of the unsuitability of selenium and tended to discourage interest in schemes for distant electric vision during the last ten years of the century. An ingenious attempt to overcome this sluggishness of selenium was suggested in 1897 by Szczepanik.[13] He proposed to construct the selenium cell in the form of a ring which was to rotate at a steady speed. By means of a pair of oscillating mirrors, the image was to be scanned and projected through an aperture onto a part of the selenium ring. As the ring rotated, it continually exposed a fresh surface to the aperture and thus became capable of responding to higher frequencies. There is no evidence that Szczepanik

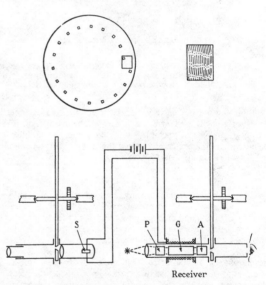

Receiver

Fig. 5.—Nipkow's Patent, 1884: a redrawing of the diagram published in the original patent specification showing his scanning disc with apertures on a spiral track. A selenium cell was proposed as the light-sensitive element at the transmitter, while rotation of the plane of polarization in a magnetic field was proposed as the light-control at the receiver.

ever attempted to construct his apparatus; indeed, the receiver he described in his patent was founded on incorrect principles and could not have worked without major modification to the light control.

It had been only natural that the tremendous success and widespread expansion of the telephone which took place in the years which followed its invention, in 1875, should have encouraged scientific thought towards the parallel subject of distant vision.

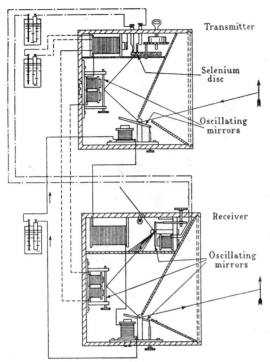

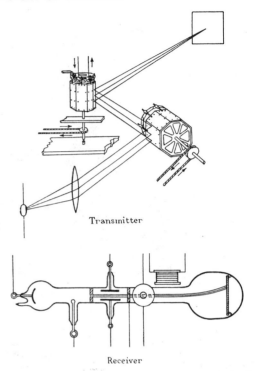

Fig. 6.—Szczepanik's Patent, 1897: an attempt to overcome the sluggishness of selenium by using a cell in the form of a rotating disc S. Note the proposed use of oscillating mirrors for scanning purposes.

Fig. 7.—Rosing's Patent, 1907: a reproduction of the main drawings from Rosing's patent in which, for the first time, the use of a cathode-ray tube was proposed as the receiver.

Within only ten or twelve years, the telephone had become the basis of a great and growing industry; the parallel problem, however, proved intractable, and when the unsuitability of selenium finally became recognized the interest in the possibility of distant vision appears to have waned.

(1.9) Boris Rosing

Early in the twentieth century, interest in the possibility of electric vision was revived by the gradual development of photoelectric elements and especially by the work of Elster and Geitel in this field. Further encouragement was provided by the introduction of the first commercial cathode-ray tubes by Braun in 1897. It was not, however, until 1907 that Boris Rosing, a teacher at the Technological Institute of St. Petersburg became the first to suggest the use of a cathode-ray tube as the receiving element in a system of remote electric vision.[14] After the primitive and impracticable schemes of the nineteenth century, Rosing's ideas seemed to possess at least the elements of practicability, and although much detailed development would have been needed to obtain even crude results, the originality of his proposals gives them a place in the history of the art.

From Fig. 7, which has been reproduced from Rosing's British patent specification of 1907, it will be seen that his proposed transmitting arrangements employed two mirror-drums revolving at right angles to each other, but at widely different rates, to scan the image. The varying current impulses from the photocell were transmitted to the receiver, where they were chased to charge a pair of deflecting plates in the Braun tube. The fluctuating charges on these plates caused the beam of electrons projected

from the cathode to be deflected away from an aperture. The beam in these early cathode-ray tubes was not sharply focused, and thus the amount of the beam which passed through the aperture was roughly proportional to the voltage of the deflecting plates and therefore to the degree of light and shade in the original scene. Having passed through the aperture, the modulated beam was caused to scan the surface of a fluorescent screen and so to produce an image of the scene presented to the transmitter.

A brief account of Rosing's work appeared in the *Scientific American* in 1911, and a careful reading of his subsequent British patent[15] suggests that he had devoted considerable time to experimental work on the subject, since he referred to certain difficulties with which he could have become familiar only by practical experience. Rosing's second British patent,[15] however, is a disappointing document; he seems to have diagnosed incorrectly certain faults in his earlier apparatus and, on the basis of this wrong diagnosis, the cure he prescribed could not have achieved the results for which he must have hoped. If anything, therefore, the second patent was a retrograde step, but it nevertheless contains two features worthy of brief comment.

First, it describes a device for interrupting or "chopping" the light falling on the photocell, an arrangement which was used by Baird and others in about 1924, but which cannot have been of practical use to Rosing, who was not confronted by any problems of amplification. Secondly, the patent contains a reference to a scanning system which later became known as "velocity modulation," in which the speed of the scanning spot contains a component which is inversely proportional to the illumination intensity, so that, in Rosing's words, "the time of action of the signal upon the eye of the observer corresponds to the intensity of the light signals at the transmitting station." Unfortunately, it does not seem to have occurred to Rosing that the light signal must also modulate the speed of the transmitting scanner if accurate reconstitution of the original scene is to be obtained at the receiver.

(1.10) A. A. Campbell Swinton

Campbell Swinton's proposals were put forward during almost exactly the same period as Rosing's, i.e. 1908–1911, but it is clear that they were made entirely independently, and indeed both men approached the problem by a very different route. A comparison is instructive. Rosing found certain devices and methods available—mechanical scanning by mirror drums and the Braun tube—and he proceeded to apply these tools in the pursuit of remote electric vision. Campbell Swinton, on the other hand, began by analysing the problem itself—somewhat superficially it is true—but nevertheless analysing it as to its elementary and minimum requirements. Having defined the conditions to be satisfied, he proceeded to consider the general means whereby it might be possible to solve such a problem. In defining the problem, he gave little or no thought to what might or might not be technically practicable at the time. Such matters he left for future development and invention—and there can surely be little doubt that this is always the wiser and more logical course when one seeks the solution to some complex scientific problem.

It was a letter in *Nature* from Shelford Bidwell[16] which inspired Campbell Swinton to make the original suggestion which was to develop into the high-definition system of television we know to-day. As we have seen, Shelford Bidwell had taken much interest in remote electric vision as early as 1881, and on the 4th June, 1908, he wrote to *Nature* to criticize a report that a Monsieur Armengaud of Paris "firmly believed that within a year, as a consequence of the advance already made by his apparatus, we shall be watching one another across distances hundreds of miles apart." Shelford Bidwell, with his thoughts still on selenium, emphasized the impracticability of such proposals on the grounds of complexity and expense. He considered the synchronizing difficulties insuperable for a high-definition system, although he thought that the problem could possibly be solved by using a multi-wire cable with 90 000 conductors and costing £1¼ million for a point-to-point system between two places 100 miles apart.

In his reply to this letter, on the 16th June, 1908, Campbell Swinton wrote[17] that although

it is wildly impracticable to effect even 160 000 synchronized operations per second by ordinary mechanical means, this part of the problem of obtaining distant electric vision can probably be solved by the employment of two beams of kathode rays (one at the transmitting and one at the receiving station) synchronously deflected by the varying fields of two electromagnets placed at right angles to one another and energized by two alternating electric currents of widely different frequencies, so that the moving extremities of the two beams are caused to sweep synchronously over the whole of the required surfaces within one-tenth of a second necessary to take advantage of visual persistence.

Indeed, so far as the receiving apparatus is concerned, the moving kathode beam has only to be arranged to impinge on a sufficiently sensitive fluorescent screen, and given suitable variations in its intensity, to obtain the desired result.

The real difficulties lie in devising an efficient transmitter which, under the influence of light and shade, shall sufficiently vary the transmitted electric current so as to produce the necessary alterations in the intensity of the kathode beam of the receiver, and further in making this transmitter sufficiently rapid in its action to respond to the 160 000 variations per second that are necessary as a minimum.

Possibly no electric phenomenon at present known will provide what is required in this respect, but should something suitable be discovered, distant electric vision will, I think, come within the region of possibility.

66, Victoria Street, S.W.   A. A. Campbell Swinton.
    12th June, 1908.

At the time when this letter was written, radiocommunication was in its infancy, radio valves were almost unknown and the amplifier had not been invented, whilst vacuum technique was primitive and photocells were inefficient. As his letter shows, Campbell Swinton realized the difficulties to be overcome and the impossibility of putting his ideas into practice at the time. Nevertheless, the vital principles for future development had, for the first time, been clearly indicated, and one can only regret that Campbell Swinton himself did not live* to see his ideas developed into the first service of high-definition television in the world.

Three years elapsed before Campbell Swinton made any attempt to elaborate his scheme further, but in November, 1911, in the course of a Presidential Address[18] to the members of the Röntgen Society, he greatly amplified his proposals and gave comprehensive details of the scheme, illustrating it with the diagram reproduced in Fig. 8.

He conceived the idea of a mosaic screen of photo-electrical elements which were to form a part of a special type of cathode-ray tube. The image of the scene to be transmitted was to be projected onto the mosaic screen by means of a lens, and the back of the screen was to be scanned by a beam of cathode rays controlled magnetically by the currents from two a.c. generators. The cathode-ray beam in the receiver was synchronized with that in the transmitter by means of deflecting coils connected by wires to the same generators as the transmitter, and a separate conductor carried the photo-electric currents for modulating the receiver beam.

Campbell Swinton never attempted to construct a working demonstration of his proposals, and he fully realized the difficulties which would have to be overcome before such a system could be made to work. With characteristic sincerity he said, "It is an idea only and the apparatus has never been constructed. Furthermore, I do not for a moment suppose that it could be got to work without a great deal of experiment and probably much modification."

Modern television owes much to the researches and achievements of many distinguished workers, but in its essence it has been developed upon the fundamental lines first put forward by Campbell Swinton.

(2) THE LOW-DEFINITION ERA

All the schemes which were put forward prior to 1920, and which have been described in the earlier part of the paper, belong to the period when they could necessarily be little more than mere ideas. Although some of the basic schemes may have been well conceived, e.g. that put forward by Campbell Swinton, there were many ancillary problems which required solution before any real progress could be made towards the development of practical television. The thermionic valve and the multi-stage amplifier were but two of the essential requirements whose development had to be largely perfected before any real progress could be made.

By the mid-twenties, however, it seemed to a number of individuals that a stage had been reached when serious attempts could be made to devise a workable system. Contributions were made by Mihály of Hungary, Jenkins of the United States, Baird of England, and by Ives and others in the Bell Telephone Laboratories.

(2.1) Dénes von Mihály

Mihály's initial experiments were confined to systems employing oscillating mirrors and were of no practical utility. Although he subsequently developed a mechanical-optical system of some elegance, it failed to meet the competition of the cathode-ray-tube receiver and will not be discussed further. A description of his

* He died in 1930.

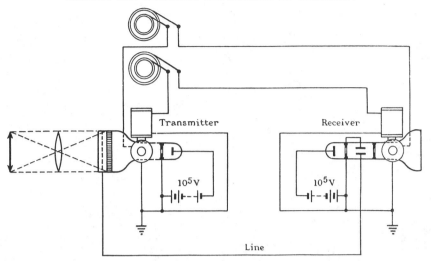

Fig. 8.—Campbell Swinton's Scheme, 1911. In his Presidential Address to the Röntgen Society, Campbell Swinton outlined the basic system upon which high-definition television was later developed. Note the use of cathode-ray tubes at transmitter and receiver. (Reproduced from the *Journal of the Röntgen Society*.)

early work together with a useful historical survey is contained in a book[43] he published in 1926.

(2.2) C. F. Jenkins

One of the earliest demonstrations of practical television was that given by Jenkins of the United States in June, 1925, his demonstration consisting of the transmission and reception by wireless of the outline of a slowly-revolving model windmill.[19] Jenkins was typical of the American inventor—homely, energetic and resourceful, but not highly scientific. He was a pioneer inventor in the field of motion pictures and was imbued with the idea of speeding-up electrical picture transmission to take motion pictures into the home by radio broadcasting, i.e. television. His scanning system employed a pair of bevel-edged glass discs, the angle of bevel changing continuously around the circumference of each disc (see Fig. 9). The bevelled edges formed prisms which deflected the beam of light as the discs rotated, and by rotating one disc many times faster than the other, the entire surface of the image could be scanned. On account of the limited speed at which the discs could be revolved, this system of scanning could be used only in a low-definition system. It was, moreover, limited to the use of transmitted, as distinct from reflected, light; thus Jenkins could transmit motion pictures faithfully, but only the outline—a shadowgraph—of a real object. In 1928, contemplating commercial exploitation, he undertook the experimental broadcasting of motion-picture film.

(2.3) J. L. Baird

The name best known among the pioneers of television in the United Kingdom is unquestionably that of John Logie Baird, who was born in 1888 at Helensburgh, Scotland, where his father was Minister of the West Parish Church. He attended classes at the Royal Technical College and at Glasgow University, but as a young man he did not enjoy good health. He was attracted to experimental work, and in 1923 decided to devote his whole time to the development of television.

In April, 1925, Baird gave a public demonstration[20] at which crude images were transmitted between two machines; a photograph of the original scanning apparatus is shown in Fig. 10.

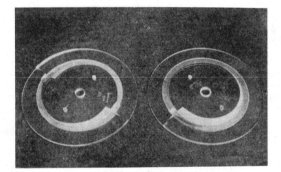

Fig. 9.—Jenkins's glass prismatic discs: used by Jenkins in June, 1925, in an early public demonstration.

Fig. 10.—The original scanning apparatus used by Baird in April, 1925, when giving public demonstrations at a London store.

Nine months later, in January, 1926, he gave a demonstration to the members of the Royal Institution, at which the images of moving human faces were shown, not as mere black-and-white outlines but having tone gradations of light and shade. At this demonstration, Baird scanned the brightly-illuminated subject by means of a lensed Nipkow disc giving a definition of 30 lines with a repetition frequency of 5 frames/sec. A gas-filled potassium photocell was used, the varying currents in which were employed to modulate a carrier wave. The picture was reconstituted at the receiver by interposing a synchronously-driven scanning disc between the eye of the observer and a neon lamp, the glow from which was made to vary in intensity according to the variations of the current in the original photocell. The image received measured barely $1 \cdot 5$ in $\times$ 2 in; it was of a dull pinkish colour, and, as was to be expected, the flicker was very pronounced. These were crude results but it was just possible to recognize individual faces, an achievement not previously attained.

Improvements were made gradually, and Baird gave frequent public demonstrations during the succeeding years. By so doing, he maintained a measure of public interest in his work and he obtained the financial support which his experiments necessitated. In May, 1927, for example, he demonstrated the transmission of low-definition images by telephone lines between Glasgow and London. In February of the following year, he transmitted similar low-definition images between London and New York, and to the *Berengaria* in mid-Atlantic. Crude colour-television and stereoscopic television were shown to the British Association at Glasgow in August, 1928, and a demonstration of large-screen television (by means of a multi-cellular-lamp screen containing 2 100 flash-lamp bulbs) was given as part of a music-hall programme at the Coliseum in London for two weeks in July, 1930.

The first stage in the emergence of television from the laboratory may be said to date from 1929, when arrangements were made between the B.B.C. and the Baird Television Co. for the regular experimental transmissions of television pictures from the London station. The transmissions were described as "experimental," but they took place for half-hour periods on five days of each week, and had a definition of 30 lines and a repetition frequency of $12 \cdot 5$ frames/sec; a receiver for these transmissions placed on the market at this time is shown in Fig. 11. No sound was broadcast to accompany the pictures until April,

1930. These early transmissions were conducted solely by the Baird Television Co., the B.B.C. having no share in the technical aspects of the television equipment or in the programme material. Nevertheless, the B.B.C. closely watched the progress being made, and, in August, 1932, they took over the responsibility for originating the transmissions in order to gain more direct experience of the problems entailed in the production of television programmes.

There can be no doubt that Baird made a notable contribution towards the development of television in the United Kingdom. He must be remembered, however, more for the public interest he created and for the way in which, indirectly, he caused others to undertake serious research, than for his own technical achievements. He was an experimenter rather than a scientist, but his enthusiasm and persistence in the development of television, at a time when its practical achievement seemed far distant, acted as a stimulus to others to commence research in this field. Thus Baird must be credited with being one of those, like Jenkins in America, whose contribution to the development of television consisted mainly in his enthusiasm, enterprise and publicity, rather than in fundamental invention or highly scientific achievement.

(2.4) Herbert Ives and others of the Bell Telephone Laboratories

Any review of the progress of television development during the "low-definition" era would be incomplete unless the detailed work on television which commenced some time in 1926 at the Bell Telephone Laboratories were mentioned. In April, 1927, the Laboratories gave an experimental demonstration of television, both by wire and by radio, at which pictures composed of 50 lines at a repetition frequency of 18 frames/sec were shown. The wire demonstration consisted of the transmission of images from Washington, D.C., to New York, a distance exceeding 250 miles. In the radio demonstration, images were transmitted from an experimental station at Whippany, New Jersey, to New York City, a distance of 22 miles. The subject was scanned by a rapidly-moving spot of light in a manner very similar to that which had also been used by Baird. The synchronizing signals were separately transmitted, and two forms of receiver were used, one giving a small image about 2 in $\times$ 2·5 in, suitable for viewing by a single person, and the other a much larger image approximately 2 ft $\times$ 2·5 ft, by means of a special multi-element neon lamp, suitable for an audience of considerable size. The smaller form of apparatus was primarily intended as an adjunct to the telephone, and it enabled individuals in New York to see their friends whom they telephoned in Washington. The larger form of receiving apparatus was designed to serve as an adjunct to a public-address system. Images of speakers in Washington, and of singers and other entertainers at Whippany, were seen by its use, simultaneously with the reproduction of their voices. A comprehensive description of this work was published by Herbert Ives and others,[21] and it is of some interest to note that, whilst we now regard the transmission of the d.c.-modulation component as an essential requirement, Ives claimed that "the transmission only of the alternating-current component of the image" was one of the chief features to which success was due.

(3) THE DEVELOPMENT PHASE OF HIGH-DEFINITION SYSTEMS

Although the regular broadcasting of low-definition television from the London transmitter of the B.B.C. was to continue almost without interruption from 1929 until September, 1935, this period is of more importance on account of the rapid development of high-definition systems, which took place during these years, than for any consolidation of the 30-line system. As early as

Fig. 11.—Baird Televisor: the first television receiver to be made commercially and offered for sale to the public. It was first offered in February, 1930, at a price of 25 guineas. It worked on a 30-line system and employed a perforated Nipkow disc with a neon lamp.

1930, it was in fact well appreciated by those closely concerned that, to the general public, television would prove acceptable only if pictures having a standard of definition far higher than 30 lines could be provided. Until high-definition systems actually became available, however, it seemed to be desirable to continue the broadcasting of the 30-line transmissions despite the fact that they were valueless as a source of entertainment.

In reviewing the progress of television development between 1930 and 1935, it is perhaps important to remember that the cathode-ray tube was itself not in an advanced state of development in 1930, at least so far as its application towards television was concerned. The art of focusing the beam to a small spot by magnetic or electrostatic lenses had not been developed and only gas-focused tubes were available. In these tubes it was difficult if not impossible to preserve a good focus with varying values of beam intensity, especially at high scanning speeds, and they possessed several other characteristics which rendered their use a matter of some difficulty. From about 1931, however, development commenced of the electrostatically focused high-vacuum-type cathode-ray tube. Experimental tubes of this type were soon manufactured in small numbers in England, and they were quickly found to be a great improvement over the gas-focused tube.

(3.1) Velocity Modulation

Among the principles which offered some promise of providing higher definition during the earlier part of this development phase was that known as "velocity modulation." As mentioned earlier in the paper, the principle had been somewhat vaguely described by Rosing in 1911, revived by Thun in 1929 and demonstrated by von Ardenne in 1931. Basically, it depends upon the idea of varying the light intensity at the receiver screen by varying the velocity of the scanning spot instead of its actual intensity. Late in 1931, Bedford and Puckle commenced a serious study of television in the course of which they considered the relative advantages of systems based upon intensity and velocity modulation, being at that time unaware of Rosing's publication on the second method. In an intensity-modulation system, the changes in picture brightness are brought about by modulating the intensity of the beam in the receiving cathode-ray tube, and the size of the fluorescent spot and its focus must not vary materially as the beam intensity is varied. These requirements were not fulfilled by the gas-focused tube, in which a good focus could be obtained only for one value of the beam intensity, and since gas-focused tubes were the only ones available at the time, Bedford and Puckle developed a velocity-modulation system using a standard of 120 lines with a repetition frequency of 25 frames/sec. Considerable success was achieved, and a description of their system was published[22] in 1934. It was later found that adequate contrast-ratio could be obtained only by the addition of a degree of intensity modulation. Since it was clearly impossible to use velocity modulation at the receiver if the original subject was being scanned at a uniform rate, the scanning spot at the transmitter had also to be velocity-modulated, and this requirement demanded, in turn, that the scanning element at the transmitter should also be a beam of cathode rays. A cathode-ray tube was in fact used as the source of light at the transmitter, but this limited the subject-matter to film material, a restriction not then considered to be of much significance.

By 1930, it had not only become evident that definition standards far higher than 30 lines would be required, but it had also become clear that the development of systems giving higher definition was a task beyond the capacity of independent investigators. Considerable laboratory and financial resources were needed to tackle the many different problems involved. Apart from the Cossor Company, for which Bedford and Puckle were

developing the velocity-modulation system just described, three other companies were interested in the development of complete systems of high-definition television, namely the Baird Television Co., the group which later became known as Electric and Musical Industries, Ltd., and Scophony, Ltd.

(3.2) Baird Television

It has already been mentioned that the Baird Company had arranged for the regular experimental transmission of 30-line television programmes from the London stations of the B.B.C. between 1929 and 1932, and, preoccupied with this task, they did not begin work seriously on higher definition until 1933. From their early days, they had concentrated on mechanical methods using Nipkow discs and mirror drums in the scanning processes. Baird had little faith in electronic devices and favoured mechanical solutions to the problems of television optics; he did not seriously encourage his staff to work on electronic methods. When the limitations of mechanical methods eventually became apparent, it was too late to retrieve the lead which had by that time been lost.

During the broadcasting of the low-definition programmes, the studio scenes and actors were scanned by a spot-light projector in which a revolving mirror-drum caused an intense spot of light to scan the subject in a series of 30 vertical lines. Light reflected from the subject in an otherwise darkened studio was picked up by banks of photocells, and the output from these constituted the initial signal currents. It was recognized that, whilst such methods might be acceptable for the televising of individuals, head-and-shoulder close-ups and announcers, the inconvenience and restrictions imposed by the need for a darkened studio would render the scheme unacceptable as a permanent arrangement for large studio scenes. Nevertheless, the system of flying-spot scanning was developed for interviews and close-ups, and when the Alexandra Palace station was opened in November, 1936, a scanner of this type was used for a time on a standard of 240 lines.

It was also recognized at an early stage that cinema films would often be required as the subject-matter for television transmissions. The high-definition scanning of an object as small as the frame of a 35-mm film by mechanical methods called for the manufacture of high-precision equipment, but this was satisfactorily developed, and by 1936 a flying-spot film scanner which worked on a standard of 240 lines was being used. Within the limitations imposed by the 240-line standard at 25 frames/sec, this equipment gave very satisfactory pictures when the Alexandra Palace station was opened.

In order to permit the televising of large studio entertainments, it was decided to develop a film process to form an intermediate link between the studio scene and the scanning unit. In this process, the scene and players were photographed as by an ordinary film camera, but after exposure and the simultaneous recording of the accompanying sound, the film was immediately developed and fixed in tanks which were situated beneath the camera. The resulting negative was then passed into the scanning unit, which also formed a part of the whole camera assembly. In practice, there was a time-delay, owing to development and fixing, of the order of 60 sec and this, combined with the inconvenience and inflexibility of the whole installation, placed it at a great disadvantage in comparison with instantaneous methods of television transmission.

As an alternative to the use of the intermediate-film process, the limitations of which were fully recognized, the Baird Company participated in the development of an electronic type of television camera under licence from the Farnsworth Television Laboratories of Philadelphia, by whom the original development had been carried out. In this type of camera, which is known as an "image-dissector tube,"[23] an image of the scene to be transmitted is focused on a large and uniform photo-electric cathode of high sensitivity. The electrons liberated from the cathode surface produce an "electron image" near the surface of the cathode, which corresponds in electron density to the optical image. The electron image is drawn forwards by appropriate fields and scanned over a tiny aperture, and the initially feeble electron currents are multiplied by a secondary-emission multiplier to give the output current from the camera. There were many difficulties in the development of this type of camera, which, unlike the mosaic type, does not employ the principle of charge storage. Although it was used for a brief period at Alexandra Palace at the end of 1936, it had not then been fully developed, and it was not found satisfactory for regular service.

(3.3) Electric and Musical Industries, Ltd.

Serious study of the problems of television was begun in 1930 by the Gramophone Company, where a team was organized under the leadership of G. E. Condliffe and C. O. Browne. They gave a successful demonstration of a mechanical system at the Physical and Optical Society's Exhibition at Imperial College in January, 1931. This was a 150-line system using five channels, each dealing with 30 lines; the picture was scanned $12\frac{1}{2}$ times a second and was demonstrated and described in some detail by Browne.[24] Transmissions were from film, and the receiver, using Kerr cells and a mirror drum, gave a picture measuring about 2 ft $\times$ 2 ft 6 in, suitable for a seated audience of about 50 people.

After the formation of Electric and Musical Industries, Ltd., by the merger of the Gramophone Co., Columbia Graphophone Co., and others, in mid-1931, the responsibility for guiding further development fell to I. Shoenberg, who had been appointed Director of Research, and the team was augmented by several ex-Columbia engineers, notably the late A. D. Blumlein and P. W. Willans. Attention was concentrated on receivers using cathode-ray tubes, mechanical scanners being retained for test transmissions since cathode-ray transmitting tubes were as yet only in an early stage of development. Mechanical scanning was raised to a high state of development, transmissions from film progressing up to 180 lines. At this time the E.M.I. engineers were working with straight scanning at 25 frames/sec. With a poorly illuminated picture at the receiver the 25-c/s flicker was not very objectionable, but as receiving tubes became brighter it was obvious that this flicker would be unacceptable in practice. The interlaced system of scanning was therefore adopted, and a 180-line interlaced Nipkow-disc studio flying-spot scanner was operated successfully in June, 1934. By the end of 1934, a 243-line picture at 50 frames/sec with interlaced scanning to give 25 pictures/sec had been achieved.

By the end of 1932, however, the limitations of mechanical systems were appreciated by the E.M.I. engineers and work was commenced on alternative methods of forming the initial vision signals. As has been noted previously in the paper, the earliest suggestion that a cathode-ray tube might be used for initiating television signals was made by Campbell Swinton in 1908, and he developed the idea in considerable detail in 1911. The two essential features of a modern television camera are the scanning of a photo-electric mosaic by a beam of cathode rays and the principle of charge storage by the elements of the mosaic. Campbell Swinton referred explicitly only to the former feature, but it seems quite clear from the context of his paper[18] that he realized that his scheme embodied the principle of charge storage, and it is significant that this fundamental invention was never claimed subsequently in the form of a patent. The first published description of a successfully operated television transmitting tube employing both these principles was by Zworykin[25]

in 1933. His tube, which he named an "iconoscope," represented a great advance on the earlier mechanical methods.

(3.4) V. K. Zworykin

Early in 1925 Zworykin began to develop electronic methods of scanning at the Research Laboratories of the Westinghouse Electric and Manufacturing Co., East Pittsburgh, and proposed various solutions of the problem. In 1930, the work was transferred to the laboratories of the R.C.A. Victor Company at Camden. From the outset, Zworykin seems to have realized the limitations of mechanical scanning and the need for using the principle of charge storage in a photo-electric mosaic. He was probably unaware of Campbell Swinton's proposals, which had been published originally in a somewhat obscure and little-known journal. Nevertheless, it is of some interest to compare Fig. 12, taken from Zworykin's United States Patent No. 1691324 of 1928, with Campbell Swinton's scheme shown in Fig. 8. Zworykin's scheme is even more remarkable when it is noted that his patent was applied for in July, 1925. Clearly, both Campbell Swinton and Zworykin were thinking along similar lines, although the ideas of the latter show, not unnaturally at this later date, a keener perception of the requirements, and there can be little doubt that he was far in advance of his contemporaries in this field. In a paper published in 1933[25] Zworykin described the development of the iconoscope, and Fig. 13 is taken from that paper. The sensitivity of the device at that date was stated to be "approximately equal to that of a photographic film operating at the speed of a motion picture camera, with the same optical system," and the author goes on to state that some of the tubes actually constructed were satisfactory up to 500 lines, with a wide margin for future improvement. A photograph of an iconoscope made in November, 1931, is shown in Fig. 14. Although only 5% to 10% of the possible increase in sensitivity due to charge storage was achieved in the tube, it was nevertheless sufficiently sensitive to enable high-definition pictures of studio and outdoor scenes to be televised.

In 1932, experiments were begun in the E.M.I. Research Laboratories[26, 27] by McGee and Tedham on the charge-storage type of transmitting tube. During the course of these experiments, many improvements were made in the technique of producing uniform and highly sensitive mosaics having a suitable colour response. The resulting tube, which had a sensitivity somewhat greater than that of the iconoscope, was named the Emitron, a photograph of which is reproduced in Fig. 15.

(3.5) Scophony, Ltd.

Mention has already been made of the development of television transmitting tubes which employ both cathode-ray scanning and the principle of charge storage. Whilst charge storage has made possible remarkable achievements on the transmission side, there has been no corresponding development in reception, and, in this field, the trend towards larger screens has been met by an increase in the size of cathode-ray tubes and an increase in their intrinsic brightness.[28] However, some believed that optical-mechanical methods were more likely to be successful in producing pictures approaching those of the home cinema in size, brightness and definition, and early in 1932, Scophony, Ltd., was formed to develop such methods. In the Scophony system,[29] the difficulties arising from mechanical scanning in a high-definition system were minimized by using the optical principle, which became known as "split focus." Increased brightness of the picture was secured by modulating a normal light source by means of a "supersonic light control," which provided a means not only of storing the picture signals, but also of projecting a series of picture elements simultaneously. Fig. 16 illustrates the use of split focus and supersonic light control.

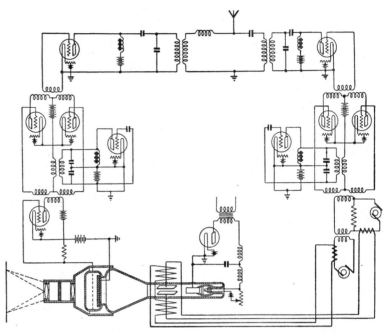

Fig. 12.—Drawing from Zworykin's Patent No. 1691324, 1928, the application for which was filed in July, 1925.

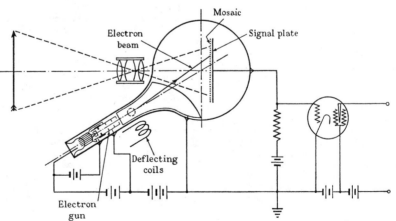

Fig. 13.—Diagram from Zworykin's paper.[31]

Fig. 14.—An early iconoscope—made in Zworykin's laboratory on the 9th November, 1931, and successfully tested by him on the following day.

In the split focus system, invented by G. W. Walton, focusing of the scanning beam was carried out in two separate planes by the use of cylindrical lenses. This gave great freedom of optical design and enabled the dimensions of the rotating mirror scanners to be greatly reduced. It thereby became possible to drive and synchronize such scanners at speeds up to 30 000 r.p.m.

The basic principle underlying the supersonic light control, invented by J. H. Jeffree, is the diffraction of light by compression waves generated in a liquid. The light control consists of·a glass-sided cell, filled with a transparent liquid, with a quartz crystal at one end. By means of a system of scanners and lenses, an image of the illuminated cell is projected onto·the viewing screen. By driving the crystal with an oscillator working at about 10 Mc/s, waves are generated in the liquid which cause it to behave as a diffraction grating, and, by modulating the amplitude of these waves by the vision signals, the amount of light diffracted at any point as it travels along the cell can be made proportional to the corresponding vision signal. Only diffracted light is permitted to pass to the screen, and the high-speed scanner is so arranged that it "follows" the supersonic waves as they travel along the cell, immobilizing an image of the train of waves as a part-line of light on the screen. In this way, the light appropriate to any point being scanned remains in its correct place on the screen during the whole time (approximately 100 microsec) which the wave takes to travel along the cell, thereby greatly increasing the brilliance of the picture. In apparatus designed for the 405-line system, approximately 200 picture elements were projected simultaneously onto the screen, giving a 200-fold increase in the amount of light at the screen.

Demonstrations were given to the press by the Scophony Company in 1936, and in June, 1937, public demonstrations were given at the Science Museum at which projected pictures measuring 5 ft × 4 ft were shown.

Fig. 15.—An Emitron camera as produced by Electric and Musical Industries, Ltd., 1937–39.

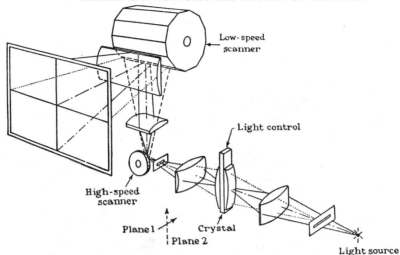

Fig. 16.—Optical schematic of Scophony receiver showing use of split focus and light control.

These pictures used a 240-line standard with 25 frames/sec, but in the following year demonstrations of large-screen television (15 ft × 12 ft) were given at two London cinemas using the B.B.C. standard of 405 lines with interlaced scanning.

(4) PROVISION OF A PUBLIC TELEVISION SERVICE FOR GREAT BRITAIN

With so many of the basic ideas of television in an advanced stage of practical development, the Postmaster General appointed, in May, 1934, a Television Committee under the chairmanship of Lord Selsdon, whose terms of reference were:

To consider the development of Television and to advise the Postmaster General on the relative merits of the several systems and on the conditions under which any public service of Television should be provided.

The Committee examined 38 witnesses representing many different interests, any interested society, firm or individual being at liberty to give evidence. They inspected the various television systems belonging to firms who were prepared to provide demonstrations, and concluded that of the systems under development in the United Kingdom the most distinctive were those of the Baird, Cossor, Marconi-E.M.I. and Scophony Companies. A delegation of the Committee visited the United States and inspected many of the chief centres of television experimental research, as well as the plant and laboratories of the principal broadcasting, telephone and telegraph authorities; they also consulted the Federal Communications Commission in Washington.

Whilst it is not possible for the authors to give a detailed and chronological review of the development of television in the United States, technical progress was comparable in America with that in England at this date.[30] Television stations were licensed only on an experimental basis, however, which did not permit the programmes to be made suitable for commercial exploitation. Consequently there were few regular programmes and very few had a regular viewing audience.

Another delegation from the Selsdon Committee visited Germany and made similar inspections of the television experimental installations belonging to the Reichspost and several private firms. Experimental transmissions were taking place on two standards, one of 30 lines and 12·5 pictures/sec, and the other of 180 lines and 25 pictures/sec; also, the interlaced system of scanning was under consideration.[33] Mechanical forms of scanning were being employed at the transmitting end, e.g. the flying spot for studio productions, and the intermediate-film system was being developed. Receivers with cathode-ray tubes were mainly used, but some with mirror drums were also produced. The general public did not participate in the reception of the transmissions.

The Committee presented their report[34] in January, 1935, and the following extracts from it are of particular interest:

51. The task of choosing a television system for a public service in this country is one of great difficulty. The system of transmission governs in a varying degree the type of set required for reception; and it is obviously desirable to guard against any monopolistic control of the manufacture of receiving sets. Further, whatever system or systems are adopted at the outset, it is imperative that nothing should be done to stifle progress or to prevent the adoption of future improvements from whatever source they may come. Moreover, the present patent position is difficult: the number of patents relating to Television is very large, and in regard to many of them there are conflicting views as to their importance and validity.

52. At the same time it is clear from the evidence put before us that those inventors and concerns, who have in the past devoted so much time and money to research and experiment in the development of Television, are looking—quite fairly—to recoup themselves and to gather the fruits of their labours by deriving revenue from the sale of receiving apparatus to the public, whether in sets or in parts, and whether by way of royalties paid by the manufacturers or by manufacturing themselves. It is right that this should be so, and that the growth of a new and important branch of industry, capable of providing employment for a large number of workers, should in every way be fostered and encouraged to develop freely and fully.

55. We have come to the conclusion that a start could best be made with a service of high definition Television by the establishment of such a service in London. It seems probable that the London area can be covered by one transmitting station and that two systems of Television can be operated from that station. On this assumption we suggest that a start be made in such a manner as to provide an extended trial of two systems, under strictly comparable conditions, by installing them side by side at a station in London where they should be used alternately—and not simultaneously—for a public service.

56. There are two systems of high definition Television—owned by Baird Television Limited and Marconi-E.M.I. Television Company Limited respectively—which are in a relatively advanced stage of development, and have indeed been operated experimentally over wireless channels for some time past with satisfactory results. We recommend that the Baird Company be given an opportunity

to supply the necessary apparatus for the operation of its system at the London station, and that the Marconi-E.M.I. Company be given a similar opportunity in respect of apparatus for the operation of its system also at that station. Besides any other conditions imposed, acceptance of offers should be subject in each case to the following conditions precedent:

* * *

(f) Transmissions from both sets of apparatus should be capable of reception by the same type of receiver without complicated or expensive adjustment.

(g) The definition should not be inferior to a standard of 240 lines and 25 pictures per second.

Because of the close relationship which must exist between sound and vision broadcasting, the Television Committee also recommended that the B.B.C. should be entrusted with the inauguration of the television service under the guidance of an advisory committee, on which the B.B.C., the Department of Scientific and Industrial Research and the Post Office were to be represented; this recommendation was accepted by the Government. The particular form of financing of broadcasting evolved in the United Kingdom may well have been the deciding factor in enabling it, in 1937, to be the first country in the world to inaugurate a regular daily service of high-definition television programmes.

By the time that the Selsdon Committee reported, early in 1935, the Emitron camera had been developed to a point which enabled E.M.I. to offer a standard of definition considerably higher than the minimum set by the Committee. With considerable courage, Shoenberg decided to offer a standard which seemed to many at the time to be beyond practical attainment, namely 405 lines with 50 frames/sec, alternate frames being interlaced to give 25 pictures/sec. Demonstrations were given using this standard, an Emitron type of camera being used in conjunction with a radio transmitter developed by Marconi's Wireless Telegraph Co. The transmitter had a peak power of 12 kW and was capable of being modulated by signals having frequencies up to some 2·8 Mc/s, although the modulator was still in need of further specialized development. Thus, when the B.B.C. ordered equipment from Marconi-E.M.I. Television, Ltd., the equipment used this standard, whilst that ordered from Baird Television Company used 240 lines and 25 frames/sec. sequentially scanned. Both the Marconi-E.M.I. and the Baird systems could be used on a single receiver by means of a simple switching operation without unduly complicating or increasing the cost of receivers. It had been hoped that a single radio-vision transmitter might be constructed which would be suitable for both systems, but this was not found practicable, as the characteristics required by the two systems were so diverse as not to permit of the use of common apparatus other than the aerial and high-frequency feeder line; the peak power of both vision transmitters was 17 kW and the carrier frequency 45 Mc/s.

On the 2nd November, 1936, the Alexandra Palace television station[35] was opened, the Marconi-E.M.I.[36] and the Baird systems being used alternately, each for a week at a time. These alternate transmissions, for two hours each day, continued until the beginning of February, 1937, when, on the recommendation of the Television Committee, the Marconi-E.M.I. system was adopted for the permanent service. The decision[37] in favour of the Marconi-E.M.I. electronic system was influenced by many factors, the more important of which were: that the quality of the picture was considerably superior because of the higher definition; that the use of 50 frames/sec (interlaced) gave much less flicker than the Baird system with its 25 frames/sec; and that the electronic camera seemed to offer the greatest scope for development, the use of the intermediate film being costly and unduly restrictive in studio productions.

Among the more important technical features of the system, the following should be noted:

(a) The transmission of the d.c. component[38] of television signals, now regarded as *sine qua non*, but not always so. In fact, at the time when it was being demonstrated and recommended to the Television Committee, a.c. transmission was used in other countries, including the United States.[44, 45] In the transmission of d.c. modulation, definite carrier values represent black, white and the synchronizing signal; there is no fixed mean carrier value, as this depends on picture brightness. The system of transmission is analogous to telegraphy rather than telephony. The establishment of a "black" datum naturally led to the problem of stabilizing this at all times, whether the transmitter was being modulated at very low or very high frequencies. At very low frequencies, the intrinsic regulation of the anode supply operates, whilst at the higher frequencies the characteristics of the smoothing-circuit impedances predominate to determine the instantaneous value of the voltage. Suitable circuits[39] were developed by Blumlein, who, until his untimely death during the 1939–45 War, was one of the foremost "circuit" engineers in the country.

(b) Positive modulation, the tips of the synchronizing pulses being of zero carrier amplitude.

(c) Line- and frame-synchronizing pulses both of equal amplitude; line synchronization employing sharp front edges of "slots" in frame pulse. Up to 1933, various schemes were tried for separating line- and frame-synchronizing pulses easily from each other in the receiver, whilst allowing line synchronization to continue satisfactorily in the presence of the frame pulse. The first simple and reliable solution[40] was the use of a series of frame pulses so located as to have sharp leading edges where the line pulses would have been in the absence of the frame pulse. This was later adopted in the United States, where more complicated arrangements were in use until 1935.

(d) Provision in the standard waveform of a datum portion at or near picture black immediately following the synchronizing pulses.[41] When effecting d.c. re-establishment, it is more important to fix the black level correctly than any other point in the range. Thus, instead of occupying the whole of the scan-return interval with a synchronizing pulse, this pulse was shortened so as to provide a short period of black before picture modulation recommenced.

(5) CONCLUSION

In this survey of the historical development of television, the authors have attempted to review those aspects which are particularly related to the formation of the vision signals; they have refrained from making any survey of such features as are common to radio practice, and they have refrained also from any discussion of such problems as frequency allocation, with which the broadcasting of television must necessarily be involved.

Many scientists and engineers have contributed to the practice of modern television. No single country can claim a monopoly of the inventive effort, and it has been the authors' endeavour to survey the field in true perspective and to give a balanced assessment of the major contributions. It is perhaps pardonable to take some small pride in the fact that this country was the first in the world to establish a regular system of high-definition television broadcasting.

(6) BIBLIOGRAPHY

(1) BECQUEREL: *Comptes Rendus*, 1839, **9**, p. 145.

(2) WILLOUGHBY SMITH: *Journal of the Society of Telegraph Engineers*, 1873, **2**, p. 31.

(3) WILLOUGHBY SMITH: *ibid.*, 1876, **5**, p. 183.

(4) SENLACQ: *English Mechanic*, 1878, **28**, p. 509.

(5) SENLACQ: *ibid.*, 1881, **32**, p. 534.

(6) AYRTON and PERRY: *Nature*, 1879–80, **21**, p. 589.

(7) CAREY: *Design and Work*, 1880, **8**, p. 569.

(8) SHELFORD BIDWELL: *Nature*, 1881, **21**, p. 344.

(9) SHELFORD BIDWELL: *English Mechanic*, 1881, **33**, pp. 158 and 180.

(10) SHELFORD BIDWELL: Report of the British Association for 1881, p. 777.

(11) NIPKOW, P.: British Patent No. 30105, 1884.

(12) WEILLER, L.: *Gén. Civ.*, 1889, **15**, p. 570.

(13) SZCZEPANIK: British Patent No. 5031, 1897.

(14) ROSING: British Patent No. 27570, 1907.

(15) ROSING: British Patent No. 5486, 1911.
See also: *Scientific American*, 1911, **105**, p. 574.

(16) SHELFORD BIDWELL: *Nature*, 1908, **78**, p. 105.

(17) CAMPBELL SWINTON: *ibid.*, 1908, **78**, p. 151.

(18) CAMPBELL SWINTON: *Journal of the Röntgen Society*, 1912, **8**, p. 1.
See also: *Wireless World*, 1924, **14**, pp. 51–56, 82–84 and 114–118.

(19) "The Jenkins System," *Wireless World*, 1926, **18**, p. 642.

(20) *Nature*, 1925, **115**, p. 505.

(21) IVES, H.: "Television," *Bell System Technical Journal*, 1927, **6**, p. 551.
GRAY, HORTON and MATHES: "The Production and Utilization of Television Signals," *ibid.*, p. 560.
STOLLER and MORTON: "Synchronization of Television," *ibid.*, p. 604.
GANNETT and GREEN: "Wire Transmission for Television," *ibid.*, p. 616.
NELSON: "Radio Transmission for Television," *ibid.*, p. 633.

(22) BEDFORD, L. H., and PUCKLE, O. S.: "A Velocity-Modulation Television System," *Journal I.E.E.*, 1934, **75**, p. 63.

(23) DIECKMANN and HILL: German Patent No. 450187, 1925.
ROBERTS: British Patent No. 318331, 1928.
FARNSWORTH: British Patents Nos. 368309 and 368721, 1930.
FARNSWORTH: *Journal of the Franklin Institute*, 1934, **218**, p. 411.

(24) BROWNE, C. O.: "Multi-Channel Television," *Journal I.E.E.*, 1932, **70**, p. 340.

(25) ZWORYKIN, V. K.: "Television with Cathode-Ray Tubes," *ibid.*, 1933, **73**, p. 437.

(26) MCGEE, J. D., and LUBSZYNSKI, H. G.: "E.M.I. Cathode-Ray Television Transmission Tubes," *Journal I.E.E.*, 1939, **84**, p. 468.

(27) MCGEE, J. D.: "A Review of Some Television Pick-up Tubes," *Proceedings I.E.E.*, 1950, **97**, Part III, p. 377.

(28) LEVY, L., and WEST, D. W.: "Fluorescent Screens for Cathode-Ray Tubes for Television and Other Purposes," *Journal I.E.E.*, 1936, **79**, p. 11.

(29) JEFFREE, J. H.: "The Scophony Light Control," *Television and Short Wave World*, May 1936.
LEE, H. W.: "The Scophony Television Receiver," *Nature*, 1938, **142**, p. 59.

ROBINSON, D. M.: "The Supersonic Light Control and its Application to Television with special reference to the Scophony Television Receiver," *Proceedings of the Institute of Radio Engineers*, 1939, **27**, p. 483.
SIEGER, J.: "The Design and Development of Television Receivers using the Scophony Optical Scanning System," *ibid.*, p. 487.
WIKKENHAUSER, G.: "Synchronization of Scophony Television Receivers," *ibid.*, p. 492.
LEE, H. W.: "Some Factors involved in the Optical Design of a Modern Television Receiver using Moving Scanners," *ibid.*, p. 496.
British Patents Nos. 328286 and 451132 (split focus), 439236 (supersonic light control), 433945, 469426 and 469427 (beam conversion).

(30) FINK, D. G.: "Television Broadcasting Practice in America: 1927–1944," *Journal I.E.E.*, 1945, **92**, Part III, p. 145.

(31) ENGSTROM, E. W.: "An Experimental Television System," *Proceedings of the Institute of Radio Engineers*, 1933, **21**, p. 1652.
ZWORYKIN, V. K.: "Description of an Experimental Television System and the Kinescope," *ibid.*, 1933, **21**, p. 1655.

(32) ZWORYKIN, V. K.: "The Iconoscope," *ibid.*, 1934, **22**, p. 16.

(33) GIBAS: "Television in Germany," *ibid.*, 1936, **24**, p. 741.

(34) Report of the Television Committee (H.M. Stationery Office, 1935. Command 4793).

(35) MACNAMARA, T. C., and BIRKINSHAW, D. C.: "The London Television Service," *Journal I.E.E.*, 1938, **83**, p. 729.

(36) BLUMLEIN, A. D., BROWNE, C. O., DAVIS, N. E., and GREEN, E.: "The Marconi-E.M.I. Television System," *ibid.*, p. 758.

(37) ASHBRIDGE, N.: "The British Television Service," *Proceedings of the Joint Engineering Conference*, 1951, Part 11, p. 491.

(38) WILLANS, P. W.: British Patent No. 422824, April 1933.

(39) BLUMLEIN, A. D.: British Patent No. 421546.

(40) PERCIVAL, W. S., BROWNE, C. O., and WHITE, E. L. C.: British Patents Nos. 425220, August 1933, and 448031, August 1934.

(41) British Patents Nos. 449242, September 1934, and 458585, March 1935.

(42) CORK, E. C., and PAWSEY, J. L.: "Long Feeders for Transmitting Wide Side-Bands with Reference to the Alexandra Palace Aerial-Feeder System," *Journal I.E.E.*, 1939, **84**, p. 448.

(43) MIHÁLY, DÉNES VON: "Das elektrische Fernsehen (Television)," *Experimental Wireless and Wireless Engineer*, 1927, **4**, p. 239.

(44) LEWIS, H. M.: "Standards in Television," *Electronics*, July 1937, **10**, p. 10.

(45) LEWIS, H. M., and LOUGHREN, A. V.: "Television in Great Britain," *ibid.*, October 1937, **10**, p. 32.

DISCUSSION ON
"THE HISTORY OF TELEVISION"
AT THE CONVENTION ON THE BRITISH CONTRIBUTION TO TELEVISION, 28TH APRIL, 1952

Sir Archibald Gill: As one of the delegation which went to Germany in 1934, I found that the Germans had in operation what might be called a high-definition system using 180 lines sequentially scanned. It was transmitting programmes every day, but very few people had receivers—I think that in those days Germany was getting ready for war and sensed that television might have some use for military purposes. In any case I believe it was controlled by the minister for propaganda even in those days. A good deal of work had been done in Germany by a company formed by Baird, Telefunken, Zeiss and the Loewe company—each of which contributed to the activities in their own sphere. The pictures were very good and so far as I know they used the intermediate-film process. It is mentioned in the paper that films exposed by the camera immediately passed into a solution, were developed, and then scanned by a disc scanner whilst still wet. Although the paper does not refer to it, the Germans carried this process a stage further in which only a short loop of film was used—this film after exposure, development and scanning was stripped, re-emulsified, re-sensitized, dried and passed through the camera again. The German authorities did not use this process much because they found it more useful to keep the picture for future use.

It is extraordinary to find how early the methods employed in television were proposed and what a number of inventors took part in the development of television. Some of them, like Campbell Swinton, put forward schemes of a very far-seeing type which could not be brought to fruition at the time because the techniques involved were insufficiently advanced.

I saw a good deal of Baird during the critical period when he was developing his high-definition system. I believe that he was working under financial difficulties, that his scientific knowledge was inadequate for the tasks he had set himself, and, in addition, that he suffered from ill health. I think that the latter was borne out by his early death; but he was a very brave, persistent and enthusiastic man, and one could not fail to admire him.

Baird's main difficulty lay in his inability to produce an electronic camera, which was essential for studio work. I think this doomed his system from the start. When it came to film transmission my recollection is that the 240-line Baird system using a mechanical-disc scanner produced a slightly better picture than electronic scanning. In those days the electronic camera was in a rather early stage of development.

Reference has been made to the system produced by Bedford and Puckle, using a cathode-ray tube in which the beam intensity was constant and the velocity of the spot was varied, and it is stated that a description of this system was published in 1934. What is much more interesting, I think, is that it was actually demonstrated in the Institution lecture theatre and it was a very good demonstration. There was another demonstration here about that time by E.M.I. in which they made up a big picture with five separate equipments, all the sections being matched together.

Mr. A. G. Jensen: The authors have discussed the difficulties of assigning an inventor to such a thing as television, and it is certainly very difficult. Perhaps I might give you an illustration of how far one can go with regard to the question of attaching a name to a system or a device. We ran into some difficulty with colour television, and one of the devices used was a little disc which is called the "Maxwell disc." When we delved into history we found, however, that the disc was first thought of by a Greek astronomer named Ptolemy in the second century.

I was surprised to learn that Nipkow did not demonstrate his ideas; I always thought that he did. It is worth recalling that as late as the 1930's, some of the best pictures we made were due to Nipkow in one form or another. I believe that in England the G.E.C. built a fine transmitter at the time; the Dutch also built one, and the first transmission over the coaxial cable in the United States in 1937 was by a Nipkow transmitter.

With regard to electronic high-definition television, we in the United States look to Dr. Zworykin as the real father of our television. Without his iconoscope we feel that television might not be what it is to-day.

Dr. J. D. McGee: First, I should like to thank the authors for having pointed out that Willoughby Smith and May were not the first to discover the electrical effects of light, and that reference is made to Edmond Becquerel. Secondly, the authors state in Section 1.10 that Campbell Swinton never attempted to make equipment of the type he advocated, and the impression is given

in Section 3.4 that, after his Presidential Address to the Röntgen Society in 1911, the whole idea was buried in the rather obscure journal of a defunct society. In actual fact, Campbell Swinton took up the matter again* and continued to advocate the system quite energetically right through from 1920 until his death in 1930. In 1924 he published a modernized version† of his 1911 proposal and in 1926 he described some experiments which he had been doing.‡ The following are extracts from the latter.

> . . . I actually tried some not very successful experiments in the matter of getting an electrical effect from the combined action of light and cathode rays incident on a selenium coated surface. . . . The transmitting apparatus consisted of a home-made Braun oscillograph in which a metal plate coated with selenium was substituted for the usual fluorescent screen, the image to be transmitted being thrown by a lens upon the selenium surface and the end of the cathode-ray beam being caused electromagnetically to traverse the projected image.

Those experiments were repeated in 1937–38 by Mr. J. Strange and Dr. H. Miller,§ who obtained pictures which were of good general quality although they were troubled by lag, which is serious in that type of tube. However, selenium was one of the light-sensitive materials used and gave successful results. It is a pity that some of those tubes which Campbell Swinton used are not still in existence, as they would be worthy of a place in the Science Museum.

During this early period Campbell Swinton carried on an energetic and lively correspondence in the non-technical Press, debating the problem of mechanical versus electronic television, and it is most instructive to anyone interested in the history of television to look up *The Times* and read those letters from Campbell Swinton advocating the electronic system as opposed to the mechanical one.

I am very glad to see that the authors, after their very exhaustive study of this problem, conclude that the storage principle was implicit in Campbell Swinton's writings. I have always been of that opinion myself, but it was never explicitly stated and one can only read it between the lines.

Dr. R. C. G. Williams: A few months ago I happened to tune in to Radio Moscow while testing a short-wave receiver and was most intrigued to hear a story in English of the origin and development of television. I must say that it deviated somewhat from the history given in the paper, the main difference being that, according to the broadcast story, the invention was largely of Russian origin. As this alternative version has no doubt been fairly widely publicized both over the air and in print, I wonder whether it would be possible for the authors— should they come across any of this in their researches—to make some reference to its existence and assess its true worth.

Mr. I. Shoenberg (*communicated*): I think the authors, in their efforts to present an unbiased account of the history of television, have perhaps not done full justice to the pioneer work carried out in Great Britain.

Although it may now seem to those with present-day knowledge that the main features of the British television system developed by E.M.I. are so self-evident that there could not have been much difficulty in choosing them, the picture becomes very different if it is examined in the light of the knowledge which existed when the system was being worked out.

In deciding the basic features of our system we frequently had to make a choice between a comparatively easy path leading to a mediocre result and a more difficult one which, if successful, held the promise of better things. Perhaps a few examples may be of interest and serve to illustrate this point.

When we started our work in 1931, the mechanically scanned

* See *Nature*, 1936, **138**, p. 674.
† See *Wireless World*, 1924, **14**, pp. 51 and 118.
‡ See *Nature*, 1926, **118**, p. 590.
§ See *Proceedings of the Physical Society*, 1938, **50**, p. 374.

receiver was the only type available and was under intensive development. Believing that this development could never lead to a standard of definition which would be accepted for a satisfactory public service, we decided to turn our backs on the mechanical receiver and to put our effort into electronic scanning. The cathode-ray tube then available used gas focusing and it was not possible to modulate the beam without losing focus. This fact did not deter one competent British organization from attempting to make use of the tube as described in Section 3.1, but, remembering the vagaries and instability of the soft valves of the 1912–16 period, I decided against the soft cathode-ray tube and directed our research towards the development of a hard type with electron focusing. The efforts in this direction, particularly by Broadway and his group, turned what was thought a very speculative development in 1931 into the commonplace of to-day.

Another problem which arose from our efforts to provide pictures of higher definition was the amplification and radiation of the greatly extended range of frequencies required. On the radiation side this involved abandoning the medium waves altogether and accepting the greatly reduced service area of the short-wave transmitter. Many thought that this limitation of the service area would be unacceptable but we saw no alternative. The transmission of a wide range of sidebands on short waves raised many new problems, but perhaps even more serious were the problems which had to be faced on the video side when one attempted to amplify wideband signals up to a level suitable for modulation of the transmitter. The amplification of the a.c. components alone would not have been so difficult and, as the authors have pointed out, this was the practice in the United States at that time. We, however, thought it essential to transmit the d.c. component of the picture signals so that the receiver would accurately reproduce the brightness values of the original picture. A great controversy raged for some months among my staff as to whether we should use d.c. amplification, or a.c. amplification with the restoration of the d.c. component. In the early days P. W. Willans was the protagonist of the restoration method and later on Blumlein advocated it and worked out the various circuits which were necessary. It seems strange now that this matter should ever have been one for controversy, but at the time there were cogent arguments on both sides and in the end I made the decision in favour of restoration.

Another difficult decision had to be faced in connection with the generation of the picture signals. In our early work we used mechanical scanners to generate our signals, but the limitations of these scanners—particularly in regard to live pick-up—were only too obvious. McGee's group started work on the Emitron pick-up tube in 1932, but the signals obtained from the early Emitrons tended to become submerged in great waves of spurious signals associated with secondary-emission effects. There was a great temptation to continue with the mechanical scanner—as, indeed, another company elected to do. Instead, we decided that the potentialities of the electronic scanning tube justified a great effort to overcome the problems it presented at that time. Improvements to the tube made the spurious signals more manageable, but although Blumlein and McGee invented at about this time the ultimate solution to the problem—namely cathode-potential stabilization—time did not allow it to be worked out in practice and we had to leave it to the circuit people to deal with the unwanted-signal components. Great credit is due to Blumlein, Browne and White for the resourcefulness with which they devised tilt, bend and suppression circuits to combat this evil.

Once the signals had become usable, the way was open for a real high-definition system. The choice as to the number of lines was no longer limited by mechanical considerations but by the bandwidths which could be dealt with at the time in the transmitter and the receivers. Great differences of opinion existed in the laboratory, but finally, early in 1935, I took my courage in both hands and chose 405 lines. This may seem a low standard now, but at the time when I made the decision there were many who thought I had taken a great risk.

At the beginning of 1935, I was thus able to submit to the Television Committee a fully detailed specification for a high-definition system operating on 405 lines. This specification was subsequently adopted, without any important modification, in the Alexandra Palace transmitter.

I hope that these remarks will serve to put into better perspective the pioneer work done on television in Great Britain.

THE AUTHORS' REPLY TO THE ABOVE DISCUSSION

Messrs. G. R. M. Garratt and **A. H. Mumford** (*in reply*): We are grateful to Dr. McGee for his correction regarding Campbell Swinton and for calling attention to his advocacy of electronic methods during the 1920's. His paper presented to the Radio Society of Great Britain had been overlooked, and we agree that it is possible that Zworykin, whose first relevant Patent was applied for in July, 1925, may have derived some inspiration from Campbell Swinton's proposals. Nevertheless, this seems to detract little from the credit due to Zworykin for his work on the development of the iconoscope, a full account of which was given in his paper* in 1933.

We cordially agree with Mr. Jensen in his comments regarding "the difficulty of assigning an inventor to such a thing as television." However convenient it may be to assign the name of some individual to a technical development, the fact remains that when this happens an injustice is often done to others who have made substantial contributions.

The communication from Mr. Shoenberg should help to dispel once and for all any suggestion that the Marconi-E.M.I. system was not a purely British development. There were, in fact, several fundamental differences between the British and United States systems—the form, shape and proportions of the synchronizing

signals, the polarity of the modulation of the vision signals and the transmission of the d.c. component—and in all these important features, British practice differed from the American. It is to be regretted that Mr. Shoenberg himself has not hitherto received wider credit for his courageous direction in the development of British television. His naturally retiring disposition is undoubtedly responsible for this, and perhaps his contribution to the discussion will serve as a record of the fact that, on many fundamental points, his was the guiding hand in the formative years of British television.

We were interested to learn from Dr. Williams that the Russians claim television as a Russian invention. Boris Rosing may certainly be credited with having been the first to propose the use of a cathode-ray tube as the receiving element, but let it not be forgotten that almost 25 years were to pass before it had been developed to the point where it could be used practically for television reception. Moreover, the cathode-ray tube is only one small link in the chain of devices which go to make up a television system.

Rosing most certainly was not the inventor of television, and neither was any other single individual. Modern television owes its achievement to the work of many different scientists and engineers, and we hope that our paper and the discussion have enabled the contributions of some of them to be viewed in true perspective.

* ZWORYKIN, V. K.: "Television with Cathode-Ray Tubes," *Journal I.E.E.*, 1933, **73**, p. 437.

THE EVOLUTION OF MODERN TELEVISION

A[xel] G. Jensen

The Evolution of Modern Television

By A. G. JENSEN

This paper first describes the gradual transition from mechanical television to the present all-electronic television. It gives a brief account of the major developments leading to the present types of pickup and reproducing equipment, and describes the growth of network facilities which has made possible the present wide distribution of television programming in the United States.

Iɴ ᴛʜɪs ᴘᴀᴘᴇʀ modern television implies the type of television now used for commercial broadcasting in the United States and elsewhere. This is an all-electronic type of television which utilizes electronic pickup tubes of one form or another at the transmitting end and cathode-ray tubes for displaying the picture at the receiving end. Such a system is in contrast to earlier experimental television systems using mechanical-optical scanning means both at the transmitting end and at the receiving end. These earlier types of systems have been described by J. V. L. Hogan in a companion paper.[1]

It is obvious that the transition from the earlier mechanical systems to our present electronic systems was a very gradual one. Roughly speaking, we may say that the period before 1930 was that of the all-mechanical television system, the period from 1930 to 1940 was the period of the partially electronic system, and the period from 1940 on was that of the all-electronic television system. It was natural that the comparatively simple cathode-ray tube at the receiving end should be perfected sooner than the more complicated pickup tubes at the transmitting end, and, up to the middle thirties, the systems experimented with were mostly of a type that utilized mechanical-optical scanning means at the transmitting end but used a cathode-ray tube as the display device at the receiving end. From 1935 to 1940 the mechanical scanning means at the transmitting end were gradually superseded by steadily improved electronic camera tubes, but it is interesting to note that even as late as 1939 and 1940 some of the best television pictures were produced with transmitting equipment using Nipkow discs. This was the case here in the United States and also in England, Germany and Holland.[2] In an art which has developed as rapidly and in as revolutionary a way as television, it is rather extraordinary that this should be the case, considering that the Nipkow disc was invented sixty years earlier.

Presented on May 7, 1954, at the Society's Convention at Washington, D.C., by A. G. Jensen, Bell Telephone Laboratories, Inc., Murray Hill, N.J.
(This paper was received on September 15, 1954.)

Early Inventions

While the experimental development of electronic pickup and reproducing equipment did not really get started until around 1930, it is interesting to note that the conception of such devices is quite a bit older.

The earliest experiments with the use of a cathode-ray tube as a television receiver were probably those of Prof. Max Dieckmann in Germany. One is described in a German patent application dated September 12, 1906, called "A Method for the Transmission of Written Material and Line Drawings by Means of Cathode Ray Tubes."[3] A description of these early experiments and several pictures of the equipment can be found in a recent article in a German magazine giving the history of television in Germany up to 1945.[4] Figure 1 is reproduced from this magazine and shows Dieckmann's experimental setup, consisting of a cathode-ray tube receiver adjacent to a Nipkow-type scanner used as a transmitter. The cathode-ray tube used deflection modulation and magnetic deflection for both directions of scanning.

Due to a particular design of the Nipkow disc, the signals obtained from the transmitter were of a facsimile nature, indicating either black or white, but no half-tones. For this reason it has been argued that Dieckmann's experiments were not true television, but rather facsimile transmission. The fact remains, however, that the cathode-ray tube receiver by itself did not have this limitation and would have been able to reproduce half-tones as well as any other cathode-ray tube of those early days. The equipment shown in Fig. 1 is still in existence and is housed in the Deutsches Museum.

Another early description of a cathode-ray tube used as a television receiver is due to Prof. Boris Rosing, who was a lecturer at the Technological Institute at St. Petersburg in Russia. Rosing called his invention the "electric eye," and it is described in two British patents of 1907 and 1911[5] and also in a couple of articles in the *Scientific American* of 1911.[6] Figure 2 is a reproduction of a drawing in Rosing's 1907 patent. The transmitter images the object (3) in the plane of the photosensitive cell (5), and by means of two rotating mirror drums this image is moved vertically and horizontally across the photosensitive element, which in this manner scans the entire picture. The receiver is a cathode-ray tube with a fluorescent screen (12), a pair of magnetic deflecting coils (14 and 15), and a

Fig. 1. Earliest experimental model of German cathode-ray tube television receiver (Dieckmann, 1906).

pair of modulating plates (16). The saw-tooth deflecting current for the coils is obtained from rheostats mounted on the mirror drums and with contacts so arranged that the current from the rheostat varies linearly with the rotation o the drums. As the signal from the photosensitive element (5) on the transmitter is impressed on the modulating plates (16), it causes the electron beam to be deflected away from a hole in a diaphragm (13), thereby changing the strength of the beam impinging on the fluorescent screen. It must be remembered that cathode-ray tubes, or Braun tubes as they were called in those days, did not then have very well focused beams, and it is doubtful whether a very satisfactory picture could be obtained with tubes of that sort. But it is interesting to note that Rosing employed the principle of deflection modulation of the electron beam, a principle which was later utilized very successfully by Dr. C. J. Davisson in a cathode-ray tube built in the late 1930's. The patent descriptions of this system indicate that Rosing did some experimental work, but there is no record of any successful demonstrations.

A still more startling invention was made by A. A. Campbell Swinton in 1911. Swinton was an outstanding British consulting engineer, who was born in 1863 and died in 1930. He was mainly connected with the design and development of electric lighting and electric traction in England, and was also associated with Sir Charles Parsons in the development of the steam turbine. However, apart from his business, he was also keenly interested in the newer developments in physics and electronics, particularly in the fields of x-rays and television. He was a Fellow of the Royal Society, and made many experiments both in the field of x-rays and in the field of radio. Swinton's invention was triggered by a letter written by Shelford Bidwell to *Nature* in 1908.[7] Bidwell had been interested in experiments in television since 1880, and he wrote his letter to comment on an earlier letter by Armengaud of Paris, in which the author stated that he "firmly believes that within a year, as a consequence of the advance already made by his apparatus, we shall be watching one another across distances hundreds of miles apart."

Armengaud's television system was purely mechanical, and Bidwell takes issue with him with respect to the accuracy of synchronization obtainable by such a system. Bidwell assumes a picture 2 in. square and estimates that, to equal the definition of a good photograph, such a picture would require 160,000 elements. With 10 pictures per second this would mean 1.6 million synchronizing operations per second which, according to Bidwell, is "wildly impracticable."

Instead of such a system, Bidwell sug-

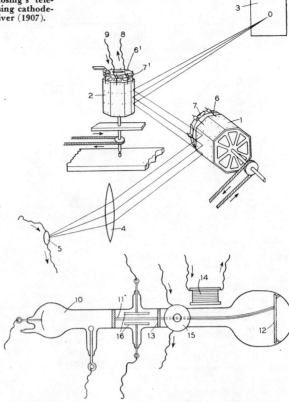

gests that one should consider a system based on the operation of the human eye, that is, a system with a large number of individual cells connected by separate wires. He estimates that for 90,000 picture elements one would require at the transmitting end a screen consisting of selenium cells and occupying a space 8 ft square. This screen would be illuminated by the object by means of a projection lens with a 3-ft aperture. He further estimates that the receiver would occupy a space of 4000 cu ft, although he does not specify what the actual receiving element would be. The 90,000-conductor cable is estimated to have a diameter of 8 to 10 in. and to cost about $1\frac{1}{4}$ million pounds for 100 mi. (It is interesting to note that this cost is at least an order of magnitude greater than the cost of present-day coaxial cables, and this is in spite of the fact that Bidwell's system would require less than one megacycle for its transmission, and that the pound of 1908 was worth a good deal more than it is today.)

In reply to Bidwell's letter, Swinton wrote as follows[8]:

"Distant Electric Vision

"Referring to Mr. Shelford Bidwell's illuminating communication on this subject published in NATURE of June 4, may I point out that though, as stated by Mr. Bidwell, it is wildly impracticable to effect even 160,000 synchronized operations per

second* by ordinary mechanical means, this part of the problem of obtaining distant electric vision can probably be solved by the employment of two beams of kathode rays (one at the transmitting and one at the receiving station) synchronously deflected by the varying fields of two electromagnets placed at right angles to one another and energised by two alternating electric currents of widely different frequencies, so that the moving extremities of the two beams are caused to sweep synchronously over the whole of the required surfaces within the one-tenth of a second necessary to take advantage of visual persistence.

"Indeed, so far as the receiving apparatus is concerned, the moving kathode beam has only to be arranged to impinge on a sufficiently sensitive fluorescent screen, and given suitable variations in its intensity, to obtain the desired result.

"The real difficulties lie in devising an efficient transmitter which, under the influence of light and shade, shall sufficiently vary the transmitted electric current so as to produce the necessary alterations in the intensity of the kathode beam of the receiver, and further in making this transmitter sufficiently rapid in its action to respond to the 160,000 variations per second that are necessary as a minimum.

"Possibly no electric phenomenon at present known will provide what is required in this respect, but should something

* Apparently Swinton lost track of the fact that Bidwell suggested 10 pictures per second, each with 160,000 elements. The figure therefore should be 1,600,000 operations per second.

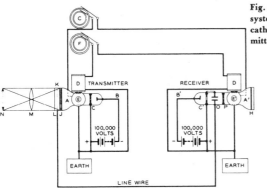

Fig. 3. Campbell Swinton's system of television using cathode-ray tubes as transmitter and receiver (1911).

witness the practical development of his ideas.

Cathode-Ray Tube Receiver

The experiments described above all dealt with the earlier types of Braun tubes which were only partially evacuated. They were generally filled with argon or some other rare gas with a pressure of 10^{-2} to 10^{-4} mm of mercury, and it was not possible in these tubes to obtain a very sharp focus. As the art progressed, however, the tubes were gradually improved, and by the middle 1920's several experimenters were working with such tubes, both here in the United States and abroad. One of the first experimenters in this field in the United States was Vladimir K. Zworykin, who came to this country in 1918 and did his early work with the Westinghouse Research Laboratories. In the years around 1910 he had been a student under Rosing at St. Petersburg and had worked with him in the early experiments with Braun tubes as television receivers. He continued this work here in America, and his first patent application describing a complete electronic television system was issued in 1923.[11] Figure 4 shows a picture of the cathode-ray receiving tube in the above-mentioned patent application. It will be noticed that the tube has a cathode (56), a modulating grid (54), an anode (57), and a fluorescent screen (60). No mention is made about the vacuum in the tube, but in a 1925 patent application a somewhat similar tube is described using low-pressure argon and it may be assumed, therefore, that this first tube also was argon-filled and depended on the argon for focusing the beam. The condenser plates (58 and 59) are used for producing electrostatic horizontal deflection, while the coils (69 and 70) produce magnetic vertical deflection.

According to Zworykin, the first demonstration of an all-electronic system using such a cathode-ray tube at the receiver and an early form of camera pickup tube at the transmitter took place in 1924. However, the pictures were scarcely more than shadow pictures with rather poor definition. It was apparent to Zworykin that the biggest problem was to produce a satisfactory pickup tube

suitable be discovered, distant electric vision will, I think, come within the region of possibility."

The foresight expressed in this letter is truly remarkable when one realizes that it was written in the early infancy of radio transmission, and at a time when the electron tube amplifier had not yet been invented and when photoelectric equipment was still very inefficient and primitive.

Three years later, Swinton elaborated on his suggestion in a presidential address to the members of the Röntgen Society in November 1911.[9] In this address he gave a detailed description of his proposal and illustrated it with a diagram which is reproduced in Fig. 3. The figure shows his conception of a camera tube involving the use of a mosaic screen of photoelectric elements which was scanned by an electron beam. The signal from the camera tube is impressed upon a pair of deflection plates in a cathode-ray tube receiver. Just as in Rosing's suggestion, Swinton makes use of deflection modulation in the receiving tube. Magnetic coils are used both at the transmitter and at the receiver for deflecting the electron beams, and synchronism is insured by using a common alternating-current supply for the two ends.

Swinton apparently never built a complete model of his system but he remained interested in it and, up to the time of his death, talks by him indicated that he had made several experiments to construct a satisfactory mosaic for the

transmitter tube. Thus in 1924 he gave a lecture before the Radio Society of Great Britain entitled "The Possibilities of Television."[10] In this lecture he dealt in detail with his earlier suggestion in 1911 and discussed his latest thoughts on the best design of the mosaic plate in the transmitting tube. He also pointed out that the advent of the electronic amplifier and of radio communication had greatly enhanced the possibilities of successful transmission of television signals. The lecture was followed by a lively discussion and, in his answer to a questioner, Swinton concluded as follows:

"...I wish to say that I agree entirely with Mr. Atkinson that the real difficulty in regard to this subject is that it is probably scarcely worth anybody's while to pursue it. That is what I have felt all along myself. I think you would have to spend some years in hard work, and then would the result be worth anything financially? If we could only get one of the big research laboratories, like that of the G.E.C. or of the Western Electric Company—one of those people who have large skilled staffs and any amount of money to engage on the business—I believe they would solve a thing like this in six months and make a reasonable job of it."

Here again Swinton shows an uncanny ability to foresee the future of television. It was indeed by the efforts of the large companies that modern-day high-definition television was finally developed, although it took a good deal longer than the six months estimated by him. It is to be regretted that he died too soon to

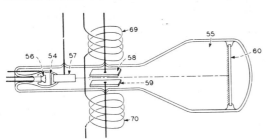

Fig. 4. Zworykin's earliest form of cathode-ray receiving tube (1923).

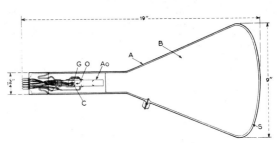

Fig. 5. Early Zworykin cathode-ray receiving tube (1933).

and, in order to achieve better results as soon as possible, he reverted for the next few years to the use of mechanical scanning means at the transmitter and concentrated on the design of better cathode-ray tubes for the receiver. The first published account of these early experiments will be found in an article in *Radio Engineering* for December 1929.[12] The article describes experiments which used a tube with a 7-in. diameter screen. It employed a simple form of electrostatic focusing, grid modulation, and both magnetic and electrostatic deflection. In this article Zworykin first coined the word "kinescope" for the receiving tube.

Around 1930 Zworykin joined the RCA Laboratories, and in the years 1931 and 1932, he and other RCA engineers constructed a complete television system utilizing a cathode-ray tube at the receiver. The transmitter still was mechanical, making use of a Nipkow disc for the scanning process, but the receiving tube had undergone a great deal of development and began to look more like present-day cathode-ray tubes.

This work was described in an article in the *Proceedings of the I.R.E.* for December 1933,[13] and Fig. 5 shows a cross-sectional picture of the cathode ray receiving tube taken from this article. Figure 6 is a diagrammatic view of the cathode and the two anodes of this tube, and it is obvious from the figure that the application of electron optics to proper focusing of the electron beam by means of electrostatic lenses was by that time quite well understood. In order to obtain such focusing the tube must have been of the same high-vacuum type as those employed in today's television receivers. The tube face was 9 in. in

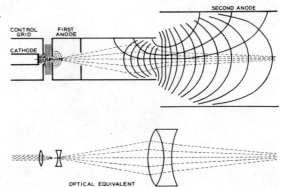

Fig. 6. Constructional details of early Zworykin cathode-ray receiving tube (1933).

diameter and the tube was 19 in. long. Magnetic deflection was used in both directions, and the pictures produced had 120 lines/frame and 24 frames/sec. Several photographs of such pictures are shown in the article mentioned above.

Another early experimenter with cathode-ray receiving tubes in this country was Philo T. Farnsworth, who did his early experimental work in San Francisco in the late 1920's and early 1930's. He too was working with an all-electronic system using cathode-ray tube devices both at the receiving and at the transmitting ends. The transmitting tube was of a different type than that conceived by Zworykin, and will be described later on. The receiving tube, however, was quite similar to the early types of receiving tubes used by Zworykin. The early experimental work by Farnsworth is described in articles appearing in *Wireless World* and *Television News* in 1931.[14] Figure 7 from the article in *Wireless World* shows a picture of the cathode-ray receiving tube which

Farnsworth called an "oscillite." A complete receiver using such a tube is shown in Fig. 8. The pictures had 200 lines/frame and 15 frames/sec, and Farnsworth stated that the bandwidth necessary for transmitting such a signal should be about 300 kc. Even in these earliest tubes, Farnsworth used magnetic deflection for both directions.

Another early experimenter with cathode-ray tubes for television was Manfred von Ardenne in Germany. His work is described in several articles in the German magazine *Fernsehen* during 1930 and 1931.[15] As will be discussed later, von Ardenne was probably the first to experiment with a cathode-ray tube as a flying-spot scanner at the transmitting end, and the experiments described in the above-mentioned article utilized this arrangement at the transmitter. Figure 9 shows a photograph of a table model receiver as taken from one of these articles, while Fig. 10 shows a photograph of a picture obtained on such a receiver. The pictures had 60 lines/frame and about 9000

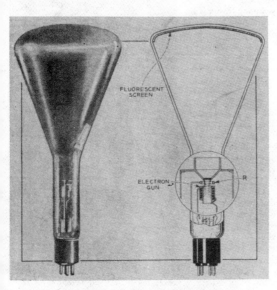

Fig. 7. Early model of Farnsworth cathode-ray receiving tube (1931).

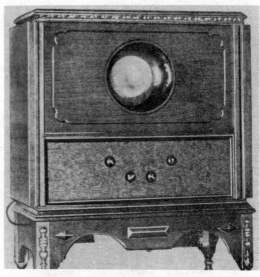

Fig. 8. Early model of American cathode-ray tube television receiver (Farnsworth, 1931).

Fig. 9. Early model of German cathode-ray tube television receiver (von Ardenne, 1930).

Fig. 10. Photograph of early German cathode-ray tube television picture (von Ardenne, 1931).

picture points, which would correspond to a bandwidth of about 75 kc. According to von Ardenne, this limitation in detail was entirely due to the sending arrangement and the receiving tube by itself was capable of producing pictures with about 20,000 to 30,000 picture elements/frame.

During the next twenty years, the early forms of cathode-ray receiving tubes described above were steadily being developed and improved in a number of laboratories and by a number of different organizations. Increased knowledge and application of electron optical principles made it possible to design tubes with higher and higher definition. By the middle 1930's pictures were being shown with as many as 300 lines/frame. During the 1940's pictures were shown with 500 or more lines and, at present, tubes can be made which are capable of resolving 1000 or more lines. At the same time manufacturing methods were developed and steadily improved to permit the production of larger diameter tubes. Up to 1940 very few tubes with diameters over 10 or 12 in. were used, while nowadays television receiver tubes are made with diameters as large as 30 in.

As a somewhat unusual phase of this development, it might be of interest to mention a cathode-ray receiving tube designed by C. J. Davisson of the Bell Telephone Laboratories for use in the first transmission of television signals over the coaxial cable from New York to Philadelphia in 1937.[16] Davisson designed this tube on the basis of his knowledge of electron optics, and at no stage would he depart from a design which would allow him accurately to predict the performance. This accounts for the "thin" lenses used in the different focusing systems, for the small deflection angles employed to insure sharp focus all over the screen, and for the extreme care with which the deflection-plate system was made to avoid either "pin cushion" or "barrel" distortion. It resulted in a very long tube (about 5 ft), as shown in Fig. 11, and an unusually complex assembly of precision mechanical parts, as shown by the diagrammatic representation of the tube in Fig. 12. It also resulted in an actual performance very close to the predicted performance and markedly superior to that of other television receiving tubes of the same period. It is interesting to note that Davisson in this tube reverted to the old method of deflection modulation in order to insure more complete control over the size and shape of the spot on the screen.

Cathode-Ray Tube Flying-Spot Scanner

The use of the light spot of a cathode-ray tube raster as a scanning means for scanning lantern slides or motion-picture film was first proposed by Zworykin in the 1923 patent application mentioned above.[11] Figure 13 shows the proposed arrangement in this patent application. The light from the scanning raster is made to fall on a lantern slide (78), and the light passing through the slide is focused by the lens (77) onto the photocell (76). While the optical arrangement proposed here is not altogether clear, the fundamental idea of obtaining a television signal in this manner is quite obvious.

The first experimenter to put this suggestion into practice was von Ardenne, and his experiments are described in the articles already mentioned above.[15] The general arrangement of the transmitter and receiver is shown in Fig. 14, taken from the article, and here the optical arrangement at the transmitting end is quite proper except that an additional condensing lens between the lantern slide and the photocell amplifier might have been advantageous. The frame frequency of 25 and the line frequency of 1500 would indicate that each picture consisted of 60 lines. Figure 15, also from the same article, shows an experimental setup of the cathode-ray tube used for scanning motion-picture film, and a sample of the pictures obtained was shown above in Fig. 10.

The use of such a cathode-ray tube flying-spot scanner even for live pickup was proposed by von Ardenne, as indicated by Fig. 16 from the same article. Here the scanning raster of the tube is

Fig. 11. C. J. Davisson cathode-ray receiving tube (Bell Telephone Laboratories, 1937).

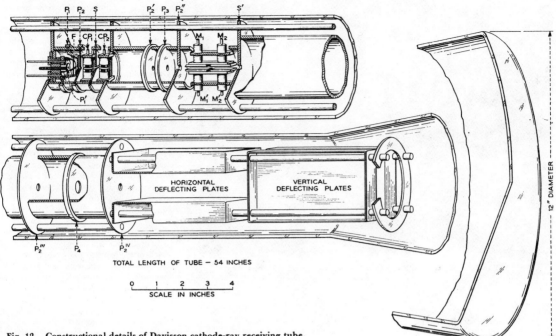

Fig. 12. Constructional details of Davisson cathode-ray receiving tube.

HORIZONTAL DEFLECTING PLATES

VERTICAL DEFLECTING PLATES

12" DIAMETER

TOTAL LENGTH OF TUBE — 54 INCHES

0 1 2 3 4
SCALE IN INCHES

imaged by means of a lens in the plane of the object, in this case the head of a person. The light reflected from the object is then picked up by photocells arranged around the object, and the signals from the photocells are amplified and sent to the transmitter.

It appears that for many years following von Ardenne's experiments no further work was done with cathode-ray tubes as spot scanners. This probably was due to the fact that, as the art progressed toward pictures with higher and higher definition, the cathode-ray tube became inadequate as a spot scanner because of the comparatively slow decay in the phosphors. For receiving tubes it is necessary only that the light decay appreciably in the time between successive field scans, i.e., in 1/60th of a second, and phosphors with this order of decay time were well known even in the 1930's. For spot-scanning tubes, on the other hand, it is necessary that the light decay

appreciably in the time corresponding to one picture element. Otherwise the afterglow would result in an intolerable smearing or blurring of the image. For modern high-definition television this means that the phosphor decay can be only a very small fraction of a microsecond, and phosphors of this type were not known until around 1940. One of the earliest discussions of the requirement for spot-scanning phosphors can be found in an article in a German magazine in 1939.[17] A description of one of the first high-definition lantern slide spot scanners used in this country will be found in an article in the *Proceedings of the I.R.E.* in 1946.[18] This particular spot scanner was used for scanning motion-picture film, a purpose for which a cathode-ray tube scanner is particularly well

adapted. This problem will be discussed in more detail later on under Film Scanners. Since that time many types of cathode-ray tube spot scanners have been developed and are now available commercially.

Camera Tubes

Electronic camera pickup tubes may be divided into two types. One type is the instantaneous or nonstorage type tube, as exemplified by the Farnsworth dissector tube, while the other is the storage type of camera tube, as exemplified by Zworykin's iconoscope and practically all of the later types of camera tube types used today. In the following we shall discuss the development of both of these types of camera tubes.

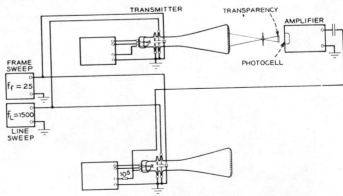

Fig. 13. Earliest picture of Zworykin's cathode-ray tube spot scanner (1923).

Fig. 14. Diagram of German proposal for cathode-ray tube spot-scanning transmitter (von Ardenne, 1930).

Jensen: **Evolution of Modern Television**

Fig. 15. Early German cathode-ray tube film scanner (von Ardenne, 1931).

A. Dissector Tube

The dissector tube was developed by Farnsworth and was first described by him in a patent application dated January 7, 1927.[19] While the fundamental principles of the tube are described in this patent application, there were still several essential improvements to be added before the tube would operate satisfactorily. These additions were incorporated in several subsequent applications, and a complete description of the method of operation of the tube was first given by Farnsworth in an article in the *Journal of the Franklin Institute* in 1934.[20]

One interesting feature of Farnsworth's earliest patent application is the fact that he does not use at the receiver a cathode-ray tube as in his later applications and as discussed above. Instead he uses a Kerr cell for controlling the intensity of a polarized light source in accordance with the received signal strength, and the light beam after leaving the Kerr cell is then deflected onto a screen by use of two galvanometer mirrors controlled by the scanning signals. This is probably one of the very few proposals, if not the only one, for a television system using electronic means at the transmitting end and mechanical scanning means at the receiving end.

Figure 17 shows a cross-sectional view of the dissector tube as described in the 1934 article. It consists of a cylindrical glass envelope with a plain circular cathode at the left and an anode enclosed in a pencil-shaped shell, called "target," at the right. The end face of the tube consists of plane glass of optical quality, and through this end plate an optical image of the object to be televised is formed on the face of the cathode. Since the surface of the cathode is photosensitive, electrons will leave each point of this surface in an amount corresponding to the illumination in the image at that point. These photoelectrons are pulled toward the anode by means of a high potential applied between the anode and the cathode. If no other arrangements were made, the electrons leaving any one point in the cathode would not all travel toward the anode in an axial direction, but would fan out in a narrow cone. However, by enclosing the entire tube in a solenoid and passing a direct current of the proper amplitude through this solenoid, it is possible to pull this cone of electrons together again in such a fashion that all electrons leaving any one point on the cathode will again come together in one point of a plane parallel to the cathode and located in the plane of the target. This focusing action is indicated by the curved electron patterns shown in Fig. 17. In other words, it is thus possible to insure that a complete electron image of the scene is reproduced in the plane of the target. The target has a small aperture facing the cathode and, by sweeping the entire electron image back and forth and up and down past this aperture, the electrons of each element of the image are made to pass successively through this aperture and reach the anode inside. That is, by the scanning of the electron image past the aperture, the image is "dissected" into its elemental areas, and a signal is produced which at each instant is proportional to the number of electrons entering the aperture at that instant, and therefore again proportional to the light falling on the corresponding point of the cathode at that same instant. The tube therefore is instantaneous in nature and no storage is involved. The sweeping of the electron image is effected by means of two sets of sweep coils placed at right angles around the cylindrical sides of the tube and energized by saw-tooth currents of the proper frequencies.

The earliest forms of dissectors built by Farnsworth were quite insensitive as compared with modern type tubes. In order to obtain television signals with any reasonable signal-to-noise ratio, it was necessary that the illumination in the optical images formed on the cathode be extremely high. This, in turn, meant that live pickup from studio scenes was out of the question since the required image illumination could be obtained only from lantern slides with high-intensity light sources. In order to remedy this situation, Farnsworth therefore replaced the original photocell with an electron multiplier capable of multiplying the original photocurrent by many orders of magnitude. The development of electron multipliers for this and other purposes will be discussed later.

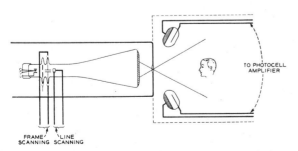

Fig. 16. Early German proposal for live pickup using cathode-ray tube spot scanner (von Ardenne, 1930).

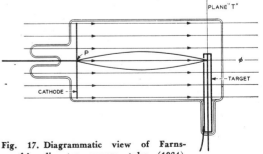

Fig. 17. Diagrammatic view of Farnsworth's dissector camera tube (1934).

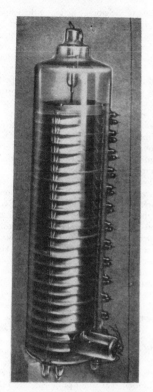

Fig. 18. Special Farnsworth dissector camera tube (Bell Telephone Laboratories, 1945).

In the late 1930's dissector tubes with electron multipliers were used in television cameras for live pickup and demonstrated by Farnsworth, but even these later type tubes required studio lighting which was uncomfortably intense and the tubes were never used in any commercial television cameras. The tube, however, has been used very successfully in many types of motion-picture film scanners, although present types of commercial film scanners all use storage-type tubes. It also has been used and still is being used for several applications of industrial television, such as the reading of instruments in inaccessible or dangerous locations. It should be mentioned that the great advantage of the dissector tube over storage-type tubes is that it is completely linear in operation so that the output signal is strictly proportional to the light falling on the cathode. This is of particular advantage when scanning motion-picture film and was the principal reason for the use of the dissector tube for this purpose until the advent of the short-decay, cathode-ray tube flying-spot scanner.

Figure 18 shows a photograph of a special type of dissector tube designed for use in an experimental motion-picture scanner which was used in the Bell Telephone Laboratories until quite recently.[21]

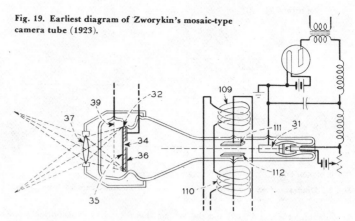

Fig. 19. Earliest diagram of Zworykin's mosaic-type camera tube (1923).

B. The Iconoscope

The storage-type camera tube was invented by Zworykin and was first described in a patent application dated December 29, 1923.[11] Figure 19 shows a cross-sectional view of the tube as taken from this application. The right end of the tube shows the cathode and a cylindrical anode (31), and through this a beam of electrons is directed toward the target (32). This target has the characteristic mosaic structure consisting of a wire mesh screen (36) facing the cathode, a thin insulating layer (34) behind the wire screen, and a layer of insulated globules of photoelectric material (35) deposited on the insulating layer. This photoelectric mosaic thus faces the camera lens (37). By means of this lens the scene to be televised is imaged on the mosaic. It will be noticed that both electrostatic and electromagnetic deflection are used in this type. The tube was argon-filled and used gas focusing of the beam and, according to the description in the patent, this gas also played a part in the collection of signal current from the mosaic onto the signal grid (39). However, the exact nature of this collection is not very clear. Tubes of this sort were constructed by Zworykin in 1924, and Fig. 20 is a photograph of one of these early tubes. It was mentioned earlier that Zworykin demonstrated the use of such a camera tube in an all-electronic system using a cathode-ray tube at the receiver sometime during 1924, but that the pictures were scarcely more than shadow pictures with rather poor definition.

After spending the next few years in improving the cathode-ray tube at the receiver, Zworykin again took up the construction of camera tubes when he joined RCA around 1930. A description of an early form of iconoscope is given in an article by Zworykin in the *Proceedings of the I.R.E.* in 1934,[22] and Fig. 21 shows a cross-section of a tube used in these experiments. By this time it had been realized that proper focusing of the electron beam could be obtained only in a high-vacuum tube with proper design of focusing electrodes, as described earlier in the section on Cathode-Ray Tube Receivers. The tube has a signal plate consisting of a thin metal plate on which is deposited a thin layer of insulating material (in modern iconoscopes this insulating layer consists of a thin sheet of mica), and on top of the insulating sheet is deposited a mosaic consisting of a large number of minute globules of photosensitive material. The globules are insulated from each other and are small enough so that the cross-section of the beam covers a large number of globules at any one time.

The method of operation of the iconoscope may be described in a somewhat simplified form as follows. An object is imaged on the mosaic by means of a camera lens and, as the globules are illuminated by this image, they give off photoelectrons which leave the mosaic and are collected by a collector electrode marked "Pa" in the figure. As the photoelectrons leave the mosaic, the globules are thus charged positive with respect to the signal plate, and the greater the illumination the more the globules are charged and, therefore, the higher the potential between the globules and the signal plate. This goes on as

Fig. 20. Earliest experimental model of Zworykin camera tube (1923).

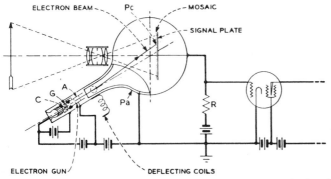

Fig. 21. Diagram of earlier model of Zworykin's iconoscope camera tube (1934).

long as the target is illuminated and the action, therefore, is cumulative. An electron beam from the cathode of the tube is made to scan over the mosaic in the usual manner by means of two sets of magnetic coils placed around the neck of the tube. As the scanning beam strikes the globules on any one point of the mosaic, these globules are thereby discharged and the discharge causes a corresponding pulse of current to flow from the signal plate to ground through the resistance, R. The varying potential across this resistance therefore constitutes the television signal. As soon as the beam has left a particular spot on the mosaic, this spot will again start to charge up in accordance with the light falling on it and will continue to do so until the beam strikes it again during the succeeding scan. In other words, energy is stored up in the mosaic by the action of light on the photoelectric globules, and this storage continues during the entire period between scans. This storage principle makes the iconoscope much more sensitive than the dissector tube, which depended for its action on the photoelectric emission from the cathode at any one instant.

The actual mode of operation of the iconoscope is somewhat more complicated than described above, due to the

fact that secondary emission takes place as the beam scans over the mosaic. The potential between the cathode and the anode is quite high, of the order of a thousand volts or so, and therefore results in a high velocity electron beam. As this beam strikes the mosaic it releases secondary electrons from the mosaic, and these secondary electrons are partly collected by the collector electrode but also partly pulled back onto the mosaic in a more or less random manner. Even if the mosaic were not illuminated by any image, such secondary emission would still take place and some of these secondary electrons would be pulled back to the mosaic and redistributed over the surface, thus giving rise to a false signal current. These spurious signals are called "shading" signals and are common to all iconoscope and other high velocity beam tubes. Since the secondary electrons are redistributed over the mosaic in a rather broad fashion, the resulting shading signals are low-frequency signals of rather simple waveforms, and they may therefore be compensated for by introducing a corresponding correcting signal from special shading signal generators incorporated as part of the camera equipment. A photograph of an early iconoscope of the type described above is shown in Fig. 22.

Due to the storage action of the iconoscope, this tube theoretically should be some 50 to 100 times more sensitive than an instantaneous-type tube like the dissector (assuming that the dissector does not incorporate the use of an electron multiplier). Actually, the iconoscope is only some 5 to 10% efficient due to the secondary emission. In the first place, the secondary electrons released from the mosaic reduce the electric field in front of the mosaic and therefore make the photoemission of the globules less efficient; also, the fact that the secondary electrons are only partly collected by the collector electrode results in lower charging potentials of the globules and, therefore, in lower signal current. Even so, the inconoscope proved to be a much more sensitive tube than any tube hitherto produced and made it possible for the first time to build live pickup cameras for studio use with studio illuminations that were not too uncomfortable for the actors.

The number of secondary electrons emitted from any part of the mosaic depends on the potential of the mosaic at the time the electron beam strikes it, and this again depends on the amount of light falling on the mosaic at that point. The result is that the overall action of the iconoscope is not linear, that is, the output current is not proportional to the amount of light falling on the mosaic. The relation between illumination on the mosaic and signal output current is more nearly a square root relation for small values of illumination, and for large values it soon reaches a saturation point, so that increased illumination results in only very little increase in corresponding signal output.

For studio use the iconoscope has long since been superseded by more modern forms of camera tubes, but it is still extensively used in motion-picture film scanners. The tube shown to the right in Fig. 27 is a photograph of a modern-type iconoscope for this use.

C. Image Iconoscope

The first step towards further improving the iconoscope was the so-called image iconoscope, which was first described in an article by Iams, Morton and Zworykin in the *Proceedings of the I.R.E.* in 1939.[23] Figure 23 shows a cross-sectional diagram of such a tube, as

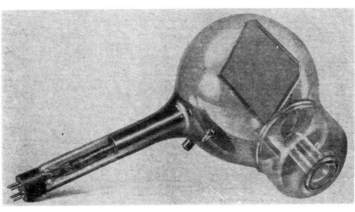

Fig. 22. Early iconoscope camera tube.

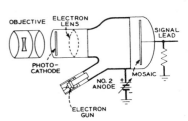

Fig. 23. Diagram of image iconoscope camera tube (1939).

taken from that article. The tube has a semitransparent photocathode on which an optical image is formed of the scene to be televised. As light falls on the photocathode, photoelectrons are emitted from this cathode and are pulled towards the mosaic at the right by means of a high electric potential. Close to the photocathode there is, in other words, an electron image corresponding to the optical image falling on the cathode, and by means of an electron lens system this electron image is focused in the mosaic in the same manner as described earlier for the dissector tube. The electron lens system may be either magnetic and consist of a coil, as in the case of the dissector, or it may be electrostatic and consist of a series of metal rings of proper potential.

Since the potential between the photocathode and the mosaic is high, the photoelectrons reach the mosaic and therefore give rise to the emission of secondary electrons from the mosaic. Since each incoming electron releases on the order of 4 or 5 secondary electrons, this in effect results in a charge image on the mosaic which is 4 or 5 times higher than if no secondary emission took place. On top of that, the high field between the photocathode and the mosaic results in a more efficient collection of current from the photocathode than in the case of the iconoscope, where the field near the photosensitive mosaic was reduced due to secondary emission. All in all, this arrangement results in a tube which is about ten times more sensitive than the ordinary iconoscope. Figure 24 shows a photograph of an image iconoscope for use in a studio camera. Tubes of this type were used here in the United States during the 1940's but were later replaced by the image orthicon. However, in England tubes of this type are used extensively for studio cameras. The English name for this type of camera tube is Super Emitron.

D. The Orthicon

It was mentioned above that the iconoscope was only 5 to 10% efficient. Due to secondary emission, only a small part of the photoelectrons emitted by the mosaic are drawn away and only a small part of the stored charge is effective in

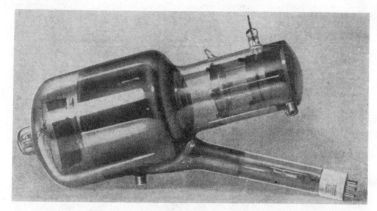

Fig. 24. Image-iconoscope camera tube (1939).

producing a signal. The image iconoscope is one attempt to improve this efficiency, but another and rather different approach consists in using a low-velocity electron beam for scanning. In this case it is possible to provide at the mosaic a field strong enough to draw away all the photoelectrons and, furthermore, no secondary electrons are emitted when the beam reaches the mosaic.

A tube of this type, called the orthicon, was developed at the RCA Laboratories, and is described by Rose and Iams in an article in *RCA Review* in 1939.[24] Figure 25 shows a cross-sectional diagram of this tube. The tube has a storage-type mosaic plate similar to that used in an iconoscope, and when light falls on this plate all the photoelectrons are drawn to the collector by means of a sufficiently high potential between the collector and the mosaic plate. In this case, however, the mosaic plate is kept at a potential equal to that of the cathode, and when the electron beam reaches the signal plate it therefore has zero velocity and does not release any secondary electrons. If the mosaic has been charged up due to light falling on it, the electron beam just resupplies the electrons that have left the mosaic due to photoemission and thus brings that part of the mosaic back to cathode potential. The remaining electrons in the beam thereafter return to the collector. The signal thus consists of the current pulses flowing in the output lead as each successive point of the mosaic is being discharged.

Since the electron beam in this tube has very low velocity, it would not stay sharply focused as it traverses the tube unless special precautions were taken. These consist in focusing coils arranged around the tube in such a manner as to produce a strong longitudinal magnetic field. This field has the effect of keeping the beam concentrated as it traverses the tube so that the cross-section of the beam at the mosaic is about the same as that leaving the gun. In order to avoid blurring near the edges of the mosaic it is further necessary to introduce correcting fields which insure that the beam is perpendicular to the mosaic over the entire surface. One of the great advantages of this type of tube is that no secondary emission takes place. For this reason the tube is nearly linear in operation and, furthermore, this also results in a complete absence of spurious shading signals. Tubes of this sort were built and used experimentally in pickup cameras. However, the tube was soon to be superseded by a still more efficient and sensitive tube and therefore never came into wide use.

E. Image Orthicon

In the description of the orthicon it was mentioned that the electron beam supplies to each part of the mosaic the electrons that have been lost by the electron image, and the remaining part of the electrons then return to the collector. This return beam, therefore, is in effect modulated by the signal current and,

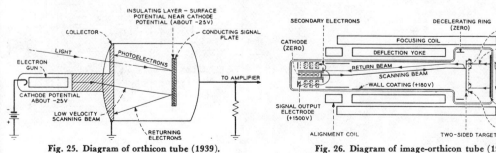

Fig. 25. Diagram of orthicon tube (1939). **Fig. 26. Diagram of image-orthicon tube (1946).**

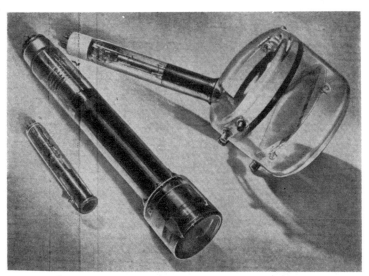

Fig. 27. Modern vidicon, image-orthicon and iconoscope tube.

if properly collected, could indeed be used as the signal. This is one of the principles employed in the image-orthicon tube which was developed by RCA and first described in an article by Rose, Weimer and Law in the *Proceedings of the I.R.E.* in 1946.[25] Figure 26 shows a cross-sectional diagram of this tube which makes use of electron image amplification, as in the image iconoscope, of the efficient low-velocity beam, as in the orthicon, and of an electron multiplier, as in the later types of dissectors. The optical image is formed on a transparent photocathode placed on the inside of the end glass wall of the tube. The electron image emitted from this cathode is focused as a second electron image in the plane of a thin two-sided insulating target. Since a high electric field exists between the photocathode and the target, the photoelectrons strike the target with high velocity and give rise to the emission of secondary electrons, with the result that the charge image on the target is some 4 or 5 times stronger than the electron image leaving the photocathode. The other side of the target is scanned by a low velocity electron beam which discharges points of the target as it scans across it, thus resulting

in a return beam which is modulated with this discharge current. As in the orthicon, the electron beam is kept focused during its traverse from electron gun to target, and the same focusing field will therefore keep the return beam focused along the same path so that the return beam strikes the plate surrounding the gun aperture very close to this aperture. It strikes these plates with a velocity high enough to release secondary electrons and, by proper electrostatic focusing, these secondary electrons are in turn guided to the first dynode of a multistage electron multiplier located behind the gun. By thus making use of all the known methods of electron magnification it has been possible to construct a camera tube which is of the order of 100 times more sensitive than any previous type of camera tube. With tubes of this sort, it is possible to make television pictures at very low light levels and, in fact, when using television cameras with this type of tube it has been possible to televise outdoor scenes at light levels so low that no satisfactory motion picture could be obtained. When used in the studio this tube produces excellent television pictures at medium light levels and is now the tube

most commonly used for all studio cameras in the United States. A photograph of a modern image orthicon is shown as the center tube in Fig. 27.

F. The Vidicon

One of the earliest discoveries in the field of photoelectricity was that of the photoconductive properties of selenium. Selenium was discovered in the early part of the 19th century and was found to have a very high electrical resistance. In 1873 Willoughby Smith in England was making experiments with submarine cables and, in connection with those tests, he had use for some very high resistances. He therefore had made up some long wires of selenium enclosed in glass tubes, with the idea of using these as resistance elements. He found, however, that he could not obtain constant results in his measurements. The results varied from day to day for no apparent reason and, in an attempt to discover the cause of the discrepancies, Smith accidentally covered one of the selenium wires, excluding the light falling on the wire, and he found that it suddenly changed its resistance materially. Smith published his discovery, which caused great excitement and gave rise to many of the early proposals for "seeing by electricity." When Swinton tried to build a mosaic for his camera tube, it was selenium he thought about as a photoelectric material, and most of the other early experimenters thought in terms of selenium when translating light energy to electrical energy. It is interesting to note that, in spite of this early discovery of the photoconductive property of selenium, it was not until 1950 that a satisfactory photoconductive camera tube was built, a tube capable of producing the high-definition television signal required for modern television. It took the technique over half a century to catch up with the ideas and visions of the early inventors.

During the late 1940's the engineers at RCA experimented with an RCA photoconductive pickup tube, and in 1950 such a tube was described in an article by Weimer, Forgue and Goodrich appearing in *Electronics*.[26] They called the tube the vidicon, and Fig. 28 shows a cross-sectional diagram of such a tube

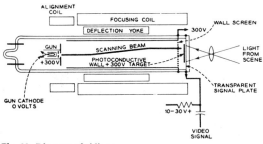

Fig. 28. Diagram of vidicon camera tube (1950).

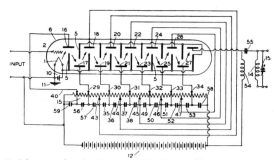

Fig. 29. Diagram of early electron multiplier amplifier tube (1926.

as taken from the article mentioned. The tube uses a low-velocity scanning beam similar to that used in the orthicon, but instead of the photoemissive surface in the orthicon the vidicon uses a photoconductive target deposited on a transparent conductive signal plate. An optical image of the scene to be transmitted is focused onto the photoconductive surface through the transparent signal plate. This signal plate is kept at a potential of about 20 v positive with respect to the cathode, and when no light falls on the photoconductive surface this surface is kept at cathode potential by means of the scanning beam. When light falls on the photoconductive surface it increases the conductivity, and a charge current flows from the signal plate to the individual elements of the photoconductive surface, the charging current being proportional to the amount of light falling on the surface. When the scanning beam strikes the charged surface it deposits sufficient electrons to neutralize this charge and thereby generates the video signal in the signal plate lead. In other words, the operation is very similar to that of the orthicon except that the positive charging effect is achieved by photoconduction through the target itself rather than by photoemission from the scanned surface.

The advantage of using a photoconducting tube rather than a photoemissive tube is that the sensitivity of photoconductive materials is such that currents of several thousand microamperes per lumen may be obtained as compared to 30 to 50 $\mu a/lm$ for the most efficient photoemissive materials. Thus a highly sensitive camera tube may be built without the additional complication of an electron image amplifier and an electron multiplier as incorporated in the image orthicon. This results in a simpler tube which is cheaper to build and can be made much smaller and more compact than any other camera tube. The photograph to the left in Fig. 27 is of a vidicon tube capable of producing satisfactory television pictures of more than 600 lines. It is about 6 in. long and 1 in. in diameter, and with tubes of this sort it has been possible to build portable field cameras that are not much bigger than an ordinary 16mm motion-picture camera.

Multiplier Phototubes

A multiplier phototube or photomultiplier is a photocell in which the photoelectrons emitted by the photosensitive surface are amplified inside the tube by making use of the phenomenon of secondary emission. The photoelectric effect itself was first discovered by Hertz in 1887, and photoelectric cells of the type we now know were first experimented with around 1900. The use of secondary emission as a means of amplification was first made use of by A. W.

Hull in a patent application dated 1915.[27] This application deals with the use of a dynatron as a device having negative electrical resistance characteristics. A second patent application by Hull, dated December 28, 1921,[28] deals with a vacuum tube amplifier which makes use of secondary emission to obtain increased amplification. Another electron multiplier amplifier is described in a patent application by Jarvis and Blair dated September 15, 1926.[29] Figure 29 shows a cross-sectional diagram of the amplifier taken from this application. In this amplifier tube, electrons from the cathode (1) are controlled by the grid (2) in the usual manner, and from there they flow to the first anode (16) which corresponds to the plate in an ordinary vacuum tube. If this anode is of the proper potential, the electrons striking it will have sufficiently high velocity so that each electron releases several secondary electrons. By means of electrostatic shields (5) of proper potential, these secondary electrons are guided to the next anode (17) which they strike, and release still more secondary electrons, and so on, through the entire tube until finally the multiplied electrons reach the collector plate at the end of the tube. With the proper potential gradient from anode to anode, and with the anodes made of the proper material, it is possible to insure that each primary electron striking an anode will release as many as 4 or 5 secondary electrons, and with enough stages of multiplication it is thus possible to obtain an electron multiplication of many thousand times.

If, in the tube shown in Fig. 29, the cathode and the grid are replaced by a photosensitive surface on which light can be projected, we have, in effect, a photomultiplier very similar to the type used nowadays. During the early development of photomultipliers some of the forms utilized external magnetic focusing or guiding of the electrons through the tube, but all of the photomultipliers used today employ electrostatic focusing similar to that described above. It is multipliers of this type that were used by Farnsworth in his later developments of the dissector tube, and it is also this type of multiplier which is employed in the image orthicon.

Motion-Picture Film Scanners

During the late 1930's it was established that future commercial television in the United States would use 30 frames/sec, that is, 60 fields interlaced. With regard to the scanning of motion-picture film, this created a problem, due to the fact that motion pictures are presented at 24 frames/sec, and some form of translation from 24 to 30 frames/sec was therefore necessary.

The problem was solved by using a storage type of camera tube as a pickup device in the scanner. The motion-

picture film was run in the usual intermittent fashion at 24 frames/sec, and the mosaic of the camera tube was illuminated by the image using very short flashes of intense light at the rate of 60/sec. These flashes were made to occur during the field blanking intervals of the television signal, and the mosaic was then scanned in the usual manner while the film was being pulled down to the next frame. Thus every other picture frame is scanned twice and the intervening frames three times in order to produce the 60/sec field scan.

The tube most commonly used for film scanners was the iconoscope and, in fact, many film scanners using the iconoscope are still in use in the broadcasting studios. As mentioned above, under the Iconoscope Section, this tube produces a spurious shading signal, the nature of which depends to some extent on the distribution of light in the image. This therefore creates an additional problem since sudden changes of scene in the film give rise to sudden changes in the shading signal. Satisfactory operation of such a film scanner calls for highly skilled personnel and, in general, requires rehearsals in order to insure prompt and accurate modifications of compensating signals when the scene changes. On top of this, the iconoscope is not capable of producing a range of tone gradations that is anywhere near that obtainable in high-grade motion-picture film.

For these reasons, television pictures from motion pictures have generally been quite inferior to those obtained with live pickup, and many attempts have been made to design a motion-picture film scanner capable of using a cathode-ray tube spot scanner as the light source, since this would result in a much better tone gradation and the absence of shading signals.

In order to accomplish this it is necessary to use a continuous or nonintermittent film drive and to incorporate in the machine a system of rotating mirrors or prisms so arranged that the scanning raster imaged onto the film via these mirrors is made to follow the motion of the film, and thus is capable of scanning the film as if the frames were stationary.

Continuous projectors with this type of optical arrangement have been known for a long time, and in one particular form were used quite extensively for commercial film projection in Germany in the 1930's. This projector was designed by Mechau, and a detailed description of its design and operation was given in 1928.[30] The optical arrangement of the projector is such that, as the film is pulled through the gate in a continuous motion, the light projected through the film passes via a series of mirrors moving in such a fashion as to keep the image of the film frame stationary on the screen. As one film frame leaves the gate, the image of this frame fades out and the image of the succeeding frame fades in. In other words, the picture on the screen consists of a continual lap-dissolve from one frame into the next.

If a machine of this sort is used backwards, that is, with a cathode-ray tube scanning raster in place of the screen and with a photomultiplier in place of the light source, then it fills all the requirements for a continuous motion-picture television scanner. In fact, quite a number of these Mechau machines were brought to England before and after the last war and are still being used by the British Broadcasting Corp. for film scanning and kinerecording.

In the United States a continuous film scanner made on a principle somewhat similar to Mechau's was designed and constructed in the Bell Telephone Laboratories around 1948.[31] It uses a system of moving mirrors like the Mechau, but the mechanical arrangement for controlling these mirrors is greatly simplified over that used in the Mechau machine. This machine is still used as a main source of television signals in the Laboratories and is capable of producing pictures with a detail corresponding to some 7 or 8 mc and with contrast ranges corresponding to theater release prints.

Lately several firms have brought on the market commercial film scanners using continuous film motion and a rotating prism as the moving optical element. Film scanners of this type are particularly significant since they greatly simplify the problem of scanning motion-picture film in color.[32]

It should be mentioned that in the last few years the advent of the image orthicon and the vidicon has made it possible to build intermittent-type film scanners with performance capabilities much higher than those of the original iconoscope film scanner, and such scanners are now being used to an increasing extent in the broadcasting studios.

In England television pictures are scanned at the rate of 25 frames/sec or 50 fields interlaced. By simply running the motion-picture film continuously at the rate of 25/sec instead of 24/sec, it has therefore been possible for the British to construct and use very high-grade cathode-ray tube spot scanning machines with very much simpler optical arrangements than are required here in the United States. A description of two different types of such film scanners was given at the Television Convention in London in 1951.[33]

Development of Television Standards in the United States

It was long ago realized that in order to insure adequate synchronization between television transmitters and television receivers it is essential to establish a "lock-and-key" relationship between the two. It was therefore also realized that a widespread commercial television service would require a set of nationally adopted systems standards insuring such a lock-and-key relationship.

As early as 1929 the Radio Manufacturers Association set up a committee on television for the purpose of coordinating and guiding the experimental television work going on at that time. By 1935 this committee was requested to investigate the possibility of setting up a set of television standards for nationwide use, and by 1936 such a set of standards had been proposed and a report submitted to the Federal Communications Commission.

In this report it was recommended that seven television channels be allocated in the region from 42 to 90 mc and that each channel should be 6 mc wide. It was proposed that the system should consist of a 441-line picture and the transmission recommended was a double sideband AM transmission of the picture signal with 3.25-mc separation between the sound and the picture carriers. By August 1936 the Federal Communications Commission permitted experimental transmission of such signals in part of the frequency band proposed but took no further action on the standards since the transmissions were still experimental only.

During the next couple of years further developments resulted in a modified set of standards which was submitted by the Radio Manufacturers Association to the Federal Communications Commission in 1938 with the proposal that they be adopted as national standards. Furthermore, the National Broadcasting Company announced that, coinciding with the opening of the New York World's Fair in 1939, they would start a limited series of programs in accordance with these standards. By this time the standards recommended made use of vestigial sideband transmission, which enabled the frequency range occupied by the picture signal to be increased from about $2\frac{1}{2}$ mc to 4 mc.

NBC maintained such a program all through 1939 without any formal approval of the standards by the FCC, but in the latter part of 1939 the FCC issued a ruling which permitted limited commercial sponsorship of such transmission and announced their intention of holding a hearing in January 1940 for the purpose of arriving at a set of standards having general industry support.

During the hearing it developed that serious objections to the RMA proposed standards were raised by several sectors of the industry, and in the report of the hearing the FCC stated that commercialization would not be permitted until the industry agreed on one common system of broadcasting.

In order to establish such a system, the National Television System Committee was set up under the sponsorship of RMA. This Committee consisted of a group of technical experts from all interested parts of industry, and during the next year and a half this Committee concentrated on the formulation of a complete set of television standards supported by the industry as a whole. The final report of the Committee was delivered to the Commission in March 1940, and in May of that year the FCC announced that the NTSC standards had been adopted officially and that commercial television broadcasting based on these standards would be permitted as of July 1941.

By this time the standards called for a picture having 525 lines and, except for minor modifications, the standards were identical to those now in force. During the war years commercial television was temporarily discontinued except for a few special transmissions, and in 1945 the prewar standards were reaffirmed by the Commission with a few minor modifications. These reaffirmed standards are the present commercial standards for black-and-white television transmission in the United States.

A very complete report on this standardization work inside RMA and NTSC was assembled by D. G. Fink and published by McGraw-Hill in book form in 1943.[34]

The Development of Television Network Facilities

The first transmission of television signals over Bell System wire facilities took place in April 1927 when signals produced by Dr. H. E. Ives' mechanical-scanning arrangements were transmitted from Washington to New York.[35] At the same time, signals were also transmitted by radio from Whippany, N.J., to New York.[36] These signals were of the low-definition type produced at that time and required a band-width of only about 20,000 cycles.

Television signals approaching the present-day wide-band variety were first transmitted over the coaxial cable from New York to Philadelphia in 1937.[2] The pictures were produced from motion-picture films by means of a 6-ft scanning disc. Two hundred and forty lines per frame were used and the band-width required to transmit the signals was 800,000 cycles. In 1940 television signals were transmitted over the coaxial cable from New York to Philadelphia and return, but now the signals corresponded to 441-line television pictures and required a frequency band of about 2.7mc.[37] Later that year scenes from the Republican National Convention in Philadelphia were transmitted over the coaxial cable to the National Broadcasting Co. studio in New York for television broadcasts. During 1941 similar television signals were transmitted over coaxial cables for a total distance of 800 mi. by looping the coaxial units in the cable between Stevens Point, Wis., and Minneapolis.

During the war years very little television transmission took place, but after the war, in 1945, television service was furnished to the broadcasters on an experimental basis between New York and Philadelphia and New York and Washington, with dropping points at Baltimore. These signals were of the 525-line variety now used for commercial broadcasting.

In 1947 the first microwave relay system was inaugurated between New York and Boston, and a round-trip television signal from New York was sent to Boston and back again and was received both in New York and, in addition, in Washington over the coaxial cable from New York.[38]

In 1948 commercial service over Bell System networks was started and since that time the number of channel miles available for transmission has grown steadily, both over the coaxial system and over the microwave relay system. The Chicago-New York radio relay link was started in 1950, and the first commercial program transmission from coast to coast over the radio relay took place in September 1951 when President Truman spoke in San Francisco, and signals were broadcast from stations in New York City.[39]

At the end of February 1954, the total number of channel miles available for television transmission in the United States was about 52,000, of which 17,000 were coaxial and the remaining 35,000 microwave relays.

The establishment of satisfactory long-distance television transmission facilities has imposed some extremely severe requirements on the reliability and stability of the individual components of such systems. The reception of a satisfactory broadcast television picture, especially for color television, requires that the studio equipment, the transmitting equipment and the receiving equipment together provide a channel which has a gain characteristic flat within a fraction of a decibel over the entire band and for which the departure from phase linearity over the band is only a few degrees. When such a television signal is transmitted over a coaxial cable from New York to Chicago, it passes through over a hundred repeaters, each with a gain of about 40 db, or a total gain of over 4000 db. In order to insure that the television signal still meets the above-mentioned requirement after transmission over such a system, it will be realized that the design of the individual components requires extreme accuracy and care in order to insure the necessary precision and stability of both the repeaters themselves and their associated passive equipment, such as amplitude and phase equalizers. Similar stringent requirements exist for the components in microwave relay systems since the signal transmitted over such a system from New York to Los Angeles again has to pass through over a hundred repeaters on the way.

Conclusion

The purpose of this paper has been to sketch the development of modern television from the early days of mechanical scanning through the transition period of partly mechanical and partly electronic scanning up to the present-day all-electronic systems of high-definition television. This paper deals only with black-and-white television; the development of color television has not been included, partly because color television has only just been made commercial and a historical paper on this subject may therefore be somewhat premature, but partly also because this entire subject by itself has a long and interesting history and therefore merits a separate paper of its own.

Apart from some of the early historical developments, the author has limited himself largely to the development here in the United States, but for those readers interested in the development of television in England and in Germany, the author would refer them to two very excellent papers published during the last couple of years, one in the *Proceedings of the Institution of Electrical Engineers* in 1952[40] and the other in a German magazine article mentioned previously.[4] Both these papers are very extensive and both have excellent bibliographies, which will prove valuable and time-saving to any reader interested in source material.

References

1. J. V. L. Hogan, "The early days of television," *Jour. SMPTE*, 63: 169–172, Nov. 1954.
2. M. E. Strieby, "Coaxial cable system for television transmission," *Bell Sys. Tech. J.*, 17: 438–457, July 1938.
   D. C. Espley and D. O. Walter, "Television film transmitters using aperture scanning discs," *J. Inst. Elec. Engrs.*, 88: 145–171, June 1941.
   Kurt Thöm, "Mechanischer Universal Abtaster für Personen-, Film-, und Diapositivübertragungen," *Fernseh Ag.*, 1: 42–48, Dec. 1938.
   H. Rinia and L. Leblans, "The Nipkow disc," *Philips Tech. Rev.*, 4: 42–47, Feb. 1939.
3. M. Dieckmann and G. Glage, German Patent No. 190102, Sept. 12, 1906.
4. Gerhart Goebel, "Television in Germany up to the year 1945," *Arch. Post- u. Fernmeldewesen*, 5: 259–393, Aug. 1953.
5. Boris Rosing, British Patent No. 27570, June 25, 1908, application filed Dec. 13, 1907; British Patent No. 5486, 1911.
6. *Sci. Amer.*, p. 384, June 17, 1911; p. 574, Dec. 23, 1911.
7. Shelford Bidwell, letter to *Nature*, 78: 105, 1908.
8. A. A. Campbell Swinton, letter to *Nature*, 78: 151, 1908.
9. A. A. Campbell Swinton, Presidential Address, *J. Röntgen Soc.*, 8: 1–15, Jan. 1912.

10. *Wireless World*, 14: 51–56, 82–84, 114–118, 1924.
11. V. K. Zworykin, U.S. Patent No. 2,141,059 Dec. 20, 1938, applic. filed Dec. 29, 1923.
12. V. K. Zworykin, "Television with cathode ray tube for receiver," *Radio Eng.*, 9: 38–41, Dec. 1929.
13. V. K. Zworykin, "Description of an experimental television system and the kinescope," *Proc. I.R.E.*, 21: 1655–1673, Dec. 1933.
14. A. Dinsdale, "Television by cathode ray. The new Farnsworth system," *Wireless World*, 28: 286–288, Mar. 1931.
15. Manfred von Ardenne, "Die Braunsche Röhre als Fernsehempfänger," *Fernsehen*, 1: 193–202, May 1930; 2: 65–68, Apr. 1931, and 173–178, July 1931.
16. A. G. Jensen, "The Davisson Cathode Ray Television Tube using deflection modulation," *Bell Sys. Tech. J.*, 30: 855–866, Oct. 1951.
17. Kurt Brückersteinkuhl, "Über das Nachleuchten von Phosphoren und seine Bedeutung für den Lichtstrahlabtaster mit Braunscher Rohre," *Fernseh Ag.*, 1: 179–186, Aug. 1939.
18. "Simultaneous all-electronic color television," *RCA Review*, 7: 459–468, Dec. 1946.
19. Philo T. Farnsworth, U.S. Patent No. 1,773,980, Aug. 26, 1930, applic. filed Jan. 7, 1927.
20. Philo T. Farnsworth, "Television by electron image scanning," *J. Franklin Inst.*, 218: 411–444, Oct. 1934.
21. A. G. Jensen, "Film scanner for use in television transmission tests," *Proc. I.R.E.*, 29: 243–249, May 1941.
22. V. K. Zworykin, "The Iconoscope—a modern version of the electric eye," *Proc. I.R.E.*, 22: 16–32, Jan. 1934.
23. Harley Iams, G. A. Morton and V. K. Zworykin, "The Image Iconoscope," *Proc. I.R.E.*, 27: 541–547, Sept. 1939.
24. Albert Rose and Harley Iams, "The Orthicon, a television pickup tube," *RCA Rev.*, 4: 186–199, Oct. 1939.
25. Albert Rose, Paul K. Weimer and Harold B. Law, "The Image Orthicon—a sensitive television pickup tube," *Proc. I.R.E.*, 34: 424–432, July 1946.
26. Paul K. Weimer, Stanley V. Forgue and Robert R. Goodrich, "The Vidicon Photoconductive Camera Tube," *Electronics*, 23: 70–73, May 1950.
27. A. W. Hull, U.S. Patent No. 1,387,984, Aug. 16, 1921, applic. filed Aug. 30, 1915.
28. A. W. Hull, U.S. Patent No. 1,683,134, Sept. 4, 1928, applic. filed Dec. 28, 1921.
29. K. W. Jarvis and R. M. Blair, U.S. Patent No. 1,903,569, Apr. 11, 1933, applic. filed Sept. 15, 1926.
30. L. Burmester and E. Mechau, "Untersuchung der mechanischen und optischen Grundlagen des Mechau-Projektors," *Kinotechnik*, 10: 395–401, 423–426 and 447–451, Aug. 5, Aug. 20 and Sept. 5, 1928.
31. A. G. Jensen, R. E. Graham and C. F. Mattke, "Continuous motion picture projector for use in television film scanning," *J. SMPTE*, 58: 1–21, Jan. 1952.
32. E. H. Traub, "New 35mm television film scanner," *J. SMPTE*, 62: 45–54, Jan. 1954.
   Victor Graziano and Kurt Schlesinger, "Continuous all-electronic scanner for 16mm color motion-picture film," *J. SMPTE*, 62: 294–305, Apr. 1954, Part I.
   Jesse H. Haines, "Color characteristics of a television film scanner," 1954 I.R.E. Convention Record, Part 7, Broadcasting and Television, pp. 100–104.
33. T. C. Nuttall, "The development of a high-quality 35mm film scanner," *Proc. Inst. Elec. Engrs.*, 99: 136–144, Apr.-May 1952.
   H. E. Holman and W. P. Lucas, "A continuous-motion system for televising motion-picture films," *Proc. Inst. Elec. Engrs.*, 99: 95–108, Apr.-May 1952.
34. Donald G. Fink, *Television Standards and Practice*, McGraw-Hill Book Co., New York and London, 1943.

35. D. K. Gannett and E. I. Green, "Wire transmission system for television," *Bell Sys. Tech. J.*, *6:* 616–632, Oct. 1927.
36. Edward L. Nelson, "Radio transmission system for television," *Bell Sys. Tech. J.*, *6:* 633–652, Oct. 1927.
37. M. E. Strieby and J. F. Wentz, "Television transmission over wire lines," *Bell Sys. Tech. J.*, *20:* 62–81, Jan. 1941.

M. E. Strieby and C. L. Weis, "Television transmission," *Proc. I.R.E.*, *29:* 371–381, July 1941.
38. G. N. Thayer, A. A. Roetken, R. W. Friis and A. L. Durkee, "The New York-Boston microwave radio relay system," *Proc. I.R.E.*, *37:* 183–188, Feb. 1949.
39. A. A. Roetken, K. D. Smith and R. W. Friis, "The TD-2 microwave radio relay system,"

Bell Sys. Tech. J., *30:* 1041–1077, Oct. 1951, Part II.

T. J. Grieser and A. C. Peterson, "A broadband transcontinental radio relay system," *Elec. Eng.*, *70:* 810–815, Sept. 1951.
40. G. R. M. Garratt and A. H. Mumford, "The history of television," *Proc. Inst. Elec. Engrs.*, *99:* 25–42, Apr.-May 1952.

TELEVISION BROADCASTING

Clure H. Owen

Television Broadcasting*

CLURE H. OWEN†, MEMBER, IRE

Summary—The story of television broadcasting may be divided into three periods of time, *i.e.*, the historical period—to about 1930, the developmental period—to the end of the Second World War, and the commercial period.

One of the first public demonstrations was given by C. Francis Jenkins in Washington, D. C. in 1925. During the 1930's the mechanical scanning equipment was replaced with all electronic equipment. The first "network" or long-distance pickup was the 1940 Republican National Convention in Philadelphia which was broadcast by the NBC station in New York and the General Electric station in Schenectady.

Commercial television was authorized by the Federal Communications Commission, effective July 1, 1941; however it was not until shortly after the end of the Second World War when the electronics industry returned to peace-time conditions that the public became aware of the magic of television in the home. In a relatively short period of time the television broadcasting industry developed from a few stations with very limited programming to a national industry with more than 600 television stations and over 50,000,000 receivers in use by the public. The impact of television has affected the lives of nearly every citizen of this country and its effects are being felt in most of the other countries of the world.

THE STORY OF television broadcasting, even when limited to the technical aspects, cannot be told by one person or contained in a single volume. Accordingly only the highlights can be touched upon in this brief article. Television broadcasting can roughly be divided into three periods of time, *i.e.*, the historical period up to about 1930, the developmental period from the early 1930's to the end of World War II, and the postwar commercial period.

HISTORICAL PERIOD

For ages man has dreamed of transmitting sight and sound from one place to another. One of the milestones along the road to television was the isolation of selenium by the Swedish scientist Baron Berzelius in 1817. While the light-sensitive properties of selenium were known for some years, little application of this property was made until 1892 when Elster and Geitel devised a photoelectric cell [1]. This cell is the basic principal upon which television, as we know it today, is based. Another milestone, that of scanning and reproducing an image, was made by Paul Nipkow, a German experimenter, when he invented the scanning disc in 1884. It was Nipkow's concept of changing the picture into electrical bits at the transmission end of a circuit, the sequential transmission of these bits and the reassembly at the receiving end into a complete picture, which makes it possible to transmit a television picture over a single telephone or radio frequency circuit.

* Received by the IRE, June 29, 1961; revised manuscript received, November 20, 1961.
† American Broadcasting Company, New York, N. Y.

The invention of the triode vacuum tube by Dr. Lee deForest in 1906 was one of the greatest achievements in the advancement of communications including, of course, both radio and television. In the early 1920's J. L. Baird in England and C. F. Jenkins in the United States carried on separate experiments using mechanical scanning. The mechanical system of scanning had very definite limitations on the amount of information that could be transmitted (picture quality) and the size of the reproduced image. The appeal of an electronic system without the above limitations was evident by the work of several of the early experimenters. M. Dieckman and G. Glace received a German patent in 1906 and B. Rosing of St. Petersburg received a patent in 1907; both for systems of television using a cathode-ray tube receiver [2]. The first suggestion of an all electronic television system was made by A. A. Campbell-Swinton in 1908 [2]. His suggestion of the camera tube included a multiplicity of photoelectric cells in parallel so that each cell could store a charge until scanned and would deliver only one electric pulse for each picture period. In 1923, V. K. Zworykin filed a U. S. patent for an electronic camera tube known as an "Iconoscope." This tube differed from earlier electronic camera tubes in that it employed beam intensity modulation by an axially symmetric grid. One of the first public demonstrations of television or radio vision, as it was then called, was given by Jenkins on June 13, 1925 when a live picture was transmitted from the Naval Air Station at Anacosta, Maryland, to the Jenkins Laboratories in Washington, D. C., a distance of some miles. A mechanical system of scanning was used in this demonstration [3].

In 1929 Zworykin attained success with an electrostatic cathode-ray picture display known as a "Kinescope." By the early 1930's most work on systems of scanning had changed over to one of the electronic systems. In the United States Philo T. Farnsworth and Vladimir Zworykin headed parallel investigations which by 1933 proposed equipment to make electronic television possible. Other persons making contributions to television about this time were Ives of AT&T, Alexanderson of GE, Goldmark of CBS, Engstrom and Goldsmith of RCA and DuMont [4]. Experimental work was also being carried on in Germany using cathode-ray tubes in receivers [5].

DEVELOPMENTAL PERIOD

It is difficult to draw a line of demarcation between the experimental period and the developmental period as there was, during the 1930's, a gradual transition from an occasional equipment test to more frequent,

and in some cases more or less regular, program operations. One of the earliest and most important stations in
the development of television was the RCA station
W2XBS in New York which went on the air with
mechanical scanning equipment in 1928.

During 1931 and 1932 field tests were made with the
transmitter located at the Empire State Building [6].
The transmitter used 120-line, 24 frames/sec mechanical
scanning. The receivers for these tests used cathode-ray
tubes instead of mechanical scanning.

During 1933 a system was built in Camden, New
Jersey which used the "Iconoscope" as a pickup device.
The system operated on 240 lines, 24 frames/sec. The
picture carrier was at 49 Mc and the sound carrier was at
50 Mc.

In 1934 the system operated with 343 scanning lines
and used interlaced scanning. The frame frequency was
30/sec and the field frequency was 60/sec. The last
mechanical element of the system was removed by the
development of an electronic synchronizing generator.

In 1936 provisions were made for live and film studios
at Radio City to be used with the Empire State Building
transmitter. The picture carrier was operated at 49.75 Mc
and the sound carrier at 52 Mc. The two transmitter
outputs were fed through coupling filters so that a common transmission line and antenna could be used. The
receivers used in these tests were superheterodynes
capable of receiving both the picture and sound simultaneously.

During the latter part of the 1930's regular studio
programs were instituted (Figs. 1 and 2). At first most of
the receivers were company-owned receivers in the homes
of RCA-NBC employees; however, by 1940 it was estimated that there were about 3000 receivers in the New
York area. On April 30, 1940 a regular public series of
television programs was inaugurated with the opening
of the New York World's Fair by President Roosevelt.

The Philco Radio and Television station W3XE-
WPTZ [7] in Philadelphia carried on extensive broadcasts with remote pickups of outdoor events at Franklin
Field. It used a high-frequency radio relay to transmit
the programs from Franklin Field to the television
station.

The General Electric Company's television station
WRGB in Schenectady [7] was one of the early television stations carrying on both equipment and program
development.

The Columbia Broadcasting System's station WCBW
in New York had its transmitter located in the Chrysler
Building and studios in the Grand Central Terminal
Building. While this station carried on public interest
programming its niche in history is on the basis of the
equipment development and color television investigations under the capable leadership of Dr. Peter Goldmark.

The DuMont Laboratories' station WABD in New
York [7] carried on both program and equipment development. The resourceful DuMont engineers, whose

Fig. 1—NBC experimental television studio.

Fig. 2—NBC experimental television transmitter.

great experience in cathode-ray electronics made many
contributions to television, produced some of the earliest
large kinescope picture tubes.

The Don Lee Company's station W6XAO in Los
Angeles [7] was instrumental in the development of
interest in television on the West Coast. Station W6XAO
carried on extensive tests of television reception in an
airplane. Self-synchronized cathode-ray television receivers demonstrated that television reception was possible using power sources other than that common to
the power source of the transmitter. A field strength
contour map was prepared based upon reception of
signals at 3000 feet [8].

The Paramount stations WBKB in Chicago and
W6XYZ in Los Angeles [7] were more concerned with
the development of programming methods and the
production of entertainment than the development of
equipment.

The Zenith station W9XZV-WTZR in Chicago [7]
carried on small-scale but intensive development of
equipment and field testing of television receivers.

The Farnsworth station W2XPF [7] originated on
the West Coast but was later moved to Philadelphia.
This organization under the guidance of the youthful

electronics genius Philo T. Farnsworth made many contributions in the field of television, most noteworthy of which was the "image dissector" camera tube.

In 1936 the British Broadcasting Corporation opened a London transmitting station at Alexander Palace. Two systems were used for television transmission. The Baird system used 240 lines, 25 pictures/sec with sequential scanning. The Marconi-E.M.I. system used 405 lines, 25 pictures/sec, interlaced, to give 50 frames/sec. The Marconi-E.M.I. system used cathode-ray-type pickup for both direct and film transmission [9].

The first "network" or long-distance television pickup [10], [11] was the 1940 Republican National Convention held in Convention Hall in Philadelphia. It was broadcast by the NBC station W2XBS in New York and the GE station W2XB in Schenectady. The pickup was made by NBC field pickup equipment and transmitted by wire lines from the Convention Hall to the Bell System carrier terminal in Philadelphia, to the RCA Building in New York by coaxial cable, to the Empire State Building by wire lines and to the Helderberg mountain site of W2XB by an off-the-air pickup of W2XBS in New York. The wire lines used from the Convention Hall to the Bell System carrier terminal in Philadelphia and from the RCA Building to the Empire State Building in New York were carefully selected telephone circuits with wide-band repeaters and equalizers installed at the terminals, as well as at intervals along the circuits. The coaxial cable used was the experimental coaxial cable installed between New York and Philadelphia by the Bell System for the study of multiple carrier telephone transmission. The off-the-air pickup at the Helderberg mountain site near Schenectady was made possible by a specially constructed receiving antenna system.

COMMERCIAL TELEVISION

Commercial television was authorized by the Federal Communications Commission, effective July 1, 1941; however, the start of the Second World War shortly thereafter effectively curtailed any rapid development of commercial television. Following the conclusion of the war the television broadcasting industry developed at an unprecedented rate. With only a few low-power television broadcasting stations on the air and a small number of receivers in use the public suddenly became aware of the magic of television in the home. The television home in each neighborhood was a popular meeting place with neighbors, both adults and children, streaming in to watch the new medium of entertainment. Although the size of the television screens was small and the technical quality and program fare inferior by latter-day comparison, the public interest generated by this new medium far exceeded the most optimistic opinions of the "experts." The neighbors soon desired receivers in their own homes and the sale of receivers sky-rocketed. The popularity of television resulted in an increased number of stations until 1948 when the Federal Com-

munications Commission stopped granting authority for additional stations pending a reallocation to minimize interference between stations. The television "freeze" imposed in 1948 was not lifted until 1952 when again new stations were permitted to start broadcasting and the total number of television stations rapidly increased. Table I shows the growth in the number of television stations authorized and the television receivers in use by the public from 1946 through 1960.

The impact of television upon the home, advertising, entertainment, etc., has been the subject of vast quantities of written material. For this article it is sufficient to say that the impact of television has affected the lives of nearly every citizen of this country and its effects are being felt in most of the other countries of the world.

TABLE I

	Commercial Television Stations Authorized* [12]	Television Receivers In Use [13]
1946	30	5,000
1947	66	150,000
1948	109	1,010,000
1949	117	3,660,000
1950	109	9,785,000
1951	109	15,590,000
1952	108	21,460,000
1953	483	26,920,000
1954	573	32,750,000
1955	582	37,360,000
1956	609	42,810,000
1957	651	46,690,000
1958	665	49,900,000
1959	667	53,290,000
1960	653	56,210,000

* At the close of each fiscal year, June 30.

TELEVISION STANDARDS

The transmission and reception of television pictures require that the image be electronically disassembled at the transmitting end, transmitted bit-by-bit and reassembled in the correct order at the receiving end. For the receiver to correctly reassemble the picture it is necessary that the transmitter and receiver use the same set of transmission standards, *i.e.*, the same number of lines per picture, the same number of pictures per second, etc. To keep the received picture in step with the transmitted picture, synchronizing signals are transmitted.

During the experimental period of television various numbers of lines per picture and pictures per second were used, with as low as 30-line pictures scanned at a frequency of 5 pictures/sec. To obtain better definition of picture detail the number of lines per picture as well as the number of pictures per second were increased from time to time. During 1936 the RMA Television Committee recommended standards which included 441 lines, frame frequency of 30/sec and field frequency of 60/sec, interlaced. During 1941 the Federal Communications Commission adopted the 525-line standards proposed by the National Television Systems Committee.

Space does not permit the detailed discussion of all the standards of television transmission. The reader is referred to the Rules and Regulations of the Federal Communications Commission for details.

Unfortunately the various countries of the world have adopted different television standards. This, as might be expected, makes for some difficulty in transmitting television programs between countries. In transmitting live or taped television programs between countries where different television standards are used, various standards converters have been employed. In general these standards converters have been based upon the photographic principle, *i.e.*, the picture presentation on a kinescope monitor operating on one set of standards would be photographed by a television camera operating on a different set of standards. The resultant pictures suffered from loss of detail and often had some visible flicker. The American Broadcasting Company has developed an all electronic standards converter which has no optics in the system and has no perceptible picture degradation (Fig. 3). Table II gives the principal television systems of the world.

In 1953 the Federal Communications Commission adopted standards for color television. The standards adopted were the result of several years' work by the National Television Systems Committee. The color standards are compatible with the monochrome standards in that a monochrome receiver is able to receive a black and white picture when the picture is transmitted in color for color-equipped receivers. The color information is transmitted on an amplitude modulated subcarrier at about 3.58 Mc above the picture carrier.

Fig. 3—All electronic standards converter.

TABLE II
PRINCIPAL TELEVISION SYSTEMS OF THE WORLD [14]

Description Used in Foreign Station Directory	British	World-Wide	Western Europe	Eastern Europe	French
Location of Principal Use	United Kingdom	Western Hemisphere, Far East	Western Europe	Soviet orbit	France and Possessions
Number of Lines Per Picture	405	525	625	625	819
Video Bandwidth (Mc)	3	4	5	6	10.4
Channel Width (Mc)	5	6	7	8	14
Sound Carrier Relative to Vision Carrier (Mc)	−3.5	+4.5	+5.5	+6.5	11.15*
Sound Carrier Relative to Edge of Channel (Mc)	+0.25	−0.25	−0.25	−0.25	0.10*
Interlace	2/1	2/1	2/1	2/1	2/1
Line Frequency (cps)	10,125	15,750	15,625	15,625	20,475
Field Frequency (cps)	50	60	50	50	50
Picture Frequency (cps)	25	30	25	25	25
Vision Modulation	+	−	−	−	+
Level of Black as Percentage of Peak Carrier	30	75	75	75	25
Sound Modulation	AM	FM	FM	FM	AM

* French standards invert the video and audio frequencies for certain channels.

TELEVISION EQUIPMENT

The experimental and developmental periods of television permitted a determination of the basic fundamentals of television equipment. However, commercial television resulted in demands for equipment which exceeded all expectations. These demands were not restricted to greater quantities of existing equipment but were also for specialized equipment to permit greater flexibility of operations, greater reliability, ease of maintenance, etc.

Competition between stations resulted in a necessity for technical quality control of broadcast signals which would permit the best possible pictures being received by the public. It is not the purpose of the paper to comment on all types of television equipment. However, comments follow on the basic types of equipment such as live cameras, projection, lighting, switching, video recording and transmitting equipment.

LIVE TELEVISION CAMERA

The live television camera, which can be considered the "eye" of television, is one of the most important single elements in the line-up of the equipment system. With the exception of film or tape reproduction of programs the live television camera serves as the "front door" for the entrance of all television signals into the broadcasting system.

The image orthicon camera is generally considered as the workhorse of the industry for studio use. Its ability to satisfactorily handle scenes with a wide range of light levels ranging from only a few foot-candles for mood scenes to some 300 foot-candles for high lights, is desirable for general studio use.

The vidicon camera has found universal use as a film camera, however, as a live camera its use has been limited. The target, being photoconductive rather than photoemissive as the image orthicon, has a higher sensitivity; as a result the size of the tube as well as the associated camera is much smaller.

Without the electron multiplier and shading control and with less critical beam current control, the operation of the vidicon camera is less critical than the image orthicon camera. The signal-to-noise ratio of the vidicon is more dependent upon light levels than the image orthicon. Accordingly, for general studio use, light levels above 200 foot-candles are required for satisfactory operation. The vidicon is being extensively used at present for unattended closed-circuit television, also for limited area scenes such as newscasts, flip-cards, etc. (Fig. 4).

Live color television cameras have three image orthicon cameras, one for each of the primary colors, red, green and blue. The light from the scene being televised is focused through a lens common to all of the tubes after which it is split three ways by means of dichroic mirrors and light filters. Each color channel has its own amplifiers and color correction circuitry.

Fig. 4—Vidicon camera.

FILM PROJECTION EQUIPMENT

Film has always played a very important part in television programming. At first the film used was restricted to motion picture films. In later years more and more of the films used were made especially for television broadcasting. Commercial television started with the use of the iconoscope film cameras. Today the vidicon film cameras have almost completely replaced the old iconoscope cameras with a marked improvement in technical quality. Both 35-mm and 16-mm film projectors are used in the telecine rooms although many of the television stations in smaller markets use only 16-mm film projectors. The film projectors are sometimes used with an individual camera; however, most stations use multiplexers for selecting, by means of movable mirrors, any one of three or four light sources, *i.e.*, film or slide projectors, for use with one film camera chain.

Color television film projection requires the use of three vidicon cameras, one for each of the primary colors, red, green and blue. The light being projected through the color film is separated by appropriate dichroic mirrors and color filters. A separate lens is used for each vidicon film camera.

LIGHTING EQUIPMENT

Artificial lighting is necessary for practically all studio television program pickups. Outdoor scenes with adequate daylight can usually be picked up with an image orthicon camera without additional lighting.

In the early days of television the relatively insensitive iconoscope camera tube required very bright illumination of the subject to be photographed. To obtain adequate pictures light levels of 800 to 1200 foot-candles were needed. There were several disadvantages to these high light levels, such as actor discomfort, high power costs, the great number of lighting fixtures, and tremendous heat loads which made it essential to air-condition the studio.

With the greater sensitivity of the image orthicon

camera tube the problem of studio lighting was greatly simplified. With the exception of mood scenes most studio illumination runs from 50 to 150 foot-candles. The lighting of television scenes is an art which involves not only the placement of the light fixtures but the regulation of the light output of the individual lights to obtain the desired artistic effect. The regulation of light output is accomplished by dimmers to control the voltage applied to the lamp filaments. The dimming circuits for the individual lamps vary from the old-fashioned series resistance, through the autotransformers, grid-controlled thyrathron rectifiers to the more modern solid-state silicon controlled rectifier dimmers with punch-card preset control (Fig. 5). The latter system permits almost instantaneous change from one lighting setup to the next.

Video Switching Equipment

In the early days of television the video switching requirements were relatively simple as it was only necessary to switch between cameras or from live to projection equipment. As great strides have been made in television programming great demands have been made upon the switching equipment. This has resulted in complex equipment for handling a large number of inputs and outputs as well as special effects and superimposition and montage of a portion of one scene as an insert within another scene.

Video Recording Equipment

The first type of video recording was accomplished by focusing a film camera on the face of a kinescope monitor. The film was then developed and the required number of prints made. This type of video recording has been used for network film distribution since the early days of commercial television. The principal users of kine recordings are stations which do not have live network connections or which desire to schedule the program at other than the time it is carried on the network.

Video tape recording equipment permits immediate playback of a program and is capable of higher technical quality than kine recordings. Video tape recording has added a third source of television programming to the previous live or filmed sources. The use of video tape equipment permits a more efficient use of studio space, as entire programs, selected scenes, or commercials can be prerecorded and inserted at the desired time.

Transmitting Equipment

Television transmitters are somewhat different from transmitters in other services as it is necessary to transmit both the aural and visual signals within the 6-Mc television channel. This is normally accomplished by two separate transmitters, one carrying the aural signals and the other carrying the visual signals. The outputs of both transmitters are generally fed into a diplexing circuit for exciting a single antenna with both aural and visual signals (Fig. 6).

The high-frequency propagation of television signals requires that the transmitting antennas be located as high as possible above the surrounding area to provide maximum service. Accordingly most transmitting antennas are located on mountain tops or tall buildings. Furthermore, due to the directional characteristics of television receiving antennas it is advantageous to have all of the television transmitters in the same area. Fig. 7 shows the transmitting antennas on the Empire State Building where all seven of the New York television stations are located.

Fig. 5—Solid-state silicon controlled rectifier dimmers with punch-card preset control.

Fig. 6—50-kw television transmitter.

Fig. 7—The Empire State Building tower with seven television transmitting antennas.

Common Carrier Transmission of Television Programs

The common carrier facilities of the telephone companies are the arteries which carry the life-blood television programs from the network control centers to the affiliated stations across the country. The transmission of television programs, as distinguished from radio programs, involves the transmission of both the sound programs and the picture programs. Technically these could be transmitted together over the same circuits. In the United States, however, due to the existence of considerable audio facilities for more than two decades of radio network transmission, the aural and visual programs are transmitted over separate circuits.

Liaison between the television networks and the telephone companies is maintained by two special committees in addition to the normal operating contacts. The Video Transmission Engineering Advisory Committee (VITEAC) promotes a common understanding of the various factors of an engineering and policy nature as related to network service. The Network Transmission Committee (NTC) is the working subcommittee of the VITEAC.

It is the function of network facilities to deliver to numerous receiving locations a signal which is essentially unchanged from the signal which is applied at the originating point. To transmit audio signals satisfactorily, the facilities must transmit all frequencies with uniform loss and phase shift and must accommodate the complete range of loudness without distortion from overloading or interference from noise. For this reason they are gain and delay equalized, and maximum and mimimum levels are established.

Similarly, video facilities are equalized for gain and delay and also for differential gain and phase, and operating levels are set. The frequency band required for video transmission extends from 0 to 4.5 Mc and the entire band is equalized essentially flat. Variations from flatness in the region from 0 to 250 kc result in streaking or smearing of the picture. In the high-frequency range different tolerances apply to color and monochrome transmission. For color television signals differential gain and phase require close control to avoid effects on the hue and saturation of the color signal. Color facilities in general are equipped with special equalizers which operate in the range of the color subcarrier and its sidebands. Circuits which are set up to handle color television transmission will also handle monochrome transmission.

In the early period of television network transmission coaxial cable was used for intercity circuits. New circuits are now using microwave radio facilities.

Community Antenna Systems

The location and service range of television stations in the United States and Canada were such that many of the people, particularly those residing in the moun-tainous areas, were unable to enjoy television reception. Due to the very great demand for television reception many community antenna systems were installed. In general these systems consisted of a high-gain television receiving system installed on a mountain top near the community, a radio frequency amplifier and a coaxial cable system for transmitting the television signals from the mountain-top location to the various homes in the community. Most of the installations have been able to receive and distribute the programs of several television stations. Some of the community antenna systems use a microwave circuit where off-the-air reception of television programs is not possible.

Television Booster and Translator Stations

Television booster and translator stations, in a manner comparable to the community antenna systems, are means of extending television service into underserved areas. In general these are low-power relay stations located on mountain tops or other areas of possible television reception. The signals are received and retransmitted, on the same television channel in the case of the booster stations, or on a different channel in the case of the translator stations, to the television receivers in the areas of desired reception.

Acknowledgment

The author wishes to express his appreciation to R. M. Morris of the American Broadcasting Company and to H. A. Ahnemann of the AT&T Long Lines Department for their assistance in the preparation of this manuscript.

Bibliography

[1] A. M. Glover, "A review of the development of sensitive photo-tubes," Proc. IRE, vol. 29, pp. 413–423; August, 1941.

[2] V. K. Zworykin, E. C. Ramberg, and L. E. Flory, "Television in Science and Industry," John Wiley and Sons, Inc., New York, N. Y., p. 9; 1958.

[3] C. F. Jenkins, "Radio vision," Proc. IRE, vol. 15, pp. 958–965; November, 1927.

[4] W. C. Eddy, "Television, The Eyes of Tomorrow," Prentice-Hall, Inc., New York, N. Y., p. 3; 1945.

[5] M. Von Ardenne, "An experimental television receiver using a cathode-ray tube," Proc. IRE, vol. 24, pp. 409–425; March, 1936.

[6] L. M. Clement and E. W. Engstrom, "RCA television field tests," RCA Rev., vol. 1, pp. 32–40; July, 1936.

[7] W. C. Eddy, "Television, The Eyes of Tomorrow," Prentice-Hall, Inc., New York, N. Y., pp. 17–25; 1945.

[8] H. R. Lubcke, "Television image reception in an airplane," Proc. IRE, vol. 20, pp. 1732–1740; November, 1932.

[9] "Radio Progress During 1936: Television and Facsimile," Proc. IRE, (Rept. by Technical Committee), vol. 25, pp. 199–211; February, 1937.

[10] O. B. Hanson, "RCA-NBC television presents a political convention as first long distance pickup," RCA Rev. vol. 2, p. 267; January, 1941.

[11] M. E. Strieby and J. F. Wentz, "Television transmission over wire lines," Bell Sys. Tech. J., vol. 20, pp. 62–81; January, 1941.

[12] Annual Reports of the Federal Communications Commission.

[13] "TV Factbook No. 32," Television Digest, Triangle Publications, Inc., Radnor, Pa., p. 24; 1961.

[14] Ibid., p. 832.

A HALF CENTURY OF TELEVISION RECEPTION

F[rank] J. Bingley

A Half Century of Television Reception*

F. J. BINGLEY†, FELLOW, IRE

Summary—The paper traces the development of television receiving, starting with a description of the earliest demonstrations of mechanical television. While the major emphasis is on reception, enough discussion of television system standards is included to support the development of the receiver story. The progression of events which led from the early mechanical systems, through the various phases of electronic television using cathode-ray picture tubes, up to the modern compatible color television receivers, is covered in some detail. The development of circuitry for IF, VF, and deflection is covered. A complete bibliography, containing some of the great historic as well as the most recent contributions, is included.

INTRODUCTION

THE DEVELOPMENT of the art of television receiving is inextricably bound up with that of television transmission; there is a "lock-and-key" relationship between the transmitter and receiver. Thus, while the emphasis of this paper will properly be placed on receiving techniques, it will be necessary in many cases to supplement the discussion with sufficient explanation of the transmitter and over-all system philosophy.

The earliest development of television was directed towards generation of television signals, and their display as television pictures without the intervention of radio transmission and reception circuits. The earliest demonstrations of television used equipment comprising mechanical devices for scanning the image. It is fitting to begin with a discussion of them.

I. EARLY MECHANICAL TV HISTORY

So far as the author is aware, the first demonstrations of actual television images were made in the year 1925 by C. F. Jenkins in Washington [1] and by J. L. Baird in London. They worked independently of each other, but it appears not to be possible to state to whom should be given the credit of priority. The pictures shown by Baird were not transmitted and received by radio, but rather over a connecting wire. Similarly he avoided the synchronizing problem by using mechanical scanning with a common drive connecting the scanning devices at both "transmitter" and "receiver." Jenkins used radio transmission, and synchronous motors for synchronization.

By the fall of 1927, Baird had progressed to the point of making experimental broadcasts from a radio station near London. His scanning equipment included an 8 foot diameter lens disk carrying 30 lenses associated with a photocell. The monitor display was an 8 foot,

Nipkow scanning disk with 30 apertures, scanning vertically. A neon lamp tube served as the light source. Synchronization at the studio end of the system was achieved by the solid shaft connecting the lens disk with the Nipkow disk. The broadcasts were received at White Plains, New York. History does not record how well the pictures were received, or how the synchronization was accomplished.

A great forward step was taken in 1927 by Bell Telephone Laboratories [2], who set up a system for transmitting television from Washington, D. C., to New York, a distance of 250 miles. The transmission could be carried alternatively by radio or by wire circuit. There were 50 scanning lines per frame and 16 frames per second. The receiver displays used were of two types; one produced a small image about $2 \times 2\frac{1}{4}$ inches, the other a large image $2 \times 2\frac{1}{2}$ ft. The smaller picture was produced by a scanning disk revolving in front of a special neon bulb having a flat cathode; the disk was driven by a synchronous motor fed and synchronized from a separately transmitted signal. The large display consisted of a set of 50 parallel tubes containing neon and a set of 2500 electrodes, arranged in a sequence of 50 along each neon tube. These electrodes were connected by 2500 wires to a 2500 segment commutator run by a synchronous motor similar to that used for the scanning disk. The video signal was used to modulate a 500 kc RF carrier, and this was distributed by the commutator which thus provided the scanning function. When the signal was received over wire facilities, the circuit was carefully corrected for amplitude and phase up to a maximum frequency of 20 kc. When it was received by radio, a special superheterodyne receiver was used for the TV signals (the transmitted carrier was 1575 kc); separate receivers were also used for speech (1450 kc) and synchronizing signals (185 kc).

Another early pioneer was E. F. W. Alexanderson who, working at General Electric, produced a projection display using a drum of mirrors associated with a modulated light source [3]. The image of the light source was formed at the screen after reflection by the mirrors of the drum; as the drum rotated the image was caused to move vertically on the screen. Because, in addition to their angular spacing around the drum, the mirrors were tilted progressively with respect to its axis of rotation, the images reflected by successive mirrors were displaced sideways on the screen as each successive mirror came into play. The result was that the receiving screen was scanned. In order to produce more scanning lines from a drum having a given number of mirrors (24 in the early experiments), he used seven light sources so that in effect each mirror scanned seven lines in parallel.

* Received by the IRE, September 21, 1961; revised manuscript received, November 16, 1961.

† Government and Industrial Division, Philco Corporation, Philadelphia, Pa.

Thus the total effective number of lines was 168. In later experiments the multispot scanning was abandoned and a simple sequential scanning of 48 lines at 15 frames per second was used; the picture was displayed on a screen about 3 ft square.

II. EARLY ELECTRONIC TELEVISION RECEIVERS

The arrival of electronic television may be dated at approximately the year 1930. Much prior to this year Dieckmann in Germany (1906), Rosing in Russia (1907) and Campbell Swinton in England (1911) [3]–[5] discussed and described electronic television receivers employing cathode-ray tubes as display devices. Dieckmann actually had a crude unit in operation in 1906; however his system transmitted only silhouettes, so there is some reluctance to give him full credit for having produced the first electronic receiver. No further work on electronic television appears to have been done until the early 1920's, when the work of Philo Farnsworth and Dr. V. K. Zworykin became known. The importance of their work was not recognized until 1930 however, perhaps because some of the mechanical approaches to television were beginning to show real promise. It was only when the overriding limitations of mechanical TV were finally, if reluctantly, admitted that electronic television received the research and development effort it merited.

In about the year 1929 Farnsworth demonstrated an all electronic TV system in which the display device was a cathode-ray tube, magnetically deflected and focused [6]–[8]. In 1931 Farnsworth joined Philco Corporation, where he continued his work for about two years during which time early experiments in broadcast TV reception on frequencies in the order of 60 Mc were made. Transmissions from W3XE (the Philco experimental station) were received in Mt. Airy, Philadelphia, a distance of about 6 miles, in 1932.

The work of Dr. V. K. Zworykin was originally carried out at Westinghouse Research Laboratories. He is reported to have demonstrated an all-electronic system in 1924. The earliest published account of his experiments appeared in *Radio Engineering* in December, 1929 [9]. The display device at the receiver was a 7-inch cathode-ray tube with ES-EM deflection, and ES focusing. For beam modulation a control grid was used. When Dr. Zworykin joined RCA in 1930, this work was continued by him and his associates.

III. TELEVISION STANDARDS

By about 1935 electronic television had progressed to the point where it became obvious that the most pressing requirement was for a set of standards, in order that the development of systems could proceed in an orderly fashion. By that year there already existed experimental transmitters in the VHF band, camera tubes such as iconoscope and dissector, and cathode-ray display tubes. The necessary ancillary circuits, such as for deflection

and video amplification, were already well known. It was obvious that vitally important factors to be decided were those of how many scanning lines should be used, whether they should be interlaced, and what field and frame rates were required. These were basic factors, but they reflected their influence widely throughout the whole system design, affecting video bandwidth (and in turn transmitter and receiver bandwidth), picture flicker, and deflection circuit frequencies. Along a parallel track lay the problems concerned with transmission and reception of the video signal and the accompanying sound. Items needing consideration here were method of modulation, polarity of transmission, polarization, the applicability of single or double sideband or other band saving means, the channel layout (spacing and relation between video and sound carrier and their locations in the channel), allocation of broadcast frequencies, and many others. Standardization of these items was achieved in the fall of 1941, and the FCC issued commercial television standards at that time. Some of the more important standards will now be discussed.

Amplitude modulation was finally selected for picture transmission. Partial suppression of the lower sideband was used to save bandwidth. Frequency modulation for picture transmission was tested early in 1941, but found to produce bad effects in the receiver in the presence of multipath; it was abandoned for broadcast use for this reason. The polarity of modulation was selected as negative (that is, more light at the scene, less carrier amplitude). On the other hand, frequency modulation was selected for sound transmission, since a sufficiently high deviation ratio could be provided to take full advantage of the FM system.

The spacing between sound and picture carriers was standardized at 4.5 Mc, with the sound carrier 0.25 Mc below the upper edge of the channel, and the total channel width 6 Mc.

It was realized quite early (1933) that the transmission of 60 fields per second, interlaced two to one, was desirable from the standpoint of elimination of flicker. The matter was briefly reexamined in 1941, but the original decision was reaffirmed. The number of scanning lines progressed from 343 in 1935, to 441 in 1937, and was finally standardized at 525 in 1941. The video bandwidth increased correspondingly from about 2 Mc in 1933 to 4.25 Mc in 1941. The horizontal deflection frequencies increased from 10.29 kc to 15.75 kc during this period.

The work of standardization was carried out by a number of industry groups. The final coordination was done by the National Television System Committee. Their work has been described in detail by Fink [10].

IV. DEVELOPMENT OF CIRCUITS FOR ELECTRONIC TELEVISION RECEIVERS

The development of television standards briefly referred to above was concurrent with a progressive development of television receiving techniques. During the period up to about 1945, superheterodyne receivers were developed capable of handling the necessary bandwidth of 6 Mc [11]–[14] and video amplifiers capable of amplifying adequate bandwidth with acceptable frequency and phase characteristics were developed and design formulas were obtained [15]–[19]. Basic factors having to do with receiver and system optimization were studied during this period, and papers were written by Kell, Bedford and Fredendall [20], Wilson [21], Wheeler [22], [23], Loughren [22] and Baldwin [24].

The design of intermediate frequency amplifiers was considered by many workers in the field, including Butterworth [11], Landon [13], Sziklai and Schroeder [14] and more recently by Avins [25], Ruth [26], Bridges [27], and Bradley [77].

Prior to 1945 the accepted method of sound reception had been to use a separate IF amplifier for this purpose. It required great stability of the local oscillator in order to maintain the sound within the narrow sound IF channel. A new method, utilizing the 4.5-Mc beat between sound and picture carriers, was developed and is now used exclusively [28]. Since the 4.5 Mc beat frequency is determined at the transmitter, receiver tuning is now not critical.

Synchronizing and deflection circuits and yokes have shown progressive development. In the 1930's deflection angles of 50° were considered large. At the present time the figure has risen to 110°, with corresponding decrease in cabinet depth [29]–[33]. This has been accomplished by increased efficiency in circuits and yokes. At the same time synchronizing circuits have been improved, notably by the introduction of AFC synchronization.

Transistors are now being used in television receivers, particularly in portable models and in IF amplifiers. Much work has been reported in this field [34]–[44].

Considerable development has been applied to picture tubes since 1930. This development has covered screens, guns and geometry. Prior to about 1935 engineers had been satisfied to do what came naturally with respect to phosphors—namely, to use the readily obtainable green fluorescing willemite as a screen phosphor on the receiver display. However, about that year it became obvious that such a color would be unacceptable for monochrome TV receivers, and development of efficient white phosphors began. By 1938 the present two component white-balanced screens began to appear [45]–[47].

In about the year 1935 it became recognized that the central regions of picture tube screens were exhibiting dead areas in the phosphor—that is to say, the picture contained a dark spot in the center. It was established that this was due to bombardment of the screen by negative ions, presumably emitted from the heated cathode. The ionic mass was so great, compared to that of the electrons in the beam, that the ions were practically undeflected by the magnetic deflection fields.

They could be focused by an electrostatic field, but not by a magnetic field. Magnetically focused tubes showed a sharp shadow of the tube neck as a boundary of the ion blemish. Electrostatically focused tubes showed a smaller more intense black spot due to the electrostatic focusing action. Research using the picture tube as a mass spectrograph identified the various ions involved.

By devising an electron gun having a bent section (the axis of the bend being normal to the tube axis) the ions became trapped in the gun. By adding an external auxiliary magnetic field the electron beam in the gun was bent so it could pass down the axis of the second section of the gun, which was coaxial with the tube neck. However, the ions, which were not deflected by the auxiliary magnetic field, remained trapped in the gun and thus could not reach the screen to damage it. The development of the aluminized tube provided protection for the phosphor from ion bombardment, and ion traps are not required with such tubes.

Phosphor screens emit light on both sides. That which is emitted on the side facing the electron gun is wasted so far as the observer is concerned, if it is permitted to be absorbed and reflected by the inside surface of the picture tube. More than this, it destroys the over-all contrast of the picture. The operation of the aluminized tube depends on the fact that an electron beam of velocity such as used in a picture tube can penetrate a film of aluminum which is at the same time thick enough to reflect essentially all the incident light. Thus, light emitted from the rear surface of the phosphor screen is reflected and appears as a useful image illumination to the viewer. It is prevented from destroying image contrast. The aluminum film also protects the phosphor screen from injury due to ion bombardment [48].

Deflection of the cathode-ray beam may be accomplished either by electrostatic or electromagnetic fields; and the same is true of focusing. During the period 1930 to 1950, ES and EM focusing were used, with the general predominance on EM focusing since it was easier to get fine bright spots and the focusing element was outside the tube, giving a freedom to adjust and experiment. In the period 1950–1955 the disadvantage of the current drain of the EM focuser was overcome by designing permanent magnet focusers with adjustable airgaps. Since 1955, with the increase in knowledge of electron optics, it has been possible to design good picture tubes with electrostatic focus, and these are now widely used in television receivers. Regarding deflection, some of the early experimenters used crossed ES-EM systems, using the ES for the high-frequency deflection. This was probably done because of the lack of knowledge at that time concerning the design of magnetic deflection circuits operating at high frequency. By about 1935 magnetic deflection circuitry was well understood and has been used almost exclusively since then. There are new developments in ES deflection cathode-ray tube which may affect the present monopoly of EM deflection [49]–[53].

V. Color Television Receivers

A. Early Devlopment

Within the period of this review, the earliest color television work appears to have been carried out in England by Baird in about June 1928. This was an all mechanical field sequential receiver display, with 15 red lines and 15 cyan lines. Naturally the results were quite crude, and only demonstrated principles [54].

The next demonstration of which the author is aware was one given by Bell Laboratories some time in June 1929, at their Laboratories in West Street, New York [55]. The system was devised by Dr. Ives, the well-known expert in optics and colorimetry. Three separate channels were used, one each for red, green and blue. The receiver display had 50 lines and 18 pictures per second, using a 20-kc band for each channel.

What may be called semi-electronic color television was first broadcast by Dr. P. C. Goldmark of CBS in 1940 [56]. The system used a field sequential receiver display, with a disk having successive red, green and blue filters revolving in front of a cathode-ray picture tube. The disk was synchronized so that red, green and blue separation images were displayed in succession. This system received much attention, particularly in the post-war years. It suffered from the compromise between resolution and flicker, and from its noncompatibility with the existing monochrome service. Since it was a sequential system, there was no way in which the bandwidths required to satisfactorily transmit red, green and blue separation images could be tailored to fit the lesser demands of red and blue as compared to those of green. It also suffered from the disadvantage of not being entirely electronic, requiring (at least at the time it was under consideration) a synchronized filter disk. In spite of these disadvantages the system was adopted as U. S. standard by FCC [57] in October 1950, the scanning specifications being 405 lines, 48 color fields, interlaced, leading to a flicker rate of 48 cps. In order to partially offset the lack of detail, ingenious "crispening" circuits were devised [58] and applied to the receiver. It was demonstrated that a receiver could be built with changeover switches to show either monochrome or color pictures, thus rendering the lack of compatibility somewhat less objectionable. The standard was rescinded in December 1953.

Most of the development of color television receivers since 1950 has centered about the development of the display device. The progress will therefore be discussed mainly in terms of the display.

B. Line Sequential Color Display

An early approach to color display was that called "line sequential," in which successive lines were scanned in red, green and blue. This was first demonstrated in 1949; but, since the display involved a triple interlace, objectionable traveling patterns intruded upon the picture. No solution being found, the system was abandoned.

C. The Chromatron Tube

This tube, developed by Dr. Lawrence, is characterized by having a screen structure in which the phosphors are laid down in horizontal stripes [59]–[61]. The phosphor lines are of three separate kinds of phosphor, fluorescing in red, green, and blue, respectively. Close to the screen is a set of deflection plates running parallel to the phosphor lines. The plates cause the electron beam to be selectively deflected according to the potential between them. In this fashion the scanned raster can be caused to fall only on the green phosphor, or only on the red or blue phosphor. The selection is made at the instants when the video signal amplitude is appropriate to control the luminance of selected phosphor. This is accomplished by signal processing at the receiver.

D. The Shadow Mask Tube

This tube is of the three gun type—that is to say, separate electron guns are assigned to exciting the red, green and blue phosphors, respectively [62]–[65]. This is achieved by the parallax existing between the three guns due to their slightly differing space positions, as they beam electrons through an apertured plate on to a phosphor screen placed parallel to it. The plate has about 300,000 holes and is of the same size as the phosphor screen. The phosphor screen has 900,000 phosphor dots, arranged in groups of three, each group consisting of a red dot, a green dot, and a blue dot (the color named refers to the fluorescent radiation). Each group of three is arranged at the corners of a very small equilateral triangle of such a size and so oriented that the electrons from one gun (call it the red gun) can only fall on the red dots; and similarly for the green gun and green dots, and the blue gun and blue dots. This relation must hold over the whole screen no matter where the three beams are deflected by a deflection yoke through which they all pass. Naturally the size of phosphor dots must be carefully controlled, as must the relation between the dots and the aperture mask, which requires great precision of alignment.

The signal processing circuits effectively feed red, green and blue separation signals between grid and cathode of each respective gun.

E. The Apple Receiving System

The name "Apple" was applied by Philco to a specific type of single gun color tube developed in its Research Laboratories [66]–[70]. Considerable signal processing is involved in utilizing the tube. Basically the tube consists of a striped phosphor screen, with the stripes running vertically. The stripes fluoresce in the additive primary colors red, green, and blue. They are arranged in the repetitive order red, green, blue, such a triplet of stripes occupying a width of about $\frac{1}{16}$ inch on a 21 inch picture tube, and there being some 320 triplets across the face of the tube. Adjacent phosphor strips are separated by an inert black material. The rear of the aluminized surface is imprinted with a set of index stripes in register with the red phosphor stripes. The index stripes provide a beam position reporting signal, which is used in the signal processing circuits to cause the video signal applied to the control grid of the gun to assume values representative of the required luminance of a given phosphor at the moment the scanning beam crosses it.

A major advantage of the Apple system is that the screen white balance is automatically determined by the screen structure itself. Absence of color carrier (whether caused by a white section of a color picture being transmitted, or by the fact that the transmitted picture is monochrome) causes the screen to fluoresce in its characteristic white. The quality of this white depends only upon the balance achieved in laying down the phosphor screen during manufacture.

VI. Miscellaneous Approaches to Television Reception

A. Supersonic Light Valve

The most sophisticated embodiment of mechanical scanning was the system employing the supersonic light valve described by J. H. Jeffrees and developed as a receiving system by Scophony Ltd. of England [71], [72]. The light valve consisted of a liquid cell about 2 cm wide by about 4.5 cm long; at one end was placed a piezoelectric crystal having a natural frequency of about 20 Mc, and at the other an acoustic absorber. The crystal was highly damped (mechanically speaking) by the liquid, so that if excited by a video modulated 10-Mc signal, its amplitude of mechanical vibration would accurately follow the modulation envelope. The crystal induced a traveling compression wave at a frequency of 10 Mc in the liquid; the wave traveled down the length of the cell to the absorbing barrier at the far end with a velocity of 10^4 cm/sec, and a corresponding wavelength of 10^{-2} cm. As the wave was formed at the sending end, it bore the instantaneous amplitude of the video signal, which it preserved throughout its journey to the end of the cell. The compression waves caused the liquid to act as a diffraction grating to light passing transversely through the cell. The energy in the zeroth order was blocked out, and that in the remaining orders was permitted to pass, by means of an optical system including a slit focused upon a complementary barrier. The greater the amplitude of the video signal modulating the carrier applied to the crystal, the greater the intensity of the compressional wave at the corresponding instant of generation; and the greater the light passed by the light valve. It took about 45 μsec for the modulated wave to travel the length of the cell.

An optical system was arranged to focus an image of the cell upon the receiver screen, with the long axis of the image (corresponding to the direction of travel of the compressional wave) horizontal. A high-speed mirror drum scanner suitably synchronized was used to immobilize the image of the compressional wave on the screen; and the light coming through the immobilized wave structure was, at any point, proportional to the corresponding video signal amplitude. Thus, the scanning lines of the image were reproduced in corresponding detail at the receiver screen. A second synchronized mirror drum was used to provide the vertical scanning component.

The system employed a 20-facet mirror drum running at 30,000 rpm for horizontal immobilization. Trouble was experienced with electronic sync generators, since their rate of change of frequency was greater than could be followed by the immobilizer. This trouble would probably not exist today when crystal masters are used for sync generators instead of 60-cycle power mains.

Using the above principles, a home television receiver giving a highlight brightness of up to 10 ft lamberts, and theater projection equipment giving 3 ft lamberts on a 12×15 foot screen, were designed.

B. Skiatron Display

The skiatron depends upon the discoloration of certain metallic salts (such as potassium chloride) by bombardment with cathode rays. It has been proposed to use this effect by causing a transparency to appear on the face of the cathode-ray tube coated with such material instead of the usual phosphor, but with a scanned and modulated beam otherwise exactly like a normal picture tube. The transparency is then projected or viewed with a light source.

Such a device was used during the war for a bright radar display. For television use it has, at the moment, the disadvantage that the effect decays slowly so that it is difficult to depict motion.

C. Eidophor Display

This is a unit primarily designed for theater projection. Its basic principle is the disturbance of the surface of a thin film of liquid brought about by electric charge. The electric charge is deposited by a cathode-ray beam scanned and modulated in the normal television fashion. The beam is focused on to the surface of the liquid, and causes a charge pattern representation of the TV picture to appear. The liquid film is formed on the surface of a mirror, and the reflection, resulting from its surface irregularities, causes the direction of the rays reflected by the mirror to be affected accordingly. A Schlieren type of optical system correspondingly varies the amount of emergent light. The surface of the liquid is focused on the screen by a projection lens. The unit was developed by Dr. Fischer at the Federal Institute of Technology in Zurich [78]. It has been adapted to field sequential color television.

Conclusion

The story of the development of television receiver techniques is a long one, characterized by contributions to its progress made by many scientists and engineers, Only the contributions made by the very earliest workers occurred long enough ago that they have acquired historic perspective so that just credit can be given them.

An appreciation of the amount of work and of the number of participating engineers and scientists involved in the recent development of color standards may be had by consulting some of the more recent texts [73]-[76]. The author does not presume to have the ability properly to weigh and apportion the credit for the more recent work.

References

[1] C. F. Jenkins, "Radio Vision," Proc. IRE, vol. 15, pp. 958–964; November, 1927.

[2] H. E. Ives, "Television," *Bell Sys. Tech. J.*, vol. 6, pp. 551–559; October, 1927.

[3] H. H. Sheldon and E. N. Grisewold, "Television," D. Van Nostrand Co., Inc., New York, N. Y.; 1929.

[4] J. D. McGee, "Distant electric vision," Proc. IRE, vol. 38, pp. 596–608; June, 1950.

[5] A. G. Jensen, "The evolution of modern television," *J. SMPTE*, vol. 63, pp. 174–188; November, 1954.

[6] P. T. Farnsworth, U. S. Patent No. 1,773,980; August 26, 1930.

[7] P. T. Farnsworth, "Television by electron image scanning," *J. Franklin Inst.*, vol. 218, pp. 411–444; October, 1934.

[8] A. Dinsdale, "Television by cathode ray—the new Farnsworth system," *Wireless World*, vol. 28, pp. 286–288; March, 1931.

[9] V. K. Zworykin, "Television with cathode ray tube for receivers," *Radio Engrg.*, vol. 9, pp. 38–40; December, 1929.

[10] D. G. Fink, "Television Standards and Practice—NTSC," McGraw-Hill Book Co., Inc., New York, N. Y.; 1943.

[11] S. Butterworth, "On the theory of filter amplifiers," *Exptl. Wireless and Wireless Engr.* (London), vol. 7, pp. 536–540; October, 1930.

[12] E. W. Herold, "Superheterodyne converter system considerations in television receivers," *RCA Rev.*, vol. 4, pp. 324–337; January, 1940.

[13] V. D. Landon, "Cascade amplifiers with maximal flatness," *RCA Rev.*, vol. 5, pp. 347–362, 481–497; January, April, 1941.

[14] G. C. Sziklai and A. C. Schroeder, "Band pass bridged T network for television intermediate frequency amplifiers," Proc. IRE, vol. 33, pp. 709–711; October, 1945.

[15] V. D. Landon, "The band-pass—low-pass analogy," Proc. IRE, vol. 24, pp. 1582–1584; December, 1936.

[16] S. W. Seeley and N. Kimball, "Analysis and design of video amplifiers," *RCA Rev.*, vol. 3, pp. 290–308; January, 1939.

[17] H. A. Wheeler, "Wideband amplifiers for television," Proc. IRE, vol. 27, pp. 429–438; July, 1939.

[18] H. L. Donley and D. W. Epstein, "Low frequency characteristics of the coupling circuits of single and multi stage video amplifiers," *RCA Rev.*, vol. 6, pp. 416–433; April, 1942.

[19] H. E. Kallman, R. E. Spencer, and C. P. Singer, "Transient response," Proc. IRE, vol. 33, pp. 169–195; March, 1945.

[20] R. D. Kell, A. V. Bedford, and G. L. Fredendall, "Determination of optimum number of lines in a television system," *RCA Rev.*, vol. 5, pp. 8–30; July, 1940.

[21] J. C. Wilson, "Channel width and resolving power in television systems," *J. Television Soc.* (London), vol. 2, pt. 2, pp. 397–420; 1938.

[22] H. A. Wheeler and A. V. Loughren, "The fine structure of television images," Proc. IRE, vol. 26, pp. 540–575; May, 1938.

[23] H. A. Wheeler, "Interpretation of amplitude and phase distortion in terms of paired echoes," Proc. IRE, vol. 27, pp. 359–385; June, 1939.

[24] M. W. Baldwin, "Subjective sharpness of simulated television images," Proc. IRE, vol. 28, pp. 458–468; October, 1940.

[25] J. Avins, "IF amplifier design for color TV receivers," IRE Trans. on Broadcast and Television Receivers, vol. PGBTR-7, pp. 14–25; July, 1954.

[26] L. Ruth, "A printed circuit IF amplifier for color TV," IRE Trans. on Broadcast and Television Receivers, vol. BTR-2, pp. 50–52; July, 1956.

[27] J. E. Bridges, "Detection of television signals in thermal noise," Proc. IRE, vol. 42, pp. 1396–1405; September, 1936.

[28] R. B. Dome, "Carrier difference reception of television sound," *Electronics*, vol. 20, pp. 102–105; January, 1947.

[29] C. E. Torsch, "High efficiency, low-copper sweep yokes with balanced transient response," IRE Trans. on Broadcast and Television Receivers, vol. PGBTR-6, pp. 17–32; April, 1954.

[30] R. A. Bloomsburg, "The measurement of yoke astigmatism," IRE Trans. on Broadcast and Television Receivers, vol. PGBTR-7, pp. 26–33; July, 1954.

[31] C. G. Torsch, "Yoke development for standardization of 110° and 90° deflection angles," IRE Trans. on Broadcast and Television Receivers, vol. BTR-1, pp. 10–15; October, 1955.

[32] R. Gethmann, "Deflection distortions contributed by the principal field of a ring deflection yoke," IRE Trans. on Broadcast and Television Receivers, vol. BTR-4, pp. 24–34; February, 1958.

[33] C. Droppa, "Improvements in deflection amplifier design," 1958 IRE National Convention Record, pt. 7, pp. 147–153.

[34] W. Palmer and G. Scheiss, "Transistorized vertical deflection system," IRE Trans. on Broadcast and Television Receivers, vol. BTR-3, pp. 98–105; October, 1957.

[35] E. M. Creamer, Jr., L. H. DeZube, and J. P. McAllister, "Transistor circuit problems in TV receiver design," 1957 IRE National Convention Record, pt. 3, pp. 205–212.

[36] A. R. Curll, "A transistorized portable television receiver," IRE Trans. on Broadcast and Television Receivers, vol. BTR-6, pp. 9–16; May, 1960.

[37] J. G. Humphrey, "A transistor TV IF amplifier," IRE Trans. on Broadcast and Television Receivers, vol. BTR-6, pp. 17–20; May, 1960.

[38] W. H. Mead, "Performance of Nuvistor small-signal tetrodes in television video IF amplifiers," IRE Trans. on Broadcast and Television Receivers, vol. BTR-6, p. 50; May, 1960.

[39] C. D. Simmons and C. R. Gray, "The video processing circuits of an all transistor television receiver," IRE Trans. on Broadcast and Television Receivers, vol. BTR-6, pp. 25–32; May, 1960.

[40] F. L. Abboud, "Transistorized vertical scan system for magnetic deflection," IRE Trans. on Broadcast and Television Receivers, vol. BTR-6, pp. 33–38; May, 1960.

[41] R. B. Ashley, "Linearization of transistorized deflection system," IRE Trans. on Broadcast and Television Receivers, vol. BTR-6, pp. 39–47; May, 1960.

[42] Z. Wiencek, "Phase splitter video amplifier for transistor TV," IRE Trans. on Broadcast and Television Receivers, vol. BTR-6, pp. 18–24; November, 1960.

[43] G. C. Hermeling, "A Nuvistor low noise VHF tuner," IRE Trans. on Broadcast and Television Receivers, vol. BTR-6, pp. 52–56; November, 1960.

[44] R. W. Ahrens, "Factors in the circuit operation of transistorized horizontal deflection," IRE Trans. on Broadcast and Television Receivers, vol. BTR-6, pp. 57–59; November, 1960.

[45] R. R. Law, "Contrast in kinescopes," Proc. IRE, vol. 27, pp. 511–524; August, 1939.

[46] H. W. Leverenz, "Optimum efficiency conditions for white luminescent screens in kinescopes," *J. Opt. Soc. Am.*, vol. 30, pp. 309–315; July, 1940.

[47] M. Sadowsky, "The double layer projection tube screen for television," Proc. IRE, vol. 38, pp. 494–498; May, 1950.

[48] D. W. Epstein and L. Pensak, "Improved cathode-ray tubes with metal backed luminescent screens," *RCA Rev.*, vol. 7, pp. 5–10; March, 1946.

[49] R. R. Law, "Factors governing the performance of electron guns in television cathode-ray tubes," Proc. IRE, vol. 30, pp. 103–105; February, 1942.

[50] G. Liebmann, "The image formation in cathode-ray tubes and the relation of fluorescent spot size and final anode voltage," Proc. IRE, vol. 33, pp. 381–389; June, 1945.

[51] J. E. Rosenthal, "Correction of deflection defocusing in cathode-ray tubes," Proc. IRE, vol. 39, pp. 10–15; January, 1951.

[52] L. C. Wimpee and L. E. Swedlund, "The development of 110° television picture tubes having an ion trap gun," 1957 IRE National Convention Record, pt. 3, pp. 150–155.

[53] J. W. Schwartz, "The annular geometry electron gun," Proc. IRE, vol. 48, pp. 1864–1870; November, 1958.

[54] A. Russell, "Book Review on 'Practical Television' by E. T. Larner," *Nature*, vol. 122, pp. 232–234; August, 1928.

[55] H. E. Ives and A. L. Johnsrud, "Television in colors by a beam scanning method," *J. Opt. Soc. Am.*, vol. 20, pp. 11–22; January, 1930.

[56] P. C. Goldmark, J. N. Dyer, E. R. Piore, and J. M. Hollywood, "Color television," Proc. IRE, vol. 30, pp. 162–182; April, 1942.

[57] P. C. Goldmark, J. W. Christensen, and J. J. Reeves, "Color television—U.S.A. standard," Proc. IRE, vol. 39, pp. 1288–1313; October, 1951.

[58] P. C. Goldmark and J. M. Hollywood, "A new technique for improving the sharpness of television pictures," Proc. IRE, vol. 39, pp. 1314–1322; October, 1951.

[59] R. Dressler, "The PDF chromatron," Proc. IRE, vol. 41, pp. 851–858; July, 1953.

[60] J. D. Gow and R. Dorr, "Compatible color picture presentation with the single gun tricolor chromatron," Proc. IRE, vol. 42, pp. 308–315; January, 1954.

[61] R. Dressler and P. Neuworth, "Brightness enhancement techniques for the single gun chromatron," 1957 IRE National Convention Record, pt. 3, pp. 220–229.

[62] R. R. Law, "A three gun shadow mask color kinescope," PROC. IRE, vol. 39, pp. 1186–1194; October, 1951.

[63] M. J. Grimes, A. C. Grum, and J. F. Wilhelm, "Improvements in the RCA three-beam shadow mask color kinescope," PROC. IRE, vol. 42, pp. 315–326; January, 1954.

[64] N. F. Fyler, W. F. Rowe, and C. W. Cain, "The CBS-colortron: A color picture tube of advanced design," PROC. IRE, vol. 42, pp. 326–334; January, 1954.

[65] M. E. Amdursky, R. C. Pohl, and C. S. Szegho, "New high-efficiency parallax mask color tube," PROC. IRE, vol. 43, pp. 936–943; August, 1955.

[66] C. F. Barnett, F. J. Bingley, S. L. Parsons, C. W. Pratt, and M. Sadowsky, "A beam indexing color picture tube—the Apple tube," PROC. IRE, vol. 44, pp. 1115–1119; September, 1956.

[67] R. G. Clapp, E. M. Greamer, S. W. Moulton, M. E. Partin, and J. S. Bryan, "A new beam indexing color television display system," PROC. IRE, vol. 44, pp. 1108–1114; September, 1956.

[68] R. A. Bloomsburgh, W. P. Boothroyd, G. A. Fedde, and R. C. Moore, "Current status of Apple receiver circuits and components," PROC. IRE, vol. 44, pp. 1120–1124; September, 1956.

[69] R. A. Bloomsburgh, A. Hopengarten, R. C. Moore, and H. H. Wilson, Jr., "An advanced color television receiver using a beam indexing picture tube," 1957 IRE NATIONAL CONVENTION RECORD, pt. 3, pp. 243–249.

[70] H. Colgate, C. Comeau, D. Kelley, D. Payne, and S. Moulton, "Recent improvements in the Apple beam indexing color tube," 1957 NATIONAL CONVENTION RECORD, pt. 3, pp. 238–242.

[71] D. M. Robinson, "The supersonic light control and its application to television with special reference to the Scophony television receiver," PROC. IRE, vol. 27, pp. 483–486; August, 1939.

[72] J. Sieger, "The design and development of television receivers using the Scophony optical scanning system," PROC. IRE, vol. 27, pp. 487–492; August, 1939.

[73] D. G. Fink, Ed., "Color Television Standards," McGraw-Hill Book Co., Inc., New York, N. Y.; 1955.

[74] J. W. Wentworth, "Color Television Engineering," McGraw-Hill Book Co., Inc., New York, N. Y.; 1955.

[75] K. McIlwain, and C. E. Dean, Eds., "Principles of Color Television," John Wiley and Sons, Inc., New York, N. Y.; 1956.

[76] D. G. Fink, Ed., "Television Engineering Handbook," McGraw-Hill Book Co., Inc., New York, N. Y.; 1957.

[77] W. E. Bradley, "Wideband amplification," in "Television Engineering Handbook, McGraw-Hill Book Co., Inc., New York, N. Y., ch. 12; 1957.

[78] E. Baumann, "The Fischer large screen projecting system," *J. SMPTE*, vol. 60, pp. 344–356; April, 1953.

PART II: MECHANICAL SYSTEMS

JOHN L. BAIRD

The Founder of British
Television

By J. D. PERCY

A Memorial issued by the Television Society, London

Revised Reprint, July, 1952

JOHN LOGIE BAIRD

Born 1888 - Died 1946

INTRODUCTION BY THE AUTHOR

JOHN LOGIE BAIRD must inevitably go down in the annals of scientific achievement as the Marconi of Television. Other men may have dreamed of ways and means of achieving it. Certainly during the early twenties, many contemporary scientists attempted by practical effort to send instantaneous moving pictures by wire and wireless. The clumsy arrangements of discs, photocells and amplifiers, which Baird assembled in 1924 at Hastings, however, proved that television was, once and for always, a solvable problem. Different methods?—but more than possible! Improved definition—certainly, and most necessary! In the meantime, here was no dream, no brave effort. Here in a dingy little room, over a small seaside shop, was television—working.

Success to this measure came to Baird only after a struggle against circumstances which would have broken many men of lesser heart. Chronic ill-health, poverty, sometimes extreme poverty, must be added as braking forces to the usual disbelief which every inventor has to face and which met his early claims with more than usual vigour. Nothing, however, could apparently daunt his staunch spirit, and although the mechanical principles on which he first achieved television were not, in the end, to be those on which the systems of the world were to be finally built, nevertheless they were the first, which, when put to work in practice achieved successful results.

Baird kept no regular notes. Indeed, it is true, I think, to say that his mind worked too quickly for the satisfactory recording of his myriads of ideas. Explanation to those of us who worked with him was amplified only by rough sketches on the backs of old envelopes, by scrawled diagrams on walls (the Long Acre Laboratories were literally covered in these) or even occasionally on the table-cloths of restaurants and inns. Since no true personal chronicle exists therefore, all that follows here has been gleaned from contemporary notes and from the memories of those of the original "Baird Boys" who are still within the author's call.

Probably the two most outstanding examples of the versatility of the remarkable man whose work for television forms our subject, are his invention and exploitation in Glasgow of a chemically treated damp-proof under-sock, and later, when forced to leave his native land for health reasons, his foray into the realms of jam manufacturing in Trinidad. This latter project, however, failed badly, caused, Baird always said, by the mosquitoes eating up any profit he might have otherwise made. Eventually, he was forced to sell the jam factory at a loss and returned to England almost penniless.

After a further attempt at commercial life his health broke down completely, and so, ill, sick at heart, and almost entirely without monetary resources, he retired to Hastings, and it was here in the little attic in a side street that the old dream of a working system of television was reborn, and here, moreover, that it first flickered into practical reality.

Of the original Hastings machine nothing now remains. Baird's own recollection was that it was sold for £2 in order to pay the rent. The indisputable fact remains, however, that the world's first practical transmission by what we now know as sequential scan television, was achieved in a Hastings garret, which in 1924 was home to a Scottish gentleman whose great courage was to defeat bodily ailments, disbelief—even hunger—and whose faith in an idea was to cause such a degree of thought to be given to its perfection that even within the span of his own lifetime, moving pictures by wireless were to be commonplace in thousands of homes. For there is little doubt that, apart from any other considerations, John Baird acted as a spur to others who, both in Europe and America, fired by his practical success, commenced work along different and independent lines. Television as we know it to-day, consists of a mass of different inventions all dovetailing together to build a whole. To John Logie Baird, however, must always remain the supreme honour of being the first man in the world to prove its achievement.

Events now moved rapidly. From Hastings to Frith Street, Soho, where money was found for a new machine (now in the Science Museum, South Kensington), where on January 27th, 1926, the results were demonstrated to 40 members of the Royal Institution, and later from Frith Street to the unforgettable third floor of No. 133, Long Acre, destined to become the home of the first British television service; all the time the definition and clarity of the pictures were rising steadily as the months flew by.

It was becoming increasingly obvious, however, that mechanical television was almost bound to be superseded by one of the swift and silent electronic methods which were already yielding surprising results, and in 1936 when a series of practical trials of each system took place at the commencement of the B.B.C.'s high definition service from the Alexandra Palace, it was soon apparent that the matter was no longer in doubt, and that mechanical television had had its day.

Although bitterly disappointed, Baird never for a single moment despaired, and at once his busy mind turned towards other uses to which his fundamental ideas could be put.

The results of his labours were to be seen during the late 1930's in the large screen demonstrations given at the Dominion Theatre in London, the colour and stereoscopic television transmissions which a few of us were privileged to see at his private laboratory at Sydenham, and the 120 line 12 feet by 9 feet colour picture of 1941. John Baird's final ambition was to be the first man to evolve a practical system of big screen colour television, and even three days before he died we spoke of it together as, weak and emaciated, he lay in his house at Bexhill where he was making desperate efforts to gather strength to renew the struggle.

The years of hardship and ill-health had taken their toll, however, and sadly I realised that the long fight was almost over now, and on June 13th, 1946, he died.

It is, of course, too early yet to put the picture of John Logie Baird into its proper historical perspective. To those of us who knew and loved him, however, one thing is clear. With his own hands he made the world's first television equipment which worked. That we will always remember—and other things too. His gentle courtesy; his magnificent sense of humour; his complete disregard for setbacks and disappointments; and above all his innate gallantry.

In the following pages mention will be made of some of the milestones which marked the progress of the mechanical television system which was its inventor's life work. The scene is greater than the available canvas, and therefore it will not be possible to do more than touch on many of the technical details as the story unfolds. Lack of space has necessitated the complete omission of much that is of secondary interest. The reader's indulgence is therefore solicited. There is so much to talk about—and so little paper.

J. D. PERCY.

The Founding of British Television

1926 One of the most important dates in the history of British television is January 27th, 1926. On this day the first demonstration of true television was given in the Baird Laboratory in Frith Street, Soho. The audience consisted of some 40 members of the Royal Institution, the transmitter and receiver were in separate rooms, and of this demonstration the *Times* wrote :—

" . . . on a portable receiver in another room the visitors were shown recognizable reception . . . of a person speaking. The image . . . was faint and often blurred but substantiated a claim that . . . it is possible to transmit and reproduce instantly the details of movement . . . on the face."

The apparatus used for this demonstration was very similar to that now resting in the South Kensington Science Museum. Scanning was by means of a lensed disc, selenium photo-cells were used, and what synchronism there was, was looked after by small linked alternators on the main driving shafts of the transmitter and receiver motors. The dummy's head and, later on, the live subject were bathed in blinding incandescent light from batteries of 100-watt lamps at very close range. Little information has come down to us about the amplifier chain, but probably a three-stage " A " with old directly-heated D.E.Ls. feeding a three-stage " B " culminating in one of the L.S.5 class of output triodes was used. Contemporary photographs of the received image argue that scant attention was paid to the frequency response of the chain, which was probably transformer coupled. Baird's main interest always lay in the mechanical and optical problems of television. An amplifier was just a necessary and rather unimportant box to him, and one amplifier was very much like another.*

Experimenting with this early machine it occurred early to Baird that the colour response of the photo-cells was bound to be an important factor in improving the results. In 1926, the best available photocells were small gas-filled ones with a high ambient noise level. By the time the signal had coped with this and the parasitics of the early dull-emitter valve amplifier, the picture/noise ratio was terrible. Accordingly, Baird early set out to try and select cells whose colour response matched the colour peak of his floodlights, which experiments led in turn to the trial of various coloured lamps and filters. It was during these trials

* This strange aloofness from the circuit side of things always persisted and during all the years I knew him. I never recall being able to detect any gleam of enthusiasm while amplifiers or circuits were being discussed. Optics, and mechanics, however, were his very life blood.

that the lamps were masked as an experiment with wafer thin ebonite sheet so that all visible light was cut off and only infra-red light played on the subject. Much to Baird's surprise, the picture was still not only visible at the receiver, but the signal/noise ratio, since he was using red-sensitive cells, was surprisingly good.

Photograph of an early 30-line television image.

"Noctovision"

Thus " Noctovision " as he christened it, came into being and engaged much of his attention for the next five years. Demonstrations of " Noctovision " commenced in December and many people who witnessed these were confident that a new and potent device for " seeing in the dark " was, indeed, in being. Baird himself certainly thought so, and years afterwards, when he lived in a house on Box Hill, experiments were still going on. The very limited range and complexity of the apparatus, however, proved formidable obstacles, and it was not until the development of Radar that another of his dreams became practical reality. Alas, however, his was not to be the hand destined to open the true door to " seeing in a fog "—or in darkness.

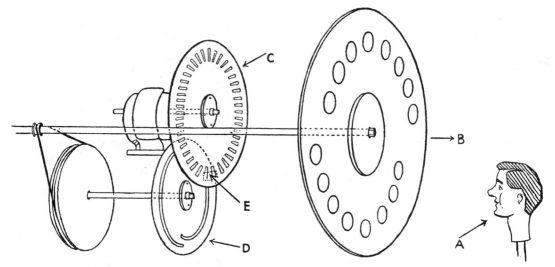

Diagram of the first television transmitter, now in the Science Museum. (A)—dummy's head, (B)—lensed disc, (C), (D)—slotted and spiral discs for scanning, (E)—aperture through which image was projected on to a photo-electric cell.

1927

Having demonstrated that television was possible, there were obviously two closely interlinked and immediate problems. The first was to explore all the ways in which the results could be readily improved. The suitability of the television signal for broadcasting, improvement in definition, increase in the size of the picture, and area transmitted—all these were necessary before the second problem, the commercial prospect of television, could be explored. The year 1927, therefore, saw many experiments, some reasonable and many otherwise. Television was being put to the test.

Distance seems to have been one of the principal worries. In April, the American Telephone and Telegraph Company staged, amid much publicity, a transmission from Washington to New York, a distance of over 200 miles. The system employed was not described in much detail, but a young English engineer of that period did not have much difficulty in making sense of the few photographs of the equipment to reach this country, and Baird himself had none. Although no acknowledgment was ever made, the system was undoubtedly fundamentally identical with that which originated two years previously in a Hastings bedroom.

Already, however, Baird was planning an even more impressive demonstration of long-distance television.

On April 22nd, the *Daily News* carried a column headline "Faces Across the Sea," and went on to describe the transmission of a television signal via an amateur short-wave station in Coulsdon, Surrey,

which was being picked up regularly in New York. The *Morning Post* in a short paragraph on April 22nd, describes the station in slightly more detail:—

" A wireless telephony transmitter capable of delivering an electrical power of one kilowatt to the aerial which is a vertical wire about 35 feet long." The wavelegnth mentioned is 45 metres and " the installation takes up the whole of a bedroom."

Although the actual resolution of an image from the transatlantic television signal was not to take place until February of the following year, another long-distance test was carried out at home. During May, demonstrations were given of a picture received in Glasgow from scanning equipment situated in London, the two stations being linked by Post Office cable. In Glasgow, an impressed reporter on the staff of *Nature* records that on the 26th " the image was amply sufficient to secure recognisability of the person being televised," and " the movement of the face or features was clearly seen."

" Noctovision " was also receiving some attention. On April 9th, London daily newspapers referred to a " Red Glare scare in the West End," and described apparatus on the roof of Motograph House, St. Martin's Lane, from which " infra-red rays were trained on Nelson's column." One account of the strange happenings in London on the night of April 8th goes on to describe how at the height of the experiments " when the red glare was at its highest, some one gave the fire alarm. The firemen, however, were stopped in time."

It was also during April that the Baird Television Development Company was formed.

1928

The year 1928 was a full and remarkable one in the history of British television. Not only was it during this year that the first home receivers were designed and built, but transmitting equipment was also for the first time standardised. All this was the work of Commander W. W. Jacomb, who during this year was appointed Chief Engineer of the Baird Television Development Company, and who probably did more than any other single individual for the rationalisation and development of the Baird principle. This remarkable man, whose association with Baird lasted

J. L. Baird and O. G. Hutchinson hoist the first television aerial of 2TV on the roof of Motograph House, St. Martin's Lane.

over seven years, displayed a brilliance and foresight, which, had mechanical television not been eventually eclipsed by electronic methods, would have placed him in a prominent position in the forefront of television history. As it is, his niche is assured by the surprising progress which was made in mechanical television under his guidance during his term of office as Baird's chief engineer.

Probably the most important development in this year was the successful application of the spotlight or flying spot principle. This enabled the heavy lensed discs which had previously been used for scanning, to be replaced by a single disc of aluminium some 14 inches in diameter, which instead of carrying heavy lenses, had a single spiral of small apertures around its periphery. In using this, Baird and Jacomb were reverting in fact to the type of disc suggested by Nipkow in his patent of 1884, and had Nipkow had the advantage of the photocells and amplifiers of 1928, there can be no doubt that he would have anticipated Baird. As it was, however, his crude selenium cells and the absence of any reasonable amplifying equipment defeated all Nipkow's efforts to make a practical success of the scheme.

During the early hours of the morning of February 9th, the signals emitted by the Coulsdon radio transmitter were finally resolved in New York via a receiver stationed outside the city in Hartsdale, and the *New York Times* for February 10th, reported :—

"Baird was the first to achieve television at all over any distance, now . . . the first to . . . flash it across the ocean . . . for American eyes."

While the *New York Herald Tribune* drew a startling comparison between the resources of American and British television pioneers. Carefully it stated—

"It is said that probably one thousand engineers and laboratorymen were involved in the American (Washington-New York) tests "—
then, acidly,

"Only a dozen worked with Baird."

Meantime, leaving the development of "straight" television to Jacomb, Baird was off once more into unexplored paths. Television in colour, daylight television and stereoscopic television were all being worked at on the third floor of a converted garage building at 133 Long Acre. Astoundingly, the first of these worked—and not too badly. It is a standing tribute to the foresight and genius of a tall fair-haired Scot who in 1928 took little sleep and wore bedroom slippers all day, that by far the most pictorially successful system of colour television demonstrated in America just 20 years later, employed the very scanning principles used on the first British experimental machine.

Colour Television

Colour television was first demonstrated on July 3rd, and on August 18th *Nature* recorded after a brief summary of the colour equipment : Delphiniums and carnations appeared in their natural colours and a basket of strawberries showed the red

7

fruit very clearly." In June, pictures using ordinary daylight had been successfully transmitted ,and the principle used for the first television transmission of an English Derby (June 3rd, 1931) was demonstrated practically for the first time.

Stereoscopic television, always a favourite Bairdian problem was being shown working. Two little images formed by concentric scanning spirals and married by an ordinary postcard type stereoscope, however, never really succeeded in giving any impression of " depth " and, indeed, even to-day three-dimensional television cannot be said to have reached any degree of practical reality.

Meanwhile, there is discernable over the exciting mechanical television scene of 21 years ago, a shadow so slight that it was not even visible to the eager and enthusiastic pioneers of the era. Nevertheless, the shade was destined to swell into the complete eclipse of a principle then looked upon as final.

For it was in 1928 that the first results were being achieved using cathode rays for scanning and reproduction, thus proving correct a remarkable prophecy made in 1903 by A. A. Campbell-Swinton who in that year forecast with astonishing accuracy modern electronic television.

By this time regular transmissions of television by radio were taking place outside normal broadcast hours by the Baird Company's Station 2TV on the roof in Long Acre. The transmitter was an entirely conventional $1\frac{1}{2}$ kilowatt one of the period, employing choke modulation and operating on a wavelength of 200 metres.

Interior of one of the 30-line disc receivers, showing motor and toothed wheel for synchronising.

1929 Things were now rapidly becoming standardised and, in view of the fact that transmissions in the broadcast frequency band (200-500 metres) were anticipated, a standard of 30 lines sequential with a picture repetition rate of $12\frac{1}{2}$ per second and vertical scanning was fixed. This low definition system lasted long after the B.B.C. had taken over television transmissions and, indeed, up to the closing down of Studio B.B. in Broadcasting House prior to the opening of Alexandra Palace and high definition in 1936.

Around this period Baird's advisers became very involved in a wordy battle with the B.B.C. Captain P. P. Eckersley, with characteristic foresight, had stated that he did not consider the Baird system final and that it would not ultimately be capable of development to the high degree of definition which, he maintained, was necessary before the home television set became really worth while. This frank, and as subsequent events proved, true prophecy, was the signal for the start of a long-drawn-out and bitter feud between the B.B.C and the Baird Company. There was right and wrong on both sides. In the

J. L. Baird and Sir Oliver Lodge at the demonstration at Leeds of Noctovision.

8

Exterior view of the same receiver known under the trade name of the "Televisor."

produced by the simple expedient of providing a "masking" signal, arranged by allowing the scanning spot to traverse a black band on the backcloth immediately after completing its scan of the picture. This sudden change in reflected light value produced a kind of pulse which was of sufficient regularity and amplitude at the receiving end to operate a somewhat complex commutator/relay arrangement, which had the effect of short-circuiting at the right intervals a minute proportion of the field resistance of the disc driving motor. Thus the speed of the receiving disc was allowed to lead that of the transmitter slightly, being dragged back when the small portion of field resistance was shorted.

The general effect of this was that although the picture could be synchronised, and, indeed, on still subjects held in step for long periods without attention, it tended to hunt very badly, the hunt becoming worse as the amount of vertical movement of the subject increased.

first place, Eckersley was undoubtedly, from a technical point of view, right in all he said. The aloof and condescending disdain with which the B.B.C. treated an all-British and quite practical if not finalised invention, however, is to their discredit, especially since during this period an early and crude form of facsimile service was being used and sponsored over the B.B.C.'s London Station.

Gradually, the situation clarified, and on September 30th, the first television transmission through a B.B.C. station took place. This was through the old and well-beloved 2LO, for years the voice of London, and situated on the rooftop of Selfridge's. The studios, of course, remained in the Baird offices in Long Acre, and among those televised on this occasion was Sir Ambrose Fleming. In October, on the 8th to be exact, a 2LO television transmission was resolved in Bradford. In spite of the limitations of 30 line television, the fact that its crude definition only required a limited sideband spread and therefore was transmittable on ordinary medium-wave broadcast frequencies, gave it an effective range greater than any subsequent high definition system.

Synchronising

The year 1929 was also that in which synchronising problems (which hitherto had been more or less ignored) received attention. A separate synchronising circuit or transmission had long been regarded as impracticable and clumsy, and the idea of radiating the synchronising impulses as an integral part of the picture signal itself was known to be the ideal method. The word "pulses" hardly applied to the curiously-shaped repetitive signal which 20 years ago was

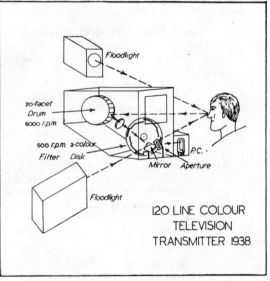

Later form of colour television using mirror drums, 1935-38.

The relay system of synchronising was later this year replaced by a simpler and more effective magnetic arrangement. Again, the masking "pulse" was used, only this time the synchronising signal at the receiving end was passed through a small electromagnetic system between which revolved a small wheel having 30 teeth around its periphery.

This tiny alternator system was quite successful, and would accelerate or retard the individual poles of the toothed wheel as they came opposite the magnets. Once again a certain amount of hunt was present, but the system was a vast improvement on the relay method.

It was also during 1929 that the Baird Company were invited by the British Association to send a complete demonstration system to South Africa, where the Association were holding their annual series of meetings. This they did and demonstrations of 30-line television were given in Cape Town, Johannesburg and Durban.

1930

At home, the ties between the B.B.C. and Baird Company were now strengthening, and most of the argument was over. Transmissions had been taking place with fair regularity all through the autumn and winter of 1929, and in March of the following year, television and sound were broadcast simultaneaously for the first time from Brookmans Park.

The old London National transmitter, working on 261 metres, radiated vision, while sound was sent out on the London Regional wavelength of 365 metres. Watching reception at " a receiving station in the centre of London," the *Times* commented : " The whole transmission was very successful, and . . . the whole programme was followed with great interest. . . . The rapidly swirling pattern was revolved into a steady head and shoulders of the speaker."

Two wireless sets were used, one for receiving television signals and the other for reception of the sound signals, and later on there is a hint of displeasure at the technique of the early sound films—" A particularly noteworthy feature . . . was that there was no lag between vision and sound, such as often destroys the illusion of the talkies."

The *Times* also reported on March 31st that " Television sets . . . are to be sold for 25 guineas," and goes on to deliver a short resumé of the qualities required in a wireless set for good television, as opposed to sound, reception.

One of the most interesting things this year was the B.B.C.'s first effort at television drama. The play was " The Man with the Flower in his Mouth " by Pirandello, and was produced in the Long Acre studio by Lance Sieveking.

The necessarily limited scenery was painted by C. R. W. Nevinson, and among the cast of four we find Miss Gladys Young, who has since become

famous as a resident member of the B.B.C. Repertory Company.

All this time, Baird himself was experimenting in new and different methods to make television better, brighter and bigger—especially bigger ; and by the end of July, he was demonstrating what surely was the world's first " large screen " television picture.

The principle he used was typical. Substituting for the receiving disc a revolving brush, he had built a mosaic of 2,100 flashlamp bulbs, arranged in 30 vertical columns of 70 each, each column representing a scanning line. The brush traversed a 2,100 segment commutator, and was synchronised to the transmitting disc. Thus each lamp became effectively a picture point, and was switched on sequentially. Current through the lamps was controlled by the television signal, and the whole effect was certainly that of a very much enlarged and infinitely brighter 30-line picture than those hitherto possible.

Large screen television made its first public bow in a theatre on Monday, July 28th, when it took part in a regular music-hall bill at the Coliseum for two weeks.

During this fortnight many celebrities were televised from Long Acre and reproduced on the Coliseum " bank of lamps." Not the least of these, and sandwiched between Sir Oswald Mosley and A. V. Alexander, was Mr. Herbert Morrison, who was then Minister of Transport.

The large screen equipment was built into a small caravan which rendered the somewhat unwieldy receiver, if not exactly portable at least mobile. Everything was built into this, and even the operator who kept continuous watch on the synchronising, sat inside and watched the reflected picture from a small mirror placed among the footlights. In the autumn the large screen equipment was despatched with its transmitter on a tour of three continental capitals, and gave regular television performances in Berlin, Paris and Stockholm. On the whole the equipment was remarkably stable and reliable. On one occasion, however, when the final stage of the television amplifier was switched on before the commutator was run up, the time taken to traverse each segment of the commutator during the first revolution was long enough to allow the lamp current to rise to a value that in a couple of seconds burned out 2,100 flashlamp bulbs. The author remembers this clearly because his was the hand that absent-mindedly caused the debacle ; his the subsequent embarrassment and humiliation, and his, too, was an evening spent replacing over 2,000 pea-lamps.

The "bank of lamps" large-screen demonstration. The mosiac of lamps was housed in the caravan, with the rotating commutator at the rear of the screen.

1931

Although true television in Baird's view could only mean the instantaneous transmission and reception of live scenes, he had always been interested in the possibilities of televising films, and in March, 1931, a part of a film was broadcast during the ordinary evening programme.

Very little information concerning this first telecine has come down to us, but it is known that a Mechau continuous motion projector, similar to those employed by the B.B.C. to-day, was used in conjunction with a normal scanning disc.

Besides telecine, improvements were steadily taking place in the normal transmission from Long Acre. Electrically maintained tuning forks, for instance, were now used to generate the synchronising impulses which led to greater stability and minimised "hunting."

This was the era too of the home constructor, and contemporary technical journals contain many examples of successful television receivers which were built by people all over the country, and which nearly all appeared to function remarkably well.

At Long Acre, some thought was being given to ways and means of transmitting television from normal B.B.C. sound studios, which meant that portable scanning equipment had to be developed.

On April 13th, a small disc scanner which, in appearance was not unlike a modern television camera, was delivered to the B.B.C. for a series of experimental trials, and another outstanding historical event of this year was the broadcasting by ordinary daylight, on Friday, May 8th, of a street scene outside the Long Acre offices.

For this experiment Baird used a mirror drum, towards which he was turning more and more from the disc as a means of increasing the optical efficiency of the scanning system, and therefore reducing his old enemy, cell and amplifier noise.

The equipment was the forerunner of that used the following month at Epsom for Cameronian's Derby, relayed by 30-line television from the course, over 15 miles of telephone wire to the Long Acre offices, and on Friday, June 4th, the *Times* reported non-committally :—

> "Yesterday afternoon the Baird Television Company, in co-operation with the B.B.C., broadcast a television transmission of scenes from the Derby, including the parade of horses before the start and the scene at the winning post during the race. This broadcast is important in that it is the first attempt that has been made in this or any other country to secure the television transmission of a topical event held in the open air, where artificial lighting is impossible."

The *Daily Telegraph* for the same morning also recorded this event, but with a little more enthusiasm

> " . . . all the Derby scenes were easily discernable. The parade of the horses, the enormous, crowd and the dramatic flash past at the winning post."

While the *Daily Mirror's* representative, who was also at Long Acre :—

> " . . . saw Cameronian, closely followed by Orpen and Sandwich, win the great summer racing classic . . . by means of the two latest and greatest marvels of science—wireless and television."

The first public showing of television on the stage. The exterior of the Coliseum with a notice-board (ringed round) announcing the event (1930)

In August, Baird made his first test on programmes from the B.B.C. studio, and it is of interest to note that the portable disc scanner had by this time been permanently abandoned for the mirror drum, which in this country held the field completely until the opening of Alexandra Palace in 1936.

The mirror drum too was being used for another system of large screen television which was to be the forerunner of the larger type of home receiver produced by the Baird Company during 1933 and 1934. This was the " modulated arc " system, which was being developed by G. B. Banks under the direction of Baird himself. The light from the arc was first polarised and then passed through a Kerr cell consisting of two metal plates immersed in nitro-benzine. The cell was followed by a second Nicol prism crossed with respect to the first, so that no light normally passed. On a signal being applied to the plates of the cell, however, a rotation of the light was secured, causing it to pass through the second prism. The refractive index of the nitro-benzine was sufficiently altered proportionately to the signal strength to provide a kind of " light valve " effect.

This system, however, was not successful for large screen work, and Baird turned his attention to the possibility of actually modulating the arc itself. This was successfully achieved over a limited band of frequencies, and low definition images of surprising brightness were obtained on a screen some 5 feet by 3 feet.

Finally, on December 29th, the *Times* printed an article by its wireless correspondent in which the case for and against low definition television for home use was discussed. The article is of interest mainly for the following prophecy which occurs at the end of it :

" . . . We have in the ultra short-wavelengths of 5 to 10 metres, transmission media admirably adapted for local television broadcasts, and it is now generally agreed that the television of the future will be accomplished by means of ultra short-wave transmitters elevated on a tower or on a hill."

But the shadow of the Thing that was to render all mechanical television obsolete, was now becoming clearer. In the early part of 1931 Manfred von Ardenne successfully transmitted and resolved a film, " using cathode ray tubes only and no mechanical contrivances whatsoever."

1932 Ultra short-wave transmission was first used by Baird on April 29th, 1932. For the first experiment a receiver was installed at Selfridge's in Oxford Street, already the scene of many of Baird's early experiments. This test is significant for the fact that as well as a very high frequency radio system being employed, a new mirror drum-type of home receiver was used in Selfridge's. This employed the Kerr cell, which, although found to be inapplicable to large screen work, was nevertheless to become for a short period a standard light modulating device for the larger type of home receiver.

Attention was also being given to the possibilities of using synchronous motors for receivers in all areas there the grid supply of electricity was available.

In order to confound the critics who argued that pictures by the Baird system on television sets using Neon tubes and discs were " too small to have any entertainment value," Baird decided to build six super home receivers giving a picture 12 inches by 6 inches, and employing the mirror-drum Kerr cell principle. The Kerr cells were actually made by Jacomb, and were built into test tubes, Jacomb and one assistant performing all the intricate assembly and glass work which was necessary. In addition to five receivers which were privately allotted in the London area, one was presented to the B.B.C. and was installed in one of their press listening halls at Broadcasting House.

These sets were probably the ultimate in 30-line receivers, and they were bright, presented a black-and-white picture, and exceedingly stable in operation.

From them was developed the well-known Bush mirror-drum receiver which was in fact a scaled-down version of the original six sets laid down in the spring of 1932.

Since the Bush mirror-drum receiver had to work in areas where mains were not synchronised, a normal phonic wheel synchronising system was fitted.

Meanwhile arrangements had been completed for the B.B.C. to take over television transmissions completely from the Baird Company, and to this end a special scanning equipment and amplifier system had been built by the Baird Company and installed in the little sub-basement studio, B.B., in Broadcasting House.

On August 22nd, the first official B.B.C. broadcast of television in this country took place. Baird himself appeared in the opening programme, and in one of the shortest speeches he ever made, said simply, " I wish to thank the B.B.C. for inviting me here to-night, and to express the hope that this new series of television transmissions will lead to developments of broadcasting, increasing its utility and adding to the enjoyment of the great listening public."

The commencement of transmissions from Studio B.B. heralded a new and important phase in the history of the Baird system. Not only was the mirror drum scanner capable of a greater light output, but improved forms of photocell enabled very much improved signal background ratio to be obtained. These two developments put together meant that instead of close-ups of head and shoulders only, a complete, though necessarily small stage, could now be televised, and dancing acts and vaudeville turns transmitted.

The sale of ordinary disc and Neon television sets was now increasing, and there can be no doubt that, although the degree of detail was very limited and the pictures still had the annoying 12½-cycle flicker, television programmes of genuine entertainment value were now being regularly radiated on a service basis in England.

This was especially so when close-ups of personalities of the day appeared in Studio B.B., for the degree of definition in the close-up and semi-extended shot had now reached its maximum possible level, and when it is remembered that the whole image was composed of only 30 lines instead of the four or five hundred line systems of to-day, some of the pictures obtained on mirror-drum/Kerr cell receivers were really quite remarkable.

The 30-line mirror drum scanner installed at the B.B.C. studio in Portland Place.

The 30-line transmissions from Studio B.B. were efficiently organised, imaginatively produced, and formed the connecting link between the years of development and the final opening in 1936 of the B.B.C.'s high definition service in London.

1933 Transmissions from Broadcasting House proceeded smoothly throughout the year, and no very great changes in technique or development marked the smooth progress of low definition television broadcasting from London by the Baird system.

One interesting event, however, was the demonstration in April of a 60-line spotlight system organised by the Television Society and held at the Imperial College, South Kensington. Contemporary accounts all bear witness to the fact that the improvement in definition was most marked. The whole equipment is of interest as it was portable in form, and a parabolic mirror was used in order to collect as much light as possible from the subject.

In April, too, the first full-dress review to be televised was produced under the direction of Eustace Robb, and attracted considerable press attention. It was during this year also that Adeline Genée televised from London her farewell performance, partnered by Anton Dolin. The transmission was received clearly in Copenhagen.

1934 This year really saw the start of practical high definition television. Already in Germany, the Fernseh Company had produced a remarkable film transmitter using disc scanning, and a standard of 120 lines 25 pictures per second. The success of this machine was to no little measure due to the micrometer precision engineering tools which the Germans had available for disc construction. Later in the year, Jacomb produced at Long Acre an improved version of this machine using German discs and operating on the same 120-line 25-frame per second standard.

Baird's first high definition telecine was monitored mechanically by means of a large aluminium lensed disc modulated by a Kerr cell and steady light source. Later on, however, a more thoroughly engineered version of this equipment was produced, into which for the first time in the history of the Baird Company a cathode ray tube monitor was built. The original tube was a standard oscilloscope one, using a green fluorescent screen, and was gas-filled.

The results from this equipment were, of course, immeasurably better in detail and scope than anything previously seen, and in order that some experience might be gained in the propagation of high definition television signals, the 120-line telecine was installed on the top floor of Broadcasting House, where there was an experimental 10-metre transmitter already operating.

The first high definition tests from the B.B.C. were between Broadcasting House and the Polytechnic Institute in Regent Street, only about 100 yards away, where was installed a television receiver using one of the new German 15-inch tubes with white screen.

Having seen the shape of television to come, Baird retired to his home, No. 3, Crescent Wood Road, Sydenham, where in a specially built private laboratory, he concentrated entirely on the problems of large screen, colour and stereoscopic television.

The Baird Company, meanwhile, having obtained a lease of part of the ground floor of the old Crystal Palace, moved at last from Long Acre, and proceeded to set up a complete transmission system for demonstration to the B.B.C., with a view to obtaining the contract for equipping the future London television station.

It was at the Crystal Palace that mechanical transmission reached its peak, here for the disc was again used for studio transmissions of limited size, while in order to cope with the demand for outdoor scenes and large scale studio shots, the Company placed its faith in the intermediate film process, whereby a film moving continuously from the camera is rapidly processed and scanned, which in this case was again by a disc.

By this time the standard had risen from 120 to 240 lines, still at 25 frames per second. There can be no doubt that for their day some of these early high definition pictures were quite remarkable, but alas, the final limitation of a mechanical system was now very definitely in sight. Such things as minute specks of dirt in the tiny apertures of the scanning discs, " hunting " between transmitter and receiver, immobility of equipment due to the weight of the mechanical parts were all conspiring to bring to an end the era of mechanical scanning. Nevertheless, it was considered improbable at the time that any electronic system likely to be offered as an alternative to the Baird one would equal in detail and picture quality the results obtainable at the Crystal Palace, and it is significant that the Television Advisory Committee were so undecided themselves about the relative methods of electronic and mechanical television, that on their advice two complete systems, the mechanical Baird and the iconoscope E.M.I., were ordered for trial side by side at the Alexandra Palace, which had been finally decided upon as the site of the B.B.C.'s London Television Station.

Baird meanwhile, having withdrawn entirely from the activities of the Company, was at Crescent Wood Road and had produced a 3-colour television system of considerable merit. This employed mechanical scanning of the transmitter, and a cathode ray tube was used as the light source in the receiver, with the usual 3-colour filter discs at both ends.

During this time he was also experimenting with high voltage cathode ray tubes for direct projection for large screen pictures, and had already achieved pictures of excellent quality and brightness on a screen which was measurable in feet instead of inches.

1935 It was during 1935 that Baird started experimenting in large screen colour television, and commenced work on a machine which ultimately took three years to build, and when finally completed, early in 1938, was installed at the Dominion Theatre in London, where pictures 12 feet by 9 feet in full colour were demonstrated by radio, being received from the Crystal Palace, about ten miles distant.

The transmitting apparatus consisted of a mirror drum 12 inches in diameter, revolving at 6,000 r.p.m., and a slotted disc revolving at 500 r.p.m. The disc was provided with 12 slots arranged at different distances from its periphery, alternate slots being fitted with red and blue-green filters. (See p. 9.)

The mirror drum was provided with 20 mirrors, which reflected the scene to be transmitted through a lens which focused an image from the disc. The combined action of the disc and drum being to produce a 120 line picture, made up of a 20 line scan interlaced six times. Images corresponding to red and blue-green were there transmitted by alternate lines by a rubidium cell placed behind the slotted disc.

The receiving apparatus in the ultimate machine used later at the Dominion was similar to that used at the transmitter, but a high intensity arc lamp of 150 amperes replaced the photo-electric cell, and a Kerr cell was used to modulate the light. Other things, too, were being worked on this year in the little laboratories in Crescent Wood Road. Colour home receivers and the old problem of stereoscopic television, were both receiving attention. Also for the very first time, Baird was giving some thought to ways and means of producing results in full colour from an all-electronic system.

At the Crystal Palace, some little way away, the Baird Company were very busy building the equipment for demonstrations to the Television Advisory Committee, which was due to be given in the summer. Success was being achieved to a remarkable degree with mechanical telecine. The spotlight direct pickup apparatus, however, was causing some headaches, and the intermediate film system was very definitely giving trouble.

1936

From a historical point of view 1936 is probably one of the most interesting in the history of British television, for it was during this year that the finest possible mechanical system was tried out, side by side with the first British all-electronic system, and at the end of a test period during which each installation took the B.B.C. programmes over during alternate weeks, it was proved beyond all possible doubt that the era of mechanical television had finally and definitely come to a close, and the discs and mirror drums for all normal domestic transmission purposes had spun to a final stop.

Baird always had the idea that in a mechanical system the degree of detail possible must always be greater than in an electronic one, since the spot size and shape could, by the use of the finest machine tools, be made finite and dead accurate.

Against this, however, of course, was the trouble into which mechanical systems got due to minute particles of dust forming in the scanning holes, but what really caused the mechanical eclipse was the utter immobility of the equipment. It did not take very long to see which future principle for television was bound to be adopted, and naturally, the decision

One of the last photographs of J. L. Baird, with his colour tube.

to use an all-iconoscope system was a severe blow to John Baird's hopes and dreams. It was probably at this time that his great quality of courage became more evident than at any other phase of his life, for he set to work at once, harder than ever, to try to prove that if normal studio transmission was not to be carried out by means of discs and mirror drums, at least the future of large screen and colour television lay with mechanical devices.

The Last Phase

It was clear now that Baird's life had but one aim, namely, to try to prove that the correct application of his original idea was practical, and indeed, the only obvious way to achieve large screen colour pictures. From now on, though, his health began to show increasing signs of deterioration, and even spending most of 1938 in the sun of Australia, where he went to lecture to the World Radio Convention, failed to restore him to full strength. When war came in 1939, he at once evacuated his family to Devonshire, and returning to London alone, worked almost single-handed throughout the years of struggle on his large screen colour system.

During the war Baird was offered an appointment as Technical Adviser to Cable & Wireless, Ltd., and devoted part of his time to the possibilities of applying television to telegraphy problems.

At the same time he continued to work on large screen receivers and colour television. At one of the National Radio Exhibitions he demonstrated what must be the world's largest direct-viewing cathode-ray tube, in which the screen was mounted inside the bulb intended for a mercury-arc rectifier—the largest he could get at the time !

He also demonstrated two noteworthy systems of colour television, both of which were subsequently developed experimentally by other rival companies. In one, the cathode-ray tube contained a serrated screen on which two powders were deposited to give green and blue fluorescence. A third screen was coated with red fluorescing powder, and when the composite screen was scanned by three independent beams a colour picture was produced. In the other, two independent pictures on the face of the cathode-ray tube were focused through red and blue filters on to a screen through a magnifying lens. The superposition of the pictures gave an accurate rendering of a two-colour picture.

Towards the end of 1945, Baird's health, never good, broke down and in the winter he caught a chill from which he never recovered. He died at his house at Station Road, Bexhill-on-Sea, on June 14th, 1946, at the age of 58.

In a tribute to his old chief, one of Baird's staff in the early days says : " Against the scepticism of current scientific thought, despite poverty and ill-health and totally inadequate resources, almost, it seemed then, against nature herself he wrested real television from a crude whirling disc, neon lamps, and selenium cells. There was no looking back or resting, and I cannot believe that there was a single day after that historic January 27th during which J. L. B. did not ponder some new aspect of his art, or invent in his mind some way to improve or embellish it."

Television as we know it to-day involves a multitude of separate ideas and patents, and is being constantly developed by the ever-active research of thousands of engineers all over the world. Had it not been for the practical success of John L. Baird's efforts in 1926 it is possible that the television age would not be fully upon us. There is little doubt that his work in England caused the beginning of an intense research such as has never been applied to any other branch of peace-time science.

Baird's name is perpetuated in the present-day company of Baird Television, Ltd., and in two memorial plaques, the first of which was unveiled by him in person at the house in Hastings where he first experimented with television.

The second plaque, at 22, Frith Street, Soho, commemorates the house where he first showed the television of living images.

At the unveiling ceremony, Sir Robert Renwick, President of the Television Society, said " Although this plaque stands in the heart of London, Baird's real memorial is in the forest of television aerials that covers the whole country. When next we look at one of these, let us remember John Logie Baird, who first showed us television."

TELEVISION

Herbert E. Ives

The Bell System Technical Journal

October, 1927

Television [1]

By HERBERT E. IVES

SYNOPSIS: The chief problems presented in the accomplishment of television are discussed. These are: the resolution of the scene into a series of electrical signals of adequate intensity for transmission; the provision of a transmission channel capable of transmitting a wide band of frequencies without distortion; means for utilizing the transmitted signals to re-create the image in a form suitable for viewing by one or more observers; arrangements for the accurate synchronization of the apparatus at the two ends of the transmission channel.

INTRODUCTION

THIS paper is to serve as an introduction to the group of papers following, which describe the apparatus and methods used in the recent experimental demonstration of television over communication channels of the Bell System. In that demonstration television was shown both by wire and by radio. The wire demonstration consisted in the transmission of images from Washington, D. C., to the auditorium of the Bell Telephone Laboratories in New York, a distance of over 250 miles by wire. In the radio demonstration, images were transmitted from the Bell Laboratories experimental station at Whippany, New Jersey, to New York City, a distance of 22 miles. Reception was by two forms of apparatus. In one, a small image approximately 2 in. by $2\frac{1}{2}$ in. was produced, suitable for viewing by a single person, in the other a large image, approximately 2 ft. by $2\frac{1}{2}$ ft., was produced, for viewing by an audience of considerable size. The smaller form of apparatus was primarily intended as an adjunct to the telephone, and by its means individuals in New York were enabled to see their friends in Washington with whom they carried on telephone conversations. The larger form of receiving apparatus was designed to serve as a visual adjunct to a public address system. Images of speakers in Washington addressing remarks intended for an entire audience, and of singers and other entertainers at Whippany, were seen by its use, simultaneously with the reproduction of their voices by loud speaking equipment.

[1] Presented at the Summer Convention of the A. I. E. E., Detroit, Mich., June 20–24, 1927.

CHARACTERISTIC PROBLEMS OF TELEVISION

The problem of television in its broad outlines is that of converting light signals into electrical signals, transmitting these signals to a distance, and then converting the electrical signals back into light signals. Given means for accomplishing these three essential tasks, the problem becomes that of developing these means to the requisite degree of sensitiveness, speed, efficiency, and accuracy, in order to re-create a changing scene at a distant point, without appreciable lapse of time, in a form satisfactory to the eye.

A convenient starting point for the discussion of television is the human eye itself. In this an image is formed upon the retina, a sensitive screen, consisting of a multitude of individual light-sensitive elements. Each of these elements is the termination of a nerve fibre which goes directly to the brain, the entire group of many million fibres constituting the optic nerve. A theoretically possible television system could be made by copying the eye. Thus a large number of photosensitive elements could be connected each with an individual transmission channel leading to a distant point, and signals could be sent simultaneously from each of the sensitive elements to be simultaneously used for the re-creation of the image at the distant point. The number of wires or other communication channels demanded in a television system of this sort would be impractically large. For practical purposes, reduction of the number of transmission channels is made possible by the fact that, while in vision all parts of the image on the retina are simultaneously and continuously acting to send nerve impulses, the inertia of the visual system is such that a sensation of continuity is obtained from discontinuous signals, provided these succeed each other rapidly enough. Due to the phenomenon of persistence of vision, it is immaterial to the eye whether the whole view be presented simultaneously or whether its various elements be viewed in succession, provided the entire image be traversed in a sufficiently brief interval, which for purposes of discussion may be taken as 1/16th of a second or less.[2] We thus have available in television the same artifice which is used in the much less exacting problem of transmission of pictures over a telephone line, that is, of *scanning*, or running over the elements of the image in sequence, instead of endeavoring to transmit all of the elementary signals simultaneously. The development of a television system therefore

[2] This figure of 1/16th of a second, commonly quoted in discussions of this sort, is a convenient one, although the frequency of image repetition necessary to extinguish "flicker" is actually proportional to the logarithm of the field brightness. A somewhat higher rate of image repetition was used in the final television apparatus.

necessitates, at an early stage, the design of some scanning system by which the image to be transmitted may be broken up into sequences of signals. In the simplest case, where one transmission channel is to be used, the whole image will be resolved into a single series of signals; if more than one transmission channel is to be utilized, the resolution may, by parallel scanning schemes, or their equivalent, be broken up into several series for simultaneous transmission.

Like the eye, an artificial television system must have some light-sensitive element or elements by means of which the light from the object shall produce signals of the sort which can be transmitted by the transmission system to be used. For a television system to operate over electrical transmission lines this means some photo-electric device. It is obvious that this photoelectric device must be extremely rapid in its response, since the number of elements of any image to be transmitted must be some large multiple of the fundamental image repetition frequency, that is 16 per second. The response should, of course, be proportional to the intensity of the light, and finally, the device must be sufficiently sensitive so that it will give an electrical signal of manageable size with the amount of light available through the scanning system. This latter requirement, that of sensitiveness, is one which, it was realized, from studies made with our earlier apparatus for the transmission of still pictures over wires,[3] would be extremely difficult to meet. In the picture transmission system a very intense beam of light from a small aperture is projected through a transparent film and on to a photoelectric cell. In practical television, the system must be arranged to handle light reflected from a natural object, under an illumination which would not be harmful or uncomfortable to a human being. Actual experiment showed that the greatest amount of light which could be collected from an image, formed by a large aperture photographic lens on the small scanning aperture of the picture transmission apparatus, was less by a factor of several thousand times than the light projected through it for still picture transmission purposes. Assuming the same kind of photoelectric cell to be used, the additional amplification required over that used in the picture transmission system, taking into account also the higher speed of response demanded, would bring us at once into the region where amplifier tube noise and other sources of interference would seriously affect the result. This indicated clearly that some more efficient method of gathering light from the object than the commonly assumed one of image formation by a

[3] "Transmission of Pictures over Telephone Lines," Ives, Horton, Parker and Clark. *Bell System Technical Journal*, Vol. IV, No. 2, April, 1925.

lens was required, unless some much more sensitive type of photoelectric cell should be found.

Assuming that means could be developed for producing an electrical signal proportional to the intensity of the light, of sufficient quickness to follow a rapid scanning device, and of sufficient strength either as directly delivered from a photosensitive device or as amplified, the next problem is that of its transmission over an electrical communication system. We may quickly arrive at an understanding of certain of the transmission problems by reviewing the requirements for the transmission of photographs. In the system of still picture transmission now in use by the American Telephone & Telegraph Company, a picture 5 in. by 7 in. in size, divided into the equivalent of 10,000 elements per square inch or 350,000 elements, is transmitted in approximately seven minutes. This requires the transmission of a frequency band of about 400 cycles per second on each side of the carrier frequency. If we plan, in the transmission of television, to transmit images of the same fineness of grain, it would mean that what is now transmitted in seven minutes would have to be transmitted in a 16th of a second, which in turn means that the transmission frequency range would have to be nearly 7000 times as great. That is, a band approximately 3,000,000 cycles wide would be required. Bearing in mind that wire circuits are ordinarily not designed to utilize frequencies higher than 40,000 cycles per second, and that with radio systems uniform transmission of wide signal bands becomes extremely difficult, it is seen at once that either an image of considerably less detail than that which we have been considering must suffice, or else some means for splitting up the image so that it may be sent by a large number of channels is indicated.

A further theoretical requirement must also be given consideration. This is that the complete television signal will consist of all frequencies up to the highest above discussed, and down to zero, that is, an essential part of the signal is the direct current component, furnished by those parts of the scene which do not change. The problem of handling the very low frequency components, presents difficulties both in the vacuum tube amplifier system adjacent to the photosensitive device, and in ordinarily available transmission channels.

In any case certain fundamental transmission requirements must be met. These are that the attenuation of the signals must be uniform over the whole frequency range and that the speed of transmission of all frequencies must be the same. Also, as in the transmission of sound signals, the amount of interference or noise must be kept down sufficiently not to impair the quality of the signal or picture.

Assuming the undistorted transmission of the signals to a distant point, the next fundamental problem of television is the reconstruction of the image, or the translation of the electrical signal back into light of varying intensity. Just as at the sending end we have seen that the production of a useful electrical signal with the amount of light available from a naturally illuminated object is a major problem, so at the receiving end the converse problem, that of securing an adequately bright light from the electrical signal, presents great difficulty. The nature of the problem may be understood by assuming that it is to be done by projecting the received image on a screen similar to an optical lantern projection screen. If the spot of light which is to build up this image scans the whole area in the same way that the object is scanned, we find that the amount of light which can be concentrated into a small elementary spot will, when distributed by the scanning operation over the whole screen, reduce the brightness of the screen in the ratio of the relative areas of the elementary spot and the whole screen. The amount of this reduction will, of course, depend upon the number of elements into which the picture is divided, but will in any event be a factor of several thousand times. It is doubtful whether any light source exists of sufficient intensity such that an image projected by it can be spread out by a scanning operation over a large screen and give an average screen brightness which would be at all adequate. It is possible to imagine optical systems by which such a thing as the crater of an arc could be projected upon the screen, but the motion of this image and its variation in intensity would involve the extremely rapid motion of lenses, mirrors and apertures of a size such as to render the operation mechanically impracticable. It appears from these considerations that the only promising means of reconstructing the image would be those in which a light source, whose intensity can be controlled with great rapidity, is directly viewed.

Another element of a television system upon whose solution success depends as much as any other is that of synchronization; the reconstruction of the image, postulated in the last paragraph, is only possible if the reconstructed elements fall in exactly the right positions at the right times, to correspond with the signals as generated at the analyzing end. The criterion for satisfactory synchronization will be expressed in terms of variation from identity of speed by figures which will depend on the fineness of grain of the image which it is planned to send. No element of the image must, of course, be out of place by a considerable fraction of the size of the element.

General Outline of Means Employed in the Present Television System

It has been pointed out above that if the goal which we set in television is the transmission of extended scenes, with a large amount of detail and hence made up of an exceedingly large number of elementary areas, we meet with the necessity for transmission channels of a character which are not now available. In the present development it was decided at the start to restrict our experiments to a size and grain of picture which, if the scanning and re-creating means were developed, would be capable of transmission over practical transmission channels, either wire or radio. This restriction fortunately leaves us with the possibility of meeting what was felt to be the typical problem of a Telephone Company, namely, the transmission of a human face in a television system used as an adjunct to a telephone system. Taking, as a criterion of acceptable quality, reproduction by the halftone engraving process, it is known that the human face can be satisfactorily reproduced by a 50-line screen. Assuming equal definition in both directions, 50 lines means 2500 elementary areas in all. 2500 elements transmitted in 1/16 second is 40,000 elements per second. The frequency range necessary to transmit this number of elements per second with a fidelity satisfactory for television cannot be calculated with assurance in advance. An approximate value can however be arrived at from a study of the results obtained in still picture transmission. In pictures transmitted by the system already referred to, individual faces contained in a square space ½ inch on a side are quite recognizable.[4] Taking the ratio of this area to the area of the whole picture, and using the frequency range figure already deduced, for a complete 5 in. by 7 in. picture, it appears that a band of 20,000 cycles would be sufficient to transmit such an image in 1/16 second.[5] These considerations led to the choice of a 50-line (2500-element) image as one which would be both satisfactory as to detail rendering, for our purposes, and as calling for frequency transmission requirements sufficiently low to give a good margin of safety in existing single communication channels.

As a method of scanning, the method which is probably mechanically simplest, namely, that of rotating disks with spirally arranged holes, proposed by Plotnow[6] in 1884, was chosen. In accordance with the

---

[4] Cf. Fig. 18 of the paper referred to (Reference 2).

[5] A factor which this analogy does not cover is that if the image is moving so that it falls on several discrete scanning elements in rapid succession a very material apparent increase in the fineness of the image structure results. This effect is similar to that by which the relatively coarse-grained individual images in a motion picture film fuse to give smooth appearing pictures.

[6] Plotnow, D. R. P. 30105, 6.1, 1884.

choice of grain above indicated, the disks were perforated with 50 apertures.

For the second element of the problem, the light-sensitive means, the alkali metal photoelectric cell was chosen as possessing the qualities of proportionality of response and quickness of reaction. The currents produced by it are at best quite small, but they lend themselves to the process of amplification by the three-electrode vacuum tube amplifier.

The problem of securing a large enough signal, which is intimately associated with that of securing enough light from the object, was, in our development work, postponed in the earlier stages, our first experimental work having been done by concentrating light through photographic transparencies.[7] The solution of the problem of securing adequate light was subsequently attained by reversing the light path and projecting a narrow beam of light through the scanning disk upon the object. By this means only the element of the object which was being scanned was illuminated at any one time, thereby reducing the average illumination enormously, and the problem of increasing the signal strength could be attacked by increasing the amount of photosensitive surface as well as by increasing the brightness of the scanning light.[8]

The problem of amplifying the photoelectric currents to sufficient value for transmission was solved by a practical compromise which at the same time met one of the transmission difficulties. This compromise consisted in amplifying and transmitting only the fluctuating or alternating current components of the signal, leaving the direct current component, which determines the general tone value of the image, for empirical reintroduction at the receiving end. By this scheme, stable amplifier constructions were made available, and the transmission channels, particularly the wire channels, could be utilized in their normal working form.

At the receiving end, the problem of securing a sufficiently bright image was solved, as indicated earlier, by the use of self-luminous surfaces of much higher intrinsic brightness than it is possible to secure by illumination of a surface by any light source which can be rapidly controlled as to its intensity. The self-luminous surfaces

[7] As one step in the development work moving picture film, projected by a commercial projector in synchronism with the scanning disks, was successfully transmitted.

[8] A still further advantage is obtained by limiting the scanning light to the region of the spectrum to which the photoelectric cells are sensitive (blue and violet). This is unnecessary where one-way transmission only is used but is of value where in two-way transmission a transmitted image is to be viewed by a person being scanned.

employed were glow lamps containing neon gas, the brightness of which changes with sufficient rapidity to follow the incoming signals.

The problem of synchronization was postponed in our earlier development work by mounting the scanning and receiving disks upon the same axle. It was later solved for the demonstration apparatus by the utilization of synchronous motors controlled by two frequencies, a low frequency, that of the image repetition period, and a high frequency, chosen of such a value that the fraction of the cycle through which transient phase displacements occurred amounted in angular displacement to less than half the angular extent of a single disk aperture. The synchronization control therefore called for the transmission of additional currents for synchronization purposes over and above the picture current.

In order to transmit and synchronize the image signals it is necessary to transmit three different frequency bands, one for the image, and two for the high and low frequency synchronization controls. In the demonstration of April 7, 1927, the images were sent in the wire demonstration over a high quality open wire line. The synchronization control was sent over two separate carrier channels of a second telephone line. In addition to these lines, another line was used for conveying the telephone conversation. In the radio demonstration two different wave-lengths were used respectively for the image signals and for the synchronization signals which were, as in the wire demonstration, carried on two different carrier frequencies. A third channel was used for the voice. In the case of both wire and radio transmission, it is quite possible to put all of the different signals upon the same transmission channel, using different carrier frequencies.

It will aid toward a clear understanding of the reasons for the success of the system of television described in the following papers if we summarize at this point the chief novel features to which that success is due. They may be listed as follows:

1. Choice of image size and structure such that the resultant signals fall within the transmission frequency range of available transmission channels.[9]

2. Scanning by means of a projected moving beam of light.

3. Transmission only of alternating current components of image.

4. Use of self-luminous surfaces of high intrinsic brilliancy for reconstruction of the image.

5. High frequency synchronization.

[9] As the succeeding papers show, the margin between the frequency range required by the scanning apparatus and that which could be made available was quite liberal. It appears in the light of our experience that apparatus with 60 or 70 scanning holes instead of 50 might be used with the transmission facilities which were at our disposal.

Applications and Future Developments

It is not easy at this early date to predict with any confidence what will be the first or the chief uses for television, or the exact lines that future development may take. It must be clearly understood that television will always be a more expensive service than telephony, for the fundamental reason that it demands many times the transmission channel capacity necessary for voice transmission. This expense will inevitably increase in proportion to the size and quality of the transmitted image.

The kinds of service which are naturally thought of upon consideration of the services now rendered in connection with sound transmission are: first, service from individual to individual, parallel in character to telephone service, and as an adjunct thereto; second, public address service, by which the face of a speaker at a distant point could be viewed by an audience while his voice was transmitted by loud speaker; third, the broadcasting of scenic events of public interest, such as athletic contests, theatrical performances and the like.

The first two types of service just mentioned lie within the range of physical practicability, with apparatus of the general type already developed. The third type, because of the uncontrolled conditions of illumination, and the much finer picture structure which would be necessary for satisfactory results, will require a very considerable advance in the sensitiveness and the efficiency of the apparatus, to say nothing of the greatly increased transmission facilities. For all three types of service, wire or radio transmission channels could be utilized, for while the problems incident to securing distortionless transmission over wide frequency bands, or multiple transmission channels, are different in detail in the two cases, they appear to be equally capable of solution by either means. However, the very serious degradation of image quality produced by the fading phenomena characteristic of radio indicates the practical restriction of radio television to fields where the much more reliable wire facilities are not available.

RADIO VISION

C[harles] Francis Jenkins

RADIO VISION*

By

C. Francis Jenkins

(Jenkins Laboratories, Washington, D. C.)

In speaking to you this evening on the subject of the electrical transmission of pictorial representations, may I say that in our laboratory we have found it convenient and informative to use the words radiogram; radiophone, and radio vision when we speak of radio-carried service; and to say telegram, telephone, or television whem we speak of wire-carried service.

Figure 1—Laboratory receiver for radio vision. The black box contains a neon gas lamp; the slotted lens-disc sweeps the light-image spot across the screen in lines while the overlapping prism-disk distributes the lines from top to bottom of the screen.

The art of electrical picture-transmission is very old, relatively, for more than fifty years ago successful demonstrations were made in sending pictures by wire.

And there have been many workers, too, but the attainment of each was given but passing notice until the stamp of approval was put thereon by the great laboratories of the Bell Telephone Company, when, in April last, they made their spectacular demonstration between Washington and New York. This demonstration gave a great impetus to the development of electrical transmission of all kinds of pictorial representation.

* Original manuscript received by the Institute, July 6, 1927.
Presented before the Philadelphia Section of the Institute of Radio Engineers, June 24, 1927.

In general there have been but two types of mechanisms employed, the cylinder and the flat surface. The cylinder has been used most often in the transmission of still pictures.

But obviously a flat receiving surface is the only possible type for radio vision and television, for the eye must seem to see the whole picture all the time, and that can only be done on a flat screen.

However, whatever type of machine is used, the only method employed to the present time consists in a linear analysis of the picture, scene, or object, and the instantaneous synthesis of each line on a distantly located receiving surface.

Figure 2—The weather map ship's receiver. The pen-box is moved across the base-map on the rotating cylinder by the screw, while the incoming radio signals touch the ink pen to the paper to build up the weather map.

The lights and darks of each successive line are changed into electrical current of corresponding strengths, which, carried to distant receivers, is there changed back into like light intensities and assembled on a suitable surface, for example, a sheet of paper, a photo film, or a flat picture screen.

It is quite evident that if this synthesis in the received picture is to result in an exact likeness to the transmitted subject, perfect synchronism of all receiving mechanisms with that of the transmitting mechanism must exist.

To attain synchronism, the synchronous motor is the simplest, but it is limited in application.

Other synchronizing methods have consisted of clock-controlled motors, where elapsed time is the standard against which all the motors were regulated.

Tuning forks for controlling the motors have been extensively employed, but this is but a modification, for it is only a smaller and more frequent division of time than the clock method.

Because pendulum clocks and most tuning forks soon stop when installed aboard a rolling ship in a rough sea, an oscillator, resembling the escapement of a ship's chronometer, has been developed, and has been found quite a successful motor controller.

A simple method of synchronizing consists in maintaining the speed of the motor constant with a governor, together with means for automatically setting the receiving cylinder at zero on the beginning of each revolution of the transmitter cylinder.

We employed this method in the "three-cornered experiment" recently conducted by the U. S. Navy, the Weather Bureau, and the writer's laboratory.

Figure 3—Fork-controlled motor unit. The vibration of the free ends of the fork arms cuts out armature resistance to keep the motor up to a definite speed. The fork continues to vibrate by reason of the current pulses in the electro-magnet between the fork arms.

Each morning the Weather Bureau made up a weather map, gave it to us, and we put it on our transmitter in the Navy building, which was connected by wire to the radio broadcast transmitters at Arlington, Virginia.

Weather map receivers were set up at certain land stations, and also on board the *U. S. S. Trenton*, and the *U. S. S. Kittery*. The latter made experimental cruises between Naval Operating Base at Hampton Roads and Caribbean Sea ports.

This territory was chosen for its well-known static disturbances, coincident with hurricane-forming zones. And it was found that weather maps could be recorded on the receivers with certainty even when weather information could not be received by code. The Florida hurricane of September provided a severe test of the system, demonstrating its worth and dependability.

While these "side-uses" are doubtless valuable, my premier ambition was radio vision, and so we do these other things only in "breathing spells" between attacks on the main problem.

Our first public demonstration of radio vision occurred on June 13, 1925, when we showed in the laboratory in Washington, in the presence of Navy Secretary Wilbur, Admirals Taylor and Robinson, and many others, what was happening at the time at the Naval Air Station at Anacostia, some miles distant. It was the first radio vision demonstration ever made, I believe, and quite an historical event to many of us.

Figure 4—The simple mechanism by which motion pictures are transmitted. The rotation of the slotted lens-disk sweeps the image of each picture "frame" on the film across the light-sensitive cell in the box shown at the extreme right above the tubes.

The possibilities of radio vision for home entertainment, and of television in business, have been told so repeatedly in the public press that I hardly need restate here the promises of views of distant inaugural ceremonies, flower festivals, and baby parades. Being technical men I rather think you are more interested in details of methods and mechanisms.

As you know, the general scheme is to analyze the object or scene by a rotating scanning disk which in the transmitter permits the light reflected from the subject to fall on a light-sensitive cell, which, just as in still pictures, changes these light values into like current values; and which at the receiver permits light from a given source to be seen, directly or by reflection from a screen.

Such a scanning disk was shown in a patent as early as 1884. It consisted of a disk with one-fiftieth of an inch (1/50 in.) holes therein, arranged in a spiral.

The holes were an inch apart in the spiral, and the ends of the spiral had an inch offset; therefore, the contemplated picture was an inch square made up of fifty lines.

The light intensity, as in a pin-hole camera, is limited to the amount which can pass through this minute aperture, that is, 1/50 of 1/50, or only 1/2,500 part of the whole light.

To overcome this limitation it was proposed by a Frenchman that an arc lamp of high-intensity be focused on these minute apertures, after passing which the spot of light is swept across the subject by the rapid rotation of the scanning disk. I hardly think

Figure 5—Strip device by which news copy, or like matter, can be transmitted by radio or by wire automatically as a continuous process. The overlapping prismatic rings in rotation sweep each typewritten line across a light-sensitive cell as the message moves longitudinally.

he expected to radio-transmit a baseball game by sweeping thereover a point of light from an arc lamp.

We, in the laboratory, think that limiting the light by passing it through these apertures is not the best plan; the available light is too limited. So, to get a greater value of the light, we usually make the openings in the scanning disk 1–1/2 inches in diameter and put lenses over the openings; and get the required tiny lightspot by focusing the light-source as a tiny flying spot on the receiving screen, to build up the moving picture.

While we have made a variety of mechanisms, this has been a fundamental principle in all of them.

Television would doubtless have been attained as early as the telephone if a suitable light-sensitive cell had been available for the transmitter, and an adequate light-source for the receiver.

Such tools are available today, and marked progress has been made in recent months toward a radio vision receiver acceptable to the public, and with the many hands and minds now engaged on the problem, I confidently believe its completion is the work of but a few months more.

In the transmitter we need more sensitive light-cells, or perhaps I should say, light-cells giving greater current output, a current output which will more dependably start the first tube of an amplifier.

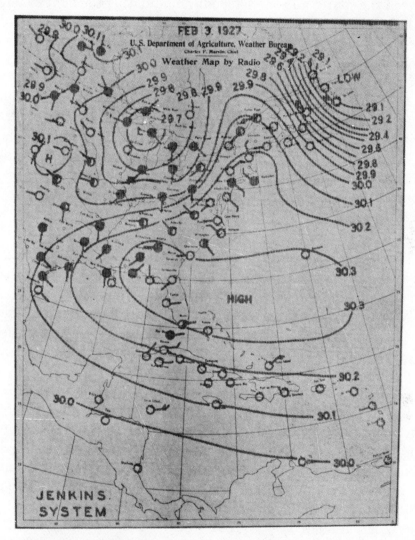

Figure 6—One of the weather maps as received aboard the *U. S. S. Kittery*. The base map is printed in brown and the isobars and other weather information received by radio is received thereon in red ink.

With light reflected from outdoor objects, the potassium cell gives but a very small current, only a few microamperes.

The present day potassium cells do not give current output proportional to the light intensity, but rather to the cell-surface covered by the light.

Resistance cells act too slowly for the required speed of light-reaction, which is of the order of 250,000 per second.

Our problem in the receiver is a light-source more intense than the neon lamp, but having approximately its high-speed light-change.

The other possibility is a steady light source and a light-valve between the source and the screen for modulating the light to build up the picture.

For this purpose many have proposed a bisulphide cell through which polarized light is passed and controlled by a potential (the Kerr effect) or a current field (the Faraday effect), but this cell requires such a large energy change to produce any useful light change as to put it out of consideration.

Of course, we have our own proposed solutions, which I shall tell you about if they are successful. Radio vision for home entertainment will revive radio interest as nothing else will, for it combines the reach of the radio and the fascination of the story told in pantomime.

Summary

To see by radio what is actually happening at a distant place is now an accomplished scientific attainment, though the mechanism is not yet a merchandising development.

When it shall have reached that point of perfection, then one may sit in one's home and see inaugural ceremonies, baseball, football, polo games, mardi gras, flower festivals, and baby parades.

Radio vision is also experimentally combined with audible radio, and when these instruments are made generally available, then radio pictures and music and speech at the fireside, sent from distant world points, will be the daily source of news; the daily instructional class, and the evening entertainment; and equally the long day of the sick and the shut-ins will be more endurable, and life in the far places less lonely, for the flight of radio is not hindered by rain or storm, or snow blockades.

The electrical transmission of still pictures of photographs, sketches, maps, and the like, is now an every day affair, applicable to newspaper illustration and other usefulness.

But for the research worker it has lost its interest for him, it is too easy. To increase the speed of picture presentation ten thousand times, as required in radio movies, makes it a sporting proposition, and really worth while.

The apparatus emplayed in transmitting pen-and-ink sketches, photographs, and movies by radio are all simple in construction, though modified to best suit each particular use.

TELEVISION WITH CATHODE-RAY TUBE FOR RECEIVER

V[ladimir K.] Zworykin

Television with Cathode-Ray Tube for Receiver[†]

Special Tube, Called the Kinescope, Eliminates Usual Scanning and Synchronizing Apparatus and Provides Larger Picture with Better Detail

By V. Zworykin*

THE problem of television has interested humanity since early times. One of the first pioneers in this field, P. Nipkow, disclosed a patent application in 1884[1] describing a mechanical scheme for television. It involved a scanning of the object and picture, for which purpose the familiar perforated disk was employed. The scanning disk is used even now, almost without alteration, in all practically-developed schemes of television apparatus. However, Nipkow's ingenious invention could not materialize in his day because of the lack of powerful modern aids—the photo-cell and radio amplification. At present, the rotating disk is giving excellent results within the mechanical possibilities of our time.

Out of a number of other methods which have been proposed for the solution of television by various inventors, the author[2] has been attracted by the application of the cathode ray for scanning purposes. This method was proposed for the first time by Boris

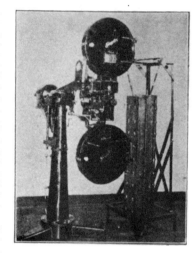

Fig. 4. A general view of the television transmitter.

working in the same direction with various degrees of success, striving to develop television reception by means of cathode-ray tubes. The cathode-ray tube presents a number of distinct advantages over all other receiving devices. There is, for example, an absence of moving mechanical parts with consequent noiseless operation, a simplification of synchronization per-

mitting operation even over a single carrier channel, an ample amount of light for plain visibility of the image, and indeed quite a number of other advantages of lesser importance. One very valuable feature of the cathode-ray tube in its application to television is the persistence of fluorescence of the screen, which acts together with persistence of vision of the eye and permits reduction of the number of pictures per second without noticeable flickering. This optical phenomenon allows a greater number of lines and, consequently, better details of the picture without increasing the width of the frequency band.

This paper will be limited to a description of an apparatus developed in Westinghouse Research Laboratories for transmission by radio of moving pictures using the cathode-ray tube for reception.

In the author's opinion, if a receiver is to be developed for practical use in private homes, it should be designed

[2] U. S. Patent application, March 17, 1924. U. S. Patent No. 1,691,324, July 13, 1925.
[3] English Patent No. 27,270, December 13, 1907.
[4] Belin et Holweck, Bull. No. 243 de la "Societé Francaise de Physique," p. 35, 8, March 1927.
[5] A. Douvillier, "Revue General de L'Electricite," p. 5, January 7, 1928.
[6] K. Takayanagi, "Jour. I. E. E.," Japan No. 482, pp. 932, Sept. 1928.

Fig. 3. A view of the projector, showing the vibrating mirror.

Rosing, professor of physics in Petrograd, in 1907[3]. The same reasons which handicapped Nipkow prevented Rosing from achieving practical results. Later Belin and Holweck,[4] Douvillier,[5] and Takayanagi[6] were

† Preprinted from a forthcoming issue of the Proceedings of the Institute of Radio Engineers.

* Engineering Department, Westinghouse Electric & Manufacturing Co., East Pittsburgh, Pa.

[1] P. Nipkow, English patent No. 30,105, January 6, 1884

Fig. 1. Details of the modified standard moving-picture projector — a part of the television transmitter — showing the location of the photo-cell, light source and vibrating mirror.

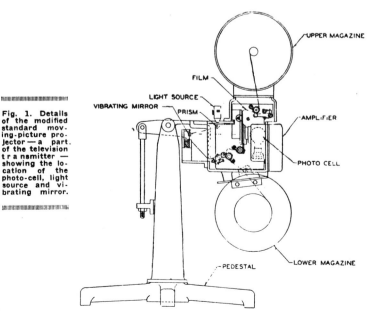

UPPER MAGAZINE

FILM

LIGHT SOURCE

VIBRATING MIRROR

PRISM

AMPLIFIER

PHOTO CELL

PEDESTAL

LOWER MAGAZINE

without any mechanically moving parts. The operation of such a receiver should not require great mechanical skill. This does not apply to the transmitter, since there is no commercial difficulty in providing a highly trained operator for handling the transmitter at a broadcasting station.

The Transmitter

The transmitter consists of a modified standard moving-picture projector. The intermittent motion device, the optical system, and the light source are dismantled. The film is caused to move with a constant speed downward, this motion providing the vertical component of scanning.

The construction of the transmitter is shown in Fig. 1. A light source is provided by an ordinary 6-volt automobile lamp. The light is focussed by a condensing lens *L* upon a diaphragm *D* with a small orifice. From there the beam of light emerging through the orifice is reflected from a vibrating mirror *M* and focussed into a sharply-defined spot on the moving film *F*. With the mirror vibrating at a

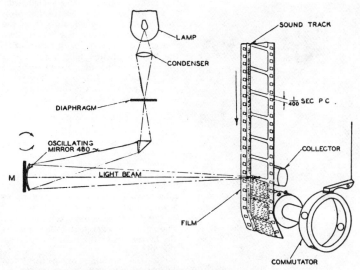

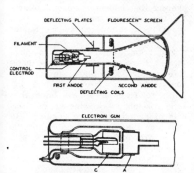

Fig. 5. Details of the special cathode-ray tube described in this article.

frequency of 480 cycles about a vertical axis, the light spot sweeps the film horizontally. This vibration of the mirror combined with the downward movement of the film to explore the whole surface of the pictures as shown in Fig. 2. After passing the film, the light enters a photoelectric cell *C* which transforms the variations of optical density in the film into a variable electric current.

The vibrating mirror is shown in Fig. 3. It consists of a small steel rod with a vane placed between the poles of an electromagnet. The poles are *U*-shaped and each leg is provided with a coil. An oscillating current of the same frequency as the natural frequency of the rod is supplied to the coils, thus causing the rod and the mirror to oscillate about the axis of the rod. In order not to depend upon the uniformity of sensitivity over the cathode area of the photo-cell, an additional lens *L₂* is provided between the film and photo-cell. This lens is so situated that the mirror and sensitive surface are at conjugate foci.

Fig. 2. Showing the manner in which the whole surface of the picture is explored by the light reflected from the vibrating mirror.

Thus, the scanning beam is always focussed upon a stationary spot in the cell.

From the fact that the horizontal scanning is produced by a sinusoidal current, it follows that the velocity of the beam across the picture is not uniform. The velocity in the center is about 57 per cent higher than that of a spot scanning at uniform rate a picture the same width. Before work was started on the machine, it was anticipated that the feature would be found objectionable and correction by optical filter was planned. Practical tests, however, indicate that the non-uniform distribution of light across the picture is not readily apparent to the eye, and, therefore, no precautions are now used. A general view of the transmitter is shown in Fig. 4.

The Receiver

The receiver consists of a cathode-ray tube especially designed for the purpose. The principles of the cathode-ray tube are well-known from their application for oscillographs. In their ordinary form, however, they cannot be used for picture reception, because although they have scanning arrangement in two dimensions they do not have means for varying the intensity of the picture. Moreover, neither of the main types of oscillographs is suited for television purposes. The high potential type which would give a sufficiently brilliant spot, is always operated in connection with a vacuum pump. Such a pump is impractical for a home television receiver. The low potential type of cathode-ray oscillographs is of the sealed-off type but the amount of light available from the screen is far too small. In order to give sufficient brilliancy for the picture of 5-in. size, the tube should operate at least at 3000 volts. For larger pictures still higher voltage is required, since the

brightness increases with the accelerating voltage. According to these requirements, a new type of cathode-ray tube was developed. This is shown in Figs. 5 and 6. An oxide-coated filament is mounted within a controlling electrode *C*. The cathode beam passes through a small hole in the front part of the controlling element and then again through a hole in the first anode *A*. The first anode accelerates the electrons to a velocity of 300 to 400 volts. There is also a second anode consisting of a metallic coating on the inside of the glass bulb. This second anode gives to the electrons a further acceleration up to 3000 or 4000 volts. The velocity of the electrons at this voltage is about one-tenth that of light. An important function of this second anode is also

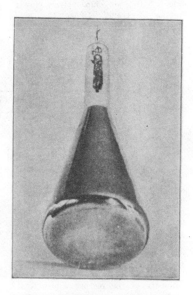

Fig. 6. The special cathode-ray tube, or Kinescope, as it is called.

Fig. 7. A view of one of the laboratory television receivers.

to focus electrostatically the beam into a sharp spot on the screen. The target wall of the bulb is about 7 in. in diameter and is covered with a fluorescent material such as willemite prepared by a special process so as to make it slightly conductive. Conductivity is required to remove the electrical charges from the screen supplied by the electron beam. This tube will be referred to hereafter in this paper as the kinescope.

The beam of electrons can easily be moved across the screen either by an electrostatic or an electromagnetic field, leaving a bright fluorescent line as it passes. For this purpose a set of deflecting plates and a set of deflecting coils are mounted on the neck of the kinescope, outside the tube. The plates and coils are adjusted in the same plane, so as to give vertical and horizontal deflection at right angles to each other. As a result of the location of the deflecting elements between first and second anode, the deflecting field is acting on comparatively slowly moving electrons. Hence the field

this mean intensity. It is evident that if we apply to this controlling electrode the amplified impulses from the transmitter and at the same time deflect the beam to synchronism with the motion of the light beam across the picture on the film, the picture will be reproduced on the fluorescent screen. Figs. 7 and 8 show a general view of two types of receivers.

Synchronization

If separate channels are available for each of the synchronizing signals, the problem of synchronization of the receiver with the transmitter is very simple. For horizontal scanning, it is necessary only to transmit the scanning frequency, operating the mirror as a sinusoidal voltage and to impress it on the deflecting coils of the kinescope. The cathode beam will follow exactly the movement of the light beam across the film.

For the framing or picture frequency, a voltage is generated at the receiving end and merely controlled by signals from the transmitter. A con-

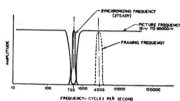

Fig. 10. The spectrum used to modulate the radio-frequency carrier.

denser is charged at constant current through a current limiting device, such as a two-electrode tube, so that the voltage at the condenser rises linearly. The deflecting plates of the kinescope are connected in parallel to this condenser, and, therefore, when the condenser is charging, this cathode beam is deflected gradually from the bottom to the top of the fluorescent screen at

Fig. 8. The television receiver built into a Radiola cabinet.

constant speed. This speed is regulated by the temperature of the filament of the charging tube to duplicate the downward movement of the film. An impulse is sent from the transmitter between pictures, which discharges the condenser, quickly returning the beam to the bottom position, ready to start upward and reproduce the next picture.

For transmission of the complete picture, three sets of signals are therefore required: picture signals, horizontal scanning frequency, and impulses for framing. It was found that it is possible to combine all of these sets of signals into one channel. In this case the photo-cell voltage of the transmitter is first amplified to a level sufficiently high for transmission. There is then superimposed upon the series of high audio-frequency impulses lasting a few cycles only and occurring when the light beam passes the interval between the pictures. (Fig. 9.)

The picture frequencies together with the framing frequencies are then passed through a band-eliminating filter, which removes the picture component of the same frequency as that of horizontal scanning. Following this, a portion of the voltage which drives the transmitter vibrator is impressed upon the signals, passed through the filter, and the entire spectrum is used to modulate the radio-frequency carrier. (Fig. 10.)

At the receiving station the output of the local radio receiver is amplified

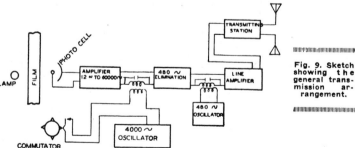

Fig. 9. Sketch showing the general transmission arrangement.

strength required is much less than that which would ordinarily be used to deflect the beam under the full acceleration of the second anode voltage.

The brightness of the line can be controlled to any desired extent by a negative bias on the controlling element. The bias controls the mean intensity of the picture whose lights and shadows are superimposed upon

Fig. 11. Sketch showing the general reception arrangement.

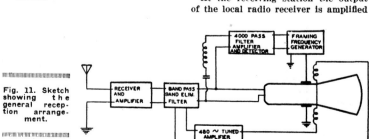

and divided by a band-pass band-elimination filter into two parts; one the synchronizing frequency, and the second the picture frequency plus the framing frequency. The synchronizing frequency is amplified by a tuned amplifier which supplies current to the deflecting coils of the kinescope. (Fig. 11.)

The picture and framing frequencies are applied directly to the control electrode of the kinescope.

The same voltage which modulates the light is impressed upon a band-pass filter, which is tuned to the frequency of the a-c. voltage used for the framing impulses. The output of this filter is amplified, rectified, and used to unbias a discharging triode which is normally biased to zero plate current, and which takes its plate voltage from the condenser which provides the vertical scanning voltage.

Thus, the picture signals and both synchronizing and framing frequencies are transmitted on one channel, and fully automatic synchronization is obtained.

The amplification problem in this case does not differ from that of the amplifier for mechanical television of the same picture frequency. The frequency band for which the amplifier should be constructed is much lower for the same number of lines due to the smaller number of pictures per second.

Conclusion

Those who are accustomed to the conventional scanning disk type of television notice a number of differences in the appearance of the picture as viewed on the end of the cathode-ray tube. The picture is green, rather than red (as when a neon glow tube is used). It is visible to a large number of people at once, for an enlargement by means of lenses is unnecessary. There are no moving parts, consequently, no noise. The framing of the picture is automatic; and it is brilliant enough to be seen in a moderately-lighted room.

Technically, the kinescope type of receiver presents added advantages. The high-frequency motor for synchronization, together with its power amplifier, is not required. The power required to operate the grid of a kinescope is no more than that for an ordinary vacuum tube.

TWO-WAY TELEVISION

Herbert E. Ives

Frank Gray

and

M. W. Baldwin

Two-Way Television
Part I—Image Transmission System

BY HERBERT E. IVES,[1]   FRANK GRAY,[1]   and   M. W. BALDWIN[1]

Fellow, A. I. E. E.   Non-member   Non-member

Synopsis.—A two-way television system, in combination with a telephone circuit, has been developed and demonstrated and is now in use between the Bell Telephone Laboratories, at 463 West Street, and the American Telephone and Telegraph Company, 195 Broadway. With this system two people can both see and talk to each other. It consists in principle of two television systems of the sort described to the Institute in 1927. Scanning is by the beam method, using disks containing 72 holes, in place of 50 as heretofore. Blue light, to which the photoelectric cells are quite sensitive, is used for scanning, with a resultant minimizing of

glare to the eyes. Water-cooled neon lamps are employed to give an image bright enough to be seen without interference from the scanning beam. A frequency band of 40,000 cycles width is required for each of the two television circuits. Synchronization is effected by transmission of a 1275-cycle alternating current controlling special synchronous motors rotating 18 times per second. Speech transmission is by microphone and loud speaker concealed in the television booth so that no telephone instrument interferes with the view of the face.

* * * * *

INTRODUCTION

DURING the past few years, since the physical possibility of television has been established, the chief problems which have received attention have been those of one-way transmission. In particular, the experimental work in radio television has had for its principal goal the broadcasting of television images, which is inherently transmission in one direction. At the time of the initial demonstration of television at Bell Telephone Laboratories in 1927,[2] one part of the demonstration consisted of the transmission to New York of the image of a speaker in Washington simultaneously with the carrying on of a two-way telephone conversation. At that time it was stated that two-way television as a complete adjunct to a two-way telephone conversation was a later possibility. It is the purpose of this paper to describe a two-way television system now set up and in operation between the main offices of the American Telephone and Telegraph Company at 195 Broadway and the Bell Telephone Laboratories at 463 West Street, New York. In principle this consists of two complete television transmitting and receiving sets of the sort used in the 1927 one-way television demonstration. In realizing this duplication of apparatus, however, a number of characteristic special problems arise, and the paper deals chiefly with matters peculiar to two-way as contrasted with one-way television.

PHYSICAL ARRANGEMENT AND OPERATION

The detailed description of the optical and electrical elements of the two-way television system will be more readily grasped if it is preceded by an account of the general arrangement of the parts and of the method of operation of the system from the standpoint of the user.

The physical arrangement of the two-way television

1. Members Technical Staff, Bell Telephone Laboratories, New York, N. Y.

2. *Bell System Technical Journal*, October, 1927, pp. 551-652.

Presented at the Summer Convention of the A. I. E. E., Toronto, Ontario, Canada, June 23-27, 1930.

system is shown by the pictorial sketch Fig. 1, and in the illustrations Fig. 2 and Fig. 3. The terminal apparatus is largely concentrated into a booth—the television booth, similar in many respects to the familiar telephone booth,—and a pair of cabinets, which contain the scanning disks and light sources. As in the 1927 demonstration, scanning is performed by the beam method, the scanning beam being derived from an arc lamp whose light passes through a disk furnished with a spiral of holes and thence through a lens on the level of the eyes of the person being scanned. The light reflected from the person's face is picked up by a group

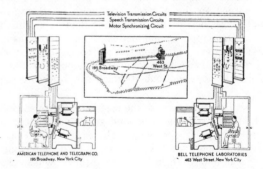

FIG. 1—PICTORIAL SKETCH OF TWO-WAY TELEVISION SYSTEM

of photoelectric cells for subsequent amplification and transmission to the distant point. The signals received from the distant point are translated into an image by means of a neon glow lamp directly behind a second disk driven by a second motor placed below the first and inclined at a slight angle to it. The two disks, which are shown in the center cabinet of Fig. 2, are of slightly different sizes; the upper one 21 in. in diameter and the lower one 30 in. They differ from the disks used in the earlier demonstration in that in place of the 50 spirally arranged holes formerly used, they carry 72 holes whereby the amount of image detail is doubled. While with the earlier "50 line" picture recognizable images of a face were obtainable, the aim in this new development was to reproduce the face so clearly that there would be no doubt of recognizability, and so that individual traits and expressions would be unmistakably transmitted. This doubled number of image elements necessarily requires, for the same image repetition frequency (18 per second), twice the transmission band, or approximately 40,000 cycles as against 20,000 for the 1927 image.

The only part of the television apparatus visible to

Fig. 2—The Three Major Cabinets of the Television-Telephone Apparatus

the user is the array of photoelectric cells which are in the television booth behind plates of diffusing glass. In addition to the photoelectric cells and their immediately associated amplifiers, the booth contains a concealed microphone and loud speaker. By means of these, the voice is transmitted to the distant station and received therefrom without the interposition of any visible telephone instrument which could obscure the face.

From the standpoint of the customer, the operation of the combined television and telephone system is reduced to great simplicity. He enters the booth, closes the door, seats himself in a revolving chair, swings around to face a frame through which the scanning beam reaches his face, and upon seeing the distant person, he talks in a natural tone of voice, and hears the image speak. Conversation is carried on as though across a table.

Optical Problems

Some of the more special of the problems encountered in two-way television are primarily optical in character. The principal one is that of regulating the intensity of the scanning light and of the image which is viewed so that the eye is not annoyed by the scanning beam, or the neon lamp image rendered difficult of observation. It has been necessary for the solution of this problem to reduce the visual intensity of the scanning beam considerably below the value

formerly used and to increase considerably the brightness of the neon lamp.

The method adopted consists first, in the use of a scanning light of a color to which the eye is relatively insensitive but to which photoelectric cells can be made highly sensitive. For this purpose blue light has been used, obtained by interposing a blue filter in the path of the arc light beam, and potassium photoelectric cells specially sensitized to blue light and more sensitive than those previously used have been developed. The number of these cells and their area has also been increased over those used in the earlier television apparatus so that the necessary intensity of the scanning beam is decreased.

The second half of the problem, namely that of securing a maximum intensity of the neon lamp, has been attained by the development of water-cooled lamps capable of carrying a high current. The net

Fig. 3—Interior of the Television Booth

result of the use of blue light for scanning, of more sensitive photoelectric cells, and of the high efficiency neon lamps is that the user of the apparatus is subjected only to a relatively mild blue light sweeping across his face, which he perceives merely as blue spot of light lying above the incoming image. Fig. 3 shows the interior of the television booth with the frame through which the observer sees the image of the distant person.

A second optical problem is the arrangement of the photoelectric cells required in order to obtain proper virtual illumination of the observer's face. As we have previously pointed out in discussing the beam scanning method,[3] the photoelectric cells act as virtual light sources and may be manipulated both as to their size and position like the lights used by a portrait photographer in illuminating the face. In the present case, it is desired to have the whole face illuminated and accordingly photoelectric cells are provided to either side and above. One practical difficulty which is encountered is that eyeglasses, which often cause annoying reflections in photography are similarly operative here. For this reason, it is important that the photoelectric cells be placed as far to either side or above as possible. The banks of photoelectric cells shown in Fig. 3 are accordingly much farther removed from the axis of the booth than were the three cells used in the first demonstration. In the position which has been chosen for the cells, reflections from eyeglasses are not annoying unless the user turns his face considerably to one side or the other.

The number of cells has been so chosen as to secure a good balance of effective illumination from the three sides and it has been found desirable to partly cover the cells on one side in order to aid in the modeling of the face by the production of slight shadows in one direction.

Another optical problem is the illumination of the interior of the booth. There must, of course, be sufficient illumination for the user to locate himself, and it is also desirable that the incoming image and the scanning spot be not seen against an absolutely black background. The illumination of the booth is by orange light, to which the cells are practically insensitive, and so arranged that the walls and floor are well illuminated. In addition to the wall and floor illumination, a small light is provided on the shelf bar in front of the observer so as to cast orange light on the front wall surrounding the viewing frame. This light contributes materially to reducing the glaring effect of the scanning beam, and to the easy visibility of the incoming image.

In addition to the optical features which are visible to the person sitting in the booth, there are very necessary optical elements which have to do with the placing of the outgoing and incoming images. A practical problem which is encountered when customers of various heights use the apparatus is that the scanning beam, if fixed in its position, would strike too high or too low upon many faces. In order to direct the beam up or down as is required, a variable angle prism, consisting of two prisms arranged to rotate in opposite directions, is interposed in the path of the scanning beam. This prism, which lies directly in front of the projection lens used with the upper disk, is shown in Fig. 4 at P. Its rotation is controlled by a knob with a numbered dial. The exact position is determined by the operator by reference to a monitoring image which will be described below.

Another optical element, which serves two purposes,

is a large convex lens lying between the receiving disk and the observing frame, shown at L, Fig. 4. This lens is used to magnify the incoming image to such a size that the image structure is just on the verge of visibility, under which condition the face of the distant person appears as though he were approximately 8 feet away. In addition to acting as a magnifier, this lens serves to position the incoming image to fit the height of the customer. For, by raising or lowering it by means of a knob, the operator, using the information as to the observer's height obtained from the position of the scanning beam, locates the lens so that the virtual image appears in the proper position.

PHOTOELECTRIC CELLS AND ASSOCIATED CIRCUITS

The photoelectric cells used in this apparatus are similar in shape to those used in the first demonstration,

FIG. 4—OPTICAL MEANS FOR CONTROLLING HEIGHTS OF SCANNING AND VIEWING BEAMS

but somewhat larger. Each cell is twenty inches long and four inches in diameter, giving it an area of approximately eighty square inches for collecting light. The anode is made in the form of a hollow glass rod wound with wire. This construction prevents the electrical oscillations that would otherwise result from mechanical vibrations of the anode. The sensitive cathode consists of a coating, covering the rear wall of the tube, of potassium sensitized with sulphur.[4] This kind of cell is considerably more sensitive than the older type of potassium-hydride cell while still having most of the sensitiveness in the blue region of the spectrum. Fig. 5 shows the response of the photoelectric cells used to the various parts of the spectrum together with the transmission of the blue filter and the brightness of the various parts of the spectrum as evaluated by the human eye. The very great efficiency of the photoelectric cells and the inefficiency of the eye to the light used are apparent.

3. *Jnl. Optical Soc. of America*, March, 1928, p. 177.

4. A. R. Olpin; *Phy. Rev.* 33, 1081, (1929).

To amplify the photoelectric current, the cells are filled with argon at a low pressure. Photoelectrons passing from the sensitive film to the anode ionize the gas atoms along their paths and thus cause a greater flow of current. The ionization of the gas does not, however, instantaneously follow sudden variations of the true photoelectric emission from the sensitive film, that is, there is a time lag in the ionization of the gas and in the disappearance of ionization. This lag results in a

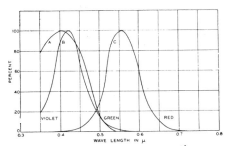

Fig. 5—A Relative Optical Transmission of the Blue Filter Through Which the Scanning Beam Passes

B Relative Sensitivity of the Photoelectric Cells to Various Parts of the Spectrum

C Relative Sensitivity of the Eye to Various Parts of the Spectrum

relative loss and phase shift of the high-frequency components of a television signal with respect to the low-frequency components which become serious in the wider frequency range utilized in the 72 line image. The relative loss in output from a single large photoelectric cell at high frequencies is shown in decibels by curve A of Fig. 6.

In the television booth, the twelve large cells mounted in the walls of the booth present an area of approximately seven square feet to collect light reflected from a

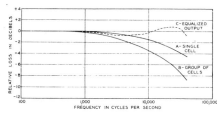

Fig. 6—Loss in Response of Photoelectric Cells at High Frequencies

subject. To secure the desired effective illumination, the cells are mounted in three groups, comprising a group of five cells in each of the two side walls of the booth and a group of two cells in the sloping front wall above the subject. The twelve cells are enclosed in a large sheet copper box, provided with doors to each group. The cells of each group are connected in parallel through the input resistance of a two stage resistance-capacity coupled amplifier similar to those

previously used. This raises the level of the signal to such a point that the output of the three amplifiers may be carried through shielded leads and connected in parallel to a common amplifier.

The metal anodes and lead wires of the cells in parallel in any one group give an appreciable capacity to ground, which results in a further loss in amplitude and phase shift of the high-frequency components of the signal. The combined loss introduced by ionization of the gas in the cells and by capacity to ground is shown by curve B of Fig. 6. This combined loss is equalized by an interstage amplifier coupling, Fig. 7. The equalized output from the photoelectric cells is shown by curve C, Fig. 6.

Two-Way Image Signal Amplifiers

The vacuum tube system used to amplify the photoelectric cell currents in two-way television is patterned closely after that used previously in one-way television, and the description here will be confined chiefly to novel features. These new features are necessary to take care of the doubled frequency band which results when the scanning is done with a 72-hole disk rather than with a 50-hole one, and to provide sufficient

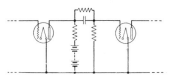

Fig. 7—Schematic of Interstage Amplifier Coupling to Equalize for the High Frequency Losses in the Photoelectric Cells

power to operate the high intensity neon lamp which is essential to two-way television. Certain other new features have been introduced in order to simplify the apparatus and to reduce the maintenance required to keep it in good working condition.

The vacuum tubes which operate at low energy levels are the so-called "peanut" type, chosen because of their freedom from microphonic action and their low inter-electrode capacities. Protection against mechanical and acoustical interference is secured by mounting these tubes in balsa wood cylinders which are loaded with lead rings and cushioned in sponge rubber. The tubes are electrically connected in cascade by means of resistance-capacity coupling, so that the whole amplifier system is stable over long periods of time and is also uniformly efficient over the required frequency band. Grid bias for the small tubes is supplied by the potential drop across a resistance in the filament circuit; the power requirements for the low level stages of the amplifier are filled by 6-volt filament batteries and 135-volt plate batteries, all located externally where they can be checked and replaced conveniently.

The amplifier system is divided into units of convenient size as shown in Fig. 8. Associated with each of the three banks of photoelectric cells is a two-stage unit known as the photoelectric cell amplifier; the combined output of these three units is carried to a four-stage unit known as the intermediate amplifier, output of which is sufficiently high level to be carried outside the copper cell cabinet to the three-stage transmitting power amplifier on the relay rack. A four-stage unit known as the receiving power amplifier is also on the relay rack, and serves to amplify the signal from the other station to a level which will yield an image of satisfactory contrast when it is translated into a light variation by means of the neon lamp. The final stage of this amplifier consists of two special 250-watt tubes in parallel. These large tubes are used because their plate impedance is of the same order of magnitude as the impedance of the neon lamp, and because they will supply the necessary direct current to the neon lamp without overheating.

Fig. 8 also shows what may be termed a voltage level diagram for the whole system. Ordinates on this

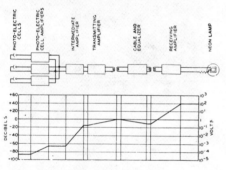

FIG. 8—SCHEMATIC DIAGRAM OF THE COMPLETE TELEVISION CHANNEL AND THE RELATIVE VOLTAGE LEVELS OF THE SIGNAL ALONG THE CHANNEL

diagram represent voltage amplitudes at the junctions between units of the system, and by themselves tell nothing at all about the power conditions in the system, since the impedances are not specified. It is interesting to observe that the signal voltage produced by the three banks of photoelectric cells has an effective value of about 50 microvolts across the 50,000 ohm input resistance; the transmitting amplifier delivers about 1 volt to the 125 ohm cable circuit, and the receiving amplifier delivers about 100 volts to the 1000 ohm neon lamp circuit. The signal current through the neon lamp has an effective value about a thousand million times greater than that of the current variation in one of the photoelectric cells.

The most outstanding contribution to the development of television amplifiers is the combination of output and input transformers whose transmission characteristics are shown in Fig. 9, *A*, and whose impedance characteristics are shown in Fig. 9, *B* and *C*. The exceptionally wide frequency range, corresponding to a ratio of limiting frequencies of 5000 to 1, trans-

mitted by these transformers is due largely to the use of chrome permalloy, a recently developed core material having very high permeability. The improved characteristics are also the result of refinements in design which involve the use of adjusted capacities and resistances to control the characteristics at the higher frequencies. Due to the fact that each terminated transformer looks like a resistance of 125 ohms over practically the entire frequency range of the image signal, it makes no difference in the form of the over-all voltage amplification characteristic of the circuit whether the transformers are connected together

FIG. 9A—VOLTAGE RATIO CHARACTERISTICS OF W-7879 INPUT TRANSFORMER AND W-7880 OUTPUT TRANSFORMERS, EACH CONNECTED BETWEEN ITS RATED IMPEDANCE

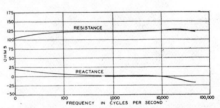

FIG. 9B—IMPEDANCE CHARACTERISTIC OF W-7880 OUTPUT TRANSFORMER WITH 1765 OHM RESISTANCE LOAD

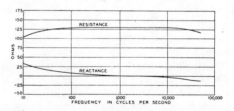

FIG. 9C—IMPEDANCE CHARACTERISTIC OF W-7879 INPUT TRANSFORMER WITH 20 MMF. CAPACITY LOAD

directly or by means of the equalized cable circuit whose characteristic is shown in Fig. 10. Advantage of this circumstance is taken in providing switching means whereby each transmitting amplifier may be connected through a resistance pad to its local receiving amplifier, enabling a person to see his own image in the television booth, which is a convenience in making apparatus adjustments.

Transformers of this type must be carefully protected against magnetizing forces which might cause polarization of the core material. In order to keep the plate current of the final tube of the transmitting power amplifier from flowing through the winding of the output transformer, the transformer winding is shunted by a battery and a resistance in series. The resistance is made high, so that the transmission loss due to bridging it across the circuit is small; the voltage of the battery is made equal to the potential drop across the resistance due to the plate current of the tube, so that the average voltage across both the battery and resistance, and hence across the transformer winding, is zero.

A vacuum thermocouple is connected in series with the line winding of the output transformer, serving as a level indicator for the transmitting amplifier. The level indicator for the receiving amplifier is a vacuum thermocouple in series with the grid resistance of the two 250-watt tubes.

The electrical control panels associated with one terminal of the television apparatus are shown in Fig. 11.

TRANSMISSION CIRCUITS

Two special requirements for the two-way television transmission circuits are to be emphasized. The first, which has already been referred to, is the wide frequency transmission band, from 18 cycles to 40,000

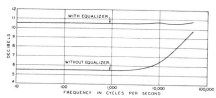

FIG. 10—INSERTION LOSS CHARACTERISTIC OF CABLE CIRCUITS WHICH TRANSMIT THE IMAGE SIGNAL, MEASURED BETWEEN 125 OHM RESISTANCES

cycles, which must have a high degree of uniformity of transmission efficiency and freedom from phase distortion. The second is the necessity for *two* circuits for the television images. This arises from the fact that the two parties to the conversation must both see and be seen at all times. There can be no interruption of one face by the other, comparable with the alternation of the role of speaker and listener in telephony which permits the use of a single circuit for ordinary speech communication.

The terminal stations of the two-way television system are connected by eight underground circuits, each consisting of 13,032 ft. of No. 19 gage and 390 ft. of No. 22 gage non-loaded cable. Two circuits are used for transmitting the image signals, two for the accompanying speech, one for the synchronizing current, two are used as order wires, and one is kept as a spare. All of the circuits have identical transmission characteristics, but equalization is necessary only on the two which carry the image signals. Fig. 10 shows the insertion loss characteristic of each circuit, and also shows the insertion loss characteristic of the image circuits when the image line equalizers are included.

Although the distance between the stations is small

the requirements of the television system from the standpoint of freedom from noise and other interference require that considerable care be given to the selection of the cable circuits used. All terminal connections are made through balanced repeating coils or transformers so that all of the circuits are balanced to ground. Also, in order to insure that the crosstalk between the various channels be unnoticeable the terminal equip-

FIG. 11—CONTROL APPARATUS PANELS ASSOCIATED WITH ONE TERMINAL OF THE TELEVISION APPARATUS

ment is so adjusted that approximately the same amount of power is transmitted by each circuit.

NEON LAMPS AND ASSOCIATED CIRCUITS

After amplification, the received television signal is impressed on the grids of two power tubes in parallel to

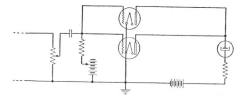

FIG. 12—SCHEMATIC OF NEON LAMP CIRCUIT

furnish current for a neon receiving lamp. The terminal lamp circuit is shown in Fig. 12. The grid bias of the two power tubes is varied by the operator to control the d-c. plate current, which replaces the original d-c. signal component suppressed at the sending end. The quality of the reproduced image is determined by the operator's control over the relative levels of the incoming a-c. signal and the restored d-c. current.

The television current from the power tubes is translated back into light by a water-cooled neon lamp designed to operate on a large current. The structural details of the lamp are shown in Fig. 13. Heavy metal bands attach the rectangular cathode to a hollow glass stem occupying the central portion of the tube. Water from a small circulating pump flows continuously through the glass stem and cools the cathode by thermal conduction through the metal bands. To reduce sputtering of the cathode and consequent blackening of the glass walls, the front surface of the cathode is coated with beryllium. This

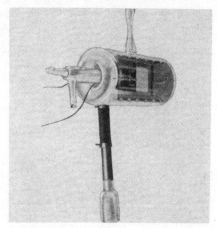

FIG. 13—WATER-COOLED NEON LAMP

metal resists the disintegrating action of the glow discharge very satisfactorily and gives the lamp a prolonged life. Other metal surfaces in the tube are shielded from the discharge by mica plates; and the discharge passes from the frame-like anode to the front surface of the cathode, covering it with a brilliant layer of uniform cathode glow.

Pure neon in a plate type of lamp gives a very inferior reproduction of an image. The impedance of the lamp is relatively high and comprises both a resistance and a reactance which vary with frequency. The variation in the impedance causes a relative loss in the frequency components of the signal and also introduces spurious phase shifts. In addition, pure neon has an after-glow; the gas continues to glow for an appreciable time after current ceases to flow. This after-glow casts spurious bands of illumination out to one side of the brighter image details.

A small amount of hydrogen in the neon prevents such an after-glow; and at the same time improves the circuit characteristics of the lamp. The total impedance of the lamp is lower, making it a less influential part of the lamp circuit; and the resistance and reactance vary in such a manner that the phase shift is more nearly proportional to frequency, (a phase shift proportional to frequency causes no distortion in the reproduction of an image). Other active gases may be used with the neon to improve the operation of a

television lamp, but one or two per cent of hydrogen is most satisfactory.

Since hydrogen is absorbed by the electrodes in a glow discharge, it slowly disappears from the neon during operation of the lamp. For this reason the lamp is provided with a small side reservoir of hydrogen. The lamp and the reservoir carry porous plugs immersed in a pool of mercury; and a flexible rubber connection permits the two plugs to be brought into contact at will. Minute quantities of hydrogen may be introduced into the lamp by simply bringing the two plugs into contact for a short time.

Even with this improvement the circuit characteristics of a lamp are not ideal. With power tubes it is usually desirable to include a fixed resistance in series with the lamp to prevent semi-arcing conditions. Such a resistance also makes the lamp a less influential fraction of the total circuit impedance.

OPTICAL MONITORING SYSTEMS

In order to insure that the incoming and outgoing images are properly positioned, no matter what the

FIG. 14—SENDING AND RECEIVING DISKS, WITH NEON LAMPS AND OPTICAL ARRANGEMENTS FOR IMAGE MONITORING

stature of the person sitting in the booth, and that the images shall be of proper quality, it is essential to have some means for the operator to observe and adjust these images. The optical monitoring system provided consists of an outgoing monitor for adjusting the scanning beam, and an incoming monitor and means for adjusting the position of the viewing lens to suit the height of the sitter.

The outgoing monitoring system is the same as that used in the one-way television apparatus which has already been described. A small neon lamp (Fig. 14, at bottom of top disk) is placed behind the sending disk but displaced several frames from the aperture through which the arc lamp beam passes. By continuing the spiral of holes part way around it is possible to see the complete image from the auxiliary neon lamp, to which the outgoing signals are also supplied. In order to see this monitoring lamp from the operator's position, a right-angle prism and a magnifying lens are placed in front of the disk and the image is observed through an opening in the side of the motor cabinet. The task of the operator is to direct the scanning beam up or down by means of the variable angle prism until the face of the person in the booth is centrally located. This adjustment is facilitated by a wire which passes across the image and is placed at the height at which the customer's eyes should appear.

The height of the observer's eyes is an indication of the position which should be taken by the large magnifying lens L, and the operator, after having properly placed the scanning beam, reads the scale on the variable angle prism dial, and then sets the magnifying lens by turning its controlling knob to the same number. When both adjustments are complete, the person in the booth will not only be properly scanned but will be in the best position to see the image.

In order to monitor the incoming image, an optical arrangement is adopted by means of which light from the water-cooled neon lamp is taken off at the side and reflected through the disk and thence reflected again, as shown in Fig. 14, (top of bottom disk), through a second, lower, observing hole on the side of the motor cabinet. Because of the small area of the side view of the neon lamp, a lens system is inserted which focuses the image of the lamp at the place to be occupied by the pupil of the operator's eye. When the eye is properly placed, the whole of the lens area is seen filled with light and exhibits the incoming image.

In addition to the monitoring means just described, an additional view of the incoming image is provided by means of a 45-deg. mirror which is carried on the back of a movable shutter which is shown at S in Fig. 4. This shutter carries an illuminated sign on the side turned to the customer with the inscription, "Watch this space for television image." The shutter with its sign covers the image until the adjustments just described are made, when it is dropped out of sight. While it is in place, the operator is provided with an additional monitoring image reflected from the 45-deg. mirror. This view is, of course, in every respect identical with that which the customer sees.

The function of the incoming monitoring system is primarily to enable the operator to set the electrical controls to give the proper quality of image. He also has another task which is that of properly framing the image. This he can do by turning the framing handle, which is described elsewhere, while watching the image from the mirror. The framing operation is preferably performed not on a person sitting in the booth but upon some suitable object such as a mirror located upon the rear door of the booth. In order to

make this framing adjustment, the operators at the two terminals set their scanning beam dials to predetermined positions such that the scanning beams place the framing mirrors at the lower edge of the scanning rectangles; the phases of the incoming disks are then shifted until the images of the mirrors are seen properly located in the incoming monitors.

SIGNALING SYSTEM

In order to coordinate operations at the two terminal stations, an order wire system is provided. There are four telephone sets at each station; one on the attendant's desk in the ante-room, one concealed inside the television booth, one in the control room, and one at the control panels for the technical operator, who operates the small switchboard which is part of the system. Two of the underground cable circuits connect the two switchboards, so that there may be not more than two separate conversations between stations at one time. Ringing is accomplished by means of standard 20-cycle ringing current furnished by the telephone company.

During a demonstration, the attendants' telephones are connected permanently over one of the cable circuits. To relieve the operators of the duty of ringing each time the attendants wish to communicate, a push button and buzzer are provided at each attendant's desk, operated by the standard ringing currents simplexed on the synchronizing circuit. This arrangement leaves the operators free to manipulate the television apparatus.

The two order wire circuits are each simplexed to provide two additional circuits which operate signal lamps indicating to both operators when either chair in the television booths is occupied and turned in position.

CONCLUSION

The primary objects in developing and installing the two-way television system have been two. The first was to obtain information on the value of the addition of sight to sound in person to person communication over considerable distances. The second was to learn the nature of the apparatus and operating problems which are involved in a complete television-telephone service. While the installation is entirely experimental, it is being maintained in practically continuous operation for demonstration to employees and guests of the telephone company, and interesting data are being gathered on all aspects of the problem.

It may be said without fear of contradiction that the pleasure and satisfaction of a telephone conversation are enhanced by the ability of the participants to see each other. This is, of course, more evident where there is a strong emotional factor, as in the case of close friends or members of the same family, particularly if these have not been seen for some time.

Were the television apparatus and required line facilities of extreme simplicity and cheapness it would be safe to predict a demand for its early use. At the present time, however, the terminal apparatus is complex and bulky, and requires the services of trained engineers to maintain and operate it. In addition to the cost of the terminal apparatus there is the unescapable item of a many-fold greater transmission channel cost. Because of the wide transmission bands required for the television images, the inherent necessity for a television channel in each direction, and the extra channels for synchronizing and signaling, the total transmission facilities used in this demonstration are those which could, according to current practise, carry about fifteen ordinary telephone conversations. It is to be expected, of course, that development work will result in some increase in the efficiency of the transmitting channels and in simplifications of the terminal apparatus. It is conceivable, therefore, that our present conception of the cost of the whole system may ultimately be materially changed.

PART II—SYNCHRONIZATION SYSTEM

By H. M. Stoller[5]

Member, A. I. E. E.

General Requirements

Television transmission requires not only synchronization of the transmitting and receiving equipment but

Fig. 15—New Television Motor and Vacuum Tube Control Circuit

such synchronization must be held to a narrow phase angle so that the scanning disks at the transmitting and receiving end will never depart more than a small fraction of a picture frame width from the desired position.[6] In the 1927 demonstration, 2125-cycle synchronous motors were employed with supplementary d-c. motors to facilitate starting. This

5. Apparatus Development Dept., Bell Telephone Laboratories, New York, N. Y.

6. These requirements are more fully discussed in a previous paper of the A. I. E. E. Journal, Vol. 46, p. 940, 1927.

plan required the use of vacuum tube amplifiers of large size in order to supply sufficient power to the synchronous motors.

Such high-frequency synchronous motors, however, are inefficient and expensive, so that when designing the new system it was desired to solve the problem of synchronization with simpler and cheaper equipment and in a manner which would require less attention in starting. It was particularly desired to employ a motor which could be operated directly from the 110-

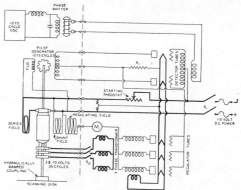

Fig. 16—Schematic Diagram of Control Circuit

volt lighting circuit without any auxiliary A, B, or C batteries for the control equipment.

Description of Motor

Fig. 15 shows a photograph of the new television motor and its associated control equipment.

The motor is a four-pole, compound-wound, d-c. motor with the following special features added:

1. An auxiliary regulating field, the current through which is controlled by the vacuum tube regulator.

2. A damping winding on the face of the field poles to prevent the field flux from shifting (Fig. 17).

3. Three slip rings are provided at points 120 electrical degrees apart for furnishing three-phase power to supply plate and filament voltage for the regulating circuit.

4. A pilot generator of the inductor type is built into the motor frame and delivers a frequency proportional to the motor speed for actuating the control circuit.

5. A hydraulically damped coupling is provided between the motor shaft and the scanning disk. (Fig. 18).

The motor frame was made from a standard 36 tooth stator punching by cutting out three teeth per pole, thus forming four polar areas of six teeth each. The shunt, series, and regulating fields enclose the entire polar areas. The damping winding consists of insulated closed turns of heavy copper wire distributed over the pole faces in the slots as shown in Fig. 17. It will be noted that this damping winding has no effect upon the flux through the poles as long as the flux density over the polar surface does not shift. In other words, the damping winding permits the total flux of the motor to increase or decrease as required by the regulating circuit but will oppose any tendency of the flux to shift back and forth across the pole face. As will be explained later on, this feature is essential in order to prevent hunting or instability of the image.

The hydraulically damped coupling between the motor shaft and the scanning disk is also essential in order to avoid hunting. It employs flexible metal bellows filled with oil and connected by a small pipe

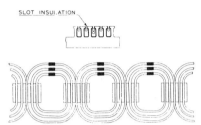

Fig. 17—Damping Winding Preventing Shifting of Field Flux

equipped with a needle valve for adjusting the amount of damping. Fig. 18 shows its construction. The scanning disk itself is centered on a ball bearing which allows the disk to rotate with respect to the shaft within approximately ± 5 degrees mechanical movement.

Control Circuit

Fig. 16 shows a schematic diagram of the control circuit. When the motor is operating at full speed the

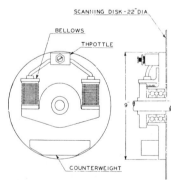

Fig. 18—Hydraulically Damped Coupling to Prevent Hunting of Motor

pilot generator delivers approximately 1 watt of power at 300 volts, 1275 cycles to the plates of a pair of push-pull detector tubes. The grids of these tubes are supplied with an e. m. f. of the same frequency from an oscillator or other source of power having a sufficiently constant frequency. The amount of power required for this grid circuit is only a few thousandths of a watt. The detector tubes rectify the plate voltage producing a potential drop across the coupling resistance R_1. If the plate and grid voltages are in phase, so that the

grids of the tubes are positive at the same instant that the plates are positive, the plate current will be a maximum. If the grid voltage is negative when the plate voltage is positive the plate current is practically zero, so that the magnitude of this current is a function of the phase relationship between the grid and plate voltages as shown in Fig. 19.

The voltage drop across the coupling resistance R_1 is applied to the grid circuits of three regulator tubes. These tubes derive their plate voltage supply from a three-phase transformer fed with power from the three slip rings provided on the motor. These tubes act as a rectifier whose output is controlled by the potential impressed upon the grids from the coupling resistance R_1. The current of the regulator tubes is passed through the regulating field provided on the motor. This field is in a direction to aid the shunt field and series fields of the motor.

The operation of the circuit is as follows: In starting switch S_1 is closed which applies direct current to the shunt field and armature circuits of the motor. The motor accelerates as an ordinary compound wound motor. Switch S_2 is then closed applying three-phase

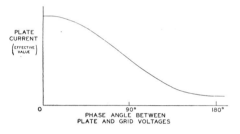

Fig. 19—Phase Detector Tube Characteristic

power from the slip rings of the motor to the transformer. As the speed of the motor approaches the operating point, the beat frequency between the pilot generator and the oscillator will cause beats in the current through the regulating field which are visible on the meter M_1. Let us assume that the field rheostat has been previously adjusted so that with the shunt field alone the motor will tend to run slightly over the desired operating speed. When the exact operating speed is obtained, the beat frequency in the regulating field will be zero and as the motor tends to accelerate, the phase relationship between the pilot generator and the oscillator will reach a point tending to give maximum strength to the regulating field. When this point is reached, the acceleration of the motor will be checked by the increased field and the speed will tend to fall until the phase of the pilot generator with respect to the oscillator has shifted sufficiently so that the regulating field current is reduced to an equilibrium value, after which the motor continues to run at constant speed in accordance with the frequency of the oscillator.

Operating tests on the circuit show that the motor will hold in step over line voltage ranges from 100 to 125 volts and will be self-synchronizing over somewhat narrower voltage limits. Thus, under normal conditions all that is necessary from an operating standpoint is to close the switch and wait for the motor to pull into step.

CONTROL OSCILLATOR

The control oscillator is a standard type of vacuum tube oscillator having a frequency precision of the order of 1 part in 1000, when delivering the negligible output of 0.005 watt to the grid circuit of the detector tubes. This frequency is delivered directly to the motor circuits at one end of the line and is transmitted over a separate cable pair to the control circuits at the other end of the line. It was found that the detector tubes would operate successfully over a considerable variation in power level, provided the minimum oscillator output was sufficient.

An interesting alternative method was developed in which the synchronizing channel between stations may be omitted entirely, but this method was not used in the present system as the additional cost was not justified. The method, however, is described as it may prove of value if television transmission over long distances is considered.

Mr. W. A. Marrison in his paper "A High Precision Standard of Frequency," *Proceedings I. R. E.*, July, 1929, described a crystal controlled oscillator which would maintain a precision as to frequency of 1 part in 10,000,000. This oscillator employs a quartz crystal as its primary means of control and by means of secondary circuits, the natural period of the crystal, which is approximately 100,000 cycles, may be stepped down to lower frequencies which are more convenient for such purposes as motor control.

By this means, a frequency of the desired value may be obtained with a precision so great that the speed of the scanning disks under control of the above described circuit will be so nearly perfect that no synchronization channel at all is required. For example, if the period of observation of the television image is 5 minutes, the scanning disk will make 5300 revolutions when operating at a speed of 1060 rev. per min. Assuming a precision of control of 1 part in 10,000,000, the maximum error during the 5 minute interval will be 5300 divided by 10,000,000 or about 1/2000 of 1 revolution. Expressed in degrees on the periphery of the disk, this is equivalent to approximately 1/6 of 1 degree or since the width of the television image with 72 holes in the scanning disk is 5 degrees, the image will drift 1/30 of a frame width during the 5 minute interval If the speed of the scanning disk at the other end drifts an equal amount in the opposite direction, the displacement of the television image will be only 1/15 of a frame width, which is a tolerable amount of drift.

From a practical standpoint, however, it is apparent that the additional cost of very precise independent oscillators would be greater than the cost of providing the synchronization channel, except possibly for transmission over long distances.

FRAMING

Referring to Fig. 16, it will be noted that a phase shifter is provided between the oscillator and the input terminals to the control circuit. This phase shifter is designed with a split phase primary member producing a rotating magnetic field. The secondary member is single phase and is mounted on a shaft provided with a handle. By rotating the handle of the phase shifter in the desired direction, the frequency delivered from the phase shifter will be the algebraic sum of the frequencies of the oscillator plus the frequency of rotation of the armature of the phase shifter. It is, therefore, a simple matter for the operator at the receiving end to momentarily increase or decrease the control frequency and thus bring the picture into frame.

CONCLUSION

During the development of the control system, one of the first difficulties encountered was hunting of the controlled motor. The problem of hunting, of course, becomes more difficult of solution the greater the precision of speed regulation desired and the greater the moment of inertia of the load connected to the motor, the latter statement applying only to controlled systems of the synchronous type. Since the moment of inertia of the scanning disk is large relative to that of the motor armature, it is seen that the conditions for securing stable rotation would be unfavorable in both the above mentioned respects if the scanning disk were mounted directly on the motor shaft. The hydraulically damped type of coupling above described was, therefore, inserted between the motor shaft and the scanning disk. It was found, however, that hunting still occurred. A further analysis of the problem showed that the axis of the field flux of the motor was shifting back and forth across the pole faces. The damping winding shown in Fig. 18 was then added with a marked improvement. It was also observed that a strong series field on the motor assisted in securing stability and it was, in fact, necessary to employ all three expedients to secure satisfactory performance. In the system as finally developed the television image, if disturbed by a momentary load such as the pressure of the hand against the disk, would come back to rest within approximately one second, there being two oscillations during this interval. In actual operation, it was found that the normal fluctuations in line voltage occurring on the commercial power supply produced no transients of sufficient magnitude to cause any objectional instability in the received image.

In conclusion, it should be pointed out that this type of control system could be equally well employed with larger motors for other applications requiring precise speed regulation. While the circuit described is applicable only to a d-c. motor, a similar system may be applied to an a-c. motor substituting a saturating reactor in place of the regulating field winding in the manner described by the author in his paper[7] presented before the Society of Motion Picture Engineers, September, 1928.

PART III—SOUND TRANSMISSION SYSTEM

By D. G. Blattner[8]   AND   L. G. Bostwick[8]
Associate, A. I. E. E.   Non-member

From the very beginning the ultimate goal of the communication engineer has been to annihilate distance and establish between remotely located parties the effect of face to face communication. Means for accomplishing this goal must of course involve the reproduction at some particular point of both visual and aural effects originated at another and it goes without saying that the two types of effects must be mutually coordinated if the result is to be of the best.

In the design of a sound transmission system to be correlated with a visual system, the requirements as to perfection of results desired are no more stringent than for other high grade sound reproducing systems[9] that have been described in the literature from time to time. Rather in this case the peculiarities of the system are largely those incidental to the adaptation of old technique to meet new conditions.

The principal limitation of the sound system imposed by the visual system is that the use be relieved of all necessity of holding a telephone in close proximity to the head. Such a limitation is highly desirable in order to secure the most natural pose of the features and the most satisfactory scanning. Obviously, the best way of meeting this limitation is by the use of telephone instruments of the type adapted for picking up and reproducing sounds at a distance. The use of such instruments has the further advantage that they can be located near the vision screen and so reproduce any peculiarities in tone quality that would result if the speaker were actually located at the position of the image. Of course, the sharpness of this perspective effect obtained is influenced by the loudness of both the original and the reproduced sounds but the matter of location of instruments is also very important.

It would thus seem that the use of distant pick-up and distant projecting instruments offers certain rather fundamental advantages but it is also true that it presents certain other disadvantages. One of the disadvantages is that the distant pick-up microphone gives less output than a close-up device because of the reduced sound pressure on the diaphram; also a sound

producing device to give suitable reception at a distance must be supplied with a higher transmission level than would a close-up instrument. It thus becomes necessary to provide for obtaining greater gain in transmission and greater electrical power capacity than would be required were the instruments held close to the head. The use of the more elaborate transmission facilities is in itself disadvantageous but it also tends to increase the feed-back from the loud speaker to the microphone, also the effect of any noise at the microphone position or at the listening position tends to interfere more seriously with the successful conduct of conversation. In the design of the two-way television system under discussion it was felt that it would be possible to overcome these technical objections to the distant type instruments and that the advantages mentioned would justify any measures necessary to do so.

The question of instruments was solved by the use of the Western Electric 394 condenser type transmitter[10] and a dynamic direct radiator loud speaker. The transmitter is one of the type generally used for phonograph and sound picture recording and for other purposes where good quality, high stability, and quietness of operation are essential. The direct radiator type of loud speaker was used instead of the usual horn type because of the limited mounting space available. It consists of a dynamic structure with a rigid duralumin diaphram about 3 in. in diameter flexibly supported at the edge and radiating directly into free air. While such a structure is not highly efficient and permits of only a small sound power output these considerations are of secondary importance in this case. The instruments were located in the front wall of the booth about 2 ft. from the position of the user and adjacent to the viewing screen in order to enhance the perspective as described above, the microphone being above and the loud speaker below as shown in Fig. 20. These instruments, in this particular case, were connected into a 4-wire circuit although in certain cases it might be desirable to use a 2-wire circuit. Such a change would of course be entirely feasible. The remainder of the apparatus used consisted of amplifiers located at the transmitting end of each channel and an attenuator at the receiving end, the two ends being connected by means of a loop of approximately 3 miles of non-loaded non-equalized cable. The amplifiers and the attenuators were each readily adjustable so that the sounds of different speakers could be reproduced at the optimum loudness. Observation of the performance of the system was made possible in each of the control rooms by means of a monitoring head type receiver bridged across the mid-point of an attenuator tying the two channels together. The attenuation used in the monitoring circuit was such as to give no audible feed-back in either booth. The results obtained with this step-up were considered satisfactory from the standpoint of

7. S. M. P. E. *Transactions*, Vol. 12, No. 35, p. 696.

8. Telephone Engineers, Bell Telephone Laboratories, New York, N. Y.

9. *Public Address Systems*, by J. P. Maxfield and I. W. Green in A. I. E. E., Feb. 14, 1923; *High Quality Recording and Reproducing of Music and Speech*, by J. P. Maxfield and H. C. Harrison, A. I. E. E., Feb. 1926.

10. E. C. Wente in *Physical Review*, May 1922.

both volume and quality. Ready recognition of familiar voices and the association of the source of the reproduced sounds with the image were the usual occurrence. Fig. 21 shows in block form the complete circuit set-up and Fig. 22 shows the combined response frequency characteristic of the microphone, amplifier, and loud speaker. The ordinates of this curve represent variations in sound pressure from the loud speaker for constant pressure on the transmitter diaphram. These data were obtained with the loud speaker located in a heavily damped room. The measurements were made on the sound axis at a distance of 2 ft., representing the relative position of the observer under conditions of actual use.

In setting up such a system the chief consideration is in regard to the acoustic feed-back from the loud

FIG. 20—MICROPHONE AND LOUD SPEAKER IN POSITION ABOVE AND BELOW TELEVISION SCANNING AND VIEWING APERTURE

speaker to the microphone and in this connection the design of the booth is an important factor. The booth must necessarily be so shaped that the user, looking at the viewing screen, can be satisfactorily scanned and the light reflected from the scanned areas will strike the banks of photoelectric cells required for the reproduction of the visual likeness. This requires that the person scanned be located in close proximity to the scanning disk and to the photoelectric cells as well as to the microphone and loud speaker. Such an arrangement is objectionable from an acoustic standpoint because in the present state of development the cells are necessarily large and poor absorbers of sound. They thus tend to cause part of the sound output from the loud speaker to reflect back into the microphone.

If the sound so reflected or fed-back is equal or greater in magnitude than the original sound picked up and is of the proper phase relation, the system will "sing" and become of no practical use. A further effect of the design of the booth is that, as a closed cavity, it tends to cause sounds of a certain pitch range to be accentuated. To reduce these effects as far as possible, the television booths were made as large as other considerations would permit and all surfaces were covered where

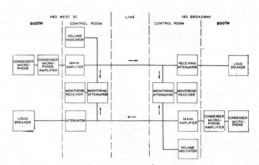

FIG. 21—CIRCUIT DIAGRAM FOR SOUND TRANSMISSION SYSTEM FOR TWO-WAY TELEVISION

possible with acoustic absorbing material. They have a floor area of about 35 sq. ft. and are about 8 ft. high. Because of the increased transmission required for the proper interpretation of sounds in the presence of noise, the booths were made of heavy masonry material to insulate the user and the microphone from the noise incidental to the rotating parts of the television apparatus. It was thus necessary to project the scanning beam and to view the illuminated image through a

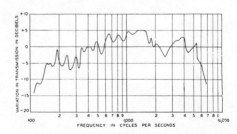

FIG. 22—RESPONSE FREQUENCY CHARACTERISTIC OF MICROPHONE, AMPLIFIER AND LOUD SPEAKER

window located in the front wall. The microphone and the loud speaker were fitted into this wall, which was then covered over with a thin screen to improve the appearance as shown in Fig. 3. These means effectively reduced the noise in the booth to an unnoticeable amount and reduced reflection effects to the extent that the voice of the average speaker talking in a conversational manner could be reproduced at a loudness best suited to the general effect. The optimum loudness seemed to be about the same as would be obtained from the speaker direct at a distance of 10 ft., the apparent distance between the image and the observer. At this loudness the gain of the amplifiers was 12 db. less than that required to cause singing.

While the system demonstrated was operated over a distance of only a few miles, it will be appreciated that the same facilities might have been used over much greater distances. Thus for the first time in the history of electrical communication it can be said that complete freedom of exchange of both visual and aural expression between distant users of the telephone has been made possible.

Discussion

N. S. Amstutz: It is with no little thrill that I arise to congratulate the Institute and Dr. Ives on a realization of the vision of one of those dreamers who thirty-six years ago wrote about "Visual Telegraphy" and assumed the role of a prophet, outlining a two-way working system to be used in combination with a telephone circuit.[1] There was described a television sending circuit, a television receiving circuit, a speech transmission circuit

1. "Electricity," (New York) Feb. 28, 1894, pp. 77-80, and March 14, 1894, pp. 110-111.

and a motor synchronizing circuit. The one-way television circuit and scanning disk device of Nipkow was described in detail, among a number of other proposals. The simultaneous sending and receiving of television currents was by means of two separate cylinders directly connected to a synchronous motor, each cylinder carried a group of openings arranged spirally.

It is indeed a great personal satisfaction to have the dreams of yesteryear come true and this opportunity is taken to congratulate Dr. Ives along with other workers, such as Jenkins at Washington, Alexanderson at Schenectady, and Baird of London, England, all of whom, thanks to sensitive light responsive devices, neon lamps, carrier currents, ultra precision apparatus, adequate financing and technical staffs, are hastening the day when television will not only be spectacular but harnessed to the service of man.

Dr. Ives, in his address referred to television as having been "in the air" for a long time, as far back as the Arabian Nights. Conjecture reverts to an incident described in Dr. Guthries "General History of the World" 1764, Vol. VII pages three and four in which portraits were made and sent in some mysterious manner during the reign of the twenty-sixth califf of the house of Al Abbas, circa A. D. 1037. This is referred to as follows "Avicenna, was obliged to fly to Forjan, where Washmakin reigned. Upon this, Mahmud, having got one of his portraits, ordered a great number of copies of it to be made and dispersed all over the east, that Avicenna might be seized." * * * "Washmakin knew him, by one of his portraits which had been sent him by Sultan Mahmud." The engineers of today may well wonder how the feat described above was accomplished.

TELEVISION IN AMERICA TO-DAY

A[lfred] Dinsdale

Television in America To-day. *

By A. DINSDALE, M.I.R.E. (Fellow).

SUMMARY.

1.—Classification of Television Workers.
2.—Two-way Television over Telephone Circuit.
3.—Television Signals along a Light Beam.
4.—Future Developments in Cathode-Ray Television.
5.—Standard 60-hole Disc Receivers.

6.—The Farnsworth Cathode-Ray System.
7.—Sanabria System designed for Projection Work.
8.—Commercial Development and Government Control
    in America.
9.—Television—a Young Man's " Game."

CLASSIFICATION OF TELEVISION WORKERS.

WHEN the first tentative plans for the 1931-32 Lecture Session were being formulated by our Lecture Secretary, television conditions in America were such that it appeared that I could most usefully serve Members of the Television Society by preparing a paper on Cathode Ray systems of television, for America then appeared to be in the lead in this line of research. In the interim, however, conditions have changed. No detailed information on cathode ray systems has been released for publication for nearly a year; hence, such cathode ray material as is included in this paper is necessarily out of date. Furthermore, European workers, notably Manfred von Ardenne, appear to me to have caught up with their American contemporaries.

Under these circumstances, therefore, it would appear that I can best serve my fellow members by endeavouring, within the limits of the time available, to present to them as comprehensive a picture as possible of the state of the art of television in the United States to-day. Even this is somewhat difficult, for during the past six months there has been a lull in published information; either no outstanding progress has been made or, what is more probable, those concerned have grown more cautious and reticent regarding the progress of their work.

Those concerned with television development may be classified under two heads: (1) Research workers, operating either independently or in association with some large organisation, whose aim is to perfect existing apparatus or develop new

ideas; and (2) persons or organisations who are interested in the operation or exploitation of existing equipment either for profit, for publicity, or for experience in related problems, so that they may be ready to take full advantage of an acceptable system of television as and when it arrives.

I may as well make my meaning clear at this stage by defining an acceptable system of television as one which will be acceptable to the general non-technical public. Prior to the advent of sound broadcasting, the only persons interested in wireless receivers were those possessed of some technical knowledge and considerable enthusiasm for the subject. Shortly after the introduction of broadcasting, wireless receivers were simplified to the point where non-technical persons could operate them— persons who had no knowledge of, or interest in the technicalities of wireless, and to whom a receiver was but a means to an end—home entertainment. An acceptable system of television will be one in which the public will interest itself as a means of entertainment in just the same manner.

Members of this Society will naturally be most interested in the first class of television workers, although I will attempt to show later on that there are certain aspects of the work of the second class which are not unworthy of attention, even by a group of technically interested persons.

Perhaps I should begin by listing the leading members of the first class of worker. They are the Bell Telephone Laboratories, Inc., whose television work is under the direction of Dr. H. E. Ives: the General Electric Company, under the direction of Dr. E. F. W. Alexanderson; the Radio Corporation of America; the Westinghouse

* A Paper received 26th January, 1932, and read at the April Meeting of the Society, at University College, London, on April 13th, 1932.

Electric and Manufacturing Company; the Jenkins Television Corporation; Philo T. Farnsworth, formerly independent, but now associated with the Philadelphia Storage Battery Co., makers of Philco radio receivers; and U. A. Sanabria, independent. There are many other individuals and organisations that might be mentioned, but since they have either done nothing outstanding, or have simply duplicated the work of others, to mention them would perhaps be superfluous for the purposes of this paper.

FIG. 1—Telephone booth used in Two-way Television between Bell System Headquarters, at 195, Broadway, New York, and Bell Laboratories. Walter S. Gifford (seated); Dr. H. E. Ives (standing).

—*Courtesy Bell Telephone Laboratories.*

TWO-WAY TELEVISION OVER TELEPHONE CIRCUIT.

I will now review briefly the work of these organisations and individuals in the order named.

My audience is probably aware of the television history of the Bell Telephone Laboratories, which is a research organisation maintained by the American Telephone and Telegraph Co., and the Western Electric Co. In 1927 they gave a demonstration of television between New York and Washington, over telephone lines, using 50-hole discs. Since 1930 they have been experimenting with a two-way television system, associated with a telephone circuit, over telephone lines between the Laboratories at 463, West Street, New York, and the A.T. and T. headquarters at 195, Broadway, a distance of about two miles.

The terminal equipment at each end consists of a specially adapted telephone booth, in which the user sits facing the scanning spot. On both sides of him and in front there are fourteen large photo cells. Immediately below the aperture through which the scanning spot appears, there is placed the receiving screen, upon which the image appears of the correspondent at the other end of the line. The user does not employ standard telephone equipment, because that would obscure his face. Instead, a high grade microphone, such as is used for broadcasting, is placed just above the scanning aperture. Below the aperture there is a small loud speaker. This acoustic system is carefully worked out so that there shall be no interaction between the microphone and the loud speaker. *Fig. 1* shows the booth.

Immediately behind the booth is the scanning mechanism. Two discs are used, the top one for transmission, and the lower one for reception. Their arrangement is shown in *Fig. 2.* Both discs have 72 holes, and 18 images are transmitted per second. The water-cooled neon tube can be seen behind the top part of the lower disc. A detailed view of the tube is given in *Fig. 3.* The portion of the cathode which glows is within the rectangular frame. Water cooling became necessary because the tube was overloaded in order to increase the available light.

Synchronism is effected by a highly complicated motor and associated control circuit, necessitating the use of a separate synchronising line between the two terminals. The complete system is illustrated diagrammatically in *Fig. 4.*

The general effect on a party using the two-way system, and the degree of realism, may best be described by comparing it to holding a conversation with another person across a room at a distance of about 12 feet. Only the head and shoulders of the distant party are seen, but the degree of detail, and the steadiness of synchronisation are good.

The A.T. and T. and its associates are not interested in the entertainment industry, although they frequently place their facilities for development at the disposal of organisations interested in the entertainment industry. They themselves are primarily interested only in communication in all its forms. Thus, we may expect that whatever developments they make in television will be adapted, wherever possible, to electrical communication, *e.g.*, to enable us to see the party we are talking to over the telephone line. Any by-products useful to the entertainment industry will be sold to that industry.

FIG. 2—Two Scanning Discs, one for sending (above), and one for receiving (below), are required at each terminal. Behind the upper edge of the Receiving Disc is the water-cooled Neon Tube used to form the Television Image.
—*Courtesy Bell Telephone Laboratories.*

Dr. Ives admitted to me frankly that he has not the remotest idea whether the public want to see the fellow at the other end of the telephone line badly enough to pay a high price for the privilege. But when the A.T. and T. started to develop the transatlantic telephone years ago, they did not know whether sufficient people would pay the necessarily high rates to make a service profitable. But the transatlantic telephone service does pay.

TELEVISION SIGNALS ALONG A LIGHT BEAM.

The only recent information which has come out of the G.E.C. laboratories at Schenectady on the

subject of television is that Dr. Alexanderson has been experimenting with the transmission of television signals along a light beam, instead of by wire or by wireless. Observers reported that the quality of the received images was as good as when sent along the more usual channels. This is perhaps a logical development of experiments with the transmission of television signals by quasi-optical wireless waves.

FUTURE DEVELOPMENTS IN CATHODE-RAY TELEVISION.

For some years past, television engineers from the G.E.C., Westinghouse Co. and R.C.A. have been concentrated at the R.C.A. Victor plant at Camden, N.J., where they have reputedly been

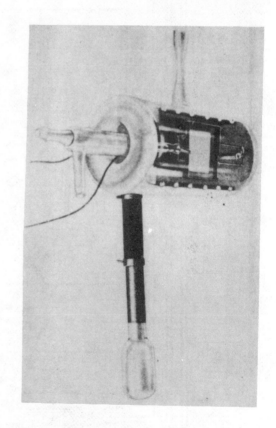

FIG. 3—Water-cooled Neon Tube.

busily engaged in secret on a cathode ray system of television. No information on this work has been officially released, unless, unknown to me at the time of preparing this paper, Dr. Zworykin,

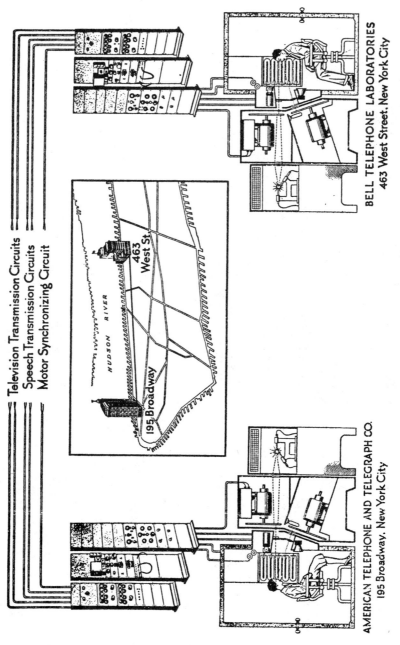

Television Transmission Circuits
Speech Transmission Circuits
Motor Synchronizing Circuit

BELL TELEPHONE LABORATORIES
463 West Street. New York City

AMERICAN TELEPHONE AND TELEGRAPH CO.
195 Broadway, New York City

FIG. 4.—Diagrammatic layout of Bell Telephone Laboratories—Two-way Television.

FIG. 5 (below)—Farnsworth Dissector—side view.

FIG. 6 (above)—Farnsworth Dissector—end view.

FIG. 10 (on right)—Farnsworth Oscillite.

At this point I must digress for a moment to mention a member of the second class of television worker, the National Broadcasting Company, which is a subsidiary of the R.C.A. For some time past the N.B.C. has been broadcasting television programmes on short wave from the roof of its Times Square studios in New York, with the object of learning (1) the effect of high buildings on short-wave television images, and other short-wave transmission problems; and (2) learning something of the problems of the studio technique necessary for television broadcasting. During these experiments, they have been using the American standard 60-hole disc, 20 frames or images per second.

Recently the N.B.C. announced that they were going to build a new television studio on the 85th floor of the Empire State building, 1,000 feet above the street, with the short-wave transmitting aerial on top of the airship mooring mast, 1,250 feet above the street. It has been generally understood in New York television circles that when this station is completed and tested, the secrets of Camden will be revealed, and sets be placed on the market at the New York Radio Show this autumn. The number of scanning lines will be 180, and the speed of transmission 24 images per second. A mechanical scanning disc will be used for transmission, but a cathode ray tube will be used for reception. The received image will be six inches square, and viewed directly on the end of the cathode ray tube. These tubes, I am informed, are short-lived and

who is in charge of the work, has forwarded a paper to the Society, as I know he has been invited to do.

expensive. The entire receiver is likely to be marketed at first at a price running into several hundreds of pounds.

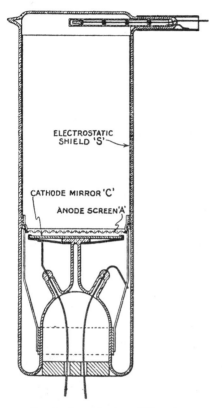

ELECTROSTATIC
SHIELD 'S'

CATHODE MIRROR 'C'

ANODE SCREEN 'A'

FIG. 7—Cross Section of Dissector.

Transmissions will be made from the Empire State studio on wave-lengths between 5 and 7·5 metres, and will be receivable only by those who, but for weather conditions, would have a direct view of the top of the Empire State Building. This embraces an area within a radius of roughly 100 miles of New York. Any obstruction whatever in the direct path will prevent reception.

The use of such a short wave-length is made necessary by the fact that a rough calculation shows that the signal frequency to be transmitted is approximately 390,000 cycles. The accompanying speech and music will be sent out on a second channel half a megacycle removed from the image wave.

I am given to understand that raising the number of scanning lines from 60 to 180 has not improved the detail of the received image to anything like

the extent that one might expect. The scope or field of view of the system has not been increased. That is to say, only head and shoulder views can be sent to advantage. The speed of transmission has been raised from 20 to 24, because the latter figure is the standard speed of talking films, and it is anticipated that the Empire State station will employ most of its transmission periods by broadcasting R.C.A. Photophone " shorts " or short talking films.

STANDARD 60-HOLE DISC RECEIVERS.

The Jenkins Television Corporation has not achieved anything outstanding for some time. It has been engaged in the manufacture and sale of standard 60-hole disc receivers and transmitting equipment, and in broadcasting regular programmes in conjunction with WGBS, one of the New York broadcasting stations. The images of artists who broadcast over the regular WGBS broadcast wave are sent out on short wave through W2XCR by means of a television " camera " placed in the broadcasting studio. Details of television transmissions will be found in an appendix to this paper.

I should perhaps state at this juncture that the Standard American transmissions all use 60-hole discs; horizontal scanning from left to right, and from top to bottom (*i.e.*, the top of the disc is used at both transmitter and receiver); 20 images are sent per second; synchronisation is achieved by driving the disc with a synchronous motor taking power from the standard 110 volt, 60 cycle, A.C. mains, or by means of a phonic wheel control in cases where non-standard power supply is available. The image is practically square.

THE FARNSWORTH CATHODE-RAY SYSTEM.

Philo T. Farnsworth, formerly of San Francisco, came east to the Philadelphia plant of the Philco concern about the middle of 1931, and since then no publicity matter concerning his activities has been released. It is generally understood, however, that he is being provided with the necessary facilities to reduce his mathematical conceptions to actual practice. In view of the sensational character of Farnsworth's past claims, the American technical world awaits interestedly and patiently, a practical demonstration of them.

Farnsworth has been using the cathode ray tube for both transmission and reception. He calls the transmitting tube, which is of the cold cathode type, a " dissector tube," and the receiving tube, which is of the hot cathode type, an " oscillite." A side view of the dissector is shown in *Fig. 5*. An end view is shown in *Fig. 6*. A cross-sectional diagram is shown in *Fig. 7*, and a skeleton sketch, with focussing coils in place, in *Fig. 8*.

The dissector tube comprises a cathode, C, coated with photo-sensitive material and mounted parallel and close to an anode screen or grid, A. At the opposite end of the tube there is a target electrode having all but a single small area shielded from the discharge from the cathode. photo-electric cell wherein provision is made for

Considered broadly, this tube is a form of forming an " electron image " of an optical image focussed through the polished target end of the tube, on to the cathode surface. In operation,

an image of the object to be transmitted is focussed through the polished window on to the cathode. Electrons are emitted from the photo-sensitive cathode mirror in proportion to the amount of light falling on it, and these electrons are accelerated by a potential of the order of 500 volts which is applied between the cathode and the anode screen, as shown in *Fig. 9*. Most of the electrons are projected into the equipotential region between the anode screen and the target, wherein they follow a helical path and re-combine to form an electron image in the plane of the target. This electron image is then shifted by the transverse magnetic field provided by the focussing deflecting coil systems, so that the entire image is caused to move over the aperture in the target shield, thus achieving the scanning of the image.

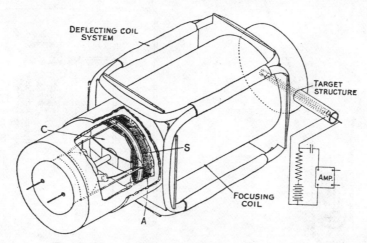

FIG. 8—Diagrammatic Sketch of Farnsworth Dissector.

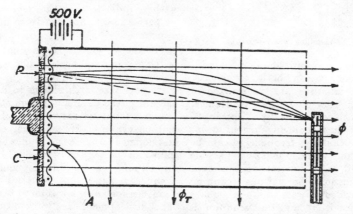

FIG. 9—Illustrating how the electron stream is deflected on to the aperture leading to the "target" or anode of the photo-electric cell.

The transverse scanning field is produced by two sets of coils, which are mounted at right angles to one another on the outside of the dissector tube. A saw-tooth wave alternating current of relatively high frequency flows through one set of coils, providing line scanning in a horizontal direction. A low frequency current of similar wave formation flows through the other set of coils and produces vertical deflection or line spacing of the image. These currents are produced by local oscillators, and it is of interest to note that by varying either or both frequencies, any number of scanning lines per inch can be transmitted, and any number

The receiving oscillite is illustrated in *Fig. 10*, and in section in *Fig. 11*. It is a hot cathode type of tube. *Fig. 12* shows the details of the " electron gun " element, which consists of a helical filament, oxide-coated only on the inside. A shield is placed over this filament having in it a hole the same diameter as the filament helix. The anode, which is tubular in form, is positioned in front of the cathode, and midway between the filament shield and the anode there is mounted a ring grid, or control element, marked 22 in *Fig. 12*.

Farnsworth has used as many as 400 scanning lines in his images, but such meagre eye-witness

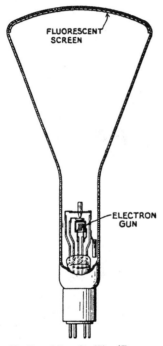

FIG. 11—Receiving Oscillite (Farnsworth).

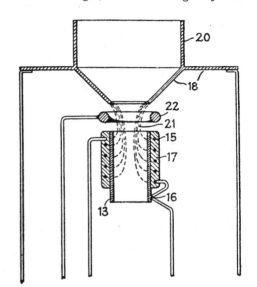

FIG. 12—Details of Filament.

of images can be transmitted per second. Arrangements are made so that both frequencies, in the form of brief impulses, are transmitted to the receiving station between images, or frames, there to control local oscillators by " locking " them into step and thus achieving synchronism. The fact that the scanning frequencies can be varied in such a simple manner makes it possible for anyone possessed of a receiver of this type to pick up transmissions of various scanning characteristics, provided always that a square picture is being transmitted, and that scanning is from left to right and from top to bottom—the American standard method.

reports of these images as I have been able to gather, indicate that the degree of image detail, or perfection, is not nearly so great as one might expect.

Of course, such a large number of image lines involves the amplification and transmission of an enormously wide frequency band, and for this purpose Farnsworth claims to have developed an amplifier (the details of which still remain secret), the characteristic curve of which is flat up to one million cycles. He had scarcely completed this development when he announced another, which he called an " image compressor," by means of which he claimed to be able to compress a 300 Kc. frequency band into something like 7 Kc. for transmission over ordinary broadcast channels. This development was worked out mathematically first, and then, Farnsworth claims, put into experimental practice.

However, none of Farnsworth's claims has yet been demonstrated publicly. One of the latest newspaper comments on his activities in Philadelphia stated that he applied to the Federal Radio Commission for a short-wave, 100 Kc. frequency band, explaining in his application that he had been mistaken in his previous conclusion that he could confine a wide frequency band to a few kilocycles.

FIG. 13—Sanabria Transmitter.

Nevertheless, Farnsworth is a young man striving earnestly to perfect a new art, and I am still of the opinion that he will ultimately add a valuable contribution to our sum of knowledge on te'evision.

Ulysses A. Sanabria is another young man who has been actively engaged in television research in Chicago for some years. Although his system is mechanical, it nevertheless accomplishes some remarkable results, and is worthy of some attention here.

SANABRIA SYSTEM DESIGNED FOR PROJECTION WORK.

At the transmitting end he uses a disc scanner, but instead of employing a single spiral of holes, he has three spirals of 15 holes each, a total of 45 holes in all. Each spiral scans the image completely, but the arrangement of the spirals is such that each spiral projects light spots which half overlap those produced by the preceding spiral. In this way " strip effect " is almost entirely eliminated, and, although the speed of transmission is only 15 images per second, there is much less flicker than is apparent at a transmission speed of 20 images per second with a single spiral disc.

Sanabria's transmitter is shown in *Fig. 13*. The light beams, after passing through the scanning disc and associated optical system, are deflected by a mirror set at an angle of 45°. In this way space is saved behind the photo cell rack, and the operator is better able to control the projector and the focussing of the triple lens turret. The eight photo-electric cells, unlike most American cells, are no larger than receiving valves, but they are set at the focus of large parabolic reflectors. The associated photo-cell amplifier is shown in *Fig. 14*. It has an output power of something like 300 watts.

Sanabria's receivers are designed and built for projection work. They vary in detail according to size, but the principle is the same. A triple-spiral disc is employed, but two-inch lenses are used instead of pin-holes. The light source is a specialiy developed neon lamp of the crater type, which might almost be described as a neon arc. Instead of operating the lamp cold, as is usually done, and forcing the signal currents to heat up the gas as well as vary the intensity of its glow, Sanabria has incorporated a heater in his lamp, which is somewhat similar in design and function to the heater element of an indirectly heated valve cathode. Thus, the gas in the vicinity of the electrodes having already been heated, the incoming signal currents have no more to do but to vary the intensity of the glowing gas. Sanabria claims that, because of this, the light source is responsive to much finer gradations and that, in consequence, much finer detail can be reproduced in the received image. In addition, because the gas is already pre-heated, its break-down voltage is reduced from the usual 175 volts for cold neon to between 15 and 25 volts for the Sanabria arc.

I have witnessed several demonstrations of the Sanabria equipment, on screens varying in size from two to twelve feet square. Even on the largest screens, the brilliance of the screen appears to be about half that of the average cinema screen; it is quite adequate for viewing purposes in a dark or very dimly lighted room.

As regard the amount of detail, this is very excellent indeed. By means of the triple lens turret at the transmitter, three-quarter length, head-and-shoulder, or close-up, head only, views can be given, all with a degree of excellence almost as good as that provided by similar views reproduced by the average small home cinema machine.

Latest reports indicate that at the tenth annual Radio-Electrical Show at Chicago, opening on January 18, Sanabria proposes to demonstrate his images to visitors to the show on a screen measuring 25 feet square. It would appear that Sanabria is concentrating on large screen projection work with a view to introducing it as a vaudeville theatre novelty. He has already met with a measure of success in this direction. *Fig. 15* illustrates a home type of receiver developed by Sanabria, which reproduces the image on a ground glass screen measuring about ten by twelve inches.

COMMERCIAL DEVELOPMENT and GOVERNMENT CONTROL IN AMERICA.

Turning now to the second group of television workers, those interested in the exploitation or development of existing equipment, a perusal of the list of stations licensed to transmit television experimentally will reveal the varied character of the interests concerned.

In Great Britain, the Postmaster-General exercises control over all radio communications. In America, a separate governmental body, known as the Federal Radio Commission acts as the controlling body, which administers international radio regulations, and assigns transmission frequencies. A North American Conference, held in Ottawa in January, 1929, set aside the following frequency bands for television experiments :—

2000	—	2100 Kilocycles.
2100	—	2200 do.
2750	—	2850 do.
2850	—	2950 do.

As will be seen from the attached list, there are 18 concerns engaged in television research work, operating a total of 22 stations, and these numbers are steadily increasing. According to the terms of their licences, the facilities may be used for experimental purposes only, the sale of " time on the air " or other commercialisation being forbidden for the present. The latter fact indicates that in the opinion of the Federal Radio Commission television has not yet reached a degree of perfection sufficient to warrant the inauguration of a public service on a commercial basis.

FIG. 14—Amplifier used in Sanabria System.

Some of the licensees are engaged in research problems, while others are interested only in gaining experience in the presentation of television entertainment. This involves research into such problems as facial make-up for actors, effective back-drops for different acts, rapid focussing on moving actors, etc.

Attached to this paper is a typical day's schedule of the Columbia Broadcasting System's station, W2XAB, which broadcasts every day except Sunday between 7 and 11 p.m., G.M.T., and between 1 and 4 a.m., G.M.T., on the 2750—2850 Kc. band, with sound synchronised over W2XE on 49 metres. From the schedule it will be seen that a wide range of subjects is tackled daily, and very good results are being obtained. Mechanical scanning is used, 60 lines, 20 images per second, scanning from left to right, top to bottom.

The problem of presenting entertainment by television differs in many respects from the problems of the stage, screen or radio, and much research will be necessary before sufficient is learned to enable an acceptable entertainment service to be inaugurated on a regular commercial basis.

FIG. 15 - De Luxe television equipment—Sanabria receiving set

These experiments are not without their amusing side. A story is told, for example, of a Chicago station which elected to televise a well-known speaker. As the speaker was conducted to the microphone a few minutes before the appointed time, the engineers, unknown to the waiting orator, opened up their television transmitter. Watchers peering into their receivers saw the gentleman sit down before the mike and arrange his papers, straighten his necktie, and prepare for his speech. The watching audience saw him as he appeared, visibly nervous, waiting for the signal to do ahead. Just one minute before the zero hour they were astonished to see him reach round to his hip pocket, from which he extracted a flask, and then, without apologies to Mr. Volstead, he fortified himself for his coming ordeal with a sizeable gulp of liquid. For the next few days the station manager was kept busy explaining that the flask contained merely cough medicine—a statement which quite possibly may have been true.

As far as public interest in television is concerned, it may be summed up as follows. An ever-increasing number of amateur enthusiasts is following the regular experimental transmissions, especially in neighbourhoods which are served by one or more transmitters. The latest burst of activity in this direction is in the vicinity of San Francisco, where a local broadcasting station has just established a short-wave television transmitter, using a cathode ray scanner with 80 lines. Local radio shops report that their telephones have been busily engaged with enquiries as to how the callers may obtain television equipment wherewith to pick up the new programmes.

So far as the non-technical public is concerned, however, it has adopted an attitude of mildly interested, watchful waiting, until such time as it can go to a neighbouring shop and buy a television receiver as cheaply as it can buy a wireless receiver to-day, and be assured that it will function as simply as the modern wireless set, and with as great a freedom from trouble. It must also be able to provide a sizeable image which the whole family can watch in comfort. And, of course, there must be a continuous service of programmes similar to existing sound programmes. Technical problems interest the general public but little; they want a perfected service, and are prepared to wait until it can be provided. This is partly due to a " once-bitten-twice-shy " experience in the development of radio, and partly because it is only a few years since the feverish development of radio killed all the technical aspirations of the average citizen.

TELEVISION—A YOUNG MAN'S " GAME."

Before closing this paper, there is one feature of American television activities which I think will prove interesting and encouraging to members of this Society, and that is that television in America, if not anywhere else in the world, is distinctly a young man's game, as the following examples will show.

Television's youngest administrative head is George Gruskin, President of the Sanabria Television Corporation. He is only 22. Sanabria himself is 24. Hollis Baird, Chief Engineer of the Shortwave and Television Laboratories, Boston, is also 24; so also is Farnsworth. Edwin L. Peterson, an experimenter well known in Hollywood, is in his late twenties. Several of the more responsible engineers at the R.C.A.-Victor plant at Camden are in their early thirties, and D. E. Replogle, Chief Engineer of the Jenkins Television Corporation is about 36 years of age. In the Columbia Broadcasting System, Edwin K. Cohan, Chief Engineer, and his assistant in television activities, Richard E. Wallace, are both in their middle thirties; and Bill Schudt, Columbia's television programme director, is in his late twenties. And in the Federal Radio Commission itself, Gerald C. Cross, the member directly responsible for the Commission's television regulations, is but 28 years old.

Older members of this audience will recall that radio, in its early beginnings, was also largely a young man's game.

To the younger members of the Television Society, therefore, I dedicate this paper. This new art in which you are interested is younger even than you are yourselves! Advance boldly on its problems, solve them, and secure for yourselves an assured future, surrounded by the satisfaction and rewards of achievement.

APPENDIX.—1.

LIST OF STATIONS LICENSED TO TRANSMIT TELEVISION EXPERIMENTALLY IN THE UNITED STATES:—

2000—2100 Kc.

Call Letters	Company and Location	Power (watts
W3Xk	Jenkins Laboratories, Wheaton, Md. ...	5000
W2XCR	Jenkins Television Corporation, Jersey City, N.J.	5000
W2XAP	Jenkins Television Corp., Portable ...	250
W2XCD	De Forest Radio Co., Passaic, N.J. ...	5000
W9XAO	Western Television Corp., Chicago, Ill.	500
W2XBU	Harold E. Smith, North Beacon, New York (School)	100
W1OXU	Jenkins Laboratories, aboard cabin monoplane	10
W1XY	Pilot Electrical & Manufacturing Co., Springfield, Mass	250
W1XAE	Westinghouse Elec. & Manufacturing Co., Springfield, Mass	20000

2100—2200 Kc.

W3XAK	National Broadcasting Company, Bound Brook, N.J.	5000
W3XAD	R.C.A.—Victor Co., Camden, N.J. ...	500
W2XBS	National Broadcasting Co., New York	5000
W2XCW	General Electric Company, South Schenectady, N.Y.	20000
W8XAV	Westinghouse Elec. & Manufacturing Co., East Pittsburg, Pa.	20000
W9XAP	Chicago Daily News, Chicago, Ill. ...	1000
W2XR	Radio Pictures Inc., Long Island City, New York	500

2750—2850 Kc.

W2XBO	United Research Corporation, Long Island City, N.Y.	20000
W8XAA	Chicago Federation of Labour, Chicago, Ill.	1000
W9XG	Purdue University, West Lafayette, Ind.	1500
W2XBA	W.A.A.M. Inc., Newark, N.J. (broadcasting station)	500
W2XAB	Columbia Broadcasting System, New York	500

2850—2950 Kc.

W1XAV	Shortwave and Television Labs., Boston, Mass.	500
W2XR	Radio Pictures Inc., Long Island City, N.Y.	500
W9XR	Great Lakes Broadcasting Co., Downers Grove, Ill.	5000

APPENDIX.—2.

COLUMBIA BROADCASTING SYSTEM EXPERIMENTAL TELEVISION PROGRAMME—3.

TUESDAY, DECEMBER 29th.

W2XAB and W2XE NEW YORK.

2·00—6·00 P.M. EXPERIMENTAL SIGHT PROGRAMMES. Card station announcements, and drawings of radio celebrities.

8·00 P.M. HEMSTREET QUARTET. All girls' novelty group with Helen Andrews, soloist. Long shot group picture. White and silver backdrop curtains.

8·15 P.M. GRACE VOSS — Pantomimes. Long shot and close-up. Silver backdrop curtain.

8·30 P.M. SENORITA SOLEDAD ESPINAL and HER PAMPEROS in a half-hour program of Spanish and Latin American music and songs. Guitar Sextette and Mezzo Soprano. Group projection with various backdrop screens.

9·00 P.M. THE TELEVISION GHOST. Mystery character in weird costume enacts the murder mysteries in the character of one risen from the grave. . . . . THE MURDERED! !

9·15 P.M. HAZEL DUDLEY — Song recital. Series of close-up pictures to be scanned.

9·30 P.M. THREE ROUND EXHIBITION BOXING BOUT. An experimental television demonstration of what the flying spot can do at a fight. Miniature ring will be used. Blow by blow description by Bill Schudt. Dark backdrops. Long shot pick-up. Scanner will follow boxers around ring.

9·45 P.M. GLADYS SHAW ERSKINE and MAJOR IVAN FIRTH PRESENT Television novelties with visual illustrations. Alternate backdrops will be used.

10·00 P.M. "TASHAMIRA" introduces new German modernistic dances and technique. Extreme long shot focus. Close-up with varied backdrop curtains.

10·15 P.M. ONE MAN NOVELTY BAND with VINCENT MONDI. Close-up against white backdrop.

10·30 P.M. ELIENE KAZANOVA—Violinist.

10·45 P.M. GRACE YEAGER, song recital. Close-up shot of an artist singing semi-classic favourites.

11·00 P.M. SIGN OFF.

Marconi "News" Television Transmitter

TELEVISION IN GERMANY

Hubert Gibas

Proceedings of the Institute of Radio Engineers
Volume 24, Number 5   *May, 1936*

TELEVISION IN GERMANY*

BY

HUBERT GIBAS

(N. V. van der Heem and Bloemsma, The Hague, Holland)

Summary—*This paper gives the status of television in Germany in the autumn of 1935. The systems used for transmission and reception of television are described. The subject of image quality and the possibilities of improvement, together with the prospects of television, are discussed.*

BERLIN'S first television transmitter was erected at the beginning of 1935 for the purpose of transmitting television program material. Two transmitters are used, one for the picture and the other for the sound. The antennas are located at the top of the broadcast tower in Witzleben, Berlin. The transmission is primarily from sound film, but there is some transmission from the studio. The

Fig. 1—Fernseh-A-G film transmitter.

intermediate film system is also used, in which case the image to be transmitted is first photographed on a film, which also carries sound. The film is then developed, fixed, washed, dried, and transmitted within one minute.

The Fernseh-A-G's film transmitter is shown in Fig. 1. The image to be transmitted is dissected by a Nipkow disk which runs in a vacuum. This provides a dust-free space for the disk and also reduces the required driving power.

* Decimal classification: R583. Original manuscript received by the Institute, November 1, 1935.

The range of the radio transmitter is a little more than the optical range, say about 100 kilometers. However, the distance depends on the location of the receiver so that very good reception from Berlin is obtained on the top of the Brocken, which is 1100 meters higher than Berlin. The distance in this case is 200 kilometers. In Neuruppin, seventy kilometers from Berlin, is located a very satisfactory receiving station. To cover a portion of Germany with television a network of about twenty-five transmitters has been planned.

The most important firms engaged in television in Germany are Fernseh-Aktiengesellschaft, Berlin—Zehlendorf; Radioaktiengesell-

Fig. 2—Manfred von Ardenne with his television equipment.

schaft, D. S. Loewe, Berlin—Steglitz; Lorenz-Aktiengesellschaft, Berlin—Tempelhof; C. F. H. Müller Aktiengesellschaft, Hamburg; Tekade, Nürnberg; and Telefunken Aktiengesellschaft, Berlin SW 11.

All of these, except Tekade, employ the cathode-ray tube at the receiver. Fig. 2 shows a picture of Manfred von Ardenne seated near his television equipment. He is well known in Germany for his cathode-ray tube work. His book "Fernseh-Empfang" gives his latest results in television reception, as well as details concerning receivers and accessories.

The elements of a cathode-ray tube manufactured by Leybold and von Ardenne, Köln-Bayenthal, are shown in Fig. 3. Two pairs of plates are provided for horizontal and vertical electrostatic deflection. Some cathode-ray tubes use only one pair of plates in combination with magnetic deflecting in one direction. For such a combination the length of the tube becomes shorter. The operation of cathode-ray tubes is well known and will not be discussed here.

Tekade uses a mirror-screw to reproduce the image at the receiver. This is shown in Fig. 4. The number of mirrors equals the number of

Fig. 3—Leybold and von Ardenne cathode-ray tube elements.

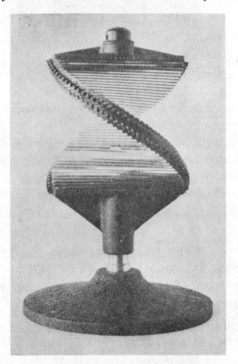

Fig. 4—Tekade mirror-screw.

lines in which the image is divided. To control the light intensity a Kerr cell is employed. The advantages of this arrangement are that no parts need be renewed from time to time, and possibly the advantage of cost over cathode-ray tube receivers. Cathode-ray tubes, in the thirty-centimeter diameter size, cost about RM. 300. No deflecting voltage supply or other equipment, including a source of high voltage, is necessary with the Tekade system. On the other hand, it is necessary to have a synchronous motor to turn the mirror-screw.

Television receiver production has proceeded to the point where large scale production is practicable. Figs. 5, 6, and 7 illustrate some of the receivers. Up to a year ago the number of control knobs was as great as twelve. These have now been reduced to not more than four,

Fig. 5—Loewe television receiver.

so that the average person can tune and operate the receivers. The present controls are: receiver tuning, regulation of light intensity (background control), volume and tone control of the sound receiver. It is planned in Germany to standardize on a constant frequency difference between the carrier waves of the picture and sound transmitters.

The size of the picture for normal home use is about twenty by twenty centimeters. Its color is now pure white—a year ago it was yellow. The brightness of the image has been considerably improved and is now about five times as great as a year ago. The pictures are sufficiently bright to be visible in slightly reduced daylight. The receivers, while designed for 180-line operation, are so built that they can be changed with slight alterations to receive 240 lines. The synchronizing is automatic. The tone quality of the sound is excellent because higher modulation frequencies are not cut off by too great selectivity, as must happen in the regular broadcast bands.

The television receiver manufactured by Loewe is shown in Fig. 5. Under the cathode-ray tube is the loud speaker, right and left are two double control knobs for receiver tuning, light intensity, volume, and tone control.

Fig. 6—Fernseh-A-G television receiver.

The receiver produced by Fernseh-A-G is shown in Fig. 6. The knobs shown in a row on the right-hand side are for adjustment of the position and sharpness of the image, as well as light intensity. They are adjusted once and then remain set. At the rear of the receiver are terminals for antenna, ground, and power.

Fig. 7—Tekade television receiver.

The Tekade television receiver is illustrated in Fig. 7. The mirror-screw is visible in the background of the shadow box, on the left of which can be seen a portion of the loud speaker. The control units in this case are for tuning the picture, the sound, and also for volume and light intensity. Tekade uses an incandescent lamp having a long incandescent wire as a light source. The light passes through the Kerr cell, with which are associated two prisms. Such a light controlling system is well known. The focusing of the light controlling equipment is carried out at the factory and must not be altered. Previously the mirror-screw system had the disadvantage that the picture was visible over a restricted angle, so that the observer had to be on its center line. With the so-called optical grid this disadvantage disappeared. This grid consists of a small glass vessel, in which are thin glass sticks, one above the other, surrounded by a fluid. In this way the light

Fig. 8—Telefunken television receiver.

beam is directed in a vertical direction and the image appears to be essentially longer, so that the observing angle is considerably larger. The size of image produced by this system can be enlarged rather simply, a feature that is not possible in ordinary cathode-ray tubes. If an arc light is used as a light source there is sufficient light for an image of large size.

Fig. 8 is a rear view of a Telefunken receiver.

The Loewe receiver without its cabinet is shown in Fig. 9. The parts of this receiver are the picture receiver, sound receiver, scanning voltage supply unit, cathode-ray tube, loud speaker, and voltage source.

In some television receivers the first radio-frequency stages of the television and sound receivers are combined. The picture receiver must have a communication band width of about one megacycle for an image of 180 lines, with each line having 240 image points, and at a picture repetition rate of twenty-five pictures per second.

We now come to the subject of synchronization. The carrier which is modulated by the picture is interrupted twenty-five times a second. The duration of each interruption is five per cent of the transmitting time for one picture. During this interruption the electron beam at the receiver must return from the bottom of the image to the top. The

Fig. 9—Loewe television receiver, removed from cabinet.

carrier wave is also interrupted 180 times per image. During these interruptions the electron beam must return from the end of one line to the beginning of the next. The interruption time in this case is about five per cent of the transmitting time for one line. During the interruptions the electron beam is not visible. These interruptions form the synchronizing impulses for the scanning oscillators at twenty-five images per second and 180 lines per image.

Most of the deflecting apparatus uses gas-filled tubes as generators. The scanning voltages in some cases must be amplified before use. The cathode-ray tubes are surrounded by metallic screens to avoid disturbances from outside. The placing of the components in the cabinet must be carefully carried out. The high voltage for the cathode-ray tube is about 4000 volts. Most of the receivers use a special antenna which can be placed anywhere in the room. If, however, the receiver is at some distance from the transmitter the antenna must be located in or near the optical path of the transmitter.

The quality of the reproduced pictures can be judged by Fig. 10. This is an unretouched photograph of the announcer at the television transmitter in Berlin. At present television images in Germany have two difficulties. First, the image flickers so that the eyes of the observer are soon tired, and second, 180-line detail is not sufficient. Many efforts are being made to eliminate these troubles. To provide more detail it is intended to use 240 lines. The Fernseh-A-G has made experiments with 320 lines.

Fig. 10—Unretouched photograph of television image.
Von Ardenne and Lorenz.

For the elimination of flicker there are two possibilities; first, the speeding up of the number of pictures per second. Thirty-five images per second have been tried, in which case no flicker was visible, but this would not be true if more brightness were available. The greater the light intensity the greater the flicker. When transmitting more than twenty-five images per second normal sound films, of course, cannot be used. Then too, the modulation frequency bands are still wider because there are more picture elements per second to be transmitted.

The second possibility of flicker elimination is the so-called "line-jump" system. Two German manufacturers, Loewe and Tekade, have demonstrated this. The principle is as follows: Suppose we have a 180-line picture, twenty-five frames per second. Now if the 1st, 3rd, 5th . . . etc., lines be transmitted during the first half of the 25th of a second interval, and then the 2nd, 4th, 6th, . . . etc., line be transmitted in the second half of the picture interval, it means that the two complete pictures are shown displaced by the width of one line. This gives the eye the same impression as though fifty images per second were transmitted. Under these conditions the picture is entirely free of flicker.

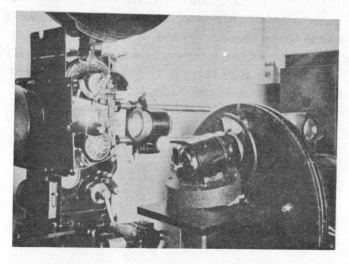

Fig. 11—Tekade film scanning equipment for interlaced pictures.

For such an interlacing system Tekade uses a small disk placed between the film and the Nipkow disk, as seen in Fig. 11. This disk, with its motor, is shown in Fig. 12. The disk has glass in one half and air in

Fig. 12—Tekade disk used in connection with interlaced film scanning.

the other half. For 180 lines the Nipkow disk has ninety holes and makes two revolutions per image. During the first revolution the air space of the small disk is between the film and the Nipkow disk. During the second revolution the glass is in place. The glass breaks up the light beam in such a manner as to shift it by the width of one line.

Looking into the future we see that before very long Germany will have a television system which will meet the highest requirements, about the same as that to which we are accustomed in cinema theatres. If it is possible to manufacture great numbers of television receivers the price, of course, will be reduced—then many more people will be able to buy television receivers. It is quite possible, therefore, that within ten years television reception will be as prevalent as sound broadcast reception is now.

PART III: ELECTRONIC TELEVISION

PRESIDENTIAL ADDRESS

A. A. Campbell Swinton

THE JOURNAL

OF THE

RÖNTGEN SOCIETY.

Vol. VIII.   JANUARY, 1912.   No. 30.

SESSION 1911—1912.

A GENERAL MEETING was held at the Institution of Electrical Engineers, Victoria Embankment, W.C., on Tuesday, 7th November, 1911, Dr. G. H. Rodman in the Chair. The minutes of the last meeting in June were read and confirmed.

Ballot was taken for Major Charles William Profeit, R.A.M.C.; Charles Fred Bailey, M.D. (Lond.), M.R.C.S.; Arthur Lloyd, Esq., and Francis Hernaman-Johnson, M.D.; these gentlemen were duly elected members of the Society.

The following were nominated for ballot at the next meeting :—G. F. Westlake, Esq., 93, Clarendon Road, Putney, Pharmacist and Assistant, Electrical Department, Cancer Hospital; John Morison, Esq., M.D., 1, Jacksons Lane, Highgate, N.

Dr. Rodman then introduced the new President, Mr. A. A. Campbell Swinton, who delivered the Presidential Address :—

PRESIDENTIAL ADDRESS

By A. A. CAMPBELL SWINTON,
November 7th, 1911.

LADIES AND GENTLEMEN,

In the first place I desire to express my appreciation of the great distinction that the Society has conferred upon me in electing me to the Presidency, a post which has been rendered honourable by the many famous men of science who have held it since the inauguration of the Society in 1897.

The choosing of a subject for my address to you this evening has been a matter of some considerable difficulty, for one reason because I have not recently been doing any experimental work suitable for exhibition before a large audience, and for another, because previous Presidents seem to me to have pretty well exhausted all there is to say at present in the way of general review of the subjects with which it has been usual for this Society to deal.

I have, therefore, found myself obliged to fall back upon some other topics, which I am, however, hopeful of being able to present to you in a sufficiently interesting manner.

I will begin with a few words as to the paramount importance that must be attached to scientific progress above all other products of human effort. All advance in the relations between man and nature whereby man gains to any greater extent the mastery, may be described as scientific progress, and in this connection we must recognise that many things which we now look upon, and have for ages regarded as entirely commonplace, were, at the time of their inception, really very remarkable indeed. Take for instance the application to human needs of fire. Animals, even of the highest types, make no use of it. There must have been a period when man also did not understand its properties, and, like the animals from which he has sprung, was afraid of it and left it severely alone. A time must next have come, and with that time the valorous man who first had the temerity to experiment with this very powerful and destructive agent. The ancient Greeks believed that Prometheus first stole fire from Heaven, and this is probably a picturesque manner of saying that advantage was taken of some combustible material having become ignited by being struck by lightning. Fire may, however, have been obtained from the glowing lava of now long-quenched volcanoes ; perhaps accidentally from spontaneous combustion ; possibly by the concentration of the rays of the sun by some crude form of lens ; from the sparks produced in shaping flint implements ; or even by friction, according to the method still used by some savage races.

Anyway, think of this prehistoric investigator into the means and effects of combustion in that far distant age ; consider his inferior mental equipment ; imagine his savage surroundings ; take into account, also, his lack of any but the most primitive appliances. Must we not laud his enterprise and admire his courage ; must we not also acknowledge the enormous advantages his investigations have gained for all his posterity ? The warming of their bodies and the cooking of both animal and vegetable nutriment would, no doubt, be the first uses to which our remote ancestors would apply the new agent, but soon would follow the firing of pottery, up to that date merely sun-baked, then the reduction and smelting of metals, and finally the whole galaxy of the arts. What is scientific progress if this is not ? And yet it leaves off where what we usually mean by science begins, namely, about the Graeco-Roman period. Look out, however, into London to-day, and recognise how little of all we see around us could have ever existed but for those early high-

temperature experiments made so many thousands of years ago. Without them could human beings even live in this northern climate ?

And here may I point out that, curiously enough, it is only when we go back to the earliest evidences of primitive human life upon this planet that we take the true philosophical course of naming the periods we are dealing with after the main material advances in scientific progress made during those periods by the human race. We talk of the Stone Age, of the Bronze Age, or the Iron Age, to denote those vast expanses of time during which the primitive inventor was discovering the means of applying new materials to what was then the great necessity of mankind, namely, weapons for the chase, for self-protection, and for war upon his enemies.

Later in history we find that this really philosophical method is abandoned. As we come to know more as regards the position, supremacy and conditions of particular races, and still further, when we become better acquainted with the deeds and achievements of particular individuals, we find that historians have a tendency to over-look the enormous influence of the results obtained by scientific investigators and discoverers, and to make it appear as though the current of events were really governed by those who, from accident of birth, official position, political influence or martial achievements, have made for themselves reputations as leaders of men.

To see that this view is wrong we have only to survey the past. Can it for an instant be doubted that the labours of the unknown prehistoric individual to whom I have just alluded, who first discovered the properties of fire, or of those who originated the smelting of metals, who launched their frail, and at that time novel coracles upon the ocean, and first applied wheels to the primitive cart, are more living factors to-day than the valour of all the warriors, the wisdom of all the statesmen, or the wiles of all the politicians that the world has seen? It is a truism indeed that the world knows little of its greatest men.

Can it be questioned that the discoveries of Archimedes and his disciples have more effect to-day than the battles of Alexander or of Hannibal? Or, if we turn to modern times, can it be gainsaid that Watt and Stephenson, Davy and Faraday, have done more to change both the course of history and the material conditions of life than did Napoleon or Wellington, Walpole or Pitt?

The fact is, as I once remember hearing lamented by no less a statesman than the late Lord Salisbury, that while the work of the politician, the statesman, the soldier or the leader of men, however great and however fortunate, is of necessity but transitory, what is accomplished by one man being undone by another—the work of the scientific discoverer and inventor is everlasting. However insignificant this work may apparently be, provided it is new, it adds something more to that great store of human knowledge and experience which is slowly accumulating, and enables man more and more to triumph over nature. Moreover, results that appear of but slender importance at the time of their discovery often turn out in the end to be of the greatest moment.

For the undue amount of influence on the progress of the world that is attributed to leaders of men, in comparison with that exerted by investigators of nature, historians are no doubt to blame. In stating this, however, one must in justice remember that, after all, most histories are written to sell, or, if not that, to bring fame to their authors. Further, we must allow that the story of scientific investigation is frequently not very interesting, at any rate to the general public, who may justly find such a story dull as compared with accounts of the stirring episodes that occur in the senate or the feuds that are settled on the battle-field. Thus, the tale of, say, Marlborough's campaigns

makes probably more picturesque reading, and is more likely to interest the average student, than would be a history of the patient scientific work that led up, about the same period, to the enunciation, say, of Boyle's law of the expansion of gases. We can admit this, though there can be no doubt that the permanent influence of Boyle's discovery on the history of the world has been in the past, and will continue to be in still greater ratio in the future, incomparably greater than was that of all the battles of the day, inasmuch as Boyle's law was an important link in the chain of discoveries that led up to the steam engine and modern industrial development, while to-day the effects of the wars of the seventeenth century have, for all practical purposes, passed away.

The fact is that there is a glamour attached to the position of those who are supposed to direct the history of nations, that prevents the real directing forces from being seen in their true proportions. The great statesmen, the great generals, leaders of mankind in general, are, after all, nothing much more than glorified policemen, whose utility to the world is only occasioned by the imperfections of human nature. As organisers they are no doubt useful, but they generally benefit particular nations at the expense of others, and as a rule they leave little behind them that will stand the test of centuries.

Another product of human endeavour which also seems to have an undue amount of importance attached to it in regard to its influence on human progress, is literature, which I am here considering apart entirely from its æsthetic claims upon us as a means of relaxation. That literature has a directive influence, and that a powerful one, no one can deny; but I fancy that all scientific men will agree that it is not to be compared with that exercised by material discoveries and inventions. In saying this, I know that it is the fashion to ascribe the beginning of all modern science to what is contained in the Novum Organon, but I rather fancy that if we could truly estimate the influence on the world's history of the two men, we should find that Roger Bacon, the inventor of gunpowder, would come before the better known Francis Bacon, who, some centuries later, wrote his great work on the new learning. Indeed, probably the chief merit of the Novum Organon was that it assisted a return to experimental methods as opposed to what had become the benumbing system of Aristotle, who, by the way, is interesting to this Society for the reason that he was the author of the immortal, if not very illuminating, phrase that "nature abhors a vacuum." Anyway, since the earliest times there has never been a better organised and more successful mutual admiration society than that formed by the writers of the world, who have always been chiefly concerned to discuss one another and one another's scripts. This, and the fact that the written word endures, has given to the wielders of the pen a prominence in history to which they are scarcely entitled by their influence on progress.

At the present time, when it is the fashion to ascribe the production of all wealth to the manual labourer and all progress to the politician, it is more than ever necessary that correct views should be insisted on. Let us, therefore, emphasise the fact that from the beginning of the world all advance has been due, not to the many, but to a few exceptional individuals, and had it not been for the genius of these we should still be naked savages, not even painted with the proverbial wode.

As an instance, take the Electric Telegraph, which has had more effect on civilization than almost anything else during the past century, and gives employment to thousands. The names of those to whom it is due, beginning with Franklin, Volta, and Galvani, going on with Morse and Cook, and ending with say Wheatstone and Kelvin, can literally be counted upon one's fingers. Nor is it very different with the Steam Engine or with the Railway itself, which, to read some of the newspapers of to-day, one would almost think had been invented by the rank and file of the railway workers.

Most really scientific workers feel that knowledge for knowledge's sake is a sufficiently worthy object for pursuit, and are content with the extension of knowledge and the satisfaction that it brings without immediately desiring precise information as to the practical results that are likely to follow from any particular line of investigation. It is well that this is so, as otherwise many of the lines of scientific research that have been most fruitful in bringing lasting benefits to mankind, would never have been begun or followed up.

As an instance of this, could there be a better example than the history of that most remarkable and important discovery in physics which was the primary cause of the foundation of this Society?

Consider for a moment the position of affairs many years ago, when Sir William Crookes first commenced his laborious experiments on the electric discharge through rarified gases. Could anything be imagined of more purely academic interest, and, at the time, seemingly less likely to lead to results of a practical nature? The small scale—the extreme delicacy of the apparatus, the uncertainty of the results, the minuteness of the forces involved, all tended to give to the investigations an air so entirely aloof from the practical concerns of everyday life, that one can scarcely wonder that for years it was only a few of the very foremost scientific intellects who had sufficient insight to take much interest in the matter.

Yet we all know how things have turned out ; how, as a direct result from these very recondite investigations we have had the discovery of the Röntgen rays with their practical applications to the investigation of the human frame, to the relief of suffering and to the cure of disease ; and also, as another result, perhaps the most momentous and far-reaching upheaval in scientific thought on the constitution of matter and the nature of electricity, that has taken place for centuries, heralding the birth of a new idea—that of radio-activity—which may in future be destined to prove the salvation of the whole human race from annihilation. Here, of course, I refer to the vast and previously unsuspected source of energy that modern investigations have shown to lie hidden away in the atoms of matter, a store which is revealed by the energy given out by radium and other radio-active substances, and one which we may hope to see made available for human use in centuries to come, when others, such as those contained in the coal and oil of the earth, at present being exploited, are exhausted.

So far as I am aware, the results of modern discovery have had no effect in weakening our belief in the truth of the great principle of the conservation of energy as defined in what is commonly called the first law of thermo-dynamics, which law is really a statement that the sum of energy in the universe, just as the amount of matter, is a constant, and cannot be either increased or diminished by any means whatever.

When, however, we come to the so-called second law, which, as stated by Clausius, is that it is impossible for a self-acting machine, unaided by external agency, to convey heat from one body to another at a higher temperature, or, as given by Lord Kelvin in a somewhat different form, that it is impossible by means of inanimate material agency to derive mechanical effect from any portion of matter by cooling it below the temperature of the coldest of the surrounding objects, we find that even the authors of these statements are prepared to admit that this second law stands on a totally different basis from the first law, and, as declared by Maxwell, can only be said to be statistically correct, or correct only when we are dealing with masses of matter and not with individual molecules.

Indeed, it was in this connection that Maxwell propounded his celebrated proposition, in which he supposes that a demon who could see individual molecules, and was possessed of super-human dexterity, could open and close an aperture in a partition dividing a vessel into two separate portions, A and B, so as to allow only the swifter molecules to pass from A to B, and only the slower ones from B to A, in which case, without the expenditure of work, the temperature of B would be raised and that of A lowered in contradiction to this second law.

It will further be observed that in the definitions quoted above, Clausius is careful to qualify his statement by words to the effect that there must be no external aid, while Lord Kelvin is even still more specific, and expressly limits the whole law to things inanimate.

Now lately, in London, we have had — I suppose as the result of Professor Bergson's remarkable writings—one of the usual periodic outbreaks of more or less meta-physical discussion, and, incidentally, there has been raised quite seriously the question as to whether living organisms are subject to the laws of thermo-dynamics or not.

So far as the first law is concerned, there seems to be complete agreement that there can be no question of any but an affirmative reply, as we can scarcely suppose living things to be capable of creating energy any more than of creating matter.

But when we come to the second law, there are apparently those who hold that it is different ; who, in fact, believe that there is good reason for doubting whether this second law, which prevents us by the use of any mere machine from getting mechanical effect from the general stock of heat, applies to living organisms at all. Indeed, on the contrary, it is contended that it is probable that in the case of certain animate bodies this is actually being accomplished all the time ; in other words, that there exist living things which in some fashion or other do very much what was the business of Maxwell's demons to do, and in this manner extract the energy that they require from the general stock. Here, obviously, is a most important matter for investigation, and one, having regard to its combined physiological and physical aspects, peculiarly adapted to be tackled by the members of the Röntgen Society, that is to say if those who put it forward can make out a sufficient case to make actual experiment worth while.

The interest of the question will be especially apparent to anyone who has seen the so-called Brownian movements which can be perceived by ultra-microscopic methods in finely divided solid matter, such as particles of colloidal gold suspended in a liquid or of tobacco smoke in a gas. These movements are now believed to be due to the actual jostling of the minute particles by the moving molecules themselves, and give the most wonderful notion of the ceaseless state of agitation that exists among the molecules of all substances at any temperature above that of the ultimate zero, and the vast amount of energy that is stored in these perpetual movements.

Indeed, so remarkable do the Brownian movements appear, that their original discoverer, who detected them in finely divided vegetable matter, came to the conclusion that the particles were alive.

To return to the main question, however, without presuming to pronounce any opinion one way or another on the very startling idea that living matter is not always subject to the second law of thermodynamics, but finds means, in some cases at all events, to evade its provisions, I desire to draw attention to the stupendous consequences that would follow could such a view be established. Here at last we should have the equivalent of the perpetually burning lamp of the story books, which consumed no oil; the perpetual fire of the burning bush, which required no fuel. We should have immediately to hand the means of producing the perpetual motion dreamed of by mediæval philosophers. We should only have to cultivate the right kind of organisms in sufficient masses, and they would do all this for us. Moreover, there would be nothing lost ; the heat that was thus accumulated locally for our needs would dissipate itself again into the common store, as would also the mechanical effects after they had done their work. The unordered molecular motions of which the Brownian movements give us an indication—motions which constitute heat — would merely be directed for a time in the particular manner needful to give us the power that we require. Life would be the directing force, but it would be a directing force only and would do no work.

It is a fascinating prospect, giving us a glimpse of what some may perhaps think is destined to take the place of fuel a few hundred years hence, when the latter is all exhausted and before means have been found to unlock the still greater stores of atomic energy that have already been alluded to. To those, however, who have been brought up to rely on the orthodox doctrines of thermo-dynamics, it seems not only very revolutionary, but also very heterodox from a physical standpoint. Personally, as one totally ignorant of biology, I am only here concerned to point out the inevitable consequence of admitting that living matter is not subject to the second thermo-dynamic law, a proposition which I venture to believe has never before now been put forward seriously in any responsible quarters.

I now propose to pass on to a subject which, though not at first sight very germane to the matters usually discussed before this Society, may, I think, prove of interest, for the reason that it supposes an entirely new application of the Crookes tube and the phenomena of Cathode Rays and fluorescence with which we are all so familiar.

Among the many scientific problems that await solution, problems which, if satisfactorily solved, would have an enormous effect on the habits of mankind, is that of distant electric vision, or the power to see objects a great way off by electrical means; in other words, to do for the sense of sight what the telephone has done for the sense of hearing. Indeed, if this extension of our sense of vision was obtained, we could well afford to dispense with any extension of our other senses, namely, those of taste, smell and touch, the senses of sight and of hearing being, for all ordinary purposes, much more important to us than are the others. The difficulties that many people, especially the inexperienced, find in carrying on satisfactory conversations over the telephone, are undoubtedly due to the fact that in ordinary conversation one is accustomed not only to listen to the sounds that proceed from the speaker, but to watch his face and gestures. Consequently, if there could be added to each telephone instrument what would indeed be a magic mirror, in which we could see even only in monochrome the faces of those with whom we were communicating, the material advantages would be very great. In addition, there would be much sentimental and other value.

The problem of distant electric vision is, however, an exceedingly difficult one, much more so than that of the telephone; for while, as we know from both the telephone and phonograph, the most diverse sounds that go to make up every description of music and articulate speech are capable of being resolved into a single more or less complicated curve, the principle involved being, in fact, the converse of the well-known mathematical conception known as Fourier's Theorem, we have, in the case of television, as we may call it, the additional factor of form or position; indeed, we can consider the transmission of diverse sounds by the telephone as analagous merely to the transmission of diverse colours by our television

apparatus. But while, in the case of the telephone, when we have transmitted the sounds we have finished what we have to do, in the case of television not only have the colours to be transmitted, or, in the case of monochrome television, the lights and shades, but these must, in addition, be distributed on a surface in their proper positions, so that the form of what we are looking at may be apparent to our eyes.

Now, in considering in what way the problem can be solved, we naturally look around to find out whether we can learn anything in the matter from nature, and just as Reis and Graham Bell, in considering how to make a telephone, received much help by taking as a model the human ear, from which, in fact, they adopted the expedient of the diaphragm, so we naturally turn our attention to the structure of the human eye.

This organ, as everybody knows, consists of a camera obscura containing a lens whereby the image of what is looked at is thrown upon the retina, just as in a photographic camera the image is thrown on the ground-glass screen or on the sensitive plate. The surface of the retina is connected with the brain through the optic nerve by means of a very large number of threads or fibres, each of which connects with a certain definite point on the retina, and thus, under the stimulus occasioned by light falling upon it, communicates to the brain, in mosaic form, a conception of the different portions of the image.

Here at once is suggested a method of transmitting the image by means of a very large number of electric wires.

As is well known to everyone who has had to do with the so-called process blocks employed for the modern printing of illustrations, it is possible to represent pictures of every kind by a sufficiently large number of sufficiently small black dots on a white surface. When looked at from a distance at which the eye cannot distinguish the dots, the picture presents itself in a satisfactory manner to the observer. Furthermore, white dots on a black surface will do just as well if properly disposed. Imagine, then, that for our television receiver we have a black surface entirely covered by very minute electric lamps, each connected by a separate wire to the distant transmitter. The latter may consist of another surface entirely covered with small silenium cells, the electrical resistance of which is reduced if light falls upon them. Now, imagine that upon the surface composed of all these cells the image that is wished to be transmitted is thrown by a lens, just as the image is thrown by the lens of the eye upon the retina. Certain of the silenium cells on which a bright light falls will have their resistance lessened to a great extent. Others on which the incumbent light is less strong will have their resistance reduced in a less degree; while those in the dark portions of the image will not be affected at all. Let us suppose, then, that these cells are connected by means of the numerous wires we have imagined each individually to one of the lamps forming the surface of the receiver, care being taken that each cell of the transmitter is connected to the correspondingly placed lamp of the receiver. Let us suppose, further, that the whole of these very numerous electric circuits are completed by a wire or by connection at both ends to earth, with a battery of suitable size interpolated. Then, with proper adjustment, we can arrive at the condition whereby the lamps which are connected with the silenium cells on which the bright portions of the image are thrown will light brightly, those lamps that are connected with the cells on which the less brightly illuminated portions of the image fall will only reach semi-incandescence, while the remainder, connected with the cells in the dark portions of the image, will not light at all. Thus, we will obtain on the surface of the receiver a fairly exact reproduction of the image thrown on the transmitter, a very granular mosaic it is true, but just as fine grained

as we like to make it by multiplying the number of the lamps and wires, and looking at it from a sufficiently great distance.

The idea is old and seems very simple, and it is only when we come to consider the fearful number of silenium cells at the one end and lamps at the other, and wires in between, to get any practically useful result, that we realise that the mere complication and cost of such an arrangement, if nothing else, renders it altogether impracticable.

For instance, even if we go as far as only employing 10,000 each of cells, lamps, and transmitting wires, it is obvious that our image would only be made up of that number of bright or dark spaces, and that there would only be one hundred of them in a row to each side of the square of the luminous surface of our receiver. This would give an image of such very coarse grain in the case of an image of any useful dimensions as to be of very little value.

The late Mr. Shelford Bidwell, who gave a great deal of attention to this subject, calculated that to give good close-grained results on a surface two inches square would require 150,000 wires with their corresponding sets of transmitting and receiving apparatus at each end, while for very coarse results, about equal to the effect that one gets in the very coarsest process-block reproductions in the newspapers, about one-tenth of this number would suffice. Taking a medium number of 90,000 wires, he calculated that the whole apparatus and cables for a one-hundred mile transmission would roughly cost £1,250,000; or if the apparatus were to be triplicated in order to give coloured results by the three-colour process, the cost would be three times that amount.

Obviously, figures such as these show that any method such as is described above is commercially out of the question, and it has therefore been the aim of a number of inventors during the past twenty years to devise some arrangement whereby the number of wires and the complexity of the whole apparatus could be reduced. One apparent method whereby the large number of wires could be done away with is by employing two synchronised revolving distributors, one at each end of the transmission line, which would simultaneously connect together each one of the transmitting cells and receiving lamps in turn through a single wire, the whole number being connected one after the other within the space of the one-tenth of a second, which is the duration of human visual persistence. Here again, when the matter is looked into, the impracticability of the method becomes apparent, as no mechanical apparatus, whether with revolving or with oscillating material parts that we can imagine, could be expected to execute with sufficient accuracy even 150,000 synchronised operations per second, which, on the basis of the above figures, is the very minimum that we could do with to obtain satisfactory results.

Some years ago it occurred to me whether one could not arrive at some better solution of the problem by the employment of an agency which, at any rate, will be of interest to members of this Society. As we know from the ingenious oscillograph, invented a few years ago by Braun, in which the curve of an alternating or oscillating electric current is delineated on a phosphorescent screen in a vacuum tube, by the effect on the screen of the impact of a thin pencil of cathode rays which is deflected by a magnetic field produced by the current or oscillation in question, cathode rays, owing to their almost imperceptible momentum, can be made to move with a rapidity and accuracy that could not be expected from more material objects.

As long ago as the year 1908, in connection with a paper published in " Nature " by the late Mr. Shelford Bidwell, from which some of the figures I have given above are taken, I wrote a letter in that journal suggesting that this difficulty of obtaining enormous numbers of synchronised operations per second could possibly be solved by the employment of two beams of cathode rays, one at the transmitting and one at the receiving station, synchronously deflected by the varying fields of two electro-magnets placed at right angles to one another and energised by two alternating electric currents of widely different frequencies, so that the moving extremities of the two beams would be caused to sweep synchronously over the whole of the required surfaces within the one-tenth of a second necessary to take advantage of visual persistence, and that so far as the receiving apparatus was concerned, the moving cathode beam would only have to be arranged to impinge on a sufficiently sensitive fluorescent screen, and given suitable variations in its intensity, to obtain the desired result. As since that date I have several times been asked to explain more in detail this idea, I now propose to do so, though it must be distinctly understood that my plan is an idea only, and that the apparatus has never been constructed. Furthermore, I would explain that I do not for a moment suppose it could be got to work without a great deal of experiment and probably much modification. It is, indeed, only an effort of my imagination, and can be useful merely as a suggestion of a direction in which experiment might possibly secure what is wanted. What, however, is claimed is, that so far as I am aware, it is the first suggested solution of the problem of distant electric vision in which the difficulty of securing the required extreme rapidity and accuracy of motion of the parts is got over by employing for these parts things of the extreme tenuity and weightlessness of cathode rays. Indeed, apart from the revolving armatures of the alternators employed for synchronisation, which present no difficulty, there is no more material moving part in the suggested apparatus than these almost immaterial streams of negative electrons. Furthermore, as will be seen, only four wires, or three wires and earth connections at each end, are required.

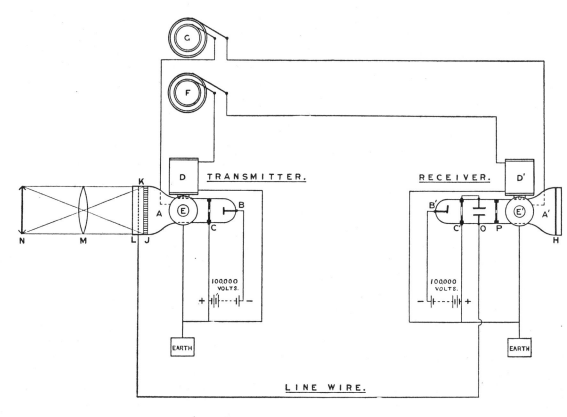

In the diagrammatic illustration the transmitter is shown on the left-hand side and the receiver on the right-hand side. The transmitter consists of a Crookes tube A fitted with a cathode B which sends a cathode ray discharge through a small aperture in the anode C, the cathode rays being produced by a battery or other source of continuous electric current giving some 100,000 volts. D and E are two electro-magnets placed at right angles to one another, which, when energised by alternating current, will deflect the cathode rays in a vertical and in a horizontal direction respectively.

The receiving apparatus consists similarly of a Crookes tube A′ fitted with a cathode B′, which, under circumstances to be further described, transmits cathode rays through an aperture in the anode C′. D′ and E′ are two electro-magnets placed at right angles, similiar to those in the transmitter, the two magnets D and D′ which control the vertical motions of the cathode ray beam being energised from the same alternating dynamo, F, which has a frequency say of 10 complete alternations per second, while the other two magnets E and E′ which control the horizontal movements of the cathode ray beam, are energised by a second alternating dynamo G having a frequency of say 1,000 complete alternations per second.

In the receiver, H is a fluorescent screen upon which, under conditions to be further described, the cathode rays impinge, and the whole surface of which they search out every tenth of a second under the combined deflecting influence of the two magnets D′ and E′, with the result that under these conditions the screen fluoresces with what appears to the eye as a uniform brilliancy.

Similarly, in the transmitting apparatus, the cathode rays fall on a screen J, the whole surface of which they search out every tenth of a second under the influence of the magnets D and E.   Further, it is to be remarked that as the two magnets D and D' and the two magnets E and E' are energised by the same currents, the movements of the two beams of cathode rays will be exactly synchronous and the cathode rays will always fall on the two screens H and J on each corresponding spot simultaneously.

In the transmitter, the screen J, which is gas-tight, is formed of a number of small metallic cubes insulated from one another, but presenting a clean metallic surface to the cathode rays on the one side, and to a suitable gas or vapour, say sodium vapour, on the other.   The metallic cubes which compose J are made of some metal, such as rubidium, which is strongly active photo-electrically in readily discharging negative electricity under the influence of light, while the receptacle K is filled with a gas or vapour, such as sodium vapour, which conducts negative electricity more readily under the influence of light than in the dark.

Parallel to the screen J is another screen of metallic gauze L, and the image to be transmitted of the object N is projected by the lens M through the gauze screen L on to the screen J through the vapour contained in K.   The gauze screen L of the transmitter is connected through the line wire to a metallic plate O in the receiver, past which the cathode rays have to pass.   There is, further, a diaphragm P · fitted with an aperture in such a position as, having regard to the inclined position of B', to cut off the cathode rays coming from the latter and prevent them from reaching the screen H unless they are slightly repelled from the plate O, when they are able to pass through the aperture.

The whole apparatus is designed to function as follows :—

Assume a uniform beam of cathode rays to be passing in the Crookes tubes A and A', and the magnets D and E and D' and E' to be energised with alternating current, as

mentioned.   Assume, further, that the image that is desired to be transmitted is strongly projected by the lens M through the gauze screen L on to the screen J.   Then, as the cathode rays in A oscillate and search out the surface of J they will impart a negative charge in turn to all the metallic cubes of which J is composed.   In the case of cubes on which no light is projected, nothing further will happen, the charge dissipating itself in the tube ; but in the case of such of those cubes as are brightly illuminated by the projected image, the negative charge imparted to them by the cathode rays will pass away through the ionized gas along the line of the illuminating beam of light until it reaches the screen L, whence the charge will travel by means of the line wire to the plate O of the receiver.   This plate will thereby be charged ; will slightly repel the cathode rays in the receiver ; will enable these rays to pass through the diaphragm P, and, impinging on the fluorescent screen H, will make a spot of light.   This will occur in the case of each metallic cube of the screen J which is illuminated, while each bright spot on the screen H will have relatively exactly the same position as that of the illuminated cube of J.   Consequently, as the cathode ray beam in the transmitter passes over in turn each of the metallic cubes of the screen J, it will indicate by a corresponding bright spot on H whether the cube in J is or is not illuminated, with the result that H, within one-tenth of a second, will be covered with a number of luminous spots exactly corresponding to the luminous image thrown on J by the lens M, to the extent that this image can be reconstructed in a mosaic fashion. By making the beams of cathode rays very thin, by employing a very large number of very small metallic cubes in the screen J, and by employing a very high rate of alternation in the dynamo G, it is obvious that the luminous spots on H of which the image is constituted can be made very small and numerous, with the result that the more these conditions are observed the more distinct and accurate will be the received image.

Furthermore, it is obvious that, by employing for the fluorescent material on the screen H something that has some degree of persistency in its fluorescence, it will be possible to reduce the rate at which the synchronised motions and impulses need take place, though this will only be attained at the expense of being able to follow rapid movements in the image that is being transmitted.

It is further to be noted that as each of the metallic cubes in the screen J acts as an independent photo-electric cell, and is only called upon to act once in a tenth of a second, the arrangement has obvious advantages over other arrangements that have been suggested, in which a single photo-electric cell is called upon to produce the many thousands of separate impulses that are required to be transmitted through the line wire per second, a condition which no known form of photo-electric cell will admit of.

Again, it may be pointed out that sluggishness on the part of the metallic cubes in J or of the vapour in K in acting photo-electrically, in no wise interferes with the correct transmission and reproduction of the image, provided all portions of the image are at rest; and it is only to the extent that portions of the image may be in motion that such sluggishness can have any prejudicial effect. In fact, sluggishness will only cause changes in the image to appear gradually instead of instantaneously.

Many modifications are of course possible in detail. For instance, the plate O of the receiver might perhaps better be replaced by an electro-magnet or solenoid so arranged as to repel the cathode beam when energised. Again, the somewhat crude form of photo-electric cell described, composed merely of insulated cubes of rubidium in contact with sodium vapour, might be improved upon.

Indeed, it is highly probable that research will reveal much more sensitive materials, the use of which would vastly improve this part of the apparatus, which, at present, is probably the one least likely to give the desired results.

Such is, however, the general idea of the scheme, which may appear complicated at first sight, but is, I believe, less so than any that has previously been suggested for the same purpose. Moreover, it is a mistake to believe that complicated apparatus cannot be successful.

Who, for instance, would have believed twenty years ago that to-day we should all be running about on vehicles which carry upon them as part of their propulsive mechanism not only an engine quite as complicated as a steam engine, but in addition a gas works and an electric generating station? Yet the modern motor car embodies all of these, as does also the modern aeroplane, while no ordinary coachman who is unable to work and keep in order all this apparatus can now easily get a situation.

My object has not, however, been to show you a really worked-out scheme for distant electric vision. So far as I am aware, no such worked-out scheme as yet exists, and my desire has been merely to put on record certain ideas that have occurred to my imagination in the hope that they may lead others to the invention of a more practicable method of arriving at what is wanted.

In conclusion, if it is objected that ideas of this kind should not be presented naked and unashamed in this manner, but should be kept to one's self till they can be put forward respectably clothed in practically worked-out garb, I can quote the example of our Honorary Member, Sir William Crookes, who, some four years prior to the first of Signor Marconi's patents, and two years previous even to Sir Oliver Lodge's celebrated Royal Institution lecture, gave in the " Fortnightly Review " a complete idea of how wireless telegraphy might be carried on over great distances by means of the Hertzian waves.

After Mr. Campbell Swinton had concluded his address from the Chair,

Professor SILVANUS P. THOMPSON, F.R.S., said: Mr. President and Gentlemen, in response to the call of your Honorary Secretary, I rise to move a vote of thanks to you, sir, for your address from the chair. It is a remarkable address to which we have listened, and we must congratulate ourselves upon the fact that we have been a privileged audience. Had other people known what was in store for them our numbers would have been larger, but this evening our President has had to contend with other attractions, even with other Presidential Addresses, which I know have drawn some away who would otherwise have been present. They do not know what they have missed. It would seem to be rather an advantage—to an audience at least—when a President-elect does not know what to discourse about in his Presidential Address; for otherwise we might expect a commonplace deliverance. But torn, as our President has been, by distracting suggestions, we find him able to maintain a kind of polar equilibrium, and the result is that we have a Presidential Address at its best. The address has contained a little bit of history, some scientific criticism, and a certain amount of prophecy; though Mr. Campbell Swinton has very modestly kept in the background those particular scientific researches with which his name has been for so long associated, and which have made him sufficiently famous to be elected President of this Society. No part of the address was more interesting than the bit of ingenious and highly speculative prophecy. We need not go back to Prometheus for the suggestion that scientific discoveries have come down from heaven. The ideas of scientific men, if they have represented any

advance at all, have always come down from heaven, in the same sense as the Promethean fire. There has always been inspiration behind them. The men of inspiration have seen that which was beyond the vision of others; and even though they may not have embodied it in any actual material form, they have realised it sufficiently to paint a word-picture, as our President has done in his suggestion with regard to a future transmitter of visible electrons. I think that the portion of Mr. Swinton's address which deals with the neglected importance of scientific discovery well deserves to be widely circulated. I wish that members of the Press had been present to listen to the remarks he made upon their shortcomings as journalists—journalists who beat the drum of the politician so loudly and largely ignore the enduring importance of science. But although we shall not see in to-morrow's *Daily X* or *Daily Y* the flying head-lines of the reporters concerning the utterances of our President, we shall carry away a very vivid recollection of that portion of his deliverance.

Mr. Swinton has touched upon one of the burning questions of to-day—one which is perhaps not directly in a line with the work of this Society, but one which was, nevertheless, quite fittingly introduced because it interests all who have any place in the borderland between physics and physiology. The question is of great interest, not because of the new chapter which is being opened up before us by a school of French thought, but because the problem is one of those which have never been plumbed to the bottom. It is a problem which brings into the consideration the two laws of thermo-dynamics. The first law—that of equivalence of work and heat—stands on a somewhat different footing from the second law—that of transformation. The first law is fundamental, unchallengeable: but the second law expresses a matter of experience, believed to be true for all cases where fluid matter in bulk is the seat of the transformation, and generally supposed to be true also in actions on the molecular scale. But it has never been rigidly proved to dominate all molecular actions ; hence, as stated both by Clausius and by Kelvin, it has been hedged round by provisos to limit the statement of it to inanimate agency. I have more than once talked on this subject with the late Lord Kelvin, and it turned up in the course of the last conversation I had with him in his own home. The second law of thermo-dynamics is more generally understood in the form in which it appears in engineering treatises. Briefly, it gives the answer to this question : What fraction of the heat that enters a steam engine is utilised by being transformed into work ? The answer is that the value of this fraction depends upon the ratio between the fall of the temperature in the engine and the absolute temperature of the source of supply. If we work with a certain temperature of source—that is to say, with a certain boiler temperature—and a certain temperature of refrigerator—that is to say, of the condenser or exhaust—we have at best a few dozen degrees between them, and the result of dividing these degrees by the larger figure gives us the utilizable fraction. The fraction of the energy that is available is simply the difference between the two temperatures—that of the boiler and that of the condenser — divided by the absolute temperature of the higher, namely, that of the boiler. The well-known formula express-

ing this is $\dfrac{T'-T}{T'}$ (T' and T being the respective absolute temperatures of boiler and condenser, and the result the proportion of the supplied heat which can be transformed into work by a perfect engine). It was once expressed thus by James Napier in some verses addressed to Lord Kelvin :—

> When you yourself once taught me
> Heat's greatest work to know,
> Wasn't it T dash minus T,
> With a T dash down below ?—

which has the merit of being poetic and at the same time a strictly scientific presentment of the second law of thermo-dynamics.

Lord Kelvin, however, saw that it was not possible to account for the operations that go on in the human machine on the supposition that a man is a thermo-dynamic engine, by comparing the amount of his mechanical output with the amount of fuel he stokes in. The simple fact is that in the case of the human machine the boiler temperature is not the boiling point and the condenser temperature is nowhere near freezing point. The two highest and lowest temperatures of the blood are very near together—so near that the ratio of the difference of temperature to the absolute temperature of supply does not represent one-thousandth part. The second law of thermo-dynamics, therefore, appears not to be applicable to the case of operations commonly called 'vital.' If man, then, is not a thermo-dynamic engine, what is he ? Since the second law of thermo-dynamics, at any rate in its mechanical form, does not apply to the vital mechanism of animals, we are left to suppose that there is some other law of thermo-dynamics pertaining to the vital operations of physiology. The thermo-dynamics of directed molecular operations has yet to be discovered. Is there a possibility of obtaining light energy and heat energy by concentrating it from the diffused energy of the universe through some process not included in the second law of thermo-dynamics ? That, at least, was not absent from Lord Kelvin's mind. A couple of years or so after the discovery of radium, the late Mons. Curie observed the singular fact that radium, left to itself, remained always a few degrees warmer than anything around it. Lord Kelvin persistently opposed the views of Rutherford and Soddy, who accounted for radio-activity by the disintegration of the atom, and until 1904, when he was constrained to abandon it, held the view that there was a mechanism unknown to us by means of which this curious material managed to concentrate the diffuse energies of the universe and to draw from them something which made it hotter than its surroundings. The atomic disintegration theory of Rutherford and others, holds the field to-day, but I do not think that it necessarily negatives the view held by Lord Kelvin as to the possibility of some substance being able to concentrate diffuse energy in itself by some process outside that second law. Our President's remarks on this part of his subject were valuable.

Lastly, we have his prediction or suggestion, which he says he has not yet worked out. It is a most interesting, beautiful and ingenious speculation, and one which is calculated to stimulate thought in many directions. The President has already anticipated my personal criticism—that it would be too complicated to work — by remarking that some things which have been worked out are far more complicated. The fact is that the great discoveries are not complicated—they never are—although it may be necessary to arrive at them by complicated means The complication is very often a step, a bridge, to attain the desired end, but when that end is attained it is found to be simple. That must not frighten us, however, from using apparently complicated things in order to arrive at the root of the matter, and such a means the President has indicated to us this evening.

THE HON. SECRETARY (Dr. Low) said that such a motion needed no formal seconding, and he put it to the meeting, when it was carried by acclamation.

THE PRESIDENT, in reply. said : I am exceedingly obliged to you for the kind way in which you have received my address and more than obliged to Professor Silvanus Thompson, one of the founders of our Society, for his remarks with regard to it. I thought that he would probably criticise the little apparatus that I brought forward, for I remember criticisms of a similar character

from his pen which appeared in the "*Times*" some years ago upon the system of another inventor. But it is a matter I have had in mind for a number of years, and it seemed to me a pity to keep it bottled up for myself when there was a possibility that its publication might stimulate someone else to bring the idea to a practical issue.

TELEVISION BY ELECTRON IMAGE SCANNING

Philo Taylor Farnsworth

TELEVISION BY ELECTRON IMAGE SCANNING.

BY

PHILO TAYLOR FARNSWORTH.

A television system is described, in which the complete optical image to be scanned is focused upon a sensitive photoelectric cathode surface, thus producing a corresponding "electron image" of photoelectrons at the surface. This electron image is brought to a sharp focus in a plane parallel to the cathode surface by means of uniform magnetic and electric fields, perpendicular to the surface. Scanning is accomplished by displacing the focused electron image with two mutually perpendicular magnetic fields (which are themselves perpendicular to the focusing field), and thus sweeping the electron image across a fixed aperture.

The current through the coils which produces the horizontal displacement of the electron image, has a saw-tooth wave form of constant frequency (line frequency) which is in the range of 5,000 to 10,000 repetitions per second. The current through the coil producing the vertical displacement has a saw-tooth wave-form of constant frequency (image frequency) which may be 16 to 30 repetitions per second. Means for producing these saw-tooth currents and means for synchronizing them between the "Image Dissector," or transmitting tube, and the "Oscillight," or receiving tube, are discussed.

A new method of amplifying very feeble electric currents by making use of secondary electron emission is described. This method of "Secondary Electron Multiplication" makes use of cold cathodes which have a high ratio (6 or greater) of secondary electrons to primary impacting electrons and uses this effect to produce an increase in an electrical current. Means are described for making use of the principle of secondary electron multiplication for producing amplification of photoelectric currents (amplification greater than 5×10^7), and of feeble directed electron streams (amplification greater than 10^3), and for the production of radio frequency oscillations (20 megacycles or greater).

The image dissector tube used without an electron multiplier requires a light intensity of 15 to 25 lumens to bring the signal impulse above the noise level inherent in the amplifier circuit and is thus suitable for use as a pick-up from motion picture film. By introducing an electron multiplier tube within the structure of the image dissector tube, placing it behind the image receiving aperture, the currents within the tube are amplified more than 10^3 times without appreciably increasing the noise level of the tube, and the tube is made sufficiently sensitive to pick up the image from a directly illuminated subject or from outdoor scenes under moderate daylight conditions.

The receiving tube or "Oscillight" consists of a cathode ray tube whose gun delivers a beam of 1 to 15 milli-amps. within an angle of 15 degrees. This beam is focused on the fluorescent screen by means of a short magnetic solenoid around the neck of the tube near the gun and coaxial with it. The horizontal (line frequency) scanning is produced by magnetic deflection coils which are at right angles to the axis of the electron gun and carry currents of the same waveform and frequency as the corresponding current in the image dissector tube. The low frequency, or picture frequency, scanning is produced by a magnetic field normal to the line frequency displacement field. This low frequency field makes use of core pole pieces of special shape.

1. ELECTRON IMAGE—DEFINITION OF THE TERM.

The fundamental idea underlying electron image scanning [1] is to receive from the optical image an electronic discharge in which each portion of the cross-section of such electronic discharge will correspond in intensity with the intensity of the light incident on that portion of the sensitive plate from which the electronic discharge originated. Then, if the electron image is bombarded against a fluorescent screen, the optical image will be reproduced there. Such a discharge is termed an "electron image."

Scanning of the electron image is accomplished by deflecting the entire electron discharge in such a manner that each elementary portion is successively picked up by a minute aperture in a suitable collector.

2. CONVERSION OF OPTICAL IMAGE TO AN ELECTRON IMAGE.

An optical image focused upon a uniform photoelectric surface produces an electron emission from each point proportional to the intensity of the light at that point. If all of the electrons emitted from the surface are drawn away from it, such a discharge is classifiable as an electron image. The image, however, is sharp only in the plane of the cathode.

In order that the electron image may be scanned by deflecting it over a small aperture, the correspondence to the optical image must be present at a suitable distance from the cathode. There are several possible methods for fulfilling this requirement. Referring to Fig. 1, if all of the electrons leaving point "P" on a cathode, could be constrained to travel parallel to one another, then any transverse section of the beam would represent a perfect electron image. For all electron paths to be parallel there must be no component of transverse velocity. The photoelectrons, however, are emitted with random velocities, ranging from zero to a volt or more. Another

---

[1] See Farnsworth's U. S. Patent No. 1,773,980.

FIG. 1.

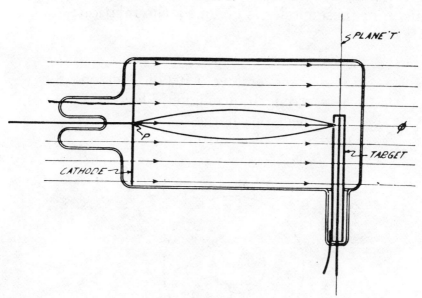

Envelope of focused electron paths from one point on the image in an image dissector tube.

component of transverse velocity is added by the non-uniformity of the electrostatic lines of force, and another large component by the microscopic irregularity of the cathode surface. Thus, the photoelectrons leave each point on the photoelectric surface as a divergent cone. We can secure something of an electron image in the plane of the anode target T, by the use of low frequency light, careful construction of the cathode of the tube, and the use of high potential between the cathode and the anode. The dissector tube, in fact, was first used in this manner.

3. MAGNETIC FOCUSING AND DEFLECTION.

In the practical form of image dissector tube now in use, the electron image is focused magnetically. To accomplish this, a uniform magnetic field of exactly the proper intensity, ϕ, is adjusted so that the lines of force are parallel to the axis of the tube. The effect of this magnetic field is to bend the electrons into helical paths, tangent to the line of magnetic force through the emitting point. Neglecting, for the present, the longitudinal velocity (see Fig. 2), and considering the helices as circles in the plane of the point "P," we see that we have a family of circles, all tangent to the point "*P*." The radius, *r*, of these circles is given by the relation:

$$r = mv/\phi e, \qquad (1)$$

where

$v =$ transverse velocity of electron,
$\phi =$ strength of magnetic field,
$e =$ electron charge,
$m =$ mass of electron.

Fig. 2.

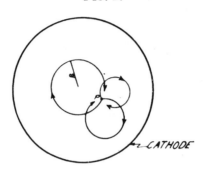

Projection of helical electron paths in the plane of the cathode.

The angular velocity ω, with which an electron travels around the circle is given by:

$$\omega = v/r. \qquad (2)$$

Substituting this value of r in equation 1, we have

$$\omega = \phi e/m; \qquad (3)$$

thus, it will be seen that the angular velocity is independent of the value of the transverse velocity of the electron. Therefore, since all of the electrons start at the same point, they will arrive back at the same point, when $\omega t = 2\pi$, that is, when they have traversed the circle once, or for any value of magnetic field intensity which gives an integral number of revolutions. In practice, the first focus is used.

The axes of the helices need not be within the electron pencil; if the longitudinal velocity is the same for all, every electron, no matter how divergent, will eventually return to intersect a line of magnetic force passing through its point of origin. It follows, therefore, that if the direction of the field be changed, the point at which the rays are focused will follow the field.   If, therefore, we superimpose a transverse magnetic field on the longitudinal field, the electrons will be deflected to follow the resultant field, as shown in Fig. 3.

FIG. 3.

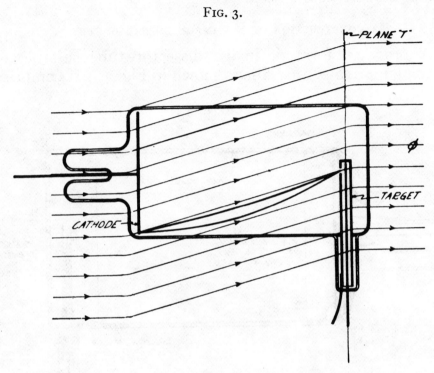

Envelope of focused electron paths when a transverse component is added to the initial magnetic focusing field of the image dissector tube.

It will be seen, therefore, that a pair of coils placed one on each side of the dissector tube, and carrying an alternating current will give a deflection along the axis of the coils, that is, in the direction of the magnetic lines produced by these coils, rather than at right angles to this direction, as would be the case if there were no longitudinal or focusing field.

4. RESOLUTION OBTAINABLE IN THE ELECTRON IMAGE.

The degree of sharpness of an electron image focused magnetically is inversely proportional to the angle of the emergent cone filled by the unfocused electrons.   The electron image may be sharper than the optical image produced by the best lens, by sufficiently limiting the angle of the emergent cone.   In the dissector tube the electron image is much sharper than required.   This result is achieved by using a cathode that is perfectly smooth (almost polished), and an accelerating potential of approximately 700 volts. The sharpness of the electron image can be directly observed by bombarding it against a fluorescent screen.

5. CONSTRUCTION OF THE IMAGE DISSECTOR TUBE.

A practical form of image dissector tube suitable for scanning motion picture film is shown in Fig. 4. It comprises

FIG. 4.

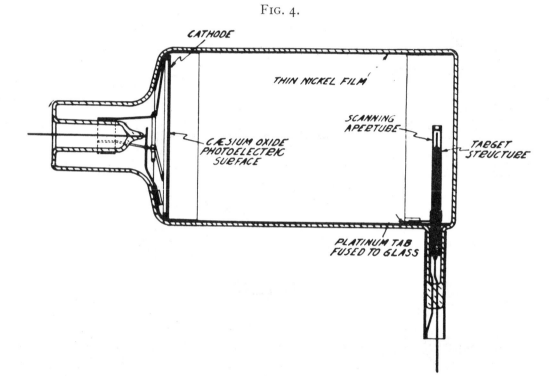

Image dissector tube used for scanning motion picture film.

an evacuated envelope in which is located a photoelectric cathode and a combination anode and target electrode. The walls of the tube are coated with a thin evaporated film of nickel. One end of this film is connected to the cathode and the other end to the anode. The electrical resistance of this metallic film may be anywhere from $\frac{1}{2}$ megohm to several megohms. The cathode is of electrolytically pure silver. The surface must be cleaned and a uniform crystal structure, and preferably having a very fine, uniformly etched finish.

Sensitization of the cathode follows very closely the usual technique [2] for such silver oxide-Cæsium cathodes.

The photoelectric cathode of the dissector tube must be uniformly sensitive over the entire surface, since any variations will appear in the image as dark or bright areas. It has been found that uniformity in color over the surface usually indicates uniformity in sensitivity. To obtain such uniform color requires not only that the silver surface be of uniform texture and thoroughly clean, but also that the oxidization process be carried out uniformly. The color of the entire surface should be the same after it has been properly oxidized. It is also important not to form the surface too rapidly, and no Cæsium must be allowed to condense directly on the cathode.

The photoelectric sensitivity of such tubes has varied in the past between 10 and 30 micro-amps. per lumen. Recently this sensitivity has been increased to 50–65 micro-amps. per lumen.

6. DEFLECTION AND FOCUSING COIL SYSTEM.

In the dissector tube shown in Fig. 4, the electron image is focused by means of the solenoid wound around the outside of the tube. Good focusing requires that the current in this solenoid be adjusted very accurately and, of course, the accuracy of the focus required will depend directly on the size of the scanning aperture which is to be used. The size of the scanning aperture we regularly use is .015 inch, allowing approximately 240 lines, since the cathode surface is 4 inches in diameter. This requires that the focusing current be adjusted to approximately plus or minus 2 per cent. The number of ampere turns used in the focusing coil is, approximately, 500 for a difference in potential between cathode and anode of 700 volts.

Deflection of the electron image for scanning is accomplished by two sets of coils at right angles to the dissector tube, and each pair of coils must be mounted at right angles to each other. These coils are rectangular, designed somewhat on the basis of the coils in the D'Arsonval Galvanometer. A

---

[2] See for example, Prescott and Kelly, *Bell System Technical Journal*, Volume XI, July 1932, p. 334.

square deflection field is obtained by spacing these coils a distance apart, approximately equal to their shortest diameter. It is an advantage also that the coils extend the full length of the dissector tube. The number of turns in the coil may be adjusted to suit the particular generator that is to feed them, but using the generators we have developed, low frequency coils have 5,000 to 10,000 turns per coil. The high frequency coils, on the other hand, have about 65 turns per coil.

In mounting these coils on the tube, they must be mounted so that mutual inductance between pairs at right angles to each other is as low as can be obtained; otherwise, difficulty is experienced with one pair resonating to harmonics of the current in the other pair. This gives a wavy appearance to the scanning field that is very troublesome. To overcome this difficulty, the two sets of coils should have widely different impedances and should be adjusted for minimum mutual inductance as stated before.

7. SENSITIVITY OF THE DISSECTOR TUBE.

The sensitivity of the dissector tube when used with a 15,000 ohm coupling resistor is not adequate for scanning directly from the subject, and if it is used in the form shown in Fig. 4, very intense light is required for illuminating the subject. This type tube is used, therefore, only for motion picture work, and for such use a light source is used which gives 15 to 25 lumens in the optical image on the cathode. With this light intensity the average signal in the output of the dissector target across 15,000 ohms, is approximately 1/10 milli-volt.

When the dissector tube is to be used for direct scanning the electrons selected from the electrical image are first multiplied by a process called "electron multiplication." In this manner, the sensitivity of the tube is increased several thousand times.

8. SECONDARY ELECTRON MULTIPLICATION.

When an electron stream is directed against a surface, it causes the emission of secondary electrons. These secondary electrons may exceed in number, those of the original or primary beam, if the primary electrons have sufficient velocity, This fact may be utilized to amplify an electron flow. Thus, an electron stream which is to be amplified is directed against a surface capable of emitting as many secondary electrons as possible by the impact of the primaries. The secondary electrons themselves are then directed against a similar surface. This process is repeated as many times as desired and the total electron flow finally collected.

9. CONSTANT POTENTIAL TYPE MULTIPLIER.

As an example of how this process may be carried out practically refer to Fig. 6. This represents an electron multiplying photocell in which the photoelectric currents from the upper end of the tube may be multiplied many times by successive secondary emission impacts.

The tube is constructed by sputtering or evaporating on the walls of a tubular bulb, a thin film of platinum or nickel, and making connection at each end of this film by means of a platinum tab fused to the glass. It is worthwhile, in addition, to evaporate a rim of heavier film around each end of the tube to insure good connection to the film. The resistance between the two platinum tabs is made of the order of 0.5 to 2 megohms. On top of this nickel film is evaporated a film of pure silver to a thickness that is nearly opaque. Down the axis of the tube is mounted an anode consisting of a fine wire or loop of wire, and having at one end a metallic collection plate as shown.

The tube is sealed onto the pump, pure oxygen admitted, and the silver surface completely oxidized so that the resistance between the platinum tabs again assumes the approximate value it had before the silver was evaporated onto it. Cæsium is then driven into the tube and the tube formed into a photoelectric cell. This process usually lowers the value of resistance somewhat, and this fact must be taken into consideration in deciding on the amount of nickel to be put on the wall. After the photoelectric surface has been formed, the tube is ready to be sealed off the pump.

In use, a potential of 500 volts or more is applied across the distributed resistance so that a potential gradient longitudinal to the tube results. At the same time the greatest positive potential in the tube is that of the anode. Photoelectrons emitted from the top of the tube are accelerated

FIG. 5.

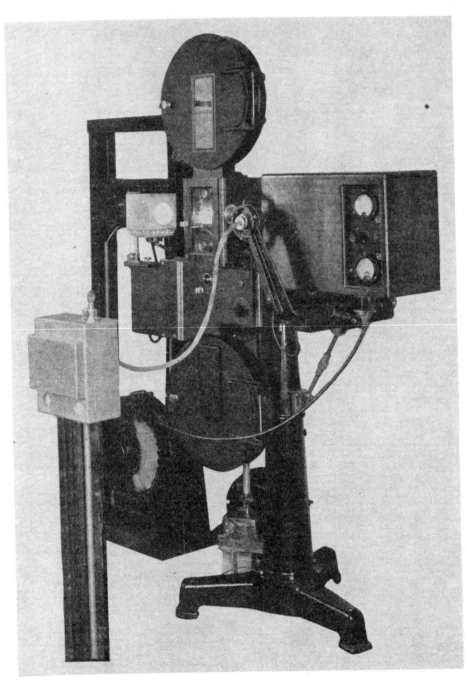

Transmitter unit used for scanning motion picture film. The image dissector tube and head amplifier are in the box on the right which is mounted on the projector unit.   The line amplifier is separately mounted in the box at the left.   The moving picture film is moved continuously past the picture aperture and no picture frequency deflector coil is used on the dissector.   The impulses for synchronizing the picture frequency of the receiving system with the film are obtained from a special impulse generator connected to the film driving sprocket.

toward the anode but due to the gradient longitudinal of the tube they also fall down the tube by a definite amount and strike the wall opposite which they were emitted with a velocity that is proportional to the potential gradient through which they have fallen. This gradient may be anywhere

FIG. 6.

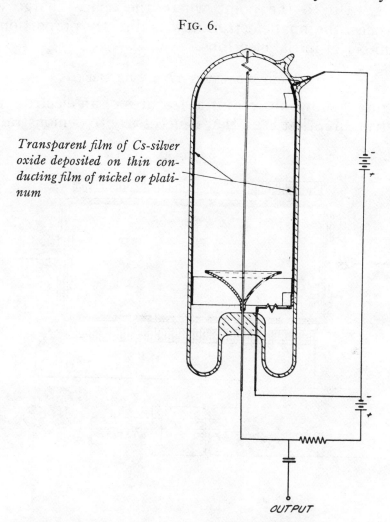

Transparent film of Cs-silver oxide deposited on thin conducting film of nickel or platinum

OUTPUT

Photocell which makes use of the electron multiplier principle.

from 50 to 100 volts, or more. The secondary electrons knocked out by impact of this primary beam are accelerated across and down the tube in a similar manner; upon reaching the opposite surface, cause an emission of electrons in increased numbers. Thus, each transverse passage results in an increase of the number of electrons. With a good Cæsium-silver oxide surface this increase amounts to a factor of three to six times for each impact.

The amount of electron multiplication obtained by this method is a function of the average potential through which the electrons fall between impacts and of the number of impacts. Up to the point where the space charge effect supervenes, the final electron flow is directly proportional to the number of initial electrons.

10. RADIO FREQUENCY TYPE MULTIPLIER.

Figure 7 shows the essential features of an electron multiplier structure, that is all that is necessary to demonstrate the

FIG. 7.

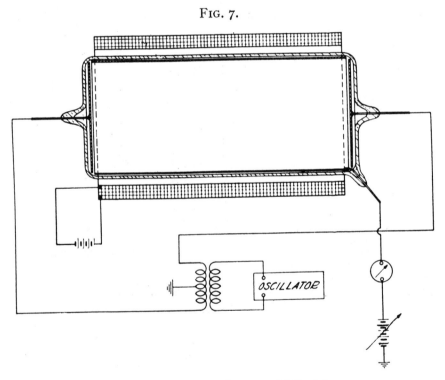

Radio frequency type electron multiplier tube.

characteristics of another method of carrying out the multiplication process. As is shown in the figure, the tube comprises a cylindrical evacuated envelope having plate like cathodes mounted in each end. These cathode plates are preferably made from pure silver, oxidized and formed with Cæsium into sensitive secondary emitting surfaces. A cylindrical anode of nickel or molybdenum is positioned between the cathode plates. This anode may fill almost the entire space between the cathodes, or may be merely a central ring.

A solenoid surrounds the envelope and is supplied with direct current for establishing a longitudinal magnetic focusing field in the space between the cathodes. For very low values of this focusing field, the output of the device is substantially proportional thereto, although saturation effects appear at higher values, as will be later explained. The cathode plates are supplied with radio frequency at a frequency of approximately 50 magacycles or higher. The anode is connected through a meter to the positive terminal of a variable potential source.

Applying, say 50 volts of radio frequency on the cathode plates, and gradually increasing the potential applied to the anode from the D.C. source, the meter will indicate currents as shown by the graph of Fig. 8. Up to a certain minimum

FIG. 8.

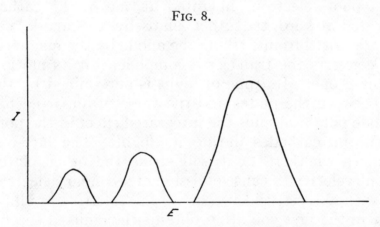

Anode current against anode voltage characteristic of the electron multiplier tube.

voltage, which depends principally on the distance between the plates, no measurable current will flow. Beyond this point, the current increases with voltage up to a definite point, then decreases again. Further increase of the voltage gives a second curve of the same general form as the first, rising to a maximum and again falling to zero, while still further increase will show a second repetition of the effect, usually of much greater magnitude.

It will be understood that the curves shown are illustrative merely, no numerical values being shown, since these will vary with the magnitudes of the oscillating potential on the cathodes, the frequency of this potential, the focusing current, and the dimensions of the device. Currents as large as a half ampere have been obtained.

Cutting off the magnetic focusing field stops the flow immediately. Increasing the oscillating potential on the cathodes broadens the range wherein the current flow occurs, eventually causing the curves to merge into a single continuous one having either multiple peaks or mere changes of slope indicating their position. Changing the frequency of the oscillator changes the position of the peaks; i.e., the voltages at which the maxima occur. The same is true as to tubes of different dimensions.

The explanation of these properties is simple in general, although the detailed analysis presents certain obscurities which render its presentation here undesirable. An electron (e.g., a photo-electron) liberated at one of the cathodes is accelerated toward the other plate by the anode voltage, being prevented from striking the anode by the magnetic field which converts the transverse component of its motion into an arcuate one. Its time of flight is determined by the distance between the plates and the velocity imparted to it by the anode potential, plus the integrated effect of the potential between the cathodes during its flight. The latter factor determines whether (1) it will strike the other plate with sufficient velocity to cause emission of secondary electrons (2) strike with a lesser velocity, or (3) fail to strike at all.

Where the first condition obtains the emitted secondaries are accelerated in the opposite direction to generate new secondaries at the cathode plate where the first electron was emitted, and if the ratio of the secondary emission be greater than unity, a multiplication by this ratio will occur at each impact. Under the second and third conditions no multiplication will occur.

It will be noted that with no R. F. in the cathodes, the anode potential contributes only to the mean velocity of the electrons through the tube, and has no effect whatever on the velocity of impact, since the acceleration it imparts to an electron leaving one of the cathodes is exactly neutralized by the deceleration imparted to the same electron approaching the other cathode.

Although the collection of any individual electron by the anode is improbable, owing to the shape and position of the latter and to the presence of a guiding field, a certain proportion of the total electrons will be collected. This proportion will depend upon the portion of the cathodes which are emitting secondaries (i.e., upon whether the electrons are striking near the center or near the periphery), and upon the transverse component of the electrostatic field within the chamber as determined by the space charge, the curvature of the lines of force between cathodes and anode, and upon any bias which may be applied within the tube.

Eventually, however, a point will be reached where the number of new secondaries emitted is equal to the number collected at each impact, and the current through the meter becomes constant. Within certain limits, therefore, the less the probability of any individual electron being collected, the greater the equilibrium current will be, and hence this current will be increased by increasing the guiding field. A limit to this, however, is the space charge developed when the number of electrons in the cloud which travels between the plates becomes very dense, causing the saturation effect above mentioned. When the space charge is the limiting factor, the largest equilibrium currents are obtained without using any magnetic field.

The peaks in the curve occur when the average time of flight or travel of the electrons is an odd number of half-cycles of the oscillating potential on the cathode, the three peaks shown in Fig. 4 representing five, three and one half-cycles respectively. With a given tube and source of oscillations and range of anode voltage it may only be possible to show one or two of the peaks. Under other circumstances still other peaks will be shown, e.g., the seven or nine half-cycle peaks. The higher the anode potential, the smaller is the time of flight, and hence the peak of output current corresponding to the highest voltage represents a time of flight of one half-cycle. This peak is usually materially higher than the others under otherwise similar conditions of operation.

Where electrons strike the cathodes without emitting

secondaries (condition 2 above); they will subtract from the anode current; where they meet condition 3 they are ineffective except as they increase the density of the electron cloud or plasma, and hence the probable number of electrons collected per half-cycle.

From the above, the reasons for the performance of the device will be clear. The total output current or equilibrium current varies with the proportionate number of electrons falling under conditions 1, 2 and 3, and this in turn varies with the potentials applied.

The uses of the device may readily be deduced from the characteristic curve, which shows alternate regions of positive and negative resistance. These regions obviously permit the modulation of the output by variation of anode potential. Operation in the negative resistance regions permits use as a generator of oscillations of any frequency, as in the case of any of the well known negative resistance devices. Rectification and frequency doubling both occur in the anode branch of the circuit, and may be put to their usual uses. Self-excited oscillations are developed if the inductance connecting the cathodes is tuned to the frequency for which the time of one-half period is approximately equal to the time of flight of the electrons. With a sensitive tube, the shock of closing the anode circuit is sufficient to start the tube oscillating. Unless these oscillations are limited by the anode supply or focusing field, their amplitude builds up until the tube is immediately destroyed.

Multiplying an Electron Flow.—Current multiplication, which is the particular concern of the present paper, may be obtained with this apparatus in several distinct ways, being dependent upon limiting the average number of impacts resulting from a single initial electron so that the total output current remains materially below the equilibrium value.

The first of these modes of operation comprises interrupting the action periodically at such short intervals that the limiting conditions cannot supervene. Since each interval will include the same number of half-cycles, and hence the same number of multiplying impacts, it is clear that the mean output current within the interval will be proportional to the number of initiating electrons liberated in the interval.

A second mode of operation depends upon a "drift" of the electron cloud toward the anode, so that if the first impact occurs on the axis of the tube, the succeeding impacts will occur successively nearer and nearer the anode, until the electrons are finally collected, the "descendants" of the successive initiating electrons reaching the collection point in their proper order, having accomplished the same number of impacts.

Electron multiplication by this "drift" method may be compared to a thermionic amplifier having many stages, and in which there is a great deal of coupling between adjacent stages, a lesser but appreciable coupling between stages farther apart in the chain, and even some coupling between the final output and input. This amount of back coupling is decreased by increasing the intensity of the magnetic guiding field. As the intensity of the field is increased, more and more amplification is possible without the tube "breaking down," that is, output becoming independent of the input. This is so because the electrons are more accurately guided and the "drift" of the electron cloud toward the anode becomes more definite the greater the magnetic guiding field. With an intense guiding field using the "drift" method, the multiplier may show complete linearity between output and input for small input variations, even after the tube has broken down and is passing considerable current that is independent of input electrons.

With a suitably intense magnetic guiding field, almost any degree of amplification is obtainable. A tube employing pure nickel cathodes has been constructed to operate as a photocell, giving an output of nearly a micro-amp. per lumen. Under these conditions by direct measurement the amplification is greater than 5×10^7 times. Whenever possible, therefore, this method is preferred. The action of the multiplier used in this manner is stable and reliable, even up to the very highest gains obtainable.

When it is not possible to employ the intense guiding field required to properly use the "drift" method of multiplication, a third mode of operation is possible which depends upon adjusting the multiplier just under its "break down" point. In this case, the "drift" of the electrons toward the anode is made as definite as possible, and the amplification of the multiplier increased slowly, either by varying the magnetic guiding field or the amount of radio frequency on the cathodes until the required multiplication is obtained.

In using the multiplier tube in this manner, the action compares somewhat to that of a gas filled photoelectric cell. The multiplication is increased slowly and is left just under the "break down" point at which the output ceases to be dependent upon the input. The amount of multiplication obtainable in this manner depends somewhat on the use to which the cell is being put, but a gain of several thousand may usually be relied upon.

11. APPLICATION OF THE ELECTRON MULTIPLIER TO THE IMAGE DISSECTOR TUBE.

Construction.—A combination of the electron multiplier and the image dissector tube for increasing the sensitivity of the latter, is carried out as shown in Fig. 9. The photoelectric

FIG. 9.

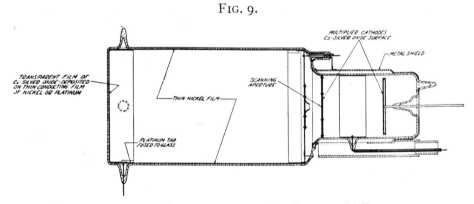

Image dissector tube having a built-in electron multiplier.

cathode onto which the optical image is focused, is formed as a thin translucent film of silver oxide on one end of the glass tube itself. The multiplier is positioned in the opposite end of the tube. The main anode of the dissector is a silver disk having the edge spun over so that it fits snugly into the glass envelope. The anode and cathode are connected together with a thin film of evaporated metal on the glass walls to insure a uniform potential gradient down the tube and to remove wall charges. In the center of the anode disk there is fastened a smaller disk with a protuberant center in which there is a rather large aperture. The anode protuberance faces the multiplier end of the tube. Behind this main anode is a tightly fitting silver cup, in the center of which the scanning aperture is punctured. This silver cup forms one of the multiplier cathodes. The second multiplier cathode is mounted from a stem and is spaced about six centimeters from the silver cup. Radio frequency is applied only to this second cathode. The output of the multiplier is taken from the anode ring which is supported from the tube walls. A metal shield covers the whole multiplier structure.

After the tube has been mechanically constructed, it is sealed onto the pump and given a good baking until a very good vacuum is obtained; oxygen is admitted and the dissector and multiplier cathodes properly oxidized for formation into photo surfaces. Cæsium pills (not shown in figure) are then flashed in both the multiplier and the dissector chambers to give a proper amount of Cæsium in each of these, after which the tube is given a heat treatment to sensitize these cathodes. The tube may then be sealed off the pump.

Fig. 10.

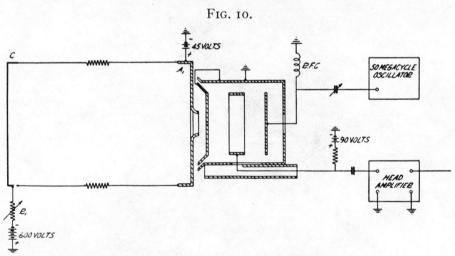

Wiring diagram of the dissector-multiplier.

Operation of the Dissector-Multiplier.—The multiplier dissector is connected as shown in Fig. 10. A potential difference of approximately 600 volts is maintained between the image cathode and the anode. This potential may be varied by the resistor in series with the cathode for adjusting the magnetic focus of the electron image. The first multiplier

FIG. 11.

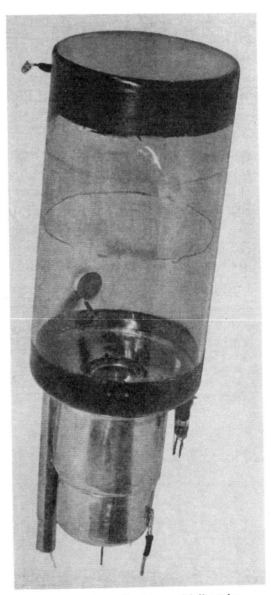

Photograph of a dissector-multiplier tube.

cathode is connected to the outside shield and to ground. The main anode of the dissector is maintained about 45 volts positively with respect to the first multiplier cathode and ground, for the purpose of removing the secondary electrons that are knocked out in the immediate neighborhood of the scanning aperture, otherwise these secondaries are drawn into the multiplier and blur the image. R.F. of about 50 megacycles is applied to the second multiplier cathode through a small variable condenser which permits the voltage to be varied from perhaps 20 to 50 volts R.M.S. The anode of the multiplier must be maintained at quite accurately the correct positive potential with respect to the cathodes. This voltage

FIG. 12.

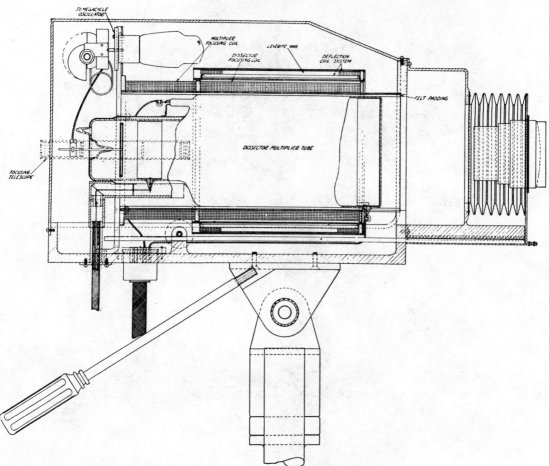

Detailed drawing of the dissector-multiplier camera.

is dependent upon the frequency of the applied R.F., but for six centimeters spacing between the cathodes and using 50 megacycles, the optimum is between 60 and 90 volts, and fair multiplication may be obtained at either of these extremes. The output of the multiplier is preferably taken across a resistor in series with the multiplier anode.

Dissector-Multiplier Camera.—The tube is positioned in a focusing and deflecting coil system as shown in Fig. 12. The

FIG. 13.

Photograph of the dissector-multiplier camera.

focusing solenoid is divided into two parts. One part extends the length of the dissector and the other part covers the multiplier chamber. Both fields have the same polarity so that they are practically equivalent to a single solenoid extending the full length of the tube, except that the current through them may be separately adjusted. Current for these focusing solenoids should be supplied from a good accumulator source through variable resistors that will give close adjustment and firm connection by the variable member. About 500 ampere turns are required in the dissector focusing coil and slightly less in the multiplier coil. The deflecting coils extend the full length of the dissector chamber and may be either inside or outside of the focusing coil. They should not extend quite to the multiplying chamber because this results in the scanning fields modulating the output of the multiplier.

In adjusting the multiplier dissector for the first time, an intense image is focused on the dissector cathode so that a picture may be received without the use of the multiplier. The approximate values of focusing current are noted for the first, second and third magnetic foci, also the effect of the multiplier focusing field on the dissector focusing field. A small amount of radio frequency is then applied to the multiplier cathode, and the multiplier focusing field slowly increased until the multiplier breaks down. The approximate amount of gain is noted just before break-down, and the intensity of the optical image reduced accordingly. This may be expected to be approximately 20 to 100 times. The amount of R.F. applied to the multiplier is then increased slightly and the process repeated. This will result in an increased sensitivity of the tube. The whole process is repeated until the optimum value for the R.F. is determined. After this, the only multiplier adjustment necessary is the focusing. The multiplier focusing also makes a very satisfactory level control.

Sensitivity.—Addition of the multiplier to the dissector in the form shown in Fig. 9 has increased the sensitivity of the latter by three or four thousand times, with no observable increase of the noise level. The sensitivity is now adequate for outdoor scanning even when there is no direct sunlight. It is not necessary to have a particularly sensitive image cathode. Good images have been transmitted direct from the subject with a tube having a cathode sensitivity considerably less than a micro-amp. per lumen.

In the multiplier used in the present combination, the "drift" of the electrons toward the collection ring is not very definite because of the necessary limitation of the intensity of the multiplier focusing field. The resulting feed-back of electrons from portions of the cathodes where the probability of collection is highest to portions near the center where the probability is a minimum, imposes a definite limit on the amount of multiplication obtainable for stable operation. The sensitivity may be improved to almost any degree by increasing the definiteness of "drift" or by using a two or three stage multiplier. It does not seem too optimistic to expect that eventually *the multiplier will eliminate the entire amplification system up to the modulator of the radio transmitter.*

11. OTHER APPLICATIONS OF THE ELECTRON MULTIPLICATION.

The multiplication principle appears applicable to many uses besides amplifying an electron flow, notably:

1. Image amplifier and translation device for such uses as an infra-red camera for fog penetration, electron microscope or telescope, and general intensification of optical images.
2. Generator of oscillations of any frequency, particularly adapted for frequencies above 20 megacycles.
3. Radio frequency amplification for transmitter use. The output may be readily modulated.
4. Short wave radio receiver.

12. GENERATION AND SYNCHRONIZATION OF THE DEFLECTING CURRENTS.

While it would be outside the scope of the present paper to give a detailed report of the method of generation and synchronization of the saw-tooth wave currents used for scanning at transmitter and receiver, nevertheless, a brief description will be included for the sake of completeness.

Modes of Scansion.—Many modes of scanning have been tried and used during the course of the present work, but two modes are especially important and will be considered here.

The first of these methods is shown in Fig. 14, and is perhaps the simplest type of linear scansion that may be used.

The picture field is scanned in the horizontal direction with a saw-tooth wave of comparatively high frequency— f_2, for example—5,000 to 10,000 repetitions per second, and in the

FIG. 14.

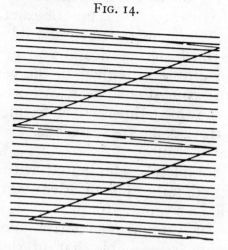

Path of cathode ray beam in simple linear scanning.

vertical direction with a saw-tooth wave of low frequency, f_1, varying between 16 and 30 repetitions per second. The number of scan lines is equal to f_2/f_1.

The second scanning mode is shown in Fig. 15, and may be

FIG. 15.

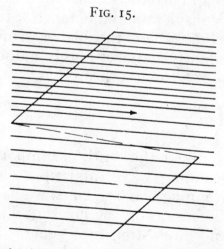

Path of cathode ray beam in interlaced linear scanning.

called interlaced scanning. This second mode is similar to the first, except that the vertical scanning is at a frequency two or three times faster than would be normally used with simpler

mode and the lines of successive scansions are displaced vertically from each other by one-half the distance between lines.

There are several methods possible for accomplishing the
required interlacing of lines. One method [3] is to use a horizontal scanning frequency that is not an exact harmonic of the
vertical scanning frequency, but preferably a harmonic of
one-half or one-third the vertical frequency. If the horizontal is commensurable with one-half the vertical frequency,
a complete picture consists of two scansions, but if commen-
surability is at one-third the vertical frequency, a complete
picture requires three separate scansions.

The principal advantage of interlaced scanning is that
flicker may be completely eliminated without too great a
sacrifice of picture detail. The disadvantage is that very accurate control of the scanning frequencies is required. Satisfactory means of such frequency control have been developed
however, and it appears probable that interlaced scanning will
be the type adopted when Television is introduced commercially.

Generation of Saw-tooth Wave Currents.—The deflection of
the electron beam at transmitter and receiver is most conveniently obtained electromagnetically. This requires that
a saw-tooth wave of current be made to flow through deflecting coils having considerable inductance. In the case of the
horizontal deflection where the higher scanning rate is required, the impedance is almost purely inductive.

The equation of a saw-tooth wave is

$$E(t) = k \left(\frac{\sin wt}{1} + \frac{\sin 2wt}{2} + \cdots + \frac{\sin nwt}{n} \right). \qquad (4)$$

Harmonics up to about the tenth are important in the horizontal (H.F.) scanning wave, and up to the twentieth in
vertical (L.F.) scanning wave. Thus, we have frequencies as
high as 70 kc. to consider in the horizontal scanning current.

Many new and interesting tube generators have been developed to generate these saw-tooth wave currents in coils.
In most of these generators a saw-tooth wave of voltage is
applied in series with a large resistor (usually the R_p of a
tube) across a network the frequency-impedance characteristic

[3] See Egerton's U. S. Patent No. 1,605,930.

of which is similar to that of the deflection coils. This results in a saw-tooth wave of current through the network, and the resulting voltage drop across the network is that required to maintain a saw-tooth wave of current through the deflection coils. The voltage across the network is amplified and the amplified voltage applied to the coils. When the coils are substantially pure inductances, the required network is simply a differentiating network and the voltage which must be supplied to the coils is simply a pulse of short duration.

In another type generator, a pulse is applied to an integrating circuit that changes the pulse into the wave form required to give a saw-tooth wave of current through the coils.

Synchronization of the Deflecting Currents at the Receiver.—Synchronization of the deflecting currents at the receiver is by means of pulses transmitted between lines and between vertical scansions. These pulses are transmitted right along with the picture frequencies, though preferably over a separate radio transmitter. They are separated from the picture frequencies at the receiver simply by amplitude selection. These pulses are applied to the receiving tube in the proper polarity to extinguish the spot during the "back time," that is, the return part of the saw-tooth wave cycle.

13. RECEIVING OSCILLIGHT AND COIL SYSTEM.

General Description.—Referring to Fig. 16, the receiving system comprises a vacuum tube having an electron gun and a fluorescent screen positioned in a focusing and deflecting coil system. The electron beam from the gun is focused into a small spot on the fluorescent screen magnetically by means of a short coil around the neck of the receiver or "oscillight" tube. This coil extends one-third to one-half the total length of the tube.

The horizontal deflection of the spot for scanning is accomplished by two small coils formed to conform closely to the neck of the tube inside of the focusing coil. Vertical deflection (at L.F.) is preferably accomplished by means of the electromagnet shown. The low frequency coil is wound to suit the impedance requirements imposed by the generator that supplies the low frequency saw-tooth wave current.

Magnetic Focus with a Short Coil.—For the best magnetic

focus the solenoid should extend the full length of the electron's path, but this gives poor deflection sensitivity. It has been found experimentally, that the use of a short focusing coil gives entirely adequate focusing of the spot, and the spot can

FIG. 16.

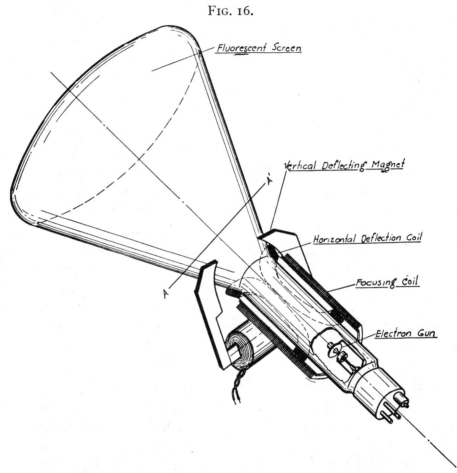

Oscillight tube and coil system.

be made smaller than necessary. The sensitivity to deflection is so greatly improved by use of the short focusing coil that its use is more than justified. With this kind of focusing, the plane of the electrons' focus is very deep; in fact an approximate focus exists between the plane of $A–A^1$ (see Fig. 16) and the fluorescent screen, and close adjustment of the current in the focusing coil is not required. Also the beam stays in focus when deflected, which is a great advantage.

Deflecting Means.—It is possible to use several different systems for deflecting the cathode ray beam, when it is focused magnetically. Perhaps the simplest system is that of an air-core coil positioned as close as possible to the walls of the oscillight tube and preferably inside the focusing coil, as shown in Fig. 16. This type of deflection system is perhaps the most sensitive that can be used, and is to be preferred, especially for use at frequencies of the order of 1,000 to 10,000 cycles per second.

Another type of deflecting system is the electromagnet shown in Fig. 16. The principal advantage in the use of the electromagnet shown, is that the deflection is concentrated at a point along the tube and minimizes the interference, which results into nearby circuits and deflection coils. The ferromagnetic type is preferable for the low frequency deflection. The disadvantages in the iron core type deflection system are that it is less sensitive than the air core, and troublesome distortion results because of the hysteresis of the core material.

When it is attempted to put two deflection systems close together deflecting at right angles to each other, several factors must be taken into consideration. Since the deflections take place simultaneously, it is necessary that all parts of deflection field of one deflecting system be equally sensitive. Otherwise, the deflection produced by this particular system will be modulated by that of the other.

In our present type of deflection system, the high frequency coils deflect the beam first. They, therefore, have to be designed simply to give linear deflection along a line. The low frequency deflection is accomplished by means of the electromagnet placed down the tube a suitable distance from the high frequency coils. These pole pieces must, therefore, produce a uniform deflection, no matter in which part of the tube the beam may be. To accomplish this requirement, the pole pieces have been given the shape shown in Fig. 16, and this suffices if the pole pieces are placed the correct distance apart.

If Keystone distortion is present, caused, for example, by the necessity of mounting the fluorescent screen oblique to the axis of the tube, such distortion is readily and completely corrected simply by suitable shaping the pole pieces of the electromagnet.

It is entirely practical to deflect the beam electrostatically, even though the focusing is obtained magnetically. The condition for satisfactory electrostatic deflection is that the deflection plates must extend practically the full length of the

FIG. 17.

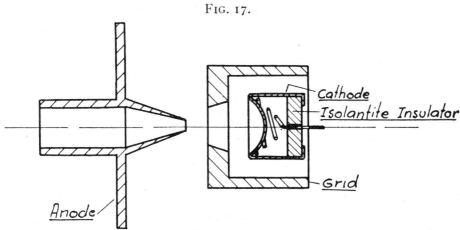

Detail of the electron gun of the oscillight.

focusing coil. A tube has been built along this line in which the plates are on the walls of the bulb itself. The sensitivity of this particular tube is approximately 100 volts per inch for an anode voltage of 4,200 volts.

FIG. 18.

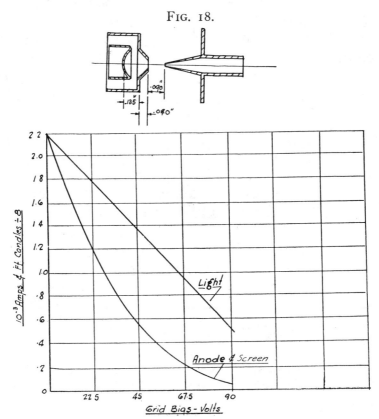

Electron Gun Design.—The electron gun used in the oscillight is shown in Fig. 17, and consists of a heater type cathode having a concave surface that faces the fluorescent screen. A grid is placed over this cathode, and usually con-

FIG. 19.

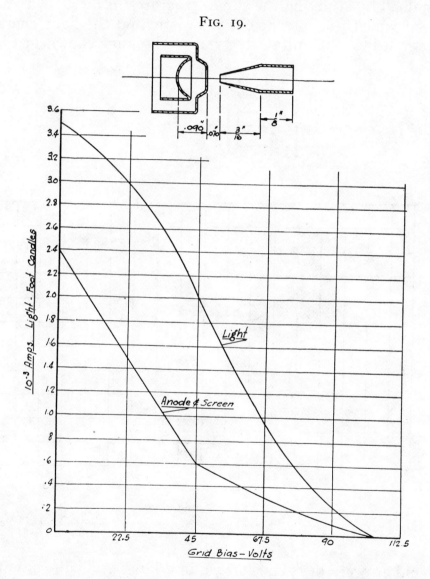

sists of a small cylinder closed at one end and having an aperture in the closed end, through which the electrostatic field of the anode may pass to the cathode. The anode in most of our work has been of the truncated type shown in Fig. 17. The purpose of the protuberant end on the anode is to simulate a point of charge from which the lines of force originate. This makes it unnecessary to carefully align the anode and cathode.

An idea as to the characteristics obtainable from electron guns built along the general line of that shown in Fig. 17, may be gained by studying the three structures in Figs. 18, 19 and 20, for which the current characteristic and the light characteristic using Willemite are given. The guns shown in Figs.

FIG. 20.

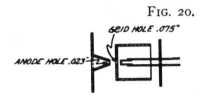

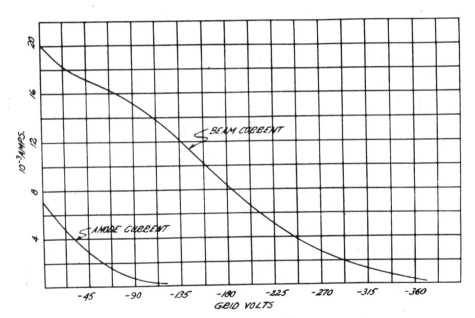

Figs. 18, 19 and 20.

Curves showing the intensity of the fluorescent spot and the anode plus screen current as a function of the control grid voltage for various electron gun structures.

18 and 19 were designed for use in a large oscillight having a nine inch end. These guns supply as much power at 4,200 volts as can be used with Willemite.

The electron gun shown in Fig. 20, was designed for projection type oscillights, in which the luminescent screen is not a fluorescent screen. It should be noted that over the useful part of the characteristic curve, practically no electrons

FIG. 21.

Photographs of received television images.

strike the anode at all, in spite of the fact that the aperture in the anode is only .020 inch in diameter, and that the maximum beam current is more than 15 milli-amps.   With a suitable magnetic focusing system the spot size obtained from this gun is less than .010 inch in diameter.

Miscellaneous Design Factors.—There are many minor factors to be considered in designing an oscillight tube, but perhaps the most important of these is the removal of the charge from the end of the tube.   Most of the systems we have used so far depend upon emission of secondary electrons from the fluorescent screen.   This requires that the fluorescent material have a secondary emission ratio that is greater than unity for the velocity of the primary electrons used.   In addition to this, some means must be provided close to the screen for drawing these secondary electrons away from the screen.   Our method of accomplishing this is to evaporate on the walls of the tube a thin film of nickel and to connect this metal coating with the anode.

Another important factor in design of oscillights is to limit the angle of divergence of the beam to approximately 15 degrees, or less.   This is important if good magnetic focusing is to be obtained.

Considerations having to do with the design of high intensity tubes using fluorescent screens must take cognizance of the fact that nearly all of the energy in the electron beam is converted directly into heat, and means must be provided for getting rid of this heat.   Most of the important fluorescent materials are spoiled completely by over-heating them, and those which are not spoiled instantly show a gradual recrystallization under heat, which decreases their efficiency.

In conclusion I want to acknowledge my indebtedness to the Staff of Television Laboratories for many suggestions and continual coöperation during the course of this development, particularly Mr. Donald K. Lippincott and Mr. R. E. Rutherford, who have contributed greatly to the present work, also Mr. B. C. Gardner who has constructed the tubes, and to Television Laboratories, Ltd. for permission to publish this material.

TELEVISION

V[ladimir] K. Zworykin

TELEVISION.*

BY

DR. V. K. ZWORYKIN, E.E., Ph.D.,

Research Division, RCA Victor Company, Inc., Camden, N. J.

SYNOPSIS.

This paper gives an outline of a new television system developed in the laboratories of RCA Victor Company, in Camden, N. J.

The system is truly electrical and employs only electronic devices, without a single mechanically moving part.

The translation of the visual image is accomplished by means of a vacuum tube called the iconoscope. This tube is a virtual electric eye and consists of a photo-sensitive mosaic corresponding to the retina of the human eye, and a moving electron beam representing the nerve of the eye. The image is projected optically on the mosaic and transformed within the tube into a train of electrical impulses, representing the illumination of individual points of the image.

The reproduction of the image is accomplished by means of another vacuum tube, the kinescope, which transforms the electrical impulses back into the variation of light intensity through the bombardment of a fluorescent screen by the moving electron beam.

The movement of the electron beams in both tubes, which is responsible for both transformations, is linear and divides the picture into a series of parallel lines. The movements are synchronized so that the instantaneous position of the beams with respect to a point in the picture is always identical. The synchronization is transmitted together with the picture signals, and operation of the receiver is completely automatic.

The sensitivity of the iconoscope, at the present time, is approximately equal to that of a photographic film operating at the speed of a motion picture camera, permitting the transmission of outdoor scenes. The resolution is high, much higher than necessary for television images of the highest quality.

The paper describes the theory of the system, its characteristics, mode of operation, and includes photographs of images obtained on the fluorescent screen of the receiver.

---

* Presented at the Stated Meeting held Wednesday, October 18, 1933.

I

INTRODUCTION.

"Television" has become a familiar word in recent times and conveys to us the idea of an artificial reproduction of transient visual images. Since it is artificial, we cannot expect to have an absolutely perfect reproduction of the original, and therefore, for better definition, the word "television" itself requires an additional adjective to identify the degree of perfection.

It is an accepted practice in reproducing pictures in printed matter, such as books and newspapers, to resolve the picture into a number of dots. The definition of the picture is then identified by the number of dots to a square inch. An ordinary picture may contain 4000 picture elements, or approximately 65 lines per linear inch, while for finer works a larger number is used. Similarly, in television, due to the requirement of transmission by one channel, the picture is dissected into a number of elements, which are transmitted in succession, one element after another in a series of parallel lines. It has, therefore, become quite common to refer to the reproduction in terms of lines. Thus we speak of 30, 60 or 120 line television, which means that the whole reproduced image is composed of 30, 60 or 120 lines, each line varying in density throughout its length.

The greater the detail is, the more difficult is the solution of the television problem; it is very important to determine accurately how far the compromise can be carried without affecting the quality of the reproduced image. In other words, what is the minimum number of lines in the television picture with which the reproduced image can be accepted as satisfactory?

A number of excellent works have already been published on the subject, both from the theoretical and experimental points of view.[1] The results of these works agree quite accurately, and predict that the increase in the number of lines at first gives a great increase in definition, then the rate of

---

[1] William H. Wenstrom, "Notes on Television Definition," *Proceedings I. R. E.*, September, 1933, Vol. 21, No. 9, p. 1317. Selig Hecht and Cornelis Verrijp, "The Influence of Intensity, Color and Retinal Location on the Fusion Frequency of Intermittent Illumination," *Proc. Nat. Academy of Science*, Vol. 19, No. 5, p. 522. E. W. Engstrom, *Proceedings of I. R. E.*, December, 1933.

increase slows down and finally approaches a certain saturation value, after which a further increase in the number of lines gives a negligible increase in the definition of the image.

FIG. 1.

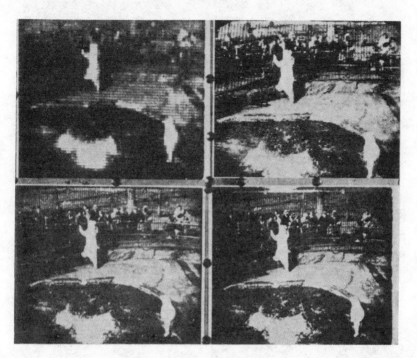

In order to give a better idea of this relation, Fig. 1 [2] shows a comparison of four reproductions of the same picture made by 60, 120, 180 and 240 lines, respectively. These pictures are made not by television, but by special optical methods, and do not contain the distortions which are possible during various transformations involved in the process of television. Therefore, they should be regarded as an optimum for television reproduction. This set of pictures indicates that for an image with a great deal of detail, we should regard 240 lines as a minimum. For a picture with less detail, these requirements can be somewhat lower, as can be seen clearly from Fig. 2. Both these examples are given for half-tone reproduction. The black and white picture requires approximately the same number of lines for the corresponding definition, as shown on Fig. 3.

[2] Courtesy of Mr. E. W. Engstrom.

FIG. 2.

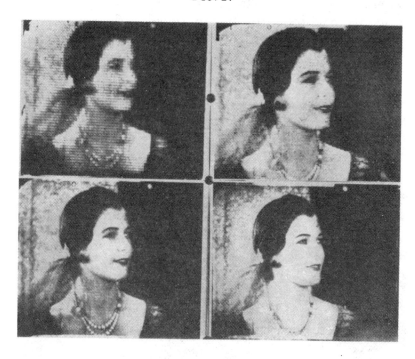

FIG. 3.

FIG. 4.

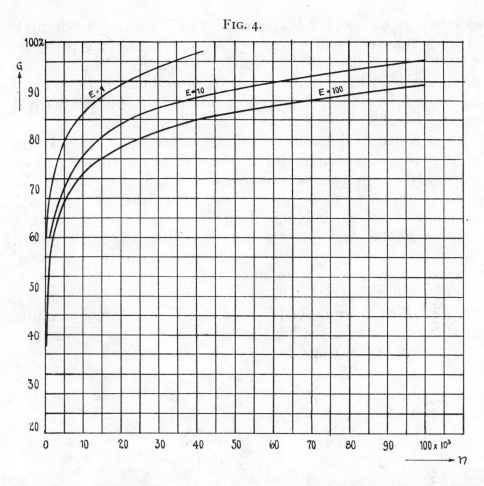

This relation between the definition and the number of lines can also be shown by a curve, such as in Fig. 4.[3] This curve is obtained theoretically from the resolution of the human eye and is represented by the expression,

$$G = \rho \log n,$$

where G is the definition of the picture,

    n—the number of picture elements,

and ρ—coefficient depending on the size of the picture and the viewing angle.

The three curves are for three different intensities of the illumination of a picture of 1, 10 and 100 lumens. All these relations are considered for a still picture. Television pictures

---

[3] J. A. Ryftin, *J. Zeitschrift für Technische Physik*, Band III, Heft 2–3, 1933.

of moving objects give better resolution for a smaller number of lines, due to the fact that television, like moving pictures, repeats many pictures per second. Therefore, some of the details missing in one image may be reproduced in others and integrated by the eye, as one combines the picture with a greater number of details.

The practical difficulties in the development of a television system are proportional to the quantity of information to be transmitted and therefore increase with the increase in the number of picture elements. These difficulties are found not only in dissecting the image, but also in the amount of light required at the transmitter and receiver, in the construction of electrical circuits, and in the limitations of the electrical channels used for transmission.

The following table illustrates the relation between the number of lines, number of picture elements, and maximum picture frequency assuming the aspect ratio of the picture as 3 to 4 with 24 repetitions per second.

Scanning Lines.	Picture Elements.	Maximum Picture Frequency.[4]	Maximum Picture Communication Band.
60	4,798	63,970	127,900
120	19,200	256,000	512,000
180	43,190	576,000	1,152,000
240	76,780	1,024,000	2,048,000

[4] 10 per cent. is added to compensate the loss of time for synchronizing signals.

The width of communication band for pictures with over 100 lines definition makes impossible the use of radio channels of frequencies utilized for sound broadcasting. The only solution for transmission of these pictures, therefore, lies in ultra-high frequencies. The communication, however, is a separate subject which will not be discussed in this paper. Also, it is not the intention of this presentation to give a history of television development. It is sufficient to say only that practically all the previous work was done with mechanical methods as a means of scanning or resolving the picture into picture elements for transmission. Likewise, mechanical methods were employed at the receiver to reconstruct the picture from the transmitted elements. This involved purely mechanical complications in construction of sufficiently precise scanners, and difficulties in increasing the number of picture elements and particularly in obtaining sufficient light. This last limitation actually introduced a stone wall which prevented the increase of resolution in the transmitted picture necessary to obtain the desired quality and practically excluded all hope of transmitting an outdoor picture—the real goal of television.

To understand the reasons for this difficulty, we should remember that the picture is scanned point by point, and, therefore, that the photo-sensitive element is affected by the light from a given point only for a very short interval of time corresponding to the time of illumination of one picture element. Assume for a picture of good quality we desire 240 lines, or 76,000 picture elements. For twenty-four repetitions per second, this means that the time of transmission of one picture element is $1/1,824,000$ of a second. On the other hand, the output of the photocell to the amplifier is proportional to the intensity of the light and time during which the light is acting on the photocell. A brief computation shows how microscopic will be the output of the photocell for this number of picture elements. If we take an average photographic camera with a lens F–4.5, the total light flux falling on the plate from a bright outdoor picture is of the order of $1/10$th of a lumen. Substituting a scanning disc suitable for 76,000 picture elements for the plate and using a photocell of 10 micro-amperes/lumen sensitivity, we will have a photo current from a single picture element

$$I_e = \frac{I \times 10^{-5}}{10 \times 76,000} = 1.3 \times 10^{-11} \text{ amp.}$$

The charge resulting from this current in the time of one picture element is

$$Q = I \times t = \frac{1.3 \times 10^{-11}}{1.824 \times 10^6} = .7 \times 10^{-17} \text{ coulombs.}$$

Comparing that with the charge of one electron,

$$e = 1.59 \times 10^{-19} \text{ coulombs,}$$

we see that only 44 electrons are collected during the scanning of one element. The amplification of such small amounts of energy involves practically insurmountable difficulties.

If we now compare this condition with that of a photographic plate during exposure, we will see that the latter operates under much more favorable conditions, since all its points are affected by the light during the whole time of exposure. This time for studio exposure is several seconds, and of the order of one hundredth of a second for outdoor exposures, or many thousand times greater than in the case of an element in the scanned television picture. The human eye, which we regard as an ideal of sensitivity, operates also under the same favorable condition.

If a television system could be devised which would operate on the same principle as the eye, all the points of the picture would affect the photo-sensitive element all the time. Then in our example of a picture with 76,000 elements, the photo-electric output for each point would be 76,000 times greater than in the conventional system. Since scanning is still necessary in order to use only one communication channel, we should have some means for storing of the energy of the picture between two successive scannings of each point.

The light requirement and necessity of avoiding the mechanically moving parts resulted in the development of an entirely electrical television system, employing special electronic devices for both transmission and reception.

ICONOSCOPE.

On the transmitting end this device took the form of a virtual artificial electric eye. The device was named iconoscope, the name being derived from two Greek words signifying "image observer." The photograph of this device is shown on Fig. 5.

It consists of two principal parts enclosed in an evacuated glass bulb. The first part is the photo-sensitive mosaic, consisting of a metal plate covered with a great number of miniature photo-electric cells, insulated from the plate and each from the other. The function of the mosaic is similar to that of the retina of the eye. It transforms the energy of the light from the image into electrical charges and stores them until they can be transformed point by point into electrical impulses and transmitted. This transformation is accomplished by an electron beam scanner, the nerve of this electric eye.

FIG. 5.

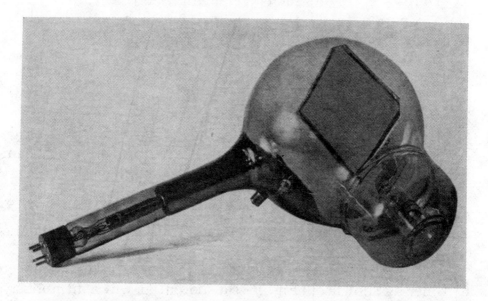

To complete the analogy of the iconoscope with the human eye, we shall mention that it possesses an electrical memory, because with a good dielectric the charges of the mosaic can be preserved for a considerable length of time.

FIG. 6.

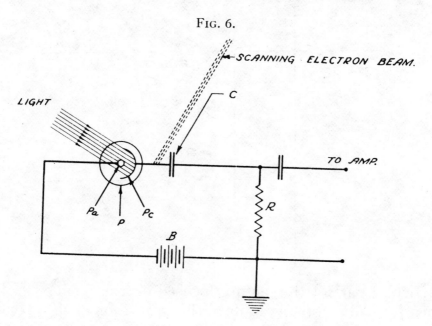

To fully understand the operation of the photo-sensitive mosaic of the iconoscope, it is best to consider the circuit of a single photo-electric element in the mosaic, as shown in Fig. 6. Here P represents such an element, and C its capacity to a plate common to all the elements, which hereafter will be called the "signal plate." The complete electrical circuit can be traced starting from the cathode P_c to C, then to resistance R, source of e.m.f. B, and back to the anode P_a. When light from the projected picture falls on the mosaic, each element P_c emits electrons, and thus the condenser element C is positively charged by the light. The magnitude of this charge is a function of the light intensity. When the electron beam which scans the mosaic strikes the particular element P_cC that element receives electrons from the beam and may be said to have become discharged. This discharge current from each element will be proportional to the positive charge upon the element.

If we plot the rise of charge of the element P_cC with respect to time, as shown on Fig. 7, the potential will continuously increase due to the light of the picture. The slope of this increase or dv/dt will depend only on the brightness of the

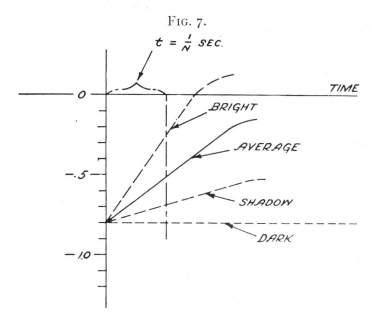

FIG. 7.

particular part of the picture focused on this element. This linearity will be preserved until the saturation of the capacity

C, which is so chosen as never to be reached at a given frequency N of repetition of the discharge. Since the scanning is constant, the interval of time, t, which is equal to $1/N$, is also constant and therefore the value of charge depends only on the brightness of this particular point of the picture. With constant intensity of the scanning beam, the impulse through R, and consequently the voltage drop V_I across R, is also proportional to the brightness of a given point of the picture. This potential V_I is the output of each single photo-element of the iconoscope, which is applied to an amplifier.

The above explanation is actually somewhat complicated by the fact that this discharging beam not only neutralizes the positive charge of the photo-element, but charges it negatively. The equilibrium potential of the element is defined by the velocity of the beam and the secondary emission from the photo-emitting substance due to bombardment by the electrons of this velocity. This equilibrium condition in the dark, for a normal iconoscope, is of the order of .5 to 1.0 volts negative. The light causes the element to gain a positive charge, thus decreasing the normal negative charge, and the scanning beam brings it back again to the equilibrium potential.

Another complication is due to the existence, besides the discharge impulses from the individual elements, of a charging current due to light on the whole mosaic. This current is constant for a stationary picture but varies when the picture, or part of it, begins to move across the mosaic. This variation, however, is very slow and does not affect an amplifier which has a cut-off below 24 cycles.

Figure 8 gives an idea of conditions on the surface of the mosaic. Here the shaded picture represents the electrical charges accumulated by the individual elements of the mosaic due to the light of the projected image. Although the background of this image is of uniform density, the corresponding charges at a given instant are not uniform but vary, as shown on the left part of this picture. The highest charge occurs just before the exploring beam has discharged the elements. After the beam has passed the charge is momentarily near its equilibrium condition and begins to increase throughout the whole scanning period, attaining its maximum value again just before the scanning by the beam.

The electron beam, in neutralizing the charge of a particular point of the picture, releases practically instantaneously the energy stored there during the whole 1/24th of the second.

FIG. 8.

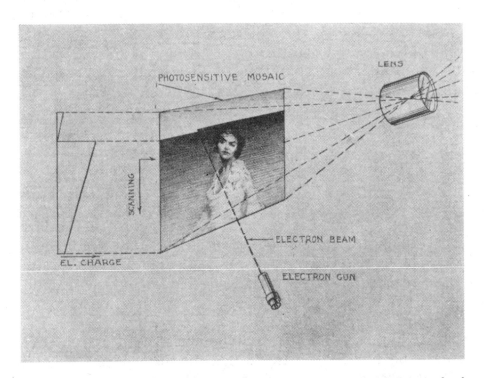

The electrical impulse created on the opposite side of the mosaic energizes the amplifier. The magnitude of this impulse is greater than the corresponding impulse, produced by a system not utilizing the storing effect, in ratio of time intervals during which the light from the picture point acts on the photo-sensitive element. In this case, it is equal to the number of picture elements, or 76,000 times as great. This is true, assuming that the device has 100 per cent. efficiency. This, however, is a practical impossibility due to various losses, and at the present we have to satisfy ourselves with only about 10 per cent. of the possible gain.

The schematic diagram of a complete electrical circuit for the iconoscope is shown in Fig. 9. Here the two parts of the photo-element P, shown on Fig. 6, are entirely separated. The cathodes are in the shape of photo-sensitive globules on the surface of the signal plate and insulated from it. The

anode or collector is common and consists of a silvered portion
on the inside of the glass bulb.

FIG. 9.

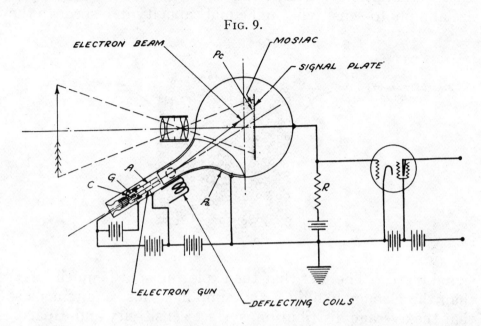

The capacity C of each individual element with respect to
the signal plate is determined by the thickness and dielectric
constant of the insulating layer between the elements and the
signal plate. The discharge of the positive charge of the
individual elements is accomplished by an electron beam
originating in the electron gun located opposite the mosaic and
inclined at 30° to the normal passing through the middle of the
mosaic. Both mosaic and electron gun are enclosed in the
same highly evacuated glass bulb. The inclined position of
the gun is merely a compromise in the construction in order
to allow the projection of the picture on the surface of the
mosaic.

The resolution of the iconoscope is determined by both the
size and number of picture elements in the mosaic, and the size
of the scanning electron beam. In practice, however, the
number of individual photo-elements in the mosaic is many
times greater than the number of picture elements, which is
determined entirely by the size of the scanning spot. This is
shown diagrammatically on Fig. 10. From the initial assumptions formulated in the analysis of the ideal circuit for
individual elements, as shown on Fig. 6, we find the qualifica-

tions which should satisfy the mosaic for the iconoscope.
These assumptions required that all the elements be of equal
size and photo-sensitivity and equal capacity in respect to the

FIG. 10.

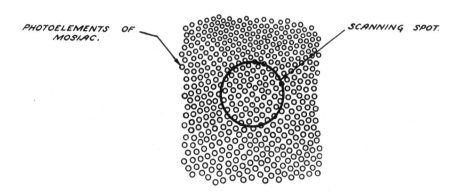

PHOTOELEMENTS OF
MOSIAC.   SCANNING SPOT.

signal plate.   The fact that the exploring spot is much larger
than the element modifies and simplifies this requirement so
that the average distribution, surface sensitivity and capacity
of elements over an area of the mosaic corresponding to the
size of the scanning spot should be uniform.   This allows
considerable tolerance in the dimensions of individual elements.

The requirement of uniformity, which at first glance
appears quite difficult to obtain, is solved by the help of
natural phenomena.   It is known that such a common material as mica can be selected in a thin sheet of practically ideal
uniform thickness and it therefore serves as a perfect insulating
material for the mosaic.   The signal plate is formed by a
metallic coating on one side of the mica sheet.   The mosaic
itself can be produced by a multitude of methods, the simplest
of which is a direct evaporation of the photo-electric metal
onto the mica in a vacuum.   When the evaporated film is
very thin it is not continuous but consists of a conglomeration
of minute spots or globules quite uniformly distributed and
insulated each from the other.   Another possible method is
that of ruling the mosaic from a continuous metallic film by a
ruling machine.

Although the initial method of formation of the photo-
sensitive mosaic was the deposition of a thin film of alkali

metal directly on an insulating plate, subsequent developments in the photocell art resulted in changes in the methods
of formation of the mosaic.   The mosaic which is used at
present is composed of a very large number of minute silver
globules, each of which is photo-sensitized by cæsium through
utilization of a special process.

Since the charges are very minute the insulating property
and dielectric losses should be as small as possible.   Mica of
good quality satisfies this requirement admirably.   However,
other insulators can also be used and thin films made of
vitreous enamels have proven to be entirely satisfactory.   The
insulation is made as thin as it can be made conveniently.

FIG. 11.

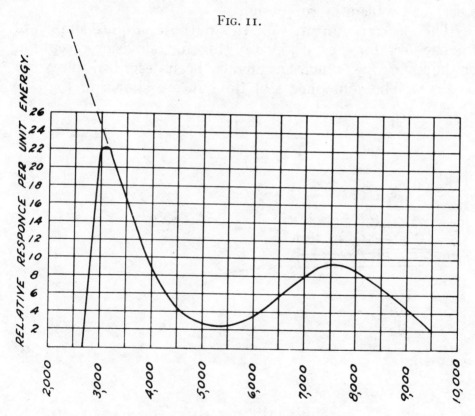

WAVELENGTH   IN   ANGSTROMS.

The sensitivity of the mosaic is of the same order as that of
corresponding high vacuum cæsium oxide photocells.   The
same is true also of the color response.   The spectral characteristic is shown on Fig. 11.   The cut-off in the blue part of the

spectrum is due to the absorption of the glass. The actual
color sensitivity of the photo-elements themselves is shown as
a dotted curve.

The electron gun producing the beam is quite an important
factor in the performance of the iconoscope. Since the resolution is defined by the size of the spot, the gun should be designed to supply exactly the size of spot corresponding to the
number of picture elements for which the iconoscope is
designed. For the given example of 76,000 picture elements
and a mosaic plate about 4″ high, the distance between two
successive lines is about .016″ and the diameter of the cathode
ray spot approximately half of this size. This imposes quite
a serious problem in gun design.

The electron gun used for this purpose is quite similar to
the one used for the cathode ray tube for television reception,
or the kinescope, which has already been described in several
papers.[5] The components of the gun are shown in Fig. 12.

FIG. 12.

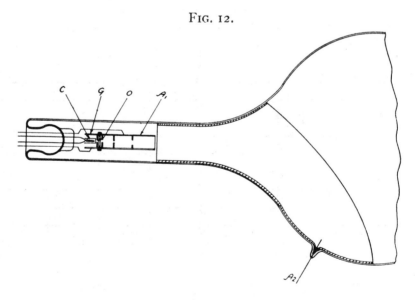

It consists of an indirectly heated cathode, C, with the
emitting area located at the tip of the cathode sleeve. The
cathode is mounted in front of the aperture O of the controlling
element, G. The anode A_1 consists of a long cylinder with
three apertures aligned on the same axis with cathode and
control element. The gun is mounted in the long narrow

[5] V. K. Zworykin, *Jour. Radio Engineering*, December, 1929.

glass neck attached to the spherical bulb housing the mosaic screen. The inner surface of the neck as well as the part of the sphere is metallized and serves as the second anode for the gun and also as collector for photo-electrons from the mosaic. The first anode usually operates at a fraction of the voltage applied to the second anode, which is approximately 1000 volts.

The focusing is accomplished by an electrostatic field set up by potential differences applied between parts of the electron gun, and between the gun itself and the metallized portion of the neck of the iconoscope.

If an electron enters the field along the lines of force, its velocity, but not the direction of motion, will be affected. If an electron enters the field with velocity, V, at an angle α, as shown on Fig. 13, both the velocity and the direction will be

FIG. 13.

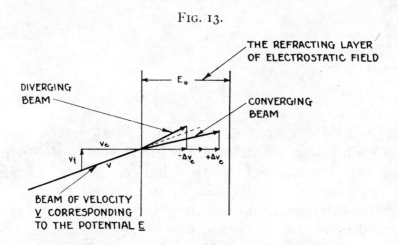

affected, as shown by the vector diagram. In case the field is accelerating, the direction of motion of the electron will be bent toward the axis of symmetry of the field, and if the field is decelerating, the electron will be deflected in the opposite direction. Thus a stream of electrons can be made to converge or diverge.

This interaction between the electrons and electrostatic field can be used to form a sort of lens, as shown on Fig. 14. The lens in this case is converging, but it can be changed to a diverging one simply by reversing the potentials.

The same effect can be accomplished by a field produced by

a difference in potential of two cylindrical electrodes or between two diaphragms. The lines of force in both cases will
force the electrons of a beam, moving inside of these electrodes,

FIG. 14.

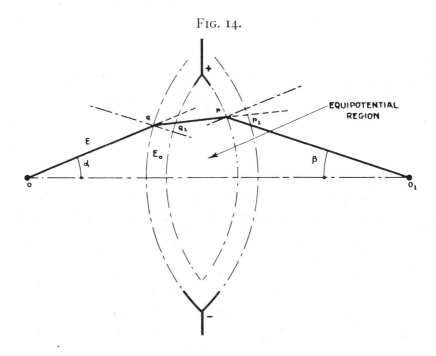

toward the axis, overcoming the natural tendency of electrons
to repel each other. This action is analogous to the focusing
of light rays by means of optical lenses. The electrostatic
lenses, however, have a peculiarity in that their index of refraction for electrons is not confined to the boundary between
the optical media, as in optics, but varies throughout all the
length of the electrostatic field. By proper arrangement of
electrodes and potentials, it is always possible to produce a
complex electrostatic lens which will be equivalent to either
positive or negative optical lenses.

The distribution of electrostatic fields in the electron gun
is shown on Fig. 15. In this particular case, the total action
of fields on electrons is approximately equivalent to a combination of two non-symmetrical lenses, as is shown in the same
figure.

The first lens forces the electrons through the aperture
of the first anode and assure the desired control of the beam by
the control element, G. The final focusing of the beam on

the mosaic is accomplished by the second lens created by the field between the end of the gun and the neck of the bulb. Thus, the final size of the spot on the mosaic, as in its optical

FIG. 15.

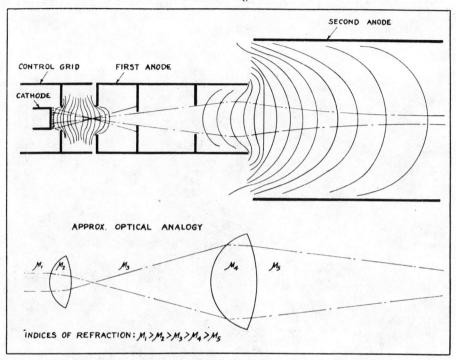

analogue, depends chiefly on the size of the active area of the cathode and the optical distances between the cathode, lenses and the mosaic.

The deflection of the electron beam for scanning the mosaic is accomplished by a magnetic field. The deflection coils are arranged in a yoke which slips over the neck of the iconoscope. The scanning is linear in both vertical and horizontal directions and is caused by saw-tooth shaped electrical impulses passing through the deflecting coils and generated by special tube generators.

From the color response curve shown on Fig. 11, it is clear that the iconoscope can be used not only for transmission of pictures in visible light but also for pictures invisible to the eye, in which the illumination is either by ultra-violet or infra-red light.

The present sensitivity of the iconoscope is approximately

equal to that of a photographic film operating at the speed of a motion picture camera, with the same optical system. The inherent resolution of the device is higher than required for 76,000 picture element transmission. Some of the actually constructed tubes are good up to 500 lines with a good margin for future improvement.

Fig. 16.

Since the iconoscope is practically a self-contained pickup unit, it is possible to design a very compact camera containing the iconoscope and a pair of amplifier stages connected with the main amplifier and deflecting units by means of a long cable.   Since the camera is portable, it can be taken to any point of interest for the transmission of a television picture. The photograph of such a camera is shown on Fig. 16.

FIG. 17.

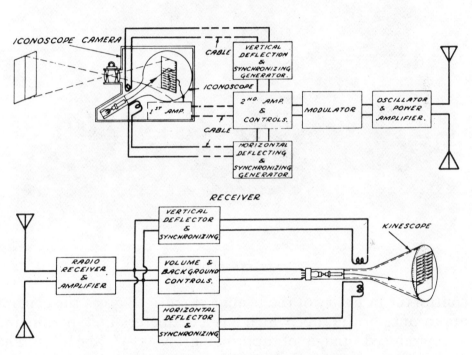

The complete block diagram of the circuit associated with the transmitting and receiving ends of the whole system is shown on Fig. 17, where the components and their location in the system are indicated.   Naming the units in order, we have for television from the studio: The iconoscope camera, the picture signal and synchronizing signal amplifiers, the control and switching equipment, the modulating and radio transmitter equipment.   The units comprising the television receiver are: A radio receiver, the cathode ray unit or kinescope, and its associated horizontal and vertical deflecting equipment.

The name "kinescope" has been applied to the cathode ray tube used in the television receiver to distinguish it from ordinary cathode ray oscilloscopes because it has several important points of difference: for instance, an added element to

FIG. 18.

THE KINESCOPE.

control the intensity of the beam. Figure 18 gives the general appearance of the tube which has a diameter of 9″, permitting a reproduced image of approximately $5\frac{1}{2}″ \times 6\frac{1}{2}″$. The electron gun is quite similar to the one used for the iconoscope, except that it operates at a higher potential for the second anode.

As in the latter, the gun is situated in the long, narrow neck attached to the large cone-shaped end of the kinescope, the inner surface of the cone being silvered or otherwise metallized to serve as the second anode. The purpose of the second anode, A_2, is to accelerate the electrons emerging from the electron gun and to form the electrostatic field to focus them into a very small, threadlike beam.

After leaving the first anode, the focused, accelerated beam impinges upon the fluorescent screen deposited upon the flat end of the conical portion of the kinescope. The fluorescent screen serves as a transducer, absorbing electrical energy and emitting light. Thus there is produced a small bright spot on the screen, approximately equal in area to the cross-section of the beam. The fluorescent screen is very thin, so a large portion of the emitted light is transmitted outside of the tube as useful illumination.

FIG. 19.

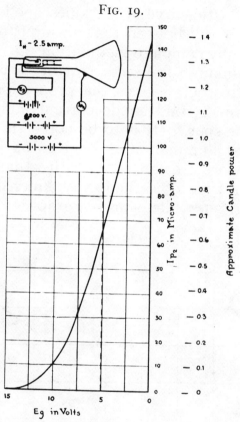

In order to reproduce the light intensity variations of the original picture, it is necessary to vary the intensity of the spot of light upon the fluorescent screen. This is accomplished by means of the control element, G, of the electron gun. For satisfactory reproduction, the control of the electron beam intensity should be a linear function of the input signal voltage. Furthermore, it is very essential that during the exercising of this control the sharp focusing of the spot shall not be destroyed. Still another requirement is that this control will not affect the velocity of the electron beam because the deflection of the beam is inversely proportional to its velocity and, therefore, a slight change due to picture modulation would disturb the image, making the bright lines shorter and the darker lines longer. As a result of careful design, the variation of velocity of the beam (from complete cut-off to full brilliancy) in the kinescope is so small as to be unnoticeable to the observer of the picture.

The characteristic curve of the kinescope is shown in Fig. 19. From this it will be seen that an input of 10 volts A.C. will give practically complete modulation (i.e., a change from maximum to minimum brilliancy) of the cathode ray beam. The shape of this curve gives the proportionality between input voltage and second anode current and corresponding brightness of the spot. (It is to be noted that the values of current, voltage, illumination, etc., given on Fig. 19, and all figures in this report, are illustrative rather than specific values for a particular type of cathode ray kinescope tube.) By referring to Fig. 20, which shows a graph of the relation of

FIG. 20.

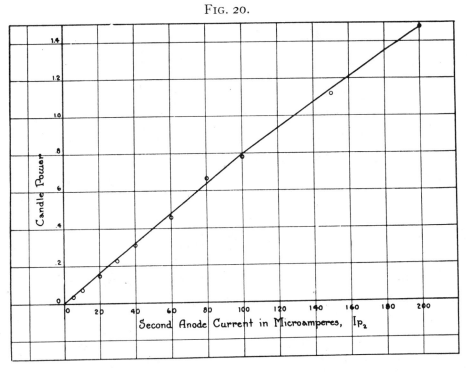

second anode current to the light emitted from the fluorescent screen, it will be noticed that a linear proportionality exists.

Therefore, we can draw the conclusion that a television picture, varying in shade from black to white, will have accurately reproduced all the intermediate shadings necessary for good half-tone pictures.

The material used for the fluorescent screen is a synthetic zinc ortho-silicate phosphor almost identical with natural Willemite. Zinc ortho-silicate was chosen because of its luminous efficiency, its short time lag, its comparative stability and its resistance to "burning" by the electron beam. The good luminous efficiency is due to the fact that the light, green in color, emitted by the zinc ortho-silicate lies in the visible spectrum in a narrow band peaked at 5230A, close to the wavelength of maximum sensitivity of the eye (5560A), as shown in Fig. 21.

FIG. 21.

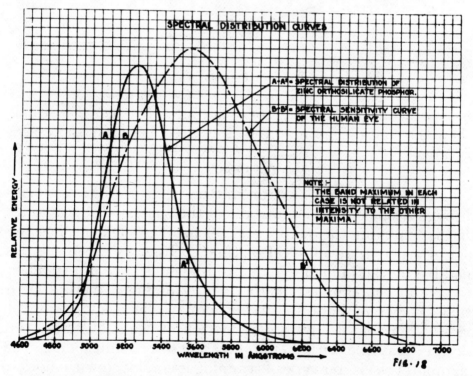

Figure 22 shows the time decay curve of the zinc-ortho-silicate luminescence. The decay curve shows that at the end of approximately 0.06 second practically all visible luminescence has ceased. For reproducing 24 pictures per second, the decay curve of the ideal phosphor should be long enough so that the phosphor just loses its effective brilliancy at the end of 1/24th of a second. If the time of decay is too long, the moving portions of the picture will "trail," as, for instance, the path of a moving baseball would be marked by a comet-like tail. If the time of decay is too short, flicker is noticeable because of

FIG. 22.

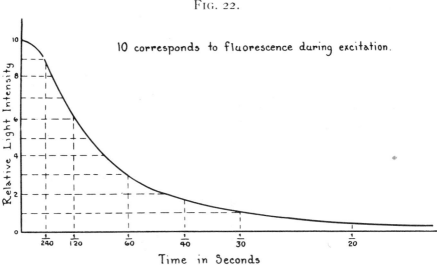

10 corresponds to fluorescence during excitation.

the space of comparative total darkness between the times when the fluorescent material is excited between successive pictures.

SCANNING.

The requirement of transmitting the picture impulses through a single channel introduces the necessity of scanning; that is, exploring the picture area, element by element, in some logical order in an interval of time so brief as not to be detectable by the human eye due to its persistence of vision.

One of the simplest methods of scanning a picture is to cause a spot of light to sweep across it in a succession of parallel horizontal lines. The motion of the spot across the picture may be either uni-directional or sinusoidal. An example of the latter type of scanning is employed with motion picture film in the system described in an earlier paper.[6] An example of uni-directional motion in scanning is that produced by the Nipkow disc widely used in television.

[6] V. K. Zworykin, "Television with Cathode Ray Tube for Receiver," *Radio Eng.*, Vol. IX, No. 12, December, 1929, pp. 38–41.

In the present system the scanning is also uni-directional and is accomplished by deflecting the electron beams, both of the iconoscope and kinescope, by means of electromagnetic fields. The beam traces a succession of equally spaced horizontal lines across the fluorescent screen, reconstructing the television picture in the identical manner that the spot at the transmitter has scanned it, beginning from the top downward and after the last line jumping back to the position at the start of a new picture.

In order to scan with a cathode ray beam in this manner, two variable magnetic fields are applied to the beam just as it emerges from the electron gun; a vertical one, pulsating N times per second, and a horizontal one, pulsating as many times faster as the number of lines in the picture.

Fig. 23.

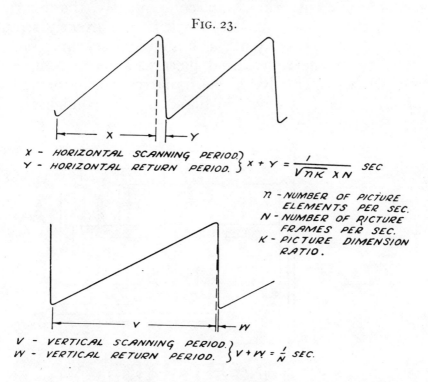

X – HORIZONTAL SCANNING PERIOD.
Y – HORIZONTAL RETURN PERIOD. $\Big\}$ $X + Y = \dfrac{1}{\sqrt{nK} \times N}$ SEC

n – NUMBER OF PICTURE ELEMENTS PER SEC.
N – NUMBER OF PICTURE FRAMES PER SEC.
K – PICTURE DIMENSION RATIO.

V – VERTICAL SCANNING PERIOD.
W – VERTICAL RETURN PERIOD. $\Big\}$ $V + W = \dfrac{1}{N}$ SEC.

In order that the cathode beam at the receiver will follow the uni-directional scanning at the transmitter, the variation of intensity of both horizontal and vertical deflecting fields plotted against time is of a "saw-tooth" shape as shown in Fig. 23. Each cycle consists of two parts; the first, linear with respect to time and lasting practically the whole cycle, and the second, or return period, lasting only a small fraction of the cycle. The picture is reproduced during the first part of the scanning period by varying the bias of the control element according to the light intensities of the transmitted picture, as described above.

There are a number of methods that will produce "saw-tooth" shaped electrical impulses. A simple one has been described in an earlier paper,[7] consisting of charging a condenser through a current limiting device such as a saturated two-electrode vacuum valve and then discharging the condenser through a thermionic or gas-discharge tube. The practical limitation of this "saw-tooth" generator lies in the fact that there is no such thing as a completely saturated thermionic tube. Therefore, the condenser cannot be charged exactly linearly with time, and, consequently, the line reproduced on the fluorescent screen will be not exactly straight.

In order to straighten the scanning lines and improve the quality of the reproduced picture, a more complicated circuit was used, involving one dynatron oscillator and two amplifying tubes, as shown in Fig. 24. The condenser, C, in

FIG. 24.

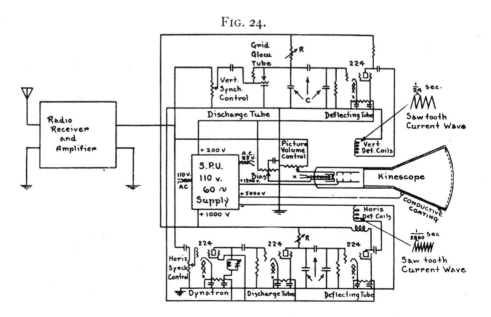

the horizontal deflecting circuit is charged continuously through the resistance, R. Periodically, at the end of pre-

[7] V. K. Zworykin, "Television with Cathode Ray Tube for Receiver," *Radio Eng.*, Vol. IX, No. 12, December, 1929, pp. 38–41.

determined intervals, the condenser is discharged.   During these intervals, the accumulated charge does not reach saturation value, for the time is insufficient.   The vacuum tube through which the discharge takes place is controlled by impulses supplied from a dynatron oscillator having a distorted wave shape.   The frequency of oscillation of the dynatron (which can be made to vary over a fairly wide range) is initially adjusted approximately to the frequency of the scanning of the transmitter, so that received synchronizing signals will have no difficulty in pulling the dynatron into step with the synchronizing impulses generated at the transmitter.   The charging and discharging of condenser, C, represent saw-tooth variation of potential, which, when applied to the grid of an amplifying tube, produce saw-tooth current impulses in deflecting coils connected in the plate of the amplifier.

The vertical deflecting circuit is similar to the horizontal circuit just described.   Both vertical and horizontal deflecting systems operate on the beam by the magnetic fields generated by coils placed about the neck of the cathode ray tube.

The choice of electro-magnetic deflection in preference to electrostatic was made more as a result of economical consideration than as a mechanical choice.   The kinescope for magnetic deflection is much cheaper to make than the one equipped with inside deflecting plates for electrostatic deflection.   On the other hand, the electromagnetic deflecting unit itself requires more power and is more costly to build than the electrostatic one.   The predominance of one or more factors depends chiefly upon the frequency of deflection and velocity of the beam.

The constants of the electrical circuits for vertical and horizontal deflection are, of course, entirely different, due to the great difference in the operating frequencies of the two deflection circuits.

SYNCHRONIZATION.

When both deflecting circuits are properly adjusted and synchronized with the transmitter, a pattern consisting of a number of parallel lines is seen on the fluorescent screen.   The sharpness of the pattern and perfection of its synchronization with the iconoscope determine to a large extent the quality of the reproduced picture. This pattern is transformed into the picture by applying the picture signal impulses from the transmitter to the control element of the kinescope, so as to momentarily vary the brilliancy of the spot.

For sending synchronizing signals to the receiver, the impulses produced by the deflecting generators of the iconoscope are fed into the amplifier and united with the picture signals and, therefore, are transmitted over the picture signal channel. They do not interfere with the picture signals, because they occur at an instant when the picture is not being transmitted.

Vertical synchronization is carried out in the same manner, synchronizing impulses for this purpose being transmitted at the completion of each frame.

Considerable advantage is gained from using a synchronizing system in which the beam at the receiver is brought into step with the transmitter at the end of each horizontal line, because momentary disturbances of the nature of static do not appreciably affect the picture.

It will be seen that the radio transmitter is modulated by picture, horizontal synchronizing and vertical synchronizing signals. The resulting composite signal which is fed to the receiver's modulator grid, therefore, appears as shown in Fig. 25, the top curve of which represents the irregular-shaped

FIG. 25.

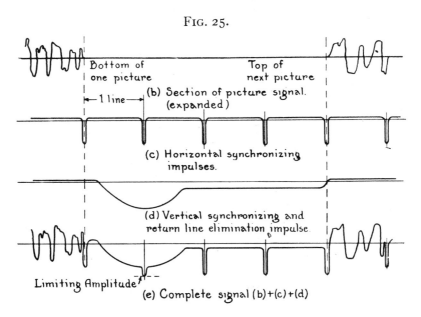

Bottom of one picture   Top of next picture

(b) Section of picture signal. (expanded)

—1 line—

(c) Horizontal synchronizing impulses.

(d) Vertical synchronizing and return line elimination impulse.

Limiting Amplitude

(e) Complete signal (b)+(c)+(d)

picture signals which are often unsymmetrical about the axis, usually more positive than negative. Both synchronizing signals are arranged to have their peaks on the negative side of the axis. The difference in shape of the horizontal and vertical impulses, of course, is due to the time of duration of both signals and the resultant difference in wave shape is utilized at the receiver for the purpose of separating these two synchronizing impulses. The three signals mentioned above differ in frequency and in amplitude. The peak picture signal amplitude is carefully adjusted to be always less than the horizontal and vertical impulses, the amplitude of the latter being approximately equal.

The separation of the three signals at the receiver is accomplished by a very simple means which is described in detail in another paper,[8] so that the fundamentals only will be mentioned here. If we trace the signals at the receiver from the antenna through the radio receiver and amplifier, shown in Fig. 24, we find that they are applied to three independent units, the vertical deflecting system, the horizontal deflecting system and the input to the kinescope. The synchronizing impulses do not affect the picture on the kinescope because they are transmitted at a time when the cathode ray beam is extinguished, that is, during its return period. The picture signals do not affect the deflecting circuits because amplitude selection is utilized; that is, the amplitude of the picture signals is never sufficient to affect the input tubes of either deflecting system. The selection between vertical and horizontal synchronizing impulses is made on the basis of wave shape selection. A simple filter in each of the input circuits of the two deflecting units gives satisfactory discrimination against undesired synchronizing impulses. The plate circuits of both dynatron input tubes contain circuits approximately resonant to the operating periods of their respective deflecting circuits, thus aiding in the matter of selectivity.

When the electron beam returns to the position from which it starts to trace a new line, and particularly when it returns from the bottom of the picture to start a new frame, an undesirable light trace, called the return line, is visible in the picture. To eliminate this the synchronizing impulses which

---

[8] R. D. Kell, *Proceedings I. R. E.*, December, 1933.

are in the negative direction are applied to the control electrode of the kinescope, so as to bias it negatively and thus eliminate the return line by extinguishing the beam during its return.

To produce a picture, the intensity of light on a fluorescent screen is varied by impressing the picture signal on the kinescope control element. If the bias adjustment on the kinescope is set so that the picture signals have the maximum swing on the characteristic curve of the kinescope (shown in Fig. 19) a picture with optimum contrast is produced. The picture background, or the average illumination of the picture, can be controlled by the operator by adjusting the kinescope bias.

REPRODUCING EQUIPMENT.

The arrangement of the television receiver built for these tests is shown in Figs. 26 and 27. The former is a photograph of the chassis containing the deflecting unit and kinescope. This chassis slides as a unit into the cabinet. Figure 27 is a photograph of the complete receiver which contains a power unit, kinescope unit, two radio receivers—one for picture and one for sound signals—and a loud speaker.

The reproduced image is viewed in a mirror mounted on the inside lid of the cabinet. In this way the lid shields the picture from overhead illumination. This method also affords a greater and more convenient viewing angle. The brilliancy of the picture is sufficient to permit observation without the necessity of completely darkening the room. Since this type of television receiver has no moving mechanical parts, it is quiet in operation.

The next pictures show the results obtained by means of the above described system. Figures 28 [9] and 29 show photographs of the fluorescent screen of the kinescope when the same pictures as used for Figs. 2 and 3 were being transmitted.

Figures 30 and 31 are actually the pictures transmitted with the iconoscope camera in the studio and outdoors.

With the advent of an instrument of these capabilities, new prospects are opened for high-grade television transmission. In addition, wide possibilities appear in the application of such

---

[9] Courtesy of Mr. R. D. Kell.

FIG. 26.

FIG. 27.

FIG. 28.

FIG. 29.

FIG. 30.

FIG. 31.

tubes in many fields as a substitute for the human eye, or for the observation of phenomena at present completely hidden from the eye, as in the case of the ultra-violet microscope.

The writer wishes gratefully to acknowledge the untiring and conscientious assistance of Messrs. G. N. Ogloblinsky, S. F. Essig, H. A. Iams, A. W. Vance and L. E. Flory, who carried on much of the theoretical and experimental work connected with the development which has been described in the foregoing, and whose ability was the major factor in the successful solution of the many problems arising in the course of this work.

REPORT OF THE
TELEVISION
COMMITTEE

Presented by the Postmaster-General to Parliament
by Command of His Majesty
January, 1935

LONDON
PRINTED AND PUBLISHED BY HIS MAJESTY'S STATIONERY OFFICE

1935

Cmd. 4793

THE TELEVISION COMMITTEE

The Right Hon. The Lord Selsdon, K.B.E. (*Chairman*).

Sir John Cadman, G.C.M.G., D.Sc. (*Vice-Chairman*).

Col. A. S. Angwin, D.S.O., M.C., B.Sc., Assistant Engineer-in-Chief, General Post Office.

Noel Ashbridge, Esq., B.Sc., Chief Engineer, British Broadcasting Corporation.

O. F. Brown, Esq., M.A., B.Sc., Department of Scientific and Industrial Research.

Vice-Admiral Sir Charles Carpendale, C.B., Controller, British Broadcasting Corporation.

F. W. Phillips, Esq., Assistant Secretary, General Post Office.

Secretary : J. Varley Roberts, Esq., M.C.,
Telegraph and Telephone Department,
General Post Office, E.C.1.

Note.—The expenses incurred in preparing this Report are estimated at £965, of which £24 represents the estimated cost of printing and publication.

INDEX

APPENDICES

THE TELEVISION COMMITTEE
REPORT

The Right Hon. Sir KINGSLEY WOOD, M.P.,
His Majesty's Postmaster General.

TERMS OF REFERENCE AND PROCEEDINGS OF THE COMMITTEE

1. The appointment of the Committee was announced in the House of Commons on the 14th of May, 1934, with the following terms of reference :—

"To consider the development of Television and to advise the Postmaster General on the relative merits of the several systems and on the conditions under which any public service of Television should be provided."

2. A notification was made in the Press on the 29th of May, 1934, and again on the 11th of June, 1934, intimating that the Committee were prepared to receive evidence on the subject of Television from any interested society, firm or individual.

3. We have examined 38 witnesses—some of them on more than one occasion—representing many different interests, on the various aspects of Television. A list of witnesses who have appeared before us is set out in Appendix I. In addition, we have had the benefit of consultation with members of various Departments of the Government, who have afforded us every facility and assistance ; and we have received numerous written statements regarding Television from various sources.

(Appendix II not printed.) 4. A note of the formal evidence given is presented in Appendix II (Volumes I to IV), but owing to the fact that much of this information, containing secrets of commercial value, was necessarily received in confidence and under promise of secrecy, we trust that this pledge may be maintained and that accordingly the record, while available for yourself and your responsible officers, will not be published. *(Appendix III not printed.)* For similar reasons we recommend that Appendix III, containing reports on developments in the United States and Germany, and *(Appendix IV not printed.)* Appendix IV, containing a description of each television system we have examined in this country, should not be published. *Appendix V (not printed.)* Appendix V, containing certain financial details, is also of a confidential nature (*see* paragraph 63).

5. We have inspected, in some cases upon several occasions, all the different television systems belonging to firms who were prepared to provide demonstrations. Of the systems under development in this country, the most distinctive are those of the Baird, Cossor, Marconi-E.M.I. and Scophony Companies.

6. Further, we despatched with your approval a delegation headed by the Chairman, to investigate and report upon progress in television research in the United States, and a delegation headed by Mr. O. F. Brown, to Germany for a similar purpose. We have also been furnished with information regarding the position in certain other countries.

7. In America, our delegation visited and inspected many of the chief centres of television experimental research, as well as the plant and laboratories of the principal Broadcasting, Telephone and Telegraph Authorities. They had also the advantage of consultation in Washington with the Federal Communications Commission. To all of these Corporations and Authorities we desire to make the fullest and most sincere acknowledgment of the kindness and courtesy extended to the delegation, who were given every possible opportunity for the fullest examination of methods and plant, and the frankest interchange of opinion. A detailed report of the proceedings and conclusions of the delegation is submitted as Appendix III, A.

8. In Germany, our delegation made a similar inspection of the television experimental installations belonging to the Reichspost and also of those of several private firms in Berlin, and they had many profitable discussions with officials of the Reichspost and others regarding various aspects of Television. As in America, so also in Germany, every facility was accorded to our delegation in their investigations, for which we likewise desire to express our sincere appreciation. A report on the visit to Germany is submitted as Appendix III, B.

BASIC PRINCIPLES OF TELEVISION

9. Television may be defined as the transmission by telegraphy and reproduction in transitory visible form of images of objects in movement or at rest. The equipment utilised usually consists of combinations of optical and electrical apparatus which at the transmitting or " pick up " end of the system convert the image of the object into electric currents, and of similar combinations at the receiving end of the system which resolve the electric currents into visible forms.

10. When an object is viewed by direct vision, light reflected from the object under observation impinges on the eye and is focussed by the lens on to the retina where it stimulates nerve cells.

Each cell communicates with the brain, and the sensation of sight and the perception of any scene result from the relative stimulation applied to the brain by the cells in the retina. As the light-sensitive cells of the retina have finite dimensions, details in an object which produce an image on the retina smaller than a single cell cannot be individually perceived. The eye, therefore, really sees a large number of infinitely small objects, which in the aggregate form the image.

11. Thus all vision is of a granular structure, as is also pictorial reproduction, and, in order to transmit pictures or images over electric circuits, a suitable granular structure is adopted, the relative brilliancy of each grain or elementary area of the picture being transmitted telegraphically to the distant point, where by suitable means an equivalent brilliancy is given to a corresponding area on the receiving screen.

12. The transmission of the relative brilliancy of each grain or elementary area of the picture must be effected in some ordered sequence, and the process by which this is achieved is termed " scanning." The usual method employed is to allow light from a selected area of the subject to impinge on a device known as a photo-electric cell, which delivers an electrical output proportional to the light stimulation it receives. This electrical output, after amplification, is used to control the output of a radio transmitter by methods similar to those in use for the transmission of speech and music. The position of the selected area of the subject is varied in a definite path so that the whole of the subject is covered in a period which should be less than the time of persistence of vision. The path of selection is usually a series of horizontal or vertical parallel lines, and the process somewhat resembles the action of the human eye in reading a page of printed matter, letter by letter and line by line. A further refinement consists in making the path of selection run first along every alternate line and then, as a second process, along the lines omitted in the first process. This is known as interlaced scanning, and it appears to be successful in reducing " flicker."

13. At the receiver, the radio signal is detected and amplified by methods similar to those used for radio-telephony. The electrical signal from the receiver thus resembles the signal from the photo-electric cell or cells at the transmitter, and is used to control the brilliancy of illumination of an elementary area of the screen on which the received picture is to be displayed.

14. It is essential to arrange that the area illuminated on the viewing screen at any given instant shall correspond in position with the area of which the illumination is then being determined by the scanning device. In other words, precise synchronism is necessary between the movements of the scanning device and the receiving device. Various methods have been proposed for

achieving this synchronism; it can, for instance, be accomplished by the sending of two series of special synchronising signals by the transmitter—one series to ensure the correct motion of the picture spot along each line and the second series to signal the instant of termination of one picture and the commencement of the next. As these series of signals occur respectively between successive lines of the picture and between successive pictures, their transmission need not interfere with the picture signals, and they can be sent on the same radio transmitter.

15. The relative brilliancy of each successive grain of the picture is transmitted with such rapidity that persistence of vision produces the effect at the receiving end of a complete picture, the degree of definition and steadiness of which is dependent upon the fineness of the individual grains composing the picture, *i.e.*, the number of lines used for scanning it, and the speed at which complete pictures are successively transmitted.

16. One of the difficulties which has been encountered in direct scanning is the small amount of light available to actuate the photo-electric cell obtained by reflection from objects which are being televised.

17. Accordingly, considerable experimental development has taken place upon a technique whereby the scene to be televised is first photographed on ordinary cinematograph film which, after being developed, is scanned by light transmitted through it. This system can be used to provide a method of delayed Television where direct scanning by a mechanical device would be difficult or impossible. In order to reduce the period of delay, equipment has now been produced in which the cinematograph camera is associated with the film scanner, and the film, after exposure, is immediately developed, fixed, washed and partially dried. It then passes through the scanner, and after further drying is stored for future use if required. In this way, the advances which have been made in photographic processes in the production of rapid and sensitive emulsions can be utilised to overcome the difficulties which are at present encountered due to the comparatively feeble sensitivity of photo-electric cells.

18. The direct scanning of open air scenes and studio subjects without abnormally powerful illuminating devices has also been made possible by the use of cathode rays in combination with photo-sensitive surfaces or minute photo-electric cells. For instance, in one such device, which is being developed in America, Germany and this country, the image to be televised is focussed by means of lenses on to a photo-electric mosaic contained in a cathode ray tube. The cathode ray beam is directed on the surface of the mosaic and by a method of magnetic control the image is scanned repeatedly.

Electrical energy is thus drawn off from the photo-electric mosaic by the cathode ray which is proportional to the light intensity of the picture and can be transmitted to operate the distant television receiver.

19. Our observations lead us to the opinion that this system of "direct pick-up" has already attained a considerable degree of effectiveness, and we should say that satisfactory reproduction of outdoor moving scenes can now be attained by this method in conditions of light, etc., approximating to those under which satisfactory cinematograph pictures can be taken, provided that the recording apparatus can be located reasonably close to 'and at a moderately constant distance from the scene to be televised. We should regard it as probable that satisfactory reproduction could, even at this stage of development, be obtained of such' scenes as a procession, a lawn-tennis match, or the actual finish of a horse race, though the transmission of a view of the whole course of a race, a cricket match, or a football match, would present much greater difficulty.

EXPERIMENTS IN TELEVISION

20. We are informed that the Post Office has always given facilities to qualified persons or firms who have applied for permission to conduct experiments in Television, but the licences issued have been restricted to purely research and experimental work and have given no authority for the conduct of any form of public service.

21. The view taken was that when any system of Television showed sufficient promise to justify its trial for public transmissions, the British Broadcasting Corporation should provide reasonable facilities for such a trial service on a limited scale at one or more of their broadcasting stations.

LOW DEFINITION TELEVISION

22. As far back as the autumn of 1929 the British Broadcasting Corporation gave the Baird Company facilities for experimental transmissions of Television from a broadcasting station. During the next two or three years a large number of experimental transmissions were carried out by the Baird Company independently, as well as in liaison with the British Broadcasting Corporation.

23. Improvements were gradually made in the system, and in August, 1932, the Corporation arranged with Baird Television Limited for public experimental transmissions from their London Station (Brookmans Park) of Television on a wavelength of

261 metres, and of the accompanying sound on a wavelength of 398 metres from the Midland Regional transmitter (Daventry). The Corporation agreed to provide special programme material and also staff for operating the television apparatus, which was installed in Broadcasting House by the Baird Company on a loan basis. These transmissions, the experimental nature of which was emphasised in a notice issued to the Press, have continued up to the present time, although their frequency has been reduced since 31st March, 1934, to two half-hour periods a week which are extended to three-quarters of an hour when circumstances permit.

24. In the case of these transmissions the size of the elements (elementary areas) composing the picture is such as to admit of transmission being effected in a series of thirty lines per picture and each picture is repeated $12\frac{1}{2}$ times per second (*see* paragraphs 12–15).

25. Any pictures built up with a structure of the order of thirty lines are, however, comparatively coarse in texture. Little detail can be given, and generally speaking the pictures are only fitted for the presentation of " close-ups "—*e.g.*, the head and shoulders of a speaker—and the quality of reproduction leaves much to be desired. Moreover, any frequency of the order of $12\frac{1}{2}$ pictures per second gives rise to a large amount of " flicker ".

26. Whilst low definition Television has been the path along which the infant steps of the art have naturally tended and, while this form of Television doubtless still affords scientific interest to wireless experimenters, and may even possess some entertainment value for a limited number of others, we are satisfied that a service of this type would fail to secure the sustained interest of the public generally. We do not, therefore, favour the adoption of any low definition system of Television for a regular public service. We refer later in our report (*see* paragraph 34) to the question of the temporary continuance of the present low definition transmissions pending the institution of a public television service of a more satisfactory type.

HIGH DEFINITION TELEVISION

27. With a view to extending the application of Television to a wider field and thereby increasing its utility and entertainment value, much attention has been given in recent years to the problem of obtaining better definition and reduced " flicker " in the received pictures.

28. The degree of definition it is essential to obtain is necessarily a matter of opinion, but the evidence received and our own observations lead us to the conclusion that it should be not less than 240 lines per picture, with a minimum picture frequency of 25 per second. The standard which has been used extensively for experimental work

is 180 lines, but we should prefer the figure of 240 and we do not exclude the possible use of an even higher order of definition and a frequency of 50 pictures per second.

29. To attain such degrees of definition and picture frequency, very high modulation frequencies are required, which in practice can only be handled by radio transmitters working on ultra-short waves the effective range of which is much more restricted than, the range of the medium waves used for ordinary sound broadcasting (*see* paragraph 47).

30. For the reception of high definition pictures the cathode ray tube is now usually employed. The cathode ray tube receiver involves no moving parts, and the picture is presented as a fluorescence at the end of the tube. A stream of electrons (particles of negative electricity) is projected along the tube, and impinges on a coating of fluorescent material at the end of the tube, the impact of the electrons on the fluorescent material causing illumination. The amount of illumination can be controlled by varying the flow of electrons, and the point of impact can be changed by deflecting the jet by means of electric or magnetic forces. The jet is modulated or controlled in amount by the received signal, and suitable electrical circuits are provided to move the point of impact in exact synchronism with the transmitter (*see* paragraph 14).

31. The size of the picture produced naturally depends upon the size of the cathode ray tube. At present the most usual size gives a picture of about 8 in. by 6 in., although good results have been seen with larger tubes. The apparent size can, of course, be increased by viewing the tube through a suitable fixed magnifying device, though with a corresponding loss of definition. Experimental work is proceeding with a view to the projection of pictures on a screen of much larger dimensions, but this is still in an early stage of development.

32. We are informed that the price to the public of a receiving set capable of producing a picture of about the first-mentioned size, with the accompanying sound, would probably at first be considerable, and various estimates have been given ranging from £50 to £80; but it is reasonable to assume that, if and when receivers were made on a large scale under competitive conditions, this price would be substantially reduced.

33. Most of the high definition television systems follow in broad outline the methods of transmission and reception referred to above, with some variations in technique. We are impressed with the quality of the results obtained by certain of these systems, and whilst much undoubtedly remains to be done in order to render the results satisfactory in all respects, we feel that a standard has now been reached which justifies the first steps being taken towards the early establishment of a public television service of the high definition type in this country.

34. As regards the existing low definition broadcasts, these no doubt possess, as we have said, a certain value to those interested in Television as an art, and possibly, but to a very minor extent, to those interested in it only as an entertainment. We feel that it would be undesirable to deprive these " pioneer lookers " of their present facilities until at least a proportion of them have the opportunity of receiving a high definition service. On the other hand, the maintenance of these low definition broadcasts involves not only some expense, but also possibly considerable practical difficulties. We can only, therefore, recommend—

(1) that the existing low definition broadcasts be maintained, if practicable, for the present ; and

(2) that the selection of the moment for their discontinuance be left for consideration by the Advisory Committee (see paragraph 41),

with the observation that, if practicable so to maintain these broadcasts, they might reasonably be discontinued as soon as the first station of a high definition service is working.

SCOPE OF TELEVISION AND ITS RELATION TO SOUND BROADCASTING

35. In our opinion there will be little, if any, scope for television broadcasts unaccompanied by sound. Television is, however, a natural adjunct to sound broadcasting and its use will make it possible for the eye as well as the ear of the listener to be reached. Associated with sound it will greatly enhance the interest of certain of the existing types of broadcast and will also render practicable the production of other types in which interest is more dependent upon sight than upon sound.

36. We are of the opinion that there are two factors which for a number of years will tend to prevent a television service being made use of to the same extent as present day sound broadcasting—

(1) The difficulties of wireless communication on ultra-short wavelengths, particularly in hilly districts, may seriously limit the extent to which the country can be effectively covered.

(2) Some time is likely to elapse before the price of an efficient television receiver will be comparable with that of the average type of receiver now in use for sound broadcasting.

Nevertheless the time may come when a sound broadcasting service entirely unaccompanied by Television will be almost as rare as the silent cinema film is to-day. We think, however, that in general sound will always be the more important factor in broadcasting. Consequently the promotion of Television must not be allowed to prevent the continued development of sound broadcasting.

37. No doubt the evolution of Television will gradually demonstrate the possibility of its application for many purposes other than those of entertainment and illustrative information. Its uses for purposes of advertisement are obvious, were such deemed desirable. We can conceive, moreover, its potential application—as distinct from existing practice in picture transmission—to public telegraphic and telephonic services, to the transmission of lists of prices, or of facsimile signatures or documents, and to its use by the police and the forces of the Crown, or as an aid to navigation.

38. We have assumed, however, that we were intended by our terms of reference to confine our attention to the question of the introduction of a public broadcast service of Television, and we do not, therefore, make any further observations regarding its other possible applications beyond expressing the earnest hope that it will be allowed the fullest possible freedom for development consonant with the public interest.

TELEVISION OPERATING AUTHORITY

39. Holding the view which we do of the close relationship which must exist between sound and television broadcasting, we cannot do otherwise than conclude that the Authority which is responsible for the former—at present the British Broadcasting Corporation—should also be entrusted with the latter. We therefore recommend accordingly ; and we have received an assurance that the Corporation is prepared fully to accept this additional responsibility and to enter whole-heartedly into the development of Television in conformity with the best interests of the licence-paying public. In discharging this task the accumulated experience of the Corporation as regards sound broadcasting cannot fail to prove of great value. Presumably a separate licence will be required from the Postmaster General specifically authorising the Corporation to undertake the broadcasting of Television.

40. We have, of course, considered the possible alternative of letting private enterprise nurture the infant service until it is seen whether it grows sufficiently lusty to deserve adoption by a public authority. This would involve the granting of licences for the transmission of sound and vision to several different firms who are pioneering in this experimental field. We should regret this course, not only because it would involve a departure from the principle of having only a single authority broadcasting a public sound service on the air, and because the subsequent process of " adoption " (which we believe would be inevitable) would be rendered costly owing to the growth of vested interests, but also because we foresee serious practical difficulties as regards the grant of licences to the existing pioneers as well as possibly to a constant succession of fresh applicants. It is therefore our considered conclusion that the

conduct of a broadcast television service should from the outset be entrusted to a single organisation, and we are satisfied that it would be in the public interest that the responsibility should be laid on the British Broadcasting Corporation.

ADVISORY COMMITTEE

41. Whilst we think that the British Broadcasting Corporation should exercise control of the actual operation of the television service to the same extent and subject to the same broad principles as in the case of sound broadcasting, we recommend that the initiation and early development of this service should be planned and guided by an Advisory Committee appointed by the Postmaster General, on which the Post Office, the Department of Scientific and Industrial Research and the British Broadcasting Corporation should be represented, together with such other members as may be considered desirable. We recommend that this Committee should be appointed forthwith, for a period of, say, five years.

42. The Committee should advise on the following :—

(a) The performance specification for the two sets of apparatus mentioned in paragraph 56, including acceptance tests, and the selection of the location of the first transmitting station.

(b) The number of stations to be built subsequently, and the choice of districts in which they should be located (see paragraph 57).

(c) The minimum number of programme hours to be transmitted from each station.

(d) The establishment of the essential technical data governing all television transmissions, such as the number of lines per picture, the number of pictures transmitted per second, and the nature of the synchronising signals.

(e) The potentialities of new systems.

(f) Proposals by the British Broadcasting Corporation with regard to the exact site of each station, and the general lines on which the stations should be designed.

(g) All patent difficulties of a serious nature arising from the operation of the service in relation to both transmission and reception.

(h) Any problem in connexion with the television service which may from time to time be referred to it by His Majesty's Government or the British Broadcasting Corporation.

Normally the Committee would not concern itself with detailed financial allocations, or with business negotiations between suppliers of apparatus and the British Broadcasting Corporation. It is further considered that the Committee should not deal with the compilation of programmes, the detailed construction of stations, or their day-to-day operation, unless specifically invited to do so under sub-paragraph (*h*).

43. It will be clear from the foregoing that the Committee would be composed of both technical and non-technical members, and it is anticipated that a part of the Committee's work would best be carried out by a technical sub-committee.

44. Such experimental work as may be necessary for the establishment of stations and the operation of the service would be carried out by the British Broadcasting Corporation in the usual course of its functions, but this would not, of course, preclude the enlistment of the co-operation of Government Departments or other organisations in technical researches.

USE OF ULTRA-SHORT WAVES FOR TELEVISION AND THEIR EFFECTIVE RANGE

45. As previously mentioned, the transmission of high definition Television is practicable only with ultra-short waves, and a wide band of frequencies is necessary. Fortunately, there should be no difficulty, at present at all events, in assigning suitable wavelengths in the spectrum—between 3 and 10 metres—for public Television in this country, although in allocating such wavelengths regard must, of course, be paid to the claims of other services. The recent experimental work has been conducted upon wavelengths around 7 metres.

46. Technically, it is desirable that the transmitting stations should be situated at elevated points, and that the masts should be as high as practicable, consistent with any restrictions which may be deemed necessary by the Government. The mast at present in use in Berlin is about 430 feet high, and the question of employing masts of greater height is under discussion in Germany. Quality of reception varies, of course, with the location of the receiving station and the nature of its surroundings. It may be observed that reception on these ultra-short waves does not seem to be materially affected by atmospherics. The most frequent sources of interference appear at present to arise from some types of electro-therapeutic apparatus, and from the ignition systems of motor cars ; but we understand that it is possible to prevent or reduce certain types of such interference by simple remedial devices.

47. Present experience both here and abroad seems to indicate that these ultra-short waves cannot be relied upon to be effective for a broadcast service much beyond what is commonly called " optical range." Generally speaking, it is at present assumed

that the area capable of being effectively covered by ultra-short wave stations of about 10 kilowatts capacity will not exceed a radius of approximately 25 miles over moderately undulating country. In more hilly districts this may be considerably reduced, and indeed in certain areas an entirely reliable service may be impracticable. It is clear, therefore, that unless and until the effective range be increased, a large number of transmitting stations would be required to provide a service covering most of the country, though we think that with 10 stations, probably at least 50 per cent. of the population could be covered from suitable locations.

PROVISION OF TELEVISION SERVICE

48. We nevertheless envisage the ultimate establishment of a general television service in this country, and in this connexion we contemplate the possibility of television broadcasts being relayed by land line or by wireless from one or more main transmitting stations to sub-stations in different parts of the country. We should observe that recent developments in cable technique render it possible for the first time to transmit, over considerable distances, frequencies such as are required for high definition Television.

49. While the establishment of such a service should be, in our opinion, the aim, we do not feel that we can advise you to proceed at once to approve the construction, at great expense, of a network of stations, intended to cover most of the country. The total number of stations required for such a purpose is as yet unknown to anyone ; and the total cost is accordingly purely speculative. Moreover, Television will be a constantly developing art, and new discoveries and improvements will certainly involve continued modifications of methods—at least during its early years. A general service will only be reached step by step ; but the steps should be as frequent as possible and in our opinion the first step should be taken now. *Solvitur ambulando.*

CHOICE OF SYSTEM AND PATENT DIFFICULTIES

50. We have been furnished with a great deal of information— much of it of a confidential character—concerning various systems of Television. Continuous progress is being made in the art ; and even during the few months of our investigations, research has brought a number of new and important discoveries. We do not think it would be right at, this early stage of development, when practical experience is small and the patent position obscure, that we should attempt to pass final judgment on the several systems of Television. A technical description of each system which we have examined in this country, indicating its distinctive features

and commenting upon its performance, is, however, submitted for your information in Appendix IV. Comments are also made in Appendix III on the systems examined in the United States and Germany.

51. The task of choosing a television system for a public service in this country is one of great difficulty. The system of transmission governs in a varying degree the type of set required for reception ; and it is obviously desirable to guard against any monopolistic control of the manufacture of receiving sets. Further, whatever system or systems are adopted at the outset, it is imperative that nothing should be done to stifle progress or to prevent the adoption of future improvements from whatever source they may come. Moreover, the present patent position is difficult : the number of patents relating to Television is very large, and in regard to many of them there are conflicting views as to their importance and validity.

52. At the same time it is clear from the evidence put before us that those inventors and concerns, who have in the past devoted so much time and money to research and experiment in the development of Television, are looking—quite fairly—to recoup themselves and to gather the fruits of their labours by deriving revenue from the sale of receiving apparatus to the public, whether in sets or in parts, and whether by way of royalties paid by the manufacturers or by manufacturing themselves. It is right that this should be so, and that the growth of a new and important branch of industry, capable of providing employment for a large number of workers, should in every way be fostered and encouraged to develop freely and fully.

53. The ideal solution, if it were feasible, would be that, as a preliminary to the establishment of a public service, a Patent Pool should be formed into which all television patents should be placed, the operating authority being free to select from this Pool whatever patents it desired to use for transmission, and manufacturers being free to use any of the patents required for receiving sets on payment of a reasonable royalty to the Pool. We have seriously considered whether we should advise you to refuse to authorise the establishment of a public service of high definition Television until a comprehensive Patent Pool of this type had been formed, on terms considered satisfactory by the Advisory Committee. From evidence we have received, however, we are convinced that, under present conditions, when the relative value of the numerous television patents is so largely a matter of conjecture, the early formation of such a Pool would present extreme difficulty. The Government would have no power to compel an owner of television patents to put them into the Pool against his will ; and, with the best will in the world, patent holders might find it exceedingly difficult to agree

among themselves on a fair basis for charging royalties and sharing the revenue so obtained. An attempt hastily to negotiate a Pool under these conditions would in all probability end in failure.

54. While, however, we have been compelled to abandon the idea that the formation of a comprehensive Patent Pool should be a condition precedent to the establishment of a public service, we are strongly of opinion that it is in the public interest, and in the interest of the trade itself, that such a Pool should be formed. In framing our recommendations we have kept this objective in mind ; and we trust that events will shape themselves in such a way as to lead to the formation of a satisfactory Patent Pool at no distant date.

START OF SERVICE

55. We have come to the conclusion that a start could best be made with a service of high definition Television by the establishment of such a service in London. It seems probable that the London area can be covered by one transmitting station and that two systems of Television can be operated from that station. On this assumption we suggest that a start be made in such a manner as to provide an extended trial of two systems, under strictly comparable conditions, by installing them side by side at a station in London where they should be used alternately—and not simultaneously—for a public service.

56. There are two systems of high definition Television—owned by Baird Television Limited and Marconi-E.M.I. Television Company Limited respectively—which are in a relatively advanced stage of development, and have indeed been operated experimentally over wireless channels for some time past with satisfactory results. We recommend that the Baird Company be given an opportunity to supply the necessary apparatus for the operation of its system at the London station, and that the Marconi-E.M.I. Company be given a similar opportunity in respect of apparatus for the operation of its system also at that station. Besides any other conditions imposed, acceptance of offers should be subject in each case to the following conditions precedent :—

(a) The price demanded should not, in the opinion of the Advisory Committee, be unreasonable.

(b) The British Broadcasting Corporation to be indemnified against any claim for infringement of patents.

(c) The Company to undertake to grant a licence to any responsible manufacturer to use its existing patents or any patents hereafter held by it, for the manufacture of television receiving sets in this country on payment of royalty.

(*d*) The terms of a standard form of such licence to be agreed upon by the Company with the Radio Manufacturers' Association, or, in default of agreement, to be settled in accordance with the provisions of the Arbitration Acts, 1889 to 1934, or any statutory modification thereof, either by a single arbiter agreed upon by the Company and the Radio Manufacturers' Association, or failing such agreement, by two arbiters—each of the parties nominating one—and an umpire nominated by the Postmaster General.

(*e*) The Company to agree to allow the introduction into its apparatus at the station of devices other than those claimed to be covered under its own patents, in the event of such introduction being recommended by the Advisory Committee.

(*f*) Transmissions from both sets of apparatus should be capable of reception by the same type of receiver without complicated or expensive adjustment.

(*g*) The definition should not be inferior to a standard of 240 lines and 25 pictures per second.

(*h*) The general design of the apparatus should be such as to satisfy the Advisory Committee, and when it has been installed, tests should be given to the satisfaction of the Committee.

DEVELOPMENT OF SERVICE

57. In the light of the experience obtained with the first station, the Advisory Committee should proceed with the planning of additional stations, until a network is gradually built up. The total number of stations and the speed at which they are provided will naturally depend upon the results obtained from the earlier stations, the popularity of the service, finance and other factors. A tentative programme for the location and provision of stations should be framed by the Advisory Committee, and reviewed by them at frequent intervals.

58. Whatever system be adopted for the second or any subsequent station, we recommend that conditions be imposed similar to those set out in paragraph 56, in so far as applicable. The Advisory Committee would, of course, endeavour to secure the incorporation in each fresh station of any improvements which had come to light, and they would also naturally consider the introduction, if possible, of such improvements into existing stations. There should be no serious difficulty in doing this, so long as the changes did not materially

affect the receiving sets, or at any rate so long as the sets already in use could be adapted, without much expense, to the modified system.

59. A more difficult situation would arise if a completely new system, requiring an entirely new type of receiving set, should be evolved and should prove on trial to be definitely superior to the systems already in use. In such a case it might be necessary to adopt the improved system, in the first instance, at new stations only, and to postpone for a time its adoption at the older stations. For it is obvious that many persons would be deterred from purchasing television sets unless they had some assurance that these sets would not be rendered useless at an early date by a complete change in the transmitting system. No radical changes should, therefore, be made in the systems serving particular areas without reasonable notice being given by the British Broadcasting Corporation of the contemplated change. In the initial stages this notice should not be less than, say, two years. The Corporation would naturally consult the Advisory Committee on this point. While giving some reasonable measure of security in this direction, the aim should be to take advantage, as far as possible, of all improvements in the art of Television, and at the same time to work towards the ultimate attainment of a national standardised system of transmission.

PROGRAMMES

60. It is scarcely within our province to make detailed recommendations on the subject of television programmes. To what extent those programmes should consist of direct transmissions of studio or outdoor scenes, or televised reproductions of films, must be determined largely by experience, technical progress and public support, as well as by financial considerations. No doubt the televising of sporting and other public events will have a wide appeal, and will add considerably to the attractiveness of the service. We regard such transmissions as a desirable part of a public television service, and it is essential that the British Broadcasting Corporation should have complete freedom for the televising of such scenes, with appropriate sound accompaniment, at any time of the day.

61. With regard to the duration of television programmes, we do not consider that it will be necessary at the outset to provide programmes for many hours a day. An hour's transmission in the morning or afternoon which will give facilities for trade demonstrations and, say, two hours in the evening, will probably suffice. As regards the future, the British Broadcasting Corporation and the Advisory Committee will doubtless be guided by experience and by financial considerations.

FINANCE

62. For reasons already explained, it will be clear that at this stage no human being can estimate the cost of constructing and working a national network—nor even of such a partial system of 10 stations, as is referred to in paragraph 47—with anything resembling accuracy. Even with all the resources at our command we have been quite unable to do so, and we confine ourselves, therefore, to giving what we hope may prove to be a fairly close estimate of the cost of providing and working the London station referred to in paragraph 55 up to 31st December, 1936. We should explain that we have taken this period, firstly on the assumption that it may be possible to start the service during the latter part of 1935, and secondly because 31st December, 1936, is the date on which the British Broadcasting Corporation's present Charter is due to expire. The relevance of this point lies in the fact that, if the television service is continued and expanded on the lines contemplated, then its finance will inevitably become bound up with the question of the Corporation's finance in general. We gather that it is probable that this will come under review in connexion with the renewal of the Charter, and accordingly we confine ourselves to the consideration of Television finance for the intervening period only, observing that within that time the Advisory Committee should be able to formulate an opinion as to the development of the service.

63. We estimate that the cost of providing the London station, including all running and maintenance expenses, programme costs and amortization charges (calculated on the basis of a comparatively rapid obsolescence), for the period up to 31st December, 1936, will be £180,000. For obvious reasons we refrain from specifying here the details upon which this estimate is built, but these are available to you in the confidential section of the Report (*see* Appendix V). Lest, however, too hasty conclusions be drawn from this figure we add the following observations. It must not be assumed that an accurate estimate of the cost of a number of stations can be reached by the simple process of multiplication. By far the largest factor in the above figure is the programme cost. On the one hand, if the service is a success, the cost of programmes will certainly rise materially, just as the cost of sound programmes has risen. We have not budgeted during this early stage for a programme comparable in duration, variety, or quality, with existing sound programmes, although the service should be amply adequate to provide interest and entertainment for the public, as well as opportunity for daily demonstrations by retailers of sets.

On the other hand, if and when a number of stations start working, it is contemplated that one programme may be relayed simultaneously to all stations (*see* paragraph **48**), and that only a small portion of the daily output will consist in each case of topical items of local interest. In the case, therefore, of each additional station, the amount to be added to other charges in respect of programme costs will be merely fractional.

64. We have carefully considered the question of providing the necessary funds. Roughly speaking, the means suggested to us for so doing may be classified under two heads :—

(*a*) Selling time for advertisements, and

(*b*) Licence revenue.

65. Advertisements may take two forms : they may be either (i) direct advertisements for which time is bought by the advertiser such as, for instance, a dress show by Messrs. Blank ; or (ii) the acceptance, as a gift, of programmes provided by an advertiser and coupled with the intimation of his name, in accordance with a standard formula, such as, for instance, " This programme comes to you through the generosity of Messrs. Dot & Dash," the latter system being usually known as that of " sponsored programmes." As regards direct advertisements, this proposal has been frequently examined in past years. In relation to sound broadcasting it was discussed and rejected by the Sykes Committee on Broadcasting in 1923 (Cmd. 1951, paragraphs 40–41). We do not differ from that Committee's view and accordingly do not recommend this course. As regards " sponsored programmes," for which the Broadcasting authority neither makes nor receives payment, the Sykes Committee saw no objection to their admission ; and they are now specifically allowed under the British Broadcasting Corporation's Licence, although the Corporation has, in fact, only admitted them on rare occasions. We see no reason why the provision concerning sponsored programmes in the existing Licence should not be applied also to the television service ; and we think it would be legitimate, especially during the experimental period of the service, were the Corporation to take advantage of the permission to accept such programmes.

66. In attempting to provide funds from licence revenue there appear to be four possible courses :—

(1) The raising of the fee for the general broadcast listener's licence.

(2) The issue of a special television looker's licence.

(3) The imposition of a licence upon retailers.

(4) The retention of the existing listener's licence at 10s. and the contribution from that licence revenue of the necessary funds during the experimental period.

67. Of these courses, the first has the merit of certainty and simplicity. It is arguable whether an additional charge would seriously diminish the number of existing listeners, or even materially abate the normal rate of growth. It would provide a definite and substantial fund to start and maintain a television service. Moreover, if the view which we have already expressed as to the future development of Television in association with sound broadcasting be well-founded, then there is considerable logical justification for treating it as an indispensable adjunct to sound broadcasting, and accordingly laying any increased consequent charge upon the broadcast licence. We, however, see no adequate answer to the inevitable complaint from country listeners " Why should we pay an increased charge for a service which only London or some other centres can receive ? "—nor even to the further complaint within such areas as are actually served, " Why should we people with restricted means pay this increased charge for a service which we cannot receive, because the necessary apparatus is at present so dear that it is only within reach of the well-off ? " We do not, therefore, recommend the adoption of this course.

68. The second course, the issue of a special licence, has also considerable logical justification. It provides a means whereby those who use—and can afford the apparatus necessary to use—this service may contribute towards the cost of it. We must, however, repeat at this juncture, that we are concerned with the means necessary to start this service—to try it out and to set it on its feet—and not with its permanent financing as part of the British Broadcasting Corporation's general system. From the former point of view the proposal, however logically justifiable, has the fatal practical defect that, if the licence fee is placed high enough even to begin to cover the cost, it will strangle the growth of the infant service—while, if it is placed low enough to encourage growth, the revenue must for some time be purely derisory as a contribution towards the cost. We do not, therefore, recommend that at the start of the service there should be any extra licence, but we think that the question should be reviewed when it is seen to what extent the use of the service has taken hold, and when the costs of further extensions of it can be more accurately estimated.

69. This conclusion naturally brought us to examine the question of the imposition of a licence upon retailers of receiving sets, based upon the number of sets sold, not wholly—nor even mainly—with a view to the collection of funds, but as providing, in the absence of a special looker's licence, the next best means of keeping a tally upon the number of users, and so measuring the extent to which the service is in demand. We regard the securing of such a tally as of great importance, and it is with some regret that we feel ourselves unable to recommend the imposition of a retailer's licence on the sale of each set. Apart, however, from the administrative difficulties and the further difficulties which would inevitably arise later on when amateur constructors become sufficiently expert to construct home-made sets, the arguments which have been put before us, and which also moved the Sykes Committee (Cmd. 1951, paragraph 39), have convinced us that the adoption of such a course would be vexatious to traders and detrimental to the development of the service. We hope, however, that it may be possible to negotiate an arrangement with the trade, whereby periodical returns may be made of the total number of television sets sold in each town or district, since this would provide some measure of the growth of the demand.

70. We are therefore left with the conclusion that, during the first experimental period at least, the cost must be borne by the revenue from the existing 10s. licence fee. The determination of the allocation of this contribution as between the British Broadcasting Corporation and the Treasury naturally presents a wide field of controversy, which we should have had to survey at length were we attempting to lay down a permanent basis. Since, however, as explained above, we are dealing only with a relatively limited sum, for a very limited period, we suggest that the best course would be for a reasonable share of the amount to be borne by each of the two parties—the Corporation and the Treasury—and we think that the matter should be considered and determined in this light by the Treasury after consultation with the Postmaster General and the Corporation.

71. We may perhaps be permitted to anticipate three different types of objection which may be raised to the course proposed in the preceding paragraph. As regards any contingent contribution from the British Broadcasting Corporation, it may be argued that the new service will, at first, enure to the benefit of a limited number of people in a limited area, and that it is unfair that the general body of licence holders should have any of their payments diverted from the ordinary programmes, to the improvement of which spare cash, if any, in the British Broadcasting Corporation's coffers

should primarily be devoted. As regards a contingent Treasury contribution, it may be maintained that this is no time to cast any fresh burden upon the taxpayer in order to make an experiment of this nature. Further, it may be said that there is no hurry, and that the start of a service can well wait until the renewal of the British Broadcasting Corporation's Charter comes to be considered, when the financial question can be fully and finally settled. We respectfully submit in answer to the first contention, that, while we have already recognised its force (*see* paragraph 67) as regards any extra levy upon the general body of licence holders, there can be no denying that the existing programmes represent amazingly good value for one-third of a penny per day and that, in these circumstances, the general body of listeners may not unreasonably be asked to help, at no extra cost to themselves, in a national experiment which, if successful, will ultimately enhance programme values for a large part of their members. As regards the second objection, we feel that the development of British Television, in addition to being of evident importance from the point of view of science and entertainment, and of potential importance from the angles of national defence, commerce and communications, will also directly assist British industries. Lastly, we are quite unable to agree that there is no urgency. On the contrary, our enquiries convince us that, apart altogether from any question of scientific prestige, any delay would be most regrettable ; and we feel that, if our conclusions are accepted, it is most desirable that the minimum amount of time should be lost in giving effect to our recommendations.

WIRELESS EXCHANGES (RADIO RELAYS)

72. We have considered the question, which has been raised in evidence, of the relaying of public television broadcast programmes by Wireless Exchanges. We see no reason why such a practice, if technically feasible, should not be allowed under the same conditions as are applicable in the case of sound broadcast programmes.

PRIVATE EXPERIMENTS AND RESEARCH

73. We hope that encouragement will continue to be given to all useful forms of experiment and research in Television by firms or private persons. It is true that much experimental work can be done by transmission from one room to another by wire without recourse to a radio link. In certain cases, however, the use of such a link is necessary; and we trust that the policy referred to in paragraph 20 will be maintained, and that adequate facilities for experimental work will continue to be given.

SUMMARY OF CONCLUSIONS AND RECOMMENDATIONS

74. Our principal conclusions and recommendations are summarised below :—

Type of Service

(1) No low definition system of Television should be adopted for a regular public service. (Paragraph 26.)

(2) High definition Television has reached such a standard of development as to justify the first steps being taken towards the early establishment of a public television service of this type. (Paragraph 33.)

Provision of Service

Operating Authority

(3) In view of the close relationship between sound and television broadcasting, the Authority which is responsible for the former—at present the British Broadcasting Corporation—should also be entrusted with the latter. (Paragraph 39.)

Advisory Committee

(4) The Postmaster-General should forthwith appoint an Advisory Committee to plan and guide the initiation and early development of the television service. (Paragraph 41.)

Ultra-short Wave Transmitting Stations

(5) Technically, it is desirable that the ultra-short wave transmitting stations should be situated at elevated points and that the masts should be as high as practicable. (Paragraph 46.)

(6) It is probable that at least 50 per cent. of the population could be served by 10 ultra-short wave transmitting stations in suitable locations. (Paragraph 47.)

Patent Pool

(7) It is desirable in the general interest that a comprehensive Television Patent Pool should eventually be formed. (Paragraphs 53 and 54.)

Initial Station

(8) A start should be made by the establishment of a service in London with two television systems operating alternately from one transmitting station. (Paragraph 55.)

(9) Baird Television, Limited, and Marconi-E.M.I. Television Company, Limited, should be given an opportunity to supply, subject to conditions, the necessary apparatus for the operation of their respective systems at the London station. (Paragraph 56.)

Subsequent Stations

(10) In the light of the experience obtained with the first station, the Advisory Committee should proceed with the planning of additional stations—incorporating any improvements which come to light in the meantime—until a network of stations is gradually built up. (Paragraphs 57 and 58.)

(11) The aim should be to take advantage, as far as possible, of all improvements in the art of Television, and at the same time to work towards the ultimate attainment of a national standardised system of transmission. (Paragraph 59.)

Finance of Service

(12) The cost of providing and maintaining the London station up to the end of 1936 will, it is estimated, be £180,000. (Paragraph 63.)

(13) Revenue should not be raised by the sale of transmitter time for direct advertisements, but the permission given in the British Broadcasting Corporation's existing Licence to accept certain types of "sponsored programmes" should be applied also to the television service. (Paragraph 65.)

(14) Revenue should not be raised by an increase in the 10s. fee for the general broadcast listener's licence. (Paragraph 67.)

(15) There should not be any separate licence for television reception at the start of the service, but the question should be reviewed later in the light of experience. (Paragraph 68.)

(16) No retailer's licence should be imposed on the sale of each television set, but arrangements should be made with the trade for the furnishing of periodical returns of the total number of such sets sold in each town or district. (Paragraph 69.)

(17) The cost of the television service—during the first experimental period at least—should be borne by the revenue from the existing 10s. licence fee. (Paragraph 70.)

---

75. In conclusion we desire to place on record our high appreciation of the services rendered by our Secretary, Mr. J. Varley Roberts. He has performed his duties with zeal and ability, and has been of the greatest assistance to us at every stage, both in the conduct of the Enquiry and in the compilation of our Report.

(*Signed*) SELSDON (*Chairman*).
JOHN CADMAN (*Vice-Chairman*).
A. S. ANGWIN.
NOEL ASHBRIDGE.
O. F. BROWN.
CHARLES D. CARPENDALE.
F. W. PHILLIPS.

J. VARLEY ROBERTS (*Secretary*).

14th January, 1935.

APPENDIX I

LIST OF WITNESSES AND ORGANISATIONS REPRESENTED

Messrs. Baird Television, Ltd.	Major A. G. Church, D.S.O., M.C. Mr. A. G. D. West, M.A., B.Sc.
Messrs. A. C. Cossor, Ltd.	Mr. W. R. Bullimore. Mr. J. H. Thomas, M.I.E.E. Mr. L. H. Bedford, M.A., B.Sc.
Messrs. Electric and Musical Industries, Ltd., and Messrs. Marconi—E.M.I. Television Co., Ltd.	Mr. Alfred Clark. Mr. I. Shoenberg. Mr. C. S. Agate. Mr. A. D. Blumlein. Mr. C. O. Browne. Mr. G. E. Condliffe. Mr. N. E. Davis. Mr. S. J. Preston.
Messrs. Ferranti, Ltd.	Mr. V. Z. de Ferranti. Mr. A. Hall.
Messrs. General Electric Co., Ltd.	Mr. C. C. Paterson, O.B.E., M.I.C.E., M.I.E.E. Mr. T. W. Heather, M.C.
Messrs. Plew Television, Ltd.	Dr. C. G. Lemon.
Messrs. Scophony, Ltd.	Mr. S. Sagall. Mr. G. W. Walton. Mr. G. Wikkenhauser.
British Broadcasting Corporation	Sir J. C. W. Reith, G.B.E.
Newspaper Proprietors' Association	Col. the Hon. F. E. Lawson. Sir Thomas McAra, J.P. Mr. A. J. Polley. Mr. F. W. Jarvis. Mr. E. J. Robertson.
Radio Manufacturers' Association	Mr. W. W. Burnham. Mr. R. Milward Ellis.
" Popular Wireless " and " The Wireless Constructor."	Dr. J. H. T. Roberts, F.Inst.P.
The Television Society	Dr. C. Tierney, F.R.M.S. Mr. Ronald R. Poole, B.Sc. Mr. W. G. W. Mitchell, B.Sc.

and

Sir William Jarratt.
Mr. W. Barrie Abbott, B.L.
Mr. J. Guibiansky.
Mr. A. B. Storrar.
Mr. R. W. Hughes.

TELEVISION IN GREAT BRITAIN

Noel Ashbridge

Proceedings of the Institute of Radio Engineers
Volume 25, Number 6   *June, 1937*

TELEVISION IN GREAT BRITAIN*

By

Noel Ashbridge

(British Broadcasting Corporation, London, England)

Summary—The development of television in Great Britain is treated in this paper and a short historical background is given, tracing the development of television in Great Britain from 1929, when the British Broadcasting Corporation first gave the Baird Television Company facilities for experimental transmissions of low definition television from an ordinary broadcast station. In May, 1934, the Postmaster-General appointed a Committee to consider the development of television, and to report on the conditions under which any public service of television should be provided. The Committee recommended in 1935 that no low definition television service should be adopted for a regular public service, and was of the opinion that high definition television had reached such a standard as to justify the first steps being taken towards the early establishment of a public television service of this type. The British Broadcasting Corporation was entrusted with the development of this service. The paper then gives a brief technical description of the London television station, which is the practical effect the British Broadcasting Corporation has given to the Television Committee's recommendations. The station has been built at Alexandra Palace, some six miles north of the center of London. Two complete television systems were installed, one by Baird Television, Limited, and the other by the Marconi-E.M.I. Television Company, Limited. Each consists of studio and control room equipment and a vision transmitter. A third transmitter common to both systems provides the sound program. An aerial mast carries two separate aerials, one for transmitting the vision and the other for transmitting the sound. The vision aerial is common to both systems. Vision is radiated on a frequency of 45 megacycles, while sound is radiated on a frequency of 41.5 megacycles. The Baird system uses 240 lines sequential scanning, 25 frames per second, while the Marconi-E.M.I. system uses 405 lines interlaced, at 50 frames per second, each of 202½ lines. The Television Advisory Committee, before approving different standards of frame frequency and definition for the two systems, was satisfied that receivers could be constructed capable of receiving both types of transmission without undue expense or complicated adjustment. The commercial receivers now on the market accomplish this by a single switch.

On November 2, 1936, a regular television service for reception by the public was opened by the Postmaster-General, regular programs being given twice a day from 3:00 to 4:00 P.M. and 9:00 to 10:00 P.M., except on Sundays.

As FAR back as the autumn of 1929, the British Broadcasting Corporation gave the Baird Television Company facilities for experimental low definition transmissions of television from an ordinary broadcast station. During the next two or three years a large number of experimental transmissions were carried out by the Baird Company independently, as well as in conjunction with the British Broadcasting Corporation.

* Decimal classification: R583. Original manuscript received by the Institute, November 12, 1936. Presented before New York Meeting, November 12, 1936.

Fig. 1—Alexandra Palace, London Television Station, Baird
studio, showing intermediate film scanner.

In August, 1932, the British Broadcasting Corporation inaugurated
a regular service of low definition television on medium wave lengths,
transmitting two or three times a week. The Baird system was used,
the number of lines was 30 and the number of pictures per second $12\frac{1}{2}$.
The service was experimental and it was terminated on September 15,
1935.

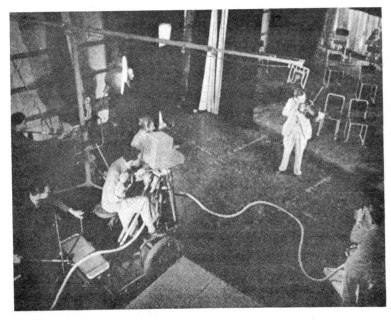

Fig. 2—Alexandra Palace, London Television Station. The Marconi-E.M.I. studio
showing two Emitron instantaneous television camers in use one transmitting
the program; the other ready to be "faded-in" for a different "shot."

In May, 1934, the Postmaster-General appointed a Committee with the following terms of reference: "To consider the development of television and to advise the Postmaster-General on the relative merits of the several systems and on the conditions under which any public service of television should be provided." Both the Post Office and the British Broadcasting Corporation were represented on this Committee. Its report was presented in January, 1935.

Fig. 3—Alexandra Palace, London Television Station. The mast and transmitting aerials (vision above) (sound below).

The Committee recommended that no low definition television service should be adopted for a regular public service, and it was of the opinion that high definition television had reached such a standard as to justify the first steps being taken towards the early establishment of a public television service of this type. The Committee recommended that the definition should not be less than that given by 240 lines per picture with a minimum picture frequency of 25 pictures per second. It also recommended that, in view of the close relationship between sound and television broadcasting, the British Broadcasting Corporation should be entrusted with the television service.

While the British Broadcasting Corporation should exercise control of the operation of the television service, it was recommended that the initiation and early development of this service should be planned and guided by an Advisory Committee appointed by the Postmaster-General, on which the Post Office, the Department of Scientific and Industrial Research, and the British Broadcasting Corporation should be represented.

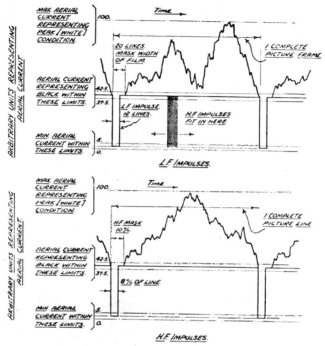

Fig. 4—Baird system. Characteristics of vision signal.

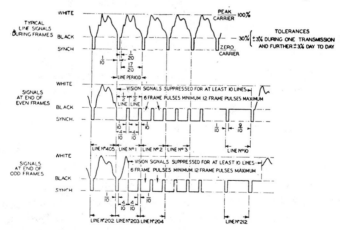

Fig. 5—Marconi-E.M.I. system. Characteristics of vision signal.

The Committee also recommended that a start should be made by the establishment of a service in the London area and that two systems, that of Baird Television, Limited and that of the Marconi-E.M.I. Television Company, Limited, should be used under conditions which admitted of a comparison being made of their respective merits.

The following is a very brief technical description of the London television station, which is the practical effect the British Broadcasting Corporation has given to the Television Committee's recommendations. Alexandra Palace is a large building situated on a hill, 300 feet above sea level and some six miles north of central London. It has been a North London landmark and a center for exhibitions and amusements for about sixty years. The British Broadcasting Corporation has leased from the Alexandra Palace trustees 31,840 square feet of floor space at the southeast corner of the building, comprising three large halls on the ground floor, the rooms over them on the first floor and the south-

Fig. 6—Baird teleciné scanners. Monitoring and control racks in background.

east tower. A further area of 24,525 square feet, comprising the theater and associated rooms, is also available.

On the ground floor there are the transmitter rooms, a film-viewing room, a restaurant, and a kitchen, while the rooms above have been treated to form two large studios with control rooms and apparatus rooms separating them. Dressing rooms and make-up rooms for band and artists have been constructed, separated from the studios by a corridor. Office accomodation has been provided in the tower at the southeast corner. This tower also carries the mast and aerial system.

The two main studios, one for use with each of the television systems, are 70 feet by 30 feet by 25 feet high. Acoustically, the studios are considerably more "dead" than is the general practice for sound broadcasting in Great Britain. The walls are covered entirely, except for door and window openings, with sheets of asbestos compound which has a high degree of sound absorption. As this material has a rather rough surface, it is covered up to about ten feet from the floor with a protective fabric which does not affect the sound-absorbing properties of the compound. The ceilings of the studios are treated with building board, as commonly used in ordinary broadcast studios. The floors are covered with black linoleum over which can be laid any type of flooring which may be required.

Several microphone points are installed in each studio, and they are arranged to allow the use of any type of microphone which may be required. Portable stands of the "lazy-arm" type are also provided.

Each studio is fitted with two stages equipped with curtains, the detailed arrangements of the stages and curtains being different in the

Fig. 7—Alexandra Palace, London Television Station Baird television radio transmitter. Control desk, foreground. Crystal drive, right background. Intermediate-radio-frequency amplifier, center background Power output stage, left background.

two studios, to take account of the different requirements of the two systems. A number of overhead battens, each of which carries several lighting circuits, has been provided in each studio. There is also a large number of wall sockets for portable lighting. In each studio a large lighting switchboard has been installed, with provision for the separate control, dimming, etc., of every circuit. In addition, there are arrangements for preselective switching and bank-dimming of any number of circuits, and the whole equipment has been designed to give the maximum possible flexibility. In addition to the above, a lighting bridge has been erected across the Marconi-E.M.I. studio to give further lighting facilities. All the lighting in both studios (a maximum of 50 kilowatts each) is at present of the incandescent lamp type, using spot- and floodlighting, on similar lines to that employed in theaters and film studios, but modifications are contemplated with developments in television technique.

Ventilation has been provided in the studios by means of extract fans situated in enclosures formed on the adjoining colonnade. Full air conditioning has not been attempted, since the studios are not regarded as permanent, but the ventilation is sufficient to keep the studios at a moderate temperature when full lighting is used, and to allow the temperature to be adjusted within normal limits.

Fig. 8—Alexandra Palace, London Television Station. Marconi-E.M.I. system, control room. Emitron camer amplifiers (right), synchronizing oscillators (left).

The southeast tower was converted to provide offices for the television staff, and the existing ornamental pylon was removed and replaced with a steel mast to carry the vision and sound aerials. The top of the mast is 300 feet above the ground, the height of the steelwork above the brick tower being 215 feet. The height of the vision aerial above sea level is thus approximately 600 feet.

Two separate aerial systems are carried by the tower, one for vision and one for sound. Both systems are similar, each consisting of a number of vertical aerial elements arranged round the mast, those for vision being above and those for sound beneath. Each aerial consists of eight push-pull end-fed vertical dipoles spaced at equal angles round the mast, together with a similar set of dipoles used as reflectors to avoid induced currents in the mast structure and to increase the radiated field. The aerials are connected to junction boxes, with which are associated a number of impedance-matching transformers to correct the aerial response. The aerial systems are connected to the transmitters by means of two five-inch concentric feeders which pass down the mast and along to the transmitting rooms, a change-over switch being provided so that either vision transmitter can be connected to the vision aerial.

The transmitter to radiate the sound accompanying the vision program is capable of operating over a band of frequencies from 35 to 50 megacycles, the working frequency being 41.5 megacycles, and the output power rating three kilowatts capable of 90 per cent peak modulation (Copenhagen rating). The frequency response of the transmitter is substantially flat between 30 and 10,000 cycles, the maximum departure being less than two decibels over this range, while the low-frequency harmonic content introduced by the transmitting apparatus is very low.

Of the two systems of television in operation, the Baird system is using 240 lines sequential scanning, 25 frames per second. Baird Television, Limited, has installed two complete teleciné scanners for film transmission and an intermediate film system in which the scene to be televised in the studio is photographed, developed, fixed and washed, all within forty seconds, the film then passing through what is, in effect, a teleciné scanner adapted to scan the film while it is wet. The sound is also recorded on the film in order that it may be synchronized with the vision. In addition, the Baird system uses mechanical spotlight scanning for direct television of head and shoulder subjects from a small studio. At the time of writing it is understood that the Baird Company is experimenting with an electron camera of the Farnsworth type.

The Marconi-E.M.I. system uses the Emitron instantaneous television camera, with electrical scanning, for studio and outside direct television as well as for film scanning. The picture is scanned with 405 lines interlaced, at 50 frames per second, each of $202\frac{1}{2}$ lines. All synchronizing impulse generators and vision and audio amplifiers are installed in a central control room, and provision is made for fading from the output of one group of cameras to another, six cameras in all being fitted.

Each system has its own vision transmitter working on a carrier frequency of 45 megacycles, the power of the transmitters in each case being 17 kilowatts peak, when giving peaks of maximum modulation which correspond in these systems to the transmission of full white. The characteristics of the radiated signals are shown in Figs. 4 and 5.

The Television Advisory Committee, before approving different standards of frame frequency and definition for the two systems, was satisfied that receivers could be constructed capable of receiving both types of transmission without undue expense or complicated adjustment. The commercial receivers now on the market accomplish this by a single switch.

The first experimental transmissions from the London television station took place just before the opening of this year's London radio exhibition. During this exhibition, demonstrations of reception of pictures from Alexandra Palace were given, and more than 100,000 people saw them. The station was then closed down for a short time

Fig. 9—Alexandra Palace, London Television Station. Marconi-E.M.I. vision transmitter, modulator units. Foreground, control desk.

for adjustment, and from October 1 test transmissions were given twice daily for periods of an hour, chiefly to allow manufacturers to test out television receivers. These test transmissions concluded on October 31, and on November 2 a regular television service was opened by the Postmaster-General, the Chairman of the British Broadcasting Corporation, and Lord Selsdon, the chairman of the original television committee. Regular programs are being given twice a day, from 3:00 to 4:00 P.M. and from 9:00 to 10:00 P.M., except on Sundays.

Some seven or eight radio manufacturers have already put, or are putting, television receivers on the market, and indeed some receivers have already been sold to members of the public. One or two London stores have installed receivers on which their customers can see the transmissions, and one of the railway companies has installed a receiver in a waiting room at one of the London termini, admission being gained by the production of a railway ticket. Public demonstrations are also given at the Science Museum in South Kensington.

The present cost of a receiver is four or five hundred dollars and it is obvious that the audience for these transmissions is not likely to become very large until this price is considerably reduced. There is reason to hope that this will be done as public interest develops.

The programs are published weekly in advance in the London edition of the *Radio Times*, the British Broadcasting Corporation's program paper, as well as daily in the morning papers. To give some idea of our initial efforts, the following items are taken from the programs for the second week:

Nov. 9 O.B.   Demonstration of the new mobile post office.
        Studio   "Picture Page." A magazine program of items of general and topical interest.
Nov. 10 O.B.   Demonstration of horsemanship and jumping.
        O.B.   Demonstration of jumping Alsations.
        Studio   A star from the entertainment world.
        Studio   A specially arranged pageant of citizen soldiers of London who formed part of the Lord Mayor's Show the previous day.
Nov. 11 Studio   Special Armistice Day program.
        Studio   Ballet program.
Nov. 12 Studio   Extracts from the new opera "Mr. Pickwick," Albert Coates, the composer, conducting, about to be produced at Covent Garden.
        Studio   Ballroom dancing demonstration.
Nov. 13 Studio   Exhibits from the International Poultry, Pigeon, and Livestock Show.
Nov. 14 O.B.   A selection of cars taking part in the Veteran Cars' Parade the following day.
        Studio   Cabaret.

News reels and one or two short films also are shown.

Up to the present, we have not attempted any outside pickups further afield than those we have done in the grounds or inside the exhibition part of Alexandra Palace itself, a maximum distance between camera and control room of about 1000 feet.

A coaxial cable has been laid between Broadcasting House and Alexandra Palace, but the terminal gear has not yet been installed. This cable has been provided largely for test purposes, and, while it may be used occasionally for actual programs, there is no intention at present of providing extensive television studio accomodation in Broadcasting House.

In this paper there has been no attempt to make any critical study of the television systems which we are using, or which we might have used, and it probably would be unwise to do so until we have obtained sufficient information based on practical daily use of the systems over a considerable period. Furthermore, in any comparison one should certainly take into account the results obtained from receivers in the hands of the general public. We are watching with great interest the reaction of the public, and further extensions of the service to other parts of the country will depend on the results of the experience we gain in London.

Author's Note:

Since this paper was read in New York on November 12, 1936, several months have elapsed, and on February 4, 1937, the Postmaster-General made the following announcement:

"The Postmaster-General announces that, as a result of the experience gained of television transmissions from the London television station at Alexandra Palace, the Television Advisory Committee have recommended that the London experimental period—during which different technical standards of transmission have been used during alternate weeks—should now be terminated, and that a single set of technical standards should be adopted for public transmissions from the London station. This recommendation, which has been approved by the Postmaster-General, provides for the adoption of standards as follows:

> Number of lines per picture = 405, interlaced.
> Number of frames per second = 50.
> Ratio of synchronising impulse to picture = 30:70.

"These standards for the television service from the London station will not be substantially altered before the end of 1938. Consequent upon this decision, television transmissions from Alexandra Palace of 240 lines with 25 frames per second will be discontinued, and all future transmissions will be on the standards set out above, which will be known as the London Television Standards."

Selections From

TELEVISION STANDARDS AND PRACTICE

Edited by

Donald G. Fink

TELEVISION STANDARDS AND PRACTICE

CHAPTER I

TELEVISION STANDARDIZATION IN AMERICA

The technical arts are usually offered to the public without benefit of standardization. The early histories of the railroads, of electric light and power, and of radio show that these public services were standardized only after bitter experience had proved such a step absolutely necessary. Each was in its first stages a gamble. Only after public approval had been won and the demand for the service had become universal did the need for standards of operation appear.

Television is a notable exception to this tendency. The need for standards had been appreciated long before television was permitted to appear commercially. Governmental authority was given for commercial broadcasting in the United States only after the standards had been debated at length (for some 12 years) in industry committee meetings and in public hearings.

The work of standardization went through a series of technical revolutions and finally culminated in the organization of the National Television System Committee. This committee, in the space of a few months, secured practical unanimity on questions that had split the industry into separate camps of opinion a year previously. The committee accomplished a distinguished piece of work thoroughly and in a short time.

Three causes contributed to the fact that standards for television received close scrutiny prior to commercialization of the art. In the first place, television is an offshoot of sound broadcasting, part and parcel of the radio industry. The radio industry has all too vivid a memory of the chaotic conditions in the early days of broadcasting. Strong industry action in 1925, under the direction of the Federal Radio Commission, produced

1

such immediate benefit for the public and so enhanced the acceptance of radio that little further proof of the basic necessity of standards was needed.

In the second place, the Federal Communications Commission, carrying on the tradition of the earlier Federal Radio Commission, made good use of its experience in the regulation of public services. The members of the commission insisted that standards for television, as well as for other services, be set only when the members of the industry were in substantial agreement as to the form they should take. The commission took the stand that, prior to such agreement, the art must remain in the experimental stage. The government and the industry agreed that standardization must precede, not follow, the initial offering of the art to the public on a commercial basis.

Finally, and perhaps most fundamentally, television transmission and reception in their very nature demand a greater degree of standardization than is required of sound broadcasting or, for that matter, of most other widespread public services. The lock-and-key relationship between television transmitter and receiver has been overemphasized at times, but its fundamental truth cannot be gainsaid. Public television service without standards is economically and socially unsound if not, in fact, practically impossible.

The fact remains that standardization must be confined to those matters that require it. The dividing line between essential matters and nonessential ones is, however, not an easy one to draw. It is hopeless, for example, to attempt to produce a receiver capable of reproducing all transmissions of any transmitter within range without standards defining the type of scanning to be employed. This is clearly an essential matter. So also is establishing the positions and widths of the channels employed by the transmitters.

The question of the numerical constants employed in scanning, on the other hand, is not so straightforward, since receivers can be built to operate over a range of scanning values. If such a range can be shown to serve a purpose sufficient to justify the cost of obtaining it, the scanning standards might well be written in terms of a range of values rather than in terms of fixed figures. If such variability in standards is shown to be undesirable, fixed values must be selected from among those proposed. In point of fact, the matter of flexibility in scanning was a matter that occupied much attention in the deliberations of the N.T.S.C.

Such decisions are best made with the judgment of many minds, guided by intelligent foresight and backed by experimental tests of the relative merits of differing proposals. It is thus no accident that even the most obvious problem in standards profits from discussion and observation in committee. Committees are, however, slow to action. Committees dealing with the quasi-permanent setting of standards are the slowest of all, because the responsibility of a wrong judgment lasts a long time.

The history of the National Television System Committee is remarkable, therefore, by virtue of the fact that it accomplished the work of drawing up standards and reporting to the FCC in less than 6 months. During this time, the 168 members of the Committee and the Panels produced reports and minutes totaling 600,000 words, devoted 4,000 man-hours to meetings and an equal time to travel, and witnessed 25 demonstrations of technical matters. This is a monumental record.

The chairman of the FCC, James Lawrence Fly, in opening the meeting on Jan. 27, 1941, paid the following tribute to the committee: "This is another example of the best that is in our democratic system, with the best in the industry turning to on a long and difficult job in an effort to help the government bodies in the discharge of their functions so that a result may be achieved for the common good of all."

THE R.M.A. BACKGROUND

Abundant as the record of the N.T.S.C. is, it is clear that the work could not have been accomplished in the time taken, nor could the unanimity of opinion have been assembled, if years of preliminary work on standards had not been carried out by a series of committees formed within the organization of the Radio Manufacturers Association.

THE R.M.A. COMMITTEE ON TELEVISION

The R.M.A. Committee on Television was active as early as 1929, before the modern cathode-ray system was fully developed. The early work was primarily one of experimentation with the rudimentary forms of mechanical scanning and occasional reports to the membership of the R.M.A. concerning the current state of the art. In 1929, the first paper on standards for television was published by Weinberger, Smith, and Rodwin.[1] These early attempts to standardize the art seem highly premature from the standpoint of present developments, as indeed were the early attempts to interest the public in a television service based on low-definition pictures.

By 1935, demonstrations of cathode-ray television employing 343 lines had been made to members of the R.M.A. and to the press. In 1935, the engineering department of the R.M.A. was instructed by its board of directors to determine when it would be advisable to adopt television standards. The investigation by the engineering group was watched with interest by the engineers of the FCC. In 1936, Commander T. A. M. Craven, then chief engineer of the commission, in a letter to the chairman of the standards section of the R.M.A., stated the philosophy that guided the commission and the industry in all the subsequent work:

The engineering department of this commission is very much interested in performance standards for visual broadcasting stations, but it is the feeling of this department that, if possible, various branches of the industry should come to an agreement among themselves prior to any commission action.

It is our present opinion that the ultimate performance standards which are adopted should be such that any receiver manufactured for the public would be capable of receiving any visual broadcasting transmitting station which may be licensed by the commission.

We are interested in information as to whether or not it will be possible for the industry to agree upon any standards in the near future if it were admitted that there has been developed a system which, from the engineering standpoint, would permit satisfactory visual broadcasting.

In the same year, 1936, Commander Craven set forth the necessity for making allocations in the ether spectrum for television stations. The commission announced that hearings would be held, beginning June 15, 1936, to determine a basis for long-time policies in the future allocations of the limited facilities, not only for television but for broadcasting and radio communication as well.

[1] The Selection of Standards for Commercial Radio Television, *Proc. I.R.E.*, **17**, 1584 (September, 1929).

Two committees of the R.M.A., one on television standards, the other on allocations, went to work in preparation for these hearings and prepared a joint report that was submitted to the FCC. Concerning allocations, their recommendation was that seven television channels, each 6 Mc/s wide, should be set up in the region between 42 and 90 Mc/s, and that experimental channels should be allocated on a band beginning at 120 Mc/s.

The standards recommended at the time constituted an incomplete list on which substantial agreement had been obtained. The principal features of the recommended system were a 441-line picture sent at a rate of 30 complete pictures per second, interlaced in two fields per frame. Double-sideband transmission was recommended, with 3.25 Mc/s separation between the sound- and picture-carrier frequencies. The aspect ratio of the picture was set at 4:3. The recommended polarity of the transmission was negative. The problem of determining a satisfactory standard for synchronization proved difficult, but it was recommended that the blanking period be one-tenth of the time required to scan a line and to scan one field, for the horizontal and vertical periods respectively. The synchronizing pulses were to occur approximately at the leading edge of the blanking pulses, but no further recommendations were made as to their form. The amplitude of the picture carrier devoted to the synchronizing signals was limited to not less than 20 per cent. The first published report[1] on these standards was prepared by A. F. Murray, then acting chairman of the R.M.A. Committee on Television.

When these proposals are compared with the commercial standards adopted 5 years later, it is clear that the workers who developed them had a clear insight into the problems of the art. The standards now officially in use differ numerically from the early proposals in every particular but the aspect ratio, but the differences are ones merely of degree. The importance of the contribution of the committee members in setting up this list of standards in 1936 can hardly be overemphasized. They determined what matters require standardization in a television system. Only three additional major items relating to picture transmission have been treated by the N.T.S.C., namely, the direction of polarization of the transmitted wave, the transmission

[1] Latest Television Standards as Proposed by R.M.A., *R.M.A. Engineer*, **1,** 1 (November, 1936).

of the d-c component, the specification of maximum percentage modulation. With a few exceptions, the members of the committee that drew up these early standards later served as members of the N.T.S.C. organization.

The broader aspects of television policy presented by the R.M.A. to the FCC in these hearings were formulated in a five-point plan by the Committee on Television, as follows:

1. One single set of television standards for the United States, so that all receivers can receive the signals of transmitters within range.

2. A high-definition picture, approaching ultimately the definition obtainable in home movies.

3. A service giving as near nation-wide coverage as possible.

4. A selection of programs—*i.e.*, simultaneous broadcasting of more than one program in as many localities as possible.

5. The lowest possible receiver cost and the easiest possible tuning.

The R.M.A. report also stated:

The commission will have the responsibility of making definite broadcasting assignments, assignments that will ensure the greatest possible service to each locality, assignments that will not lead to any monopoly, assignments that will preserve the American system of competition but which will prevent the creation of so many competitive stations that none will have enough revenue to provide fine programs.

In August, 1936, the FCC announced regulations that opened up the region from 42 to 56 Mc/s and from 60 to 86 Mc/s for experimental transmissions on channels 6 Mc/s wide. The commission took no action on the proposed standards of transmission, however, since the transmissions were to be of experimental nature, and since the standards were then in an incomplete state. Experimental transmissions were authorized also in any 6 Mc/s band above 110 Mc/s, excluding the amateur region from 400 to 401 Mc/s.

In the following years, 1937 and 1938, the R.M.A. Committees worked intensively to round out the list of standards and to obtain agreement on them in all quarters of the industry. A "recommended practice" specifying the horizontal direction of polarization for the transmitted carriers was agreed upon. An upper limit of 25 per cent was placed on the percentage of the picture carrier devoted to synchronizing. D-c transmission of the background brightness of the picture was specified. Standards for transmission of the sound signal were drawn up. The uncertainty regarding the synchronizing-signal waveform was removed by the adoption of R.M.A. standard T-111, which specified the constant-amplitude system, the vertical pulses being of the serrated form with additional equalizing pulses introduced to improve the interlacing of the frames. In July, 1938, Acting Chairman A. F. Murray was able to report[1] that the complete system of standards had been agreed upon by his committee. These standards were submitted to the FCC, and their official adoption was urged by the R.M.A. Board of Directors, early in 1938.

Only one important change impended. By 1938, it was evident that the single-sideband system of transmission would soon become a practical reality. Lack of a sufficient test of this system in the field had prevented the writing of a definite standard, but in 1938 the frequency separation between sound and picture carriers was increased to 4.5 Mc/s in anticipation of the event, superseding the 3.25 Mc/s separation that had previously been recommended for double-sideband transmission.

The synchronization waveform standard was passed by the Subcommittee on Television Standards, by a vote of 6 to 1, in May, 1938. This standard, together with that specifying negative transmission and that specifying 4.5 Mc/s bandwidth for the picture-signal sideband, was approved by the Television Committee and reported, together with the other standards previously agreed upon, to the R.M.A. Board.

In June, the R.M.A. Board approved the submission of the standards to the FCC, but suggested additional study by the Television Committee for simplification and clarification. In July the Television Committee met for this purpose, carried out the board's recommendations, and then went on to discuss three new standards: the specification of 25 per cent as the maximum carrier amplitude for maximum white in the picture; the specification of rated transmitter power as one-fourth its peak power; and the specification of approximately equal power for the sound and picture transmitters. Since no board meeting was scheduled

---

[1] R.M.A. Completes Television Standards, *Electronics*, **11**, (7), 28 (July, 1938).

at which these additions might be approved, the entire list of standards was submitted to the whole membership of the association in August. No objections to the standards were received, and accordingly on Sept. 10, 1938, all the standards were submitted to the FCC.

On the twenty-eighth of that month, the FCC requested the names of the members of R.M.A. and the individuals who participated in forming the standards, and asked whether the R.M.A. believed the FCC should call formal hearings with respect to the adoption of standards. The R.M.A. complied with this request and suggested that formal hearings should be held. In February, 1939, a committee of engineers of the R.M.A. met informally with the engineering department of the FCC and the technical background of each of the standards was presented. The hearings were not held, however, until January, 1940.

In the meantime activity had commenced in the direction of a public television program service, based on the standards recommended by the R.M.A. In October, 1938, just after the R.M.A. standards were submitted to the FCC, the Radio Corporation of America announced its intention to begin a limited production of television receivers for the public and to start a limited program service based on the R.M.A. standards from the National Broadcasting Company's transmitter in New York. The inauguration of this service was to coincide with the opening of the New York World's Fair, on Apr. 30, 1939. The service planned was limited to a minimum of 2 hr. a week, but it was the first occasion on which the public was to be invited to participate in the development of high-definition television. As such it occasioned considerable notice in the press.

Not all members of the industry considered the RCA plans wise. The Zenith Corporation gave immediate opposition to the plan. The Philco Corporation also opposed the immediate introduction of the art to the public, on the ground that an insufficient number of homes could be served. Other members of the industry, however, made plans to offer receivers to the public in the New York area on or soon after the announced date of the beginning of public service.

Before the eventful day arrived, however, a new standard was approved, which permitted a substantial improvement in the quality of the reproduced picture. This was the standard specifying vestigial sideband transmission. Throughout the latter part of 1938 it had become increasingly clear that the effective frequency range occupied by the television signal might be increased from 2.5 Mc/s to more than 4 Mc/s, with a proportionate increase in the picture detail, if vestigial sideband transmission were adopted.

Laboratory tests had proved feasible two systems, both of which involved attenuating the picture-carrier signal to one-half its normal voltage. The question was whether the attenuation should occur in the transmitter ("transmitter attenuation," or TA system) or in the receiver ("receiver attenuation," or RA system). After long and active discussion over a period of several months, the RA system was finally adopted as standard, approval being voted by the R.M.A. Committee on Television, Jan. 19, 1939. The opening of the RCA service was then less than 4 months away. The engineers of the NBC transmitter, as well as those preparing receivers for the public, worked over-time to bring the equipment in line with the new standard. The change-over was accomplished within the remaining time, and the NBC transmitter went on the air on schedule. The first program specifically intended for the public (although it had been preceded by many hours of test programs) was appropriately the speech of President Roosevelt as he opened the New York World's Fair.

Throughout 1939 an augmented schedule of programs was maintained by the NBC transmitter in New York. The transmissions conformed in all particulars to the R.M.A. standards then in the hands of the FCC. No official action on these standards had been taken, but it was assumed by most observers that they would eventually be adopted by the commission. This feeling was strengthened by the action of the FCC, in December, 1939, when the commission tentatively adopted new rules governing television broadcasting. These rules modified the then existing prohibition against commercialism and permitted a limited form of program sponsorship under certain conditions. The facilities and funds contributed by sponsors were to be used primarily for experimental development of television program service. Other forms of commercial sponsorship were prohibited. Two classes of stations were set up: Class II offered a scheduled public program service; Class I broadcast on an unscheduled experimental basis. The rules were adopted pending a public hearing to be held Jan. 15, 1940, at which all those interested were invited to state their views.

At this hearing a break in the ranks of the industry became for the first time clearly apparent. The standards that had been submitted by the R.M.A. the previous year were subject to sharp attack from two organizations. The Allen B. DuMont Laboratories, which was not a member of the R.M.A. and hence had not participated in the forming of the standards, objected that the standards were too inflexible. This organization argued that 441 lines did not provide sufficient detail and urged that a variable number of lines be employed to meet future contingencies. Mr. DuMont urged also that a lower rate of frame repetition than 30 per second was feasible and would permit greater detail in the pictures. He suggested a flexible frame rate between 15 and 30 frames per second. A cathode-ray screen designed to minimize flicker, when a low frame rate was used, had previously been demonstrated by the DuMont organization to the R.M.A. Committee on Television in December, 1939, but the results were deemed inconclusive by the committee.

The second attack came from the Philco Radio and Television Corporation. The recommended practice specifying horizontal polarization was criticized as a wrong choice that would operate to the disadvantage of receivers employing self-contained antennas. Philco also objected to the 441-line standard, suggesting a much higher value (800 lines) and a slightly lower frame rate (24 per second).

Witnesses from other organizations gave full support to the R.M.A. proposed standards. RCA engineers urged that the low frame rate of 15 pictures per second suggested by the DuMont Laboratories would intensify the problem of flicker and blurred motion in the image.

This lack of unanimity on the proposed standards made a deep impression on the members of the commission. On Feb. 29, 1940, the commission issued a report adopting with minor changes the rules concerning commercial operation which had previously been adopted tentatively. Sponsored program service, on a limited basis, was to be permitted on and after Sept. 1, 1940. But the report made no decision concerning the proposed standards; in fact it warned against "freezing" the standards.

The commission's attitude on standardization was summed up in its report:

The commission therefore recommends that no attempt be made by the industry or its members to issue standards in this field for the time being. In view of the possibilities for research, the objectives to be obtained, and the dangers involved, it is the judgment of the commission that the effects of such an industry agreement should be scrupulously avoided for the time being. Agreement upon standards is presently less important than the scientific development of the highest standards within reach of the industry's experts.

Whether or not this attitude was wise is a matter for argument. In any event, with it came a problem of interpretation which the industry found difficult to solve.

The question of interpretation arose from the fact that limited commercial sponsorship of programs was to be permitted before the end of the year, but the standards on which the service was to be operated were not specified. Two courses of action were open: The industry could assume the standards to be in a state of such rapid development that no serious commercial operation, despite the official permission, was possible. Or it could be assumed that the standards were satisfactory as they stood, despite the lack of official approval, and could be used as the basis for plans to commercialize the art.

The second course was adopted by RCA, which announced early in 1940 plans to step up production of television receivers, to reduce prices, and to enlarge the broadcasting schedule. But on Mar. 22, 1940, the FCC announced that it had decided to reconsider, at a hearing to be held Apr. 8, its intention to permit limited commercial operation of television broadcasting stations. The reason given for this action was the commercial activity of the RCA in marketing television receivers. The FCC viewed this action as one tending to freeze the standards on which the receivers and transmitters were then operating (the R.M.A. standards). Such freezing of the standards would, the commission stated, tend to discourage research and experimentation with systems based on, or requiring, other standards.

At the hearing, the DuMont and Philco organizations maintained the positions they had taken previously regarding the R.M.A. standards and agreed that the action of the RCA had caused them to abandon research with systems outside the scope of the R.M.A. standards.   The RCA witnesses stated that they had adopted the R.M.A. standards because they felt they represented the majority opinion of the industry's engineers and were the best standards with which to begin commercial operation. They noted that they were willing, however, to adopt any other standards that might be specified by the commission in licensing transmitters for public-program service.   Other witnesses stated that the R.M.A. standards would admit within their scope a large degree of progress and improvement without causing obsolescence of equipment previously sold to the public.

In the FCC report on this hearing, issued in May, the commissioners promised commercialization of the art when the industry was prepared to agree on any one system of broadcasting. The commissioners agreed that the plan of limited commercialization without previously setting the standards was not feasible. The commercial class of station (Class II) set up in the rules was therefore eliminated.   In its report the commission stated:

It is, therefore, the conclusion of the commission that in order to assure to the public a television system which is the product of comparative research on known possibilities, standards of transmission should not now be set.   It has further been decided that there should be no commercial broadcasting with its deterring effects upon experimentation until such time as the probabilities of basic research have been fairly explored.

In eliminating "limited" commercialization, the report stated:

As soon as the engineering opinion of the industry is prepared to approve any one of the competing systems of broadcasting as the standard system, the commission will consider the authorization of full commercialization.

No time limit can now be set for the adoption of standards.   The progress of the industry will largely determine this matter.   The commission will continue its study and observation of television developments and plans to make a further inspection and survey in the early fall.   Meanwhile the commission stands ready to confer with the industry and to assist in working out any problems concerned with television broadcasting.

This was in May, 1940.   Fourteen months later, on July 1, 1941, full commercial operation of television stations began with the approval of the commission.   In the short intervening time the plan for the National Television System Committee had been formulated, the committee assembled, its meetings held, its minutes recorded, technical reports compiled, and its final report delivered to the commission.

The N.T.S.C. standards resulted from a thorough reexamination of every phase of the television art relating to public service. The front displayed by the industry at the conclusion of the committee work was, if not wholly solid, as uniform as any that can be expected to result from a democratic process.   The authority of the committee's judgment was manifest at the hearing, held on Mar. 20, 1941, at which their recommendations were made to the FCC.   By that time the complexion of the industry had changed from a discord of counterclaims to a concord of expert opinion which left the commission no choice but to acknowledge its value and to proclaim the art open to the public.

The concept of the N.T.S.C. arose in a meeting between Dr. W. R. G. Baker, director of engineering for the R.M.A., and Chairman Fly of the FCC.   It was decided to open the deliberations to all members of the industry technically qualified to participate, whether they were members of the R.M.A. or not. A plan designed to segregate the technical work into panels of specially qualified experts was drawn up, and a parliamentary procedure was agreed upon which would reveal clearly the responsibility for the actions taken.

The detailed planning of the N.T.S.C. organization was carried out by three men: Dr. Baker, L. C. F. Horle, in charge of the R.M.A. Data Bureau, and I. J. Kaar, then chairman of the R.M.A. Committee on Television.   The soundness of the plan was proved by the efficiency with which the Committee and its Panels functioned.

The organization of the committee members was completed, and its first meeting was held July 31, 1940.   At that time the following statement concerning the committee and its organization was issued:

Reason and Purpose.— Because of the inadequacy of the various suggested standards for television, it is proposed to establish a committee for the purpose of developing and formulating such standards as are required for the development of a suitable national system of television broadcasting.

This project, sponsored by the R.M.A. in cooperation with the FCC, will be maintained independent of any other organization and will be truly representative of the majority opinion of the industry.

Committee and Membership.—Members of the N.T.S.C. will be appointed by the president of the R.M.A., subject to the approval of the executive committee of R.M.A., and will consist of representatives of those organizations broadly interested and experienced in the television field. In addition, there will be included representatives of such national technical organizations as are vitally interested in the research and development of television as well as individuals not associated with any organization, association, or company.

The N.T.S.C. and its component panels will be responsible for the investigation and study of all phases of a national television system. It will concern itself not only with projects that have reached the engineering and developmental stage but also with such research and experimentation as has a bearing on the system aspects and on the broad psychological and physiological aspects of television picture reproduction.

The following organizations have been requested to appoint one representative to the N.T.S.C.:

Bell Telephone Laboratories.
Columbia Broadcasting System.
Don Lee Broadcasting System.
DuMont Laboratories, Inc.
Farnsworth Television and Radio Corporation.
General Electric Company.
Hazeltine Service Corporation.
John V. L. Hogan.
Hughes Tool Company.
Institute of Radio Engineers.
Philco Corporation.
Radio Corporation of America.
Stromberg-Carlson Telephone Mfg. Company.
Television Productions.
Zenith Radio Corporation.

A quorum for the N.T.S.C. will comprise 75 per cent of its membership, and a majority vote of those present will be required for the approval of any proposal. The chairman of the N.T.S.C. will have no vote.

Panel Purpose and Functions.—The various projects may be assigned to individual members of the N.T.S.C. or to panels appointed by the chairman of the N.T.S.C.

In the operations of the panels, a quorum for any working meeting will comprise 50 per cent of the membership of the panel, and a majority vote of those present will be required for the approval of any proposal. The chairman of a panel may vote only in case of a tie.

In the event of the lack of a quorum present at any announced meeting of either the N.T.S.C. or any of its panels, action may be taken by a majority vote of those present, but such action will not be valid unless, and until, three quarters of the total membership of the N.T.S.C. or panel, as the case may be, approves such action by correspondence.

The members of the panels will be drawn from any company, association, or organization regardless of affiliation and may also include individuals not associated with any organization. The only requirement for membership on any panel is recognized skill, interest, and ability in the assigned project.

The titles and scopes of the initial panels are

1. *System Analysis.*—The analysis of foreign and proposed American television systems.

2. *Subjective Aspects.*—The influence of physiological and psychological factors in the determination of television system characteristics.

3. *Television Spectra.*—Consideration of sound and picture channel widths and locations.

4. *Transmitter Power.*—The consideration of transmitter output ratings, modulation capabilities and the relation between power requirements of picture and sound channels.

5. *Transmitter Characteristics.*—Consideration of essential systems. Characteristics of the transmitter (signal polarity, black level, etc.).

6. *Transmitter-receiver Coordination.*—Consideration of the essential factors requiring coordination in the design of receivers and transmitters (sideband distribution, audio pre-emphasis, etc.).

7. *Picture Resolution.*—Consideration of the factors influencing picture detail (aspect ratio, frame frequency, interlace, line density, etc.).

8. *Synchronization.*—Consideration of methods and means of accomplishing synchronization.

9. *Radiation Polarization.*—Consideration of the factors influencing a choice of the polarization of the radiated wave.

As the work proceeds, additional panels will be appointed as necessity arises.

Panel Reports.—Upon the completion and approval of the assigned project by a panel, a final report stating both the majority and minority opinions, together with a complete record of all meetings, will be submitted to the N.T.S.C. for its approval.

Meetings and Records.—Meetings of the N.T.S.C. and its panels shall be called at the discretion of the respective chairmen with notification to the members at least 1 week prior to the meeting date.

The chairman of the N.T.S.C. will appoint secretaries for the N.T.S.C. and its component panels.

Detailed minutes will be kept of all meetings and will record the names and votes of all voting, together with a clear statement of any minority opinion.

The minutes of all meetings shall be circulated to those attending the meeting and when approved shall constitute "official minutes."

Official minutes will then be distributed to all members of the N.T.S.C. and its component panels, the engineering department of the FCC and such others as may be approved by the executive committee of the R.M.A.

Approval and Transmission of Proposed Standards.—As standards are approved by the N.T.S.C. they will be submitted to the FCC by the board of directors of the R.M.A.

Following the initial meeting of the National Committee, the panel membership was completed, and panel meetings began early in September. By the end of December, the final reports of all the panels had been transmitted to the National Committee. During the first 2 weeks of January, 1941, a coordinating and editing committee prepared a report embodying the standards and other recommendations made by the panels. The National Committee, at its fourth meeting, on Jan. 14, considered these standards at length, modified the wording in certain particulars, and prepared and approved a "progress report," which was presented to the FCC on Jan. 27. The chairmen of the panels took the stand before the commission at that meeting and outlined in detail the substance of each panel's work and the technical background of the standards under their jurisdiction.

At this progress report meeting it became clear that the members of the FCC were satisfied that substantial agreement had been obtained on all the standards except that specifying 441 as the number of lines and that specifying amplitude modulation for the synchronization signals. Chairman Fly, in summing up the meeting, stated that the question of the synchronization signals and the question of color television were subjects on which further deliberation seemed advisable.

A special subcommittee was then set up to consider the synchronization-signal problem, which had by that time been definitely reopened by experimental tests of a system employing frequency modulation for the synchronizing pulses, which seemed to have merit. Acting on the recommendations of this subcommittee, the National Committee in its final meeting, on Mar. 8, rewrote the standard concerning synchronization, so as to admit the use of frequency modulation. At the same meeting, the committee approved the value of 525 lines as the standard for the number of lines per frame period. The change was based on demonstrations made by the Bell Telephone Laboratories of the small effect on picture quality of the relative resolution in the vertical and horizontal dimensions, as well as on the fact that the larger number of lines would provide a more uniform field of illumination in the reproduced picture.

The final report of the N.T.S.C., delivered to the commission at the hearings on Mar. 20, recommended the standards reported at the progress-report meeting, with the two changes mentioned above. The only opposition given to the standards at that time was put forward by the DuMont Laboratories, which urged that a variable number of lines and frames per second should be used.

Early in May the FCC announced that the N.T.S.C. standards had been adopted officially and that commercial television broadcasting based on these standards would be permitted on and after July 1, 1941.

CHAPTER II

THE NATIONAL TELEVISION SYSTEM STANDARDS

The work of the National Television System Committee[1] culminated in the hearing held before the Federal Communications Commission, Mar. 20, 1941. At this hearing the N.T.S.C. presented a report consisting of 22 specific standards and explanatory notes. The text of this report is reproduced in the following paragraphs. Immediately following the report, a brief outline of the standards is presented to correlate the following chapters in which the standards are treated in greater detail.

Report of the N.T.S.C. to the FCC.—The text of the N.T.S.C. report is as follows:

The National Television System Committee herewith submits transmission standards for commercial television broadcasting. The N.T.S.C. recognizes the coordinate importance of standardization and the commercial application of technical developments now in the research laboratories. These standards will make possible the creation, in the public interest, of a nationally coordinated television service and at the same time will ensure continued development of the art.

The N.T.S.C. recommends that monochromatic transmission systems other than those embodied in these standards be permitted to operate commercially, when a substantial improvement would result, provided that the transmission system has been adequately field tested and that the system is adequately receivable on receivers responsive to the then existing standards.

[1] The members and alternates of the National Television System Committee were W. R. G. Baker, chairman; V. M. Graham, secretary; Ralph Bown, D. A. Quarles, alternate; Adrian Murphy, P. C. Goldmark, alternate; A. B. DuMont, T. T. Goldsmith, alternate; B. R. Cummings, P. J. Herbst, alternate; E. F. W. Alexanderson, I. J. Kaar, alternate; D. E. Harnett, W. A. MacDonald, alternate; A. I. Lodwick, A. F. Murray, alternate; A. N. Goldsmith, H. A. Wheeler, alternate; John V. L. Hogan, L. C. Smeby, alternate; D. B. Smith, F. J. Bingley, alternate; E. W. Engstrom, C. B. Jolliffe, alternate; R. H. Manson, G. R. Town, alternate; Paul Raiborn, K. Glennan, alternate; John R. Howland, J. E. Brown, alternate.

The N.T.S.C. has broadened its standards on synchronization to permit field tests of several interchangeable systems. It is anticipated that some one of these systems will be found to be superior to the others, and it is, therefore, recommended that at that time the commission's standards be narrowed to require the commercial use of that particular and superior system (see Note A, page 21).

The N.T.S.C. believes that, although color television is not at this time ready for commercial standardization, the potential importance of color to the television art requires that

(*a*) a full test of color be permitted and encouraged, and that

(*b*) after successful field test, the early admission of color transmissions on a commercial basis coexistent with monochromatic television be permitted employing the same standards as are herewith submitted except as to lines and frame and field frequencies. The presently favored values for lines, and for frame and field frequencies for such a color system are, respectively, 375, 60, and 120.

The proposed standards are as follows:

I. The Television Channel

1. The width of the standard television broadcast channel shall be 6 Mc/s.

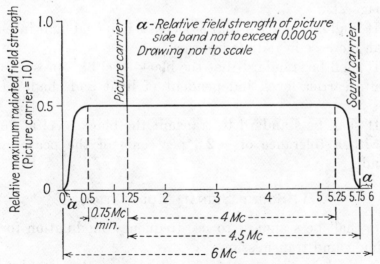

Fig. 1.—Idealized picture transmission amplitude characteristic.

2. It shall be standard to locate the picture carrier 4.5 Mc/s lower in frequency than the unmodulated sound carrier.

3. It shall be standard to locate the unmodulated sound carrier 0.25 Mc/s lower than the upper frequency limit of the channel.

4. The standard picture-transmission amplitude characteristic shall be that shown in Fig. 1.

II. Scanning Specifications

5. The standard number of scanning lines per frame period in monochrome shall be 525, interlaced two to one.

6. The standard frame frequency shall be 30 per second, and the standard field frequency shall be 60 per second in monochrome.

7. The standard aspect ratio of the transmitted television picture shall be 4 units horizontally to 3 units vertically.

8. It shall be standard, during the active scanning intervals, to scan the scene from left to right horizontally and from top to bottom vertically, at uniform velocities.

III. Picture Signal Modulation

9. It shall be standard in television transmission to modulate a carrier within a single television channel for both picture and synchronizing signals, the two signals comprising different modulation ranges in frequency or amplitude or both (see Note A, 1, page 21).

10. It shall be standard that a decrease in initial light intensity cause an increase in radiated power.

11. It shall be standard that the black level be represented by a definite carrier level, independent of light and shade in the picture.

12. It shall be standard to transmit the black level at 75 per cent (with a tolerance of ± 2.5 per cent) of the peak carrier amplitude.

IV. Sound Signal Modulation

13. It shall be standard to use frequency modulation for the television sound transmission.

14. It shall be standard to pre-emphasize the sound transmission in accordance with the impedance-frequency characteristic of a series inductance-resistance network having a time constant of 100 microseconds.

V. Synchronizing Signals

15. It shall be standard in television transmission to radiate a synchronizing waveform that will adequately operate a receiver which is responsive to the synchronizing waveform shown in Fig. 2.

16. It shall be standard that the time interval between the leading edges of successive horizontal pulses shall vary less than 0.5 per cent of the average interval.

17. It shall be standard in television studio transmission that the rate of change of the frequency of recurrence of the leading edges of the horizontal synchronizing signals be not greater than 0.15 per cent per second, the frequency to be determined by an averaging process carried out over a period of not less than 20, nor more than 100, lines, such lines not to include any portion of the vertical blanking signal (see Note B, page 24).

VI. Transmitter Ratings

18. It shall be standard to rate the picture transmitter in terms of its peak power when transmitting a standard television signal.

19. It shall be standard in the modulation of the picture transmitter that the radio-frequency signal amplitude be 15 per cent or less of the peak amplitude, for maximum white (see Note C, page 24).

20. It shall be standard to employ an unmodulated radiated carrier power of the sound transmission not less than 50 per cent nor more than 100 per cent of the peak radiated power of the picture transmission.

21. It shall be standard in the modulation of the sound transmitter that the maximum deviation shall be ± 75 kc per sec.

VII. Polarization

22. It shall be standard in television broadcasting to radiate horizontally polarized waves.

Note A: 1. Practical receivers of the RA type (those which attenuate the carrier 50 per cent before detection) designed for the synchronizing signals shown in Fig. 2 will also receive interchangeably any of the following:

   a. Amplitude-modulated synchronizing and picture signals of the 500-kc vertical synchronizing pulse type (Drawing IV, Fig. 1), Doc. 321R. [Fig. 80, page 282. Editor.]

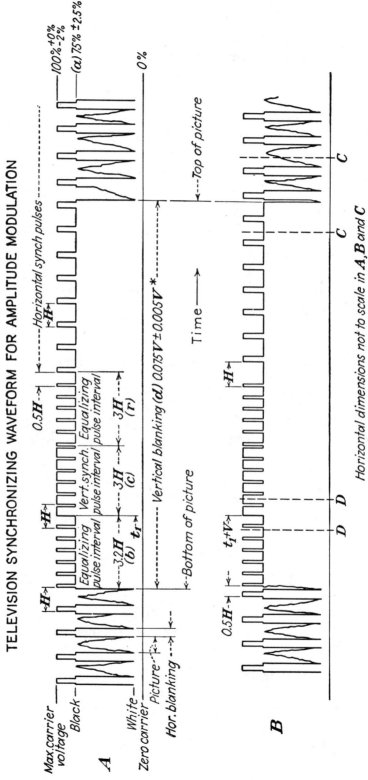

FIG. 2.—Television synchronizing waveform for amplitude modulation.

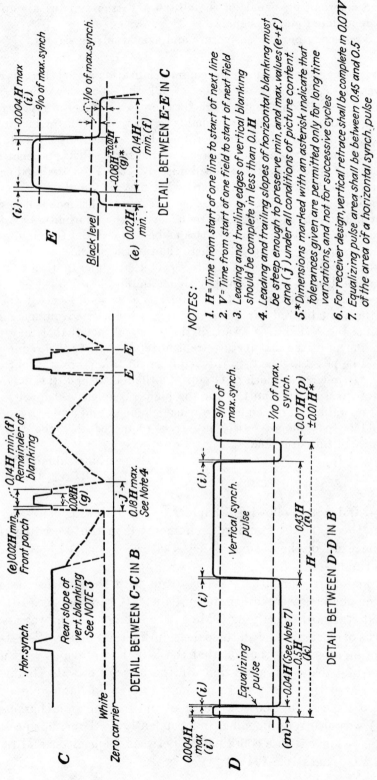

NOTES:

1. H = Time from start of one line to start of next line
2. V = Time from start of one field to start of next field
3. Leading and trailing edges of vertical blanking should be complete in less than 0.1H
4. Leading and trailing slopes of horizontal blanking must be steep enough to preserve min. and max. values (e+f) and (j) under all conditions of picture content.
5. *Dimensions marked with an asterisk indicate that tolerances given are permitted only for long time variations, and not for successive cycles
6. For receiver design, vertical retrace shall be complete in 0.07V
7. Equalizing pulse area shall be between 0.45 and 0.5 of the area of a horizontal synch. pulse

FIG. 2 (continued).

b. Synchronizing signals of the alternate carrier type with amplitude-modulated picture signals.

c. Frequency-modulated picture and synchronizing signals.

2. It is proposed that during the commercialization period there be carried out systematic, comparative tests, of all the above-mentioned signals including transmissions from a single location to a number of representative receiving locations and over a reasonable test period. It is further proposed that these tests be directed and coordinated by an accredited committee of the R.M.A. or some other committee suitable to the FCC and that on the completion of these tests there be submitted to the FCC any modifications or additions to the N.T.S.C. standards that may be found desirable.

Note B: It is recommended that as progress in the art makes it desirable, the maximum rate of change of frequency of the transmitted horizontal synchronizing signals for studio programs be reduced and that limits be set for transmissions originating elsewhere than in the studio.

Note C: It is the opinion of the N.T.S.C. that a picture transmitter not capable of a drop in radio-frequency signal amplitude to 15 per cent or less of the peak amplitude would not be completely satisfactory since it would not utilize the available radio-frequency power to the best advantage. At the same time the N.T.S.C. is aware of the practical fact that it may not be possible for all of the first picture transmitters to satisfy this requirement. It should be possible to satisfy this requirement in picture transmitters for the lower frequency channels of Group A, although, at first, this may not be possible in picture transmitters for the higher frequency channels. After the first operation on the higher frequency channels and as designs progress it should be possible to satisfy it. It is requested that the FCC take cognizance of this situation.

Respectfully submitted,

W. R. G. Baker, Chairman

Cross Index to Following Chapters.—The following notes serve to connect the N.T.S.C. standards, enumerated in the preceding report, with the work of the various panels that were engaged specifically in their formulation.

The first four standards, having to do with the dimensions of the television channel, were drawn up by Panel 3 (Chap. V, pages 101–133). The first standard, specifying 6 Mc as the channel width, was assumed as a basis of discussion by all the panels during their deliberations. This assumption arose from the fact that the FCC had previously set up the experimental television allocations on the basis of a 6-Mc channel and had indicated that no increase in this figure could be accommodated. Standard 4 specifying the vestigial sideband transmission characteristic (Fig. 1) was also recommended independently by Panel 6, since this is a matter requiring close coordination between transmitter and receiver (Chap. VII, pages 162–194).

The second group of standards, having to do with the scanning specifications, were drawn up by Panel 7 (Chap. VIII, pages 195–235) with the exception of standard 5, specifying 525 lines as the number of scanning lines per frame period. The recommendation of Panel 7 in this respect was 441 lines. The National Committee, at the meeting of Mar. 8, 1941, considered this matter at some length and finally adopted the recommendation of 525 lines. A brief recommending the latter figure, prepared for the National Committee by D. G. Fink, is printed at the conclusion of Chap. VIII (page 225).

The next group of four standards, 9 to 12 inclusive, were based on recommendations of several panels. Standard 9 was composed by the National Committee, following the recommendation of Panel 8, but allowing the alternative use of frequency modulation for the synchronizing pulses. This latter action followed demonstrations that frequency-modulated signals might be employed with benefit to the performance of this system, and that such signals could be received adequately on receivers designed for amplitude modulation, provided only that the picture-carrier signal was attenuated in the receiver before final detection, so that the frequency-modulation signals would be converted into amplitude modulation. Note A in the report explains this situation further.

The following three standards, numbered 10 to 12 (negative transmission, d-c transmission, and black level fixed at 75 per cent $\pm$ 2.5 per cent), are based on the recommendations of Panel 5 (Chap. VI, pages 134–161), together with that of Panel 8 (Chap. IX, pages 236–315), which specified the black-level percentage in Fig. 2.

Standards 13 and 14, specifying frequency modulation and audio pre-emphasis for the sound transmissions were developed jointly by Panels 5 and 6 (Chaps. VI and VII, pages 134 and 162). The standards on the synchronizing signals (numbered 15 to 17) were developed by Panel 8 (Chap. IX, pages 236–315).

The four standards on transmitter ratings (numbered 18 to 21) were based on the recommendations of Panel 4 (Chap. VI, pages 134–161). Panel 5 also took action on the relative power of the sound and picture signals. The final standard, recommending horizontal polarization is based on the study conducted by Panel 9.

In addition to the specific standards, two panels were engaged in more general investigations. Panel 1 investigated systems in general, including color television. Material abstracted from their report appears in Chap. III. The findings of Panel 2, which investigated the subjective aspects of television system performance, are reproduced in Chap. IV.

CHAPTER III

TELEVISION SYSTEMS

To Panel 1[1] of the N.T.S.C. was assigned the task of analyzing the existing systems of television, both American and foreign, and considering proposed systems. Three studies were undertaken by the Panel in carrying out this assignment.

The first was a general study of television systems as such. Included in this work was the compilation of information on existing television systems. An elaborate outline for analyzing television systems was also prepared by the Panel. This outline describes alternate methods of scanning, synchronization, video transmission, and audio transmission. The advantages and disadvantages of each method are listed, together with the controlling advantage or disadvantage in each proposal.

The conclusions reached in this outline are directly in line with the standards finally adopted by the National Committee. For example, unidirectional linear scanning in the horizontal direction, interlaced two to one, is found preferable to the many alternative methods of image analysis. Amplitude separation of the synchronization information, and waveform separation of the vertical and horizontal components are also recommended. The full text of this outline is printed on pages 33-38. It will repay careful study, since it is the most comprehensive comparison of television system proposals ever prepared.

The second study conducted by Panel 1 was concerned with color television. This subject assumed a prominent place in the discussions of the Panel as a result of the excellent demonstrations made by Dr. P. C. Goldmark at the Columbia Broadcasting System Laboratories. Since reports of these demonstrations have been published, only a brief outline of the system is presented here.

The interlaced fields are scanned at the transmitter through red, green, and blue optical filters, in succession, at a rate of 120 fields per second. At the receiver the white light from the picture

---

[1] The members of Panel 1 were P. C. Goldmark, chairman; J. E. Brown; D. G. Fink; D. E. Harnett; I. E. Lempert; H. B. Marvin; Pierre Mertz; Adrian Murphy; R. E. Shelby; D. B. Smith; and R. F. Wild.

tube is viewed through corresponding filters that rotate synchronously in succession before the fluorescent screen. Every point in the image is thus scanned in the three colors in the space of $\frac{1}{20}$ sec. This rate of frame repetition is sufficiently high to produce smooth blending of the colors.

Since the field rate is twice as great as in the corresponding black-and-white system, the picture detail is reduced to one-half the value present in the black-and-white image, for a given channel width. The resolutions in the vertical and horizontal directions are maintained approximately equal in the color system by dividing the number of lines by the square root of two. The value proposed for the color system is approximately 375 lines, corresponding to 525 lines in black and white. The reduction in image detail is compensated by the introduction of color values, which reveal contrasts not present in monochrome reproduction.

The minutes of Panel 1 reveal lengthy discussion of the merits of color-television systems, attempting to evaluate the relative importance of the loss of detail and the gain in color values, as well as other technical questions. Demonstrations of the system were viewed by a majority of the N.T.S.C. membership. The opinions of Panels 1, 6, 7, and 8 were assembled in questionnaire form, a summary of which is presented in the report of Panel 1. Near the conclusion of its work, the Panel voted without dissent that "It is the consensus of Panel 1 that color television is not yet ready for the establishment of standards for commercial operation." The Panel also took formal cognizance of the color-television experiments carried on by Dr. Goldmark.

The importance of color television for the future was reflected in the report of the National Committee.

The N.T.S.C. believes that, although color television is not at this time ready for commercial standardization, the potential importance of color to the television art requires that:

a) A full test of color be permitted and encouraged, and that

b) After successful field test, the early admission of color transmissions on a commercial basis coexistent with monochromatic television be permitted employing the same standards as are herewith submitted except as to lines and frame and field frequencies. The presently favored values for lines, and for frames and field frequencies for such a color system are, respectively, 375, 60, and 120.

TABLE I.—ANALYSIS OF AMERICAN
(Systems for band

Num-ber	Designation	Period of operation	When demon-strated	Scanning pattern			Synchronization system (H = horizontal, V = vertical)	
				Lines/frames-fields	Aspect ratio	Type of motion	Per cent of carrier	Description of waveform
1	R.M.A. standards	1939–1940	1939–1940	441/30–60	4:3	Linear	20–25	H: Single rectangular pulse V: Serrated with equalizing pulses. See R.M.A. standard M9-211
2	DuMont A (500-kc burst vertical synchronizing pulse).	1939–1940	December, 1939 to date	Variable (see Note A)	4:3	Linear	20–25	H: Rectangular pulse. V: r-f burst during V blank with H superposed
3	DuMont B (Transmission of scanning waveforms).	1938–1939	1939	Variable	4:3	Any (note B)	....	Synchronizing inherent in transmission of scanning waveforms
4	Hazeltine (FM for synchronizing).	1940	November, 1940	441/30–60	4:3	Linear	100	Any synchronizing wave shape frequency-modulated during H and V blank
5	RCA (507 lines, 495 lines later suggested).	1940	July, 1940	507/30–60 495/30–60	4:3	Linear	20–25	Waveform same as R.M.A.
6	RCA (FM for sound).	1940	November, 1940	441/30–60	4:3	Linear	20–25	R.M.A. M9-211
7	RCA (FM sound, quasi-FM for picture).	1940	October, 1940	441/30–60	4:3	Linear	....	R.M.A. M9-211
8	RCA (Long integration synchronizing pulse).	1940	October, 1940	441/30–60	4:3	Linear	20–25	Slots in vertical pulse at line frequency. V pulse = 9H
9	Philco (525 lines)	1938–1940	1938–1940	525/30–60	4:3	Linear	Over 25	Waveform same as R.M.A. M9-211
10	Philco (605 lines)	1938–1940	1938–1940	605/24–48	4:3	Linear	Over 25	Waveform same as R.M.A. M9-211
11	Philco (Narrow vertical synchronizing pulse).	1935–1938	Up to 1938	441/30–60	4:3	Linear	20–25	H: Rectangular pulses. V: single narrow rectangular pulse, same level
12	Farnsworth (Narrow vertical synchronizing pulse).	1935–1938	Up to 1938	441/30–60	4:3	Linear	20–25	H: Rectangular pulse. V: single narrow pulse of higher amplitude
13	G.E. (Picture carrier 6 db attenuated).	1938	1938	441/30–60	4:3	Linear	20–25	Same as R.M.A. M9-211
14	Kolorama (225 lines).	1935–1940	May, 1939	225/12–24	6:5	Linear	....	By transmission pulse and power line
15	"Sound-during-blanking." Single carrier system.	Prior 1940	....	Note C	Note C	Note C	Note C	Note C. Carrier frequency modulated during horizontal blanking interval with audio frequencies

AND FOREIGN TELEVISION SYSTEMS
width of 6 Mc or less)

Transmitter characteristics							Other characteristics, apparatus employed, advantages claimed by sponsor, etc. Notes
Polarity of modulation	Carrier attenuation	D-c or a-c transmission	Direction of polarization	Total bandwidth	Picture sideband width	Carrier separation	
Negative	None	D-c	Horizontal	6 Mc	4–4.5 Mc	4.5 Mc	Sound pre-emphasized. Sound and picture carriers of equal power. Standards used sometime by RCA, G.E., Farnsworth, Philco, CBS, Don Lee, General Television Corporation, Zenith, and others
Negative	None	D-c	Not specified	6 Mc	4–4.5 Mc	4.5 Mc	Note A: Designed for flexibility (continuous variability) in line and frame rates, including 15–30 as lower limit
Negative (note B)	None (note B)	D-c (note B)	Not specified	6 Mc	4–4.5 Mc	4.5 Mc	Note B: Receiver automatically follows changes in scanning motion
Negative	None	D-c	Not specified	6 Mc	4–4.5 Mc	4.5 Mc	FM modulator required for synchronizing. High synchronizing amplitude developed. Improves picture-modulation capability
Negative	None	D-c	Horizontal	6 Mc	4–4.5 Mc	4.5 Mc	
Negative	None	D-c	Horizontal	6 Mc	4–4.5 Mc	4.5 Mc	FM modulator for sound only. 75 kc maximum deviation
....	None	D-c	Horizontal	6 Mc	4–4.5 Mc	4.5 Mc	75 kc maximum deviation for sound. 0.75 Mc maximum deviation for picture. Improved transient response reported
Negative	None	D-c	Horizontal	6 Mc	4–4.5 Mc	4.5 Mc	V pulse integrated in R-C circuit of long-time constant. No equalizing pulses used
Negative	None	D-c	Vertical	6 Mc	4–4.5 Mc	4.5 Mc	Sound carriers staggered
Negative	None	D-c	Vertical	6 Mc	4–4.5 Mc	4.5 Mc	Greater horizontal definition due to lower frame rate
Negative	None	D-c	Vertical	6 Mc	4–4.5 Mc	4.5 Mc	
Negative	None	D-c	Not specified	6 Mc	4–4.5 Mc	4.5 Mc	
Negative	6 db	D-c	Horizontal	6 Mc	4–4.5 Mc	4.5 Mc	Narrower band at transmitter, wider band at receiver
Positive	None	D-c	Vertical	100 kc/300 kc	300 kc	....	Mechanical scanners. Transmitter on band at 2,000 kc
Negative	Note C	D-c	Note C	6 Mc or less	4–4.5 Mc	....	Note C: Not specified. Audio frequencies substantially higher than one-half line frequency suffer distortion. Investigated by RCA, G.E., Philco. Suggested 1940 by Kallman

TABLE I.—ANALYSIS OF AMERICAN AND

Number	Designation	Period of operation	When demonstrated	Scanning pattern			Synchronization system (H = horizontal, V = vertical)	
				Lines/frames-fields	Aspect ratio	Type of motion	Per cent of carrier	Description of waveform

American Color

Number	Designation	Period of operation	When demonstrated	Lines/frames-fields	Aspect ratio	Type of motion	Per cent of carrier	Description of waveform
16	CBS (3-color system No. 3).	1940	August, 1940	343/60–120	4:3	Linear	Note D	Adaptable to any synchronizing signal that can control filter disks
16a	CBS combination color and black and white......			*This system same as CBS No. 3, for color transmissions. For black-and-white*				
17	CBS (3-color system No. 4).	1940	....	430/45–180	4:3	Linear	Note D	Note D. Quadruple interlace
18	CBS (3-color system No. 5).	1940	....	550/30–120	4:3	Linear	Note D	Note D. Quadruple interlace
19	G.E. (2-color system).	1940	November, 1940	441/30–60	4:3	Linear	20–25	R.M.A. M9-211

Foreign

Number	Designation	Period of operation	When demonstrated	Lines/frames-fields	Aspect ratio	Type of motion	Per cent of carrier	Description of waveform
20	British (BBC) Standard.	1936–1939	1936	405/25–50	5:4	Linear	30	Rectangular H pulses. Serrated V pulse; no equalizing pulses
21	Scophony (British) (In U.S.A. also, in 1940).	....	1937	405/25–50	5:4	Linear	30	BBC Standard
22	Early Baird (British).	1936–1937	1936	240/25	4:3	Linear	40	8 per cent line synchronizing pulses, single V pulse of 12 lines duration
23	Baird 2-color (British).	....	March, 1938	120/16.6–100	3:4	Linear Vertical	40	6 to 1 interlace by nonperiodic pulses
24	Baird 3-color (British).	....	January, 1938	120/16.6–100	3:4	Linear Vertical	40	Same as Baird 2-color system
25	Velocity modulation (Br.).	....	May, 1934	60–400/25	4:3	....	30	10 per cent line pulses; frame by ratchet circuit from line
26	French (PTT) standards.	....	June, 1937	440–445/25–50	5:4	Linear	30	....
27	Barthelemy (French).	....	June, 1937	450/25–50	5:4	Linear	17 H 34 V	6 lines paired to form V pulse
28	German standards.	....	1939–1940	441/25–50	5:4	Linear	30	10 per cent line pulses; single 35 per cent V
29	German proposal	....	1937–1938	441/25–50	5:4	Linear	30	Burst of 1.1 Mc for 20 per cent line duration as V pulse. Square H
30	Italian standards	....	1939	441/21–42	5:4	Linear	30	Same as German standards
31	Russian standards.	....	1940	441/25–50	11:8	Linear	20–25	Essentially R.M.A. M9-211

FOREIGN TELEVISION SYSTEMS.—*(Continued)*

Television Systems

Polarity of modulation	Carrier attenuation	D-c or a-c transmission	Direction of polarization	Total bandwidth	Picture sideband width	Carrier separation	Other characteristics, apparatus employed, advantages claimed by sponsor, etc. Notes
Note D	Note D	Note D	Note D	6 Mc	4–4.5 Mc	4.5 Mc	Note D: Not specified. Mechanical filter disks or drums at transmitter and receiver

transmissions, no preference as to specifications is indicated

Note D	Note D	Note D	Note D	6 Mc	4–4.5 Mc	4.5 Mc	Line scan frequency approximately 19,400 p.p.s.
Note D	Note D	Note D	Note D	6 Mc	4–4.5 Mc	4.5 Mc	Line scan frequency 15,750 p.p.s. Color coincidences 10 p.s.
Negative	None	D-c	Horizontal	6 Mc	4–4.5 Mc	4.5 Mc	Dichromatic filter disks, synchronizing on power line. Odd lines always one color, even lines always other color

Systems

Positive	None	D-c	Vertical	6 Mc	2.5 Mc	3.5 Mc	Double sideband. 2 c.p.s./sec. tolerance in line frequency for mechanical receivers
Positive	None	D-c	Vertical	6 Mc	2.5 Mc	3.5 Mc	Mechanical scanners, supersonic light valve. Requires BBC tolerance on line synchronizing
Positive	None	D-c	Vertical	5 Mc	1.5 Mc	3.5 Mc	Mechanical scanner at transmitter. Live pickup by film only
Positive	None	D-c	Vertical	5 Mc	2 Mc	3 Mc	Flying spot pickup with rotating color disks. Projected large-screen image
Positive	None	D-c	Vertical	5 Mc	2 Mc	3 Mc	Same but three colors. Laboratory demonstration
Positive	None	A-c	Vertical	3 Mc	1.5 Mc	No sound	CR tube light source on film. Amplitude component added
Positive	....	D-c	....	....	....	4 Mc	Standards embrace four systems
Positive	....	....	....	....	....	4 Mc	Barthelemy interlace; rotating disk synchronizing
Positive	None	D-c	Vertical	5 Mc	2.0 Mc	2.8 Mc	Picture carrier on high side of channel
Positive	None	D-c	Vertical	5 Mc	2 Mc	2.8 Mc	
Positive	None	D-c	Vertical	5 Mc	2 Mc	2.8 Mc	
Negative	None	D-c	Horizontal	6 Mc	4–4.5 Mc	4.5 Mc	Sound carrier lower in frequency than picture

In its order commercializing the art, issued May 3, 1941, the FCC required licensees to submit data on color television. The specific instructions were: "It is further ordered that on or before January 1, 1942, the licensees of television broadcast stations shall submit to the commission complete comparative test data on color transmissions, with recommendations as to standards that may be adopted by the commission for color television." The same order permitted the submission of these data through a committee or organization representing the licensees.

To implement the collection and submission of this information, the R.M.A. appointed in May, 1941, the Subcommittee on Color Television. This subcommittee held seven meetings during 1941. On Nov. 10 of that year, the subcommittee issued a report to the N.T.S.C. Among opinions in the report was the following: "The subcommittee is of the opinion that the present knowledge of the art does not justify the recommending of standards for color television at the present time." Although no specific values of standards then in use were given, the following list of items requiring standardization was presented:

1. Color characteristics of the transmitted signal.
2. Color sequence.
3. Line and frame frequencies and interlace.
4. Phasing pulse.
5. Synchronizing-pulse constants.
6. Transmitted signal output vs. light input (generally referred to as the gamma characteristic).
7. Review of the present black-and-white transmission standards, to determine their relationship to color television.

The third study conducted by Panel 1 concerned flexibility in the scanning specifications. The DuMont Laboratories, Inc., proposed a variable number of lines and a variable number of fields and frames per second. It was argued by the DuMont organization that it might be possible in the future to employ a lower frame rate than was then practical and that the detail of the picture might then be proportionately increased without increasing the width of the channel employed by the system, provided that the system had sufficient inherent flexibility to permit changing the number of lines. This proposal received close scrutiny by the membership of Panel 1. A report on the subject, printed in the following pages, was drawn up by members of the Panel. A motion was passed by a vote of 5 to 3 that "It is the consensus of Panel 1 that the best standards for black-and-white pictures on the channels below 108 Mc referred to in Motion 6 (a motion previously passed by the Panel referring to experimental tests of color television during hours not devoted to black-and-white transmissions) would be standards using a single value for line frequency and a single value for frame frequency." This vote against flexibility was confirmed by the standards adopted by the National Committee, which recommended the single value of 525 lines and the single value of 30 frames per second.

Table of Television Systems.—Table I, an analysis of American and foreign television systems, was prepared by Panel 1.

Outline for Analysis.—The "Outline for Analysis of Television Systems" prepared by Panel 1 is as follows:

A single asterisk * indicates the preferred method, type, or system.
A dagger † indicates the controlling advantage or disadvantage.

I. Methods of scanning
   A. Type of motion
      *1. Unidirectional linear scanning (uniform velocity)

Advantages	*Disadavantages*
†a. Even distribution of detail for a given maximum video frequency	a. Precautions necessary to ensure linearity (true also of bidirectional scanning)
b. Minor departures from linearity of scanning waveform do not destroy detail (*cf.* bidirectional scanning)	b. Retrace times not available for transmitting picture detail
c. Retrace time available for auxiliary signals	

      2. Bidirectional linear scanning (uniform velocity)

Advantages	*Disadvantages*
a. More efficient use of bandwidth due to utilization of retrace interval(s)	†a. Extremely critical with respect to synchronization and linearity
b. In magnetic scanning, peak voltage associated with rapid retrace is avoided	

      3. Sinusoidal scanning

Advantages	*Disadvantages*
a. Simple, inexpensive deflection circuits	a. Difficult to synchronize
	†b. Does not make equal use of

b. Relatively free from geometric distortion

bandwidth in all portions of picture

c. Spurious signal greater in storage type camera tubes

d. Compensation required to obtain uniform brilliance

4. Velocity scanning (brightness inversely proportional to scanning velocity)

Advantages	*Disadvantages*
a. Synchronization is inherent in waveform (no separate synchronizing system necessary)	†*a.* Wasteful of amplitude and frequency characteristic
b. Brighter high lights	*b.* Excessive vertical blanking time required
	c. Has not yet been used for direct pickup

5. Spiral scanning

Advantages	*Disadvantages*
a. Utilizes time devoted to horizontal retrace in unidirectional scanning	*a.* Wasteful of screen area in transmission of rectangular pictures
b. Make maximum use of screen of circular cathode ray tube	*b.* Difficult to synchronize (for same reason as in sinusoidal scanning)
	†*c.* Does not make equal use of bandwidth in all portions of picture

B. Direction of motion of scanning lines (assuming linear scanning)

   *1. Horizontal

Advantages	*Disadvantages*
†*a.* Horizontal motion (most commonly encountered) best reproduced without disintegration of raster	*a.* Lower audio-frequency limit in single carrier system (sound during blanking) for aspect ratio greater than unity
b. Lower line frequency for pictures having greater width than height	

   2. Vertical

Advantages	*Disadvantages*
a. Higher audio-frequency limit in single carrier system (sound during blanking) for aspect ratio greater than unity	†*a.* Disintegration of raster with horizontal motion (most commonly encountered)
	b. Higher line frequency with picture having greater width than height

C. Interlace

   1. General

Advantages	*Disadvantages*
†*a.* Permits use of higher flicker rate without loss of picture	*a.* Requires more accurate vertical synchronization

information, with a given bandwidth. (Conversely, allows greater detail with a given frame rate and bandwidth)

b. Introduces optical pairing with motion across the lines (due to motion of eye following action)

c. Possibility of interline flicker being apparent

d. Jagged edges on objects moving in direction of lines

2. Simple two to one

Advantages	*Disadvantages*
†a. Easiest form of interlacing to obtain in practice (odd-line system)	a. Does not make fullest possible use of general advantages. (See *C*-1-*a*)

3. Multiple (higher than two to one)

Advantages	*Disadvantages*
a. Greater detail available with given minimum flicker rate and given bandwidth	†a. Difficult to obtain sufficiently accurate synchronizing to avoid pairing
	b. Tendency to introduce crawling with certain sequences
	c. Interline flicker more apparent (due to lower frequency and greater separation of lines in successive frames)

D. Aspect ratio

*1. 4:3 (standard for projected picture in motion pictures)

Advantages	*Disadvantages*
†a. Has all advantages found in motion-picture practice	a. Does not make fullest use of area of circular fluorescent screen
b. Permits scanning of motion-picture film without waste of screen area or distortion of the aspect ratio	

II. Synchronization

A. Methods of conveying synchronizing information

*1. Discrimination by amplitude separation of synchronization from video (composite synchronizing-video signal)

Advantages	*Disadvantages*
†a. Synchronizing clipping inexpensive and practical in infrablack region	a. Makes inefficient use of channel for transmitting synchronizing
b. Identical transmission time for video and synchronizing	b. Lowers picture-modulation capability
c. Single signal to be handled for both functions	

2. Discrimination by type of modulation (example, FM for synchronizing)

Advantages	*Disadvantages*
a. More efficient use of product of bandwidth and time	*a.* More complex modulation and demodulation circuits
b. Greater carrier amplitude available for picture modulation	
c. Substantial increase in developed synchronizing-signal amplitude	

NOTE: No adequate field test as yet.

3. Transmission of entire scanning waveform by separate carrier or subcarrier

Advantages	*Disadvantages*
a. Scanning pattern at receiver automatically follows congruently that at transmitter, with synchronism inherent	†*a.* Picture size affected by signal-strength variations line voltage, interference (not completely overcome by a-v-c)
b. Permits changes in definition to suit subject matter, provided receiver is adequately designed	*b.* Noise affects position of received picture elements by altering scanning wave shape

4. Transmission of synchronizing pulses on separate carrier or subcarrier

Advantages	*Disadvantages*
a. Less probability of cross modulation between synchronizing and video	†*a.* Separate modulator, r-f and demodulator circuits required
b. Permits full modulation capability for picture signal and synchronizing	*b.* Possibility of improper phase displacement between synchronizing and video

B. Separation of vertical and horizontal synchronizing information

*1. Frequency or waveform separation

Advantages	*Disadvantages*
†*a.* Does not require additional carrier amplitude for vertical-from-horizontal separation	NOTE: Disadvantages under examination by Panel 8
b. Will tolerate amplitude distortion in transmission and reception	
c. Permits amplitude limiters to be used to discriminate against noise	

2. Amplitude separation

Advantages	*Disadvantages*
NOTE: Advantages under examination by Panel 8	†*a.* Requires additional carrier amplitude for vertical synchronizing

b. Less tolerant of amplitude distortion

c. Requires more critical adjustment of clipping circuit

C. Method of controlling scanning

1. Self-oscillating scanning generators

Advantages

a. Scanning maintains itself in absence of synchronizing signals, thus protecting picture tube

b. Flywheel effect permits synchronization with weak lock-in

c. Less sensitive to noise during a portion of cycle

Disadvantages

a. Frequency hold-in controls required with weak lock-in

b. Oscillators subject to loss of synchronism

2. Driven scanning waveform generator

Advantages

a. No frequency hold-in control required

b. Recovers quickly from momentary noise interference

c. May be made to follow changes in lines and frames over greater frequency range without adjustment

Disadvantages

a. Loss of scanning in absence of synchronizing signal; hence protective circuit required for picture tube

b. More susceptible to random noise and to noise impulses recurring many times per frame

3. Direct amplification of scanning waveforms
   See II-A-3 above

III. Method of transmitting video information

A. Methods of modulation

1. Amplitude modulation, double sideband

Advantages

a. No sideband filter required at transmitter

b. Greater tolerance with respect to distortion produced by detuning

c. Essentially free of amplitude and phase distortion even with high percentage modulation

Disadvantages

†a. Wasteful of channel space, i.e., less detail for given bandwidth

b. Wasteful of transmitter if receiver operated single sideband

*2. Amplitude modulation, vestigial sideband

Advantages

†a. More efficient use of channel, i.e., greater detail for given bandwidth

b. Ether spectrum conserved, for a given detail

Disadvantages

a. Sideband filter required

b. Power wasted in filter, if high level modulation

c. Some amplitude and phase distortion, dependent on percentage of modulation

3. Frequency modulation, vestigial sideband (fractional deviation ratio)

Advantages	*Disadvantages*
a. Provides more useful transmitter power	*a.* More critical as to frequency characteristic of receiver
	b. Partial amplitude modulation produced, prevents use of limiter

NOTE: No adequate field test as yet.

4. Frequency modulation, **double** sideband (fractional deviation ratio)

Advantages	*Disadvantages*
a. No amplitude modulation produced; hence limiters might be used to reduce multipath effects	†*a.* More channel space required for given picture detail

IV. Method of transmission of sound

  A. Separate carrier, amplitude modulated

Advantages	*Disadvantages*
a. Simplicity and ease of separating sight and sound signals at receiver	*a.* More susceptible to noise from ignition and other impulse sources
	b. Channel space required for carrier

  B. Separate carrier, frequency modulated

Advantages	*Disadvantages*
a. Improved signal-to-noise ratio for given transmitter power when signal is greater than noise	*a.* Channel space required
	b. Somewhat more expensive receiver
b. More efficient transmitter	*c.* More critical frequency characteristic and tuning
c. Simplifies television f-m broadcast combination receivers	*d.* Oscillator stability must be better than for AM

  C. Single carrier, sound frequency modulated during horizontal blanking

Advantages	*Disadvantages*
a. Conservation of ether spectrum for given picture detail	†*a.* Audio frequency limited to approximately one-half line frequency
b. Single transmitter (two modulators, however)	*b.* Additional circuits required in receiver
	c. Possibility of cross modulation between sight and sound

Questionnaire on Color Television.—The questionnaire on color television and an abstract of the answers summarizing opinions of members of Panels 1, 6, 7, and 8 on color television are as follows:

NOTE: Qualified answers to some of the questions were given by many members.

Questions	Answers		
	Yes	No	No answer
A. COLOR ASPECT			
1. Do you prefer color television as demonstrated by the Columbia Broadcasting System to black and white?..	30	4	4
2. Will the addition of color increase the entertainment value of televised pictures?......................	34	2	3
3. Do you think that color means more to televised pictures than to moving pictures in theaters?.......	17	18	5
4. Do you think the color quality of the color television demonstrated would be acceptable to the public?...	32	3	4
5. Do you think color adds to the apparent resolution of a black-and-white picture?.....................	31	7	3
6. Do you think that the apparent resolution of color television as demonstrated is greater than R.M.A. black and white?................................	10	19	8
7. Do you think that the apparent resolution of color television as demonstrated is less than R.M.A. black and white?...................................	14	14	9
8. Is the apparent resolution of color television as demonstrated satisfactory?......................	20	12	5
9. Assuming that black-and-white transmission and color transmission will exist simultaneously, do you think that the resolution of color television received in black and white as demonstrated would be acceptable to the public?.............................	10	25	6
10. Was the brightness of the color demonstrated in your estimation acceptable?.........................	33	4	3
B. RECEIVER REQUIREMENTS			
11. Do you believe the public will pay more for color receivers than for black-and-white receivers?........	33	2	2
12. How much more do you estimate the public will pay for color receivers than for black-and-white receivers? (Answer in percentage.)*			4

* Average estimate 25 per cent.

Questions	Answers		
	Yes	No	No answer
13. Should any commercial television receivers be marketed for reception of black-and-white transmission exclusively?....................................	31	7	1
14. Should all commercial receivers be able to receive color television as well as black and white?.........	3	34	0
15. Should all commercial black-and-white television receivers be able to receive color television transmission as well as black and white?..............	16	20	2
16. Should all commercial black-and-white receivers be so designed as to be readily convertible into color receivers by the addition of only the color filter disk and its synchronizing gear?......................	7	30	2
17. Are you in favor of having the N.T.S.C. recommend to the R.M.A. that black-and-white receivers should be flexible so as to be able to receive color transmissions in black and white?....................	13	28	0
18. Will the introduction of experimental transmission of color television make the sale of black-and-white receivers more difficult if such receivers are not able to receive such transmission in black and white?....	28	10	3
19. Will the introduction of experimental transmission of color television make the sale of black-and-white receivers more difficult if such receivers are able to receive such transmission in black and white?......	19	18	4
20. Will the introduction of experimental transmission of color television hamper the commercial progress of television in general?.........................	11	24	3
C. Transmission Standards			
21. Should color transmission be allowed in Group A channels?......................................	24	11	3
22. Do you believe the present bandwidth of 6 Mc in Group A channels is adequate for color transmission?	17	16	5
23. Should special r-f channels outside of Group A channels, with a wider bandwidth (8 to 10 Mc) be allotted to color transmission?...........................	27	5	6
24. Should any transmission standards for color television be considered at this time?................	16	25	0
25. Should any transmission standards for color television be adopted at this time?...................	7	32	1
26. Should transmission standards for black-and-white television be influenced by color television considerations?......................................	12	26	0

Questions	Answers		
	Yes	No	No answer
27. Should the transmission standards for black-and-white television be fixed independent of color television considerations?..........................	27	12	1
28. Should transmission standards for black-and-white television be chosen in accordance with standards for color television (by maintaining the same number of frames and lines, *e.g.*, 2:1 interlaced, 60 frames and 343 lines, or 4:1 interlaced, 30 frames and 550 lines)?	1	36	3
29. Should transmission standards for black-and-white and for color television be so chosen that the change of black-and-white receivers to color reception is made as easy as possible? (By maintaining approximately the same line scanning frequency, *e.g.*, 60 frames, about 300 lines for color, and 30 frames, about 600 lines for black and white)?..............	7	28	3

Report on Flexibility.—The "Report on Flexibility" prepared by Panel 1 is as follows:

The term "flexibility," as applied in connection with a national television system, refers to the ability of the system to operate in more than one manner.   The different manners of operation may involve the use of fundamentally different principles of operation, the use of different types of equipment, or the optional use of different numerical values for characteristics that are quantitative in nature.

During the last few years the word "flexible" has been used in various ways to characterize various television systems, proposed transmission standards, and apparatus.   Some of the kinds of flexibility that have been proposed are

1. Variation in scanning rates for lines and fields.
2. Wide choice of methods for utilization of signal by various receiver circuits and picture-reproducing devices.
3. Other variations in transmission without obsolescence of receivers, such as flexibility to provide for:
   a. Amplitude modulation or frequency modulation of picture signal, synchronizing signal, and/or sound signal.
   b. Transmission of color television (in addition to black and white).
   c. Vertical or horizontal polarization.
   And others.

Complete flexibility, from one point of view, is obtained by eliminating all standards, and it might be said that in general flexibility tends to be the opposite of standardization in actual practice.   The ultimate ideal in flexibility would be attained if a set of transmission standards could be set up permitting the maximum economy and performance in all types of receivers and at the same time accommodating all future improvements in the television art without rendering obsolete any receiver. Since as a practical matter this is not possible, it follows that "flexible" has only a relative meaning as applied to a television system.   If all the kinds of flexibility listed above are considered, then none of the systems analyzed by this panel is completely inflexible. Furthermore, it is pointed out that the various kinds of flexibility are not, in general, independent, because an increase in one kind of flexibility often results in a relative decrease in flexibility of another kind. Economic factors, particularly as they affect the placing of an undue burden on the consumer, as well as technical factors are of importance in determining the optimum flexibility for a national television system.

The kind of flexibility that has been most widely discussed is the one permitting variation in scanning rates (*i.e.*, variable number of lines and fields per second). In this connection the word "flexible" has come to have a definite meaning in the minds of some people. However, the word "flexible" alone specifies no technical standard and, therefore, is of no value to the design engineer. Numerical limits must be set, and it is these limits that should be specified if a standard is to be variable.

Two types of flexibility in scanning rate standards are possible: one type is a continuous flexibility of scanning rates between any two limits of line and field frequencies (known as the continuous type); the other type has fixed predetermined discrete values of scanning rates (known as the discontinuous type).

The DuMont Laboratories, Inc., has proposed that the first type, or continuous flexibility, be adopted to allow for future use of the lower frame rates for black-and-white transmission and to provide for the CBS No. 3 color system. The variation in scanning rates required for such continuous flexibility is as follows:

Line frequency............... 7,875 to 20,580 per second (corresponds to
                              525/30 to 343/120)
Field frequency............... 30 to 120 per second

Some of the advantages and disadvantages of this degree and kind of continuous flexibility, provided all receivers are adapted to operate with such variable standards equally well over the entire range, are

Advantages.

1. Provides for the possibility of utilizing a frame rate lower than is at present considered feasible (*i.e.*, 15 frames proposed by DuMont) to allow substantially greater definition by a considerably greater number of lines, if in the future the present technical difficulties with flicker and brightness are overcome and if it is found that the increased definition is of greater importance than increased trailing with motion.

2. Provides for variation in line or field rate, or both, if this should become desirable for any other reason, without obsolescence of receivers.

3. Provides for the introduction of color television by the sequential color field method described as CBS No. 3 color system.

4. Allows for the use of optimum combination of definition and field rate for different program conditions.

Disadvantages.

1. Increases cost of all receivers (added components, additional tests at factory, arrangements for minimizing scanning distortion and nonsynchronous effects over wide range of frequencies).

2. Cost increase probably greatest in magnitude for cheapest receiver (owing to close proximity of components in small cabinet and probably necessary additional shielding for certain frame frequencies).

3. Consumer required to adjust receiver when scanning rates are changed unless fully automatic receiver is used and then, owing to such receiver's greater vulnerability to noise, its use is restricted to a smaller service area.

4. Reduces flexibility in choice of methods for signal utilization circuits and picture-reproducing devices.

5. Certain mechanical receivers (Scophony) are inoperable over wide range of scanning rates.

6. At frame frequencies in the lower part of the proposed range, flicker seems inevitable on presently available picture-reproducing devices at acceptable brightness levels.

7. Does not allow for the best engineering design of a receiver having optimum performance at a single line frequency and a single field frequency.

All the above advantages and disadvantages given for the continuous type of flexibility apply also to the flexibility of the discontinuous type, where the variation is in predetermined discrete steps, if the number of steps is large and covers a large range of line and field frequencies.

The Columbia Broadcasting System has proposed that flexibility of the second type (discontinuous type) be adopted, so that receivers designed for the black-and-white standards that may be adopted will be provided with a control giving fixed predetermined discrete values for the scanning rates, so that the CBS color system No. 3 may be receivable on such a receiver in black and white. This requires, in addition to the fixed predetermined values of the scanning rates for the black-and-white standards another setting on said control for adjusting the scanning rates to the additional predetermined values for the CBS No. 3 color system, namely,

Line frequency........................ 20,580 per second
Field frequency....................... 120 per second

Some of the advantages and disadvantages of this degree and kind of discontinuous flexibility, in which two predetermined discrete values of scanning rates are employed, provided all receivers are adapted to operate equally well at both of such two predetermined values of scanning rates, are

Advantages.

1. Provides for the introduction of the CBS No. 3 color system.
2. Allows for more faithful reproduction of motion in black and white at higher field rate.

Disadvantages.

1. Increases cost of all receivers (added components, additional tests at factory, arrangements for minimizing scanning distortion, and necessity for elimination of impaired resolution due to field rate being 120).
2. Cost increase probably greatest in magnitude for cheapest receiver (owing to close proximity of components in small cabinet

and probably necessary additional shielding for the higher frame frequency).

3. Consumer required to adjust receiver when scanning rates are changed.

4. Certain mechanical receivers (Scophony) may be inoperable on both scanning rates.

5. Does not allow for the best engineering design of a receiver having optimum performance at a single line frequency and a single field frequency.

6. When the transmission is in color, some flicker may be observed when receiving the transmission as a black-and-white picture.

NOTE: These disadvantages apply to a lesser degree in this special case where only one fixed step is proposed as compared with the type of flexibility mentioned previously.

TECHNICAL PAPERS
IN THE N.T.S.C. PROCEEDINGS

(Papers marked with an asterisk are contained in this book)

THE BIRTH OF A HIGH DEFINITION TELEVISION SYSTEM

S. J. Preston

The E.M.I. Research Laboratories at Hayes, Middlesex

THE BIRTH OF A
HIGH DEFINITION TELEVISION SYSTEM

By S.J. Preston, M.A., A.M.I.E.E.*

A Paper read to the Society on Friday, October 24th, 1952

I often wonder what the young people of today think about the origins of the television system used for broadcasting in this country. They probably think that those responsible for this system have no reason to be particularly proud of themselves, because the basic features of the system are obvious enough. They probably also think that the choice of only 405 lines was a great mistake. I expect they also believe that everything of any importance in the system came from the U.S.A.

It is easy to see how this mistaken impression has come about. American publications on tele-

vision greatly outnumber our own, and anyone studying television is bound to gather his information to a large extent from American sources. We cannot expect the American writers to tell the world what Britain has done—we must do that ourselves. The need for a proper account of the great pioneer work in television done in this country has now been recognized. The recent I.E.E. Convention on the British Contribution to Television was most successful and I am sure anyone who reads the paper on the History of Television by Messrs. Garratt and Mumford[1] will feel proud of his countrymen. As my modest contribution to this good cause I should like to-night to tell you something about the pioneer

*E.M.I. Ltd. Hayes.

work done by the British company which developed the British system, namely, E.M.I.

First let us cast our minds back to 1930 and consider television as it was then. The basic principles of television had been known for some 50 years and, although many detailed proposals for putting these principles into practice had been made, only certain mechanical scanning methods

Browne engineered equipment on these lines and gave demonstrations at the Physical and Optical Societies' Exhibition in January, 1931, which created quite a stir. The reproduced picture scanned in 150 lines was projected on to a screen 24" × 20" and divided vertically into five equal 30-line strips, each strip being controlled by a separate channel of about 20 Kc/s bandwidth. Trans-

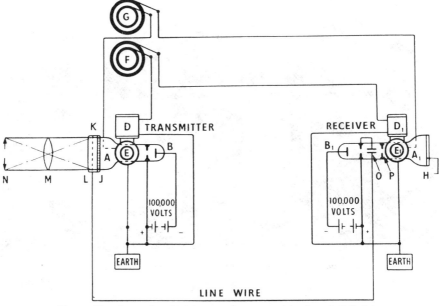

Fig. 1. The electronic television system proposed by Campbell Swinton

had been made to work, notably by Baird. The pictures obtained with this early equipment were of considerable technical interest but it was clear that much improvement in definition was necessary. Many people were actively engaged on the further development of mechanical scanning to obtain increased definition.

Electronic television was also known in principle from Campbell Swinton's publications. Fig. 1 which is taken from his address to the Rontgen Society in 1911[2] clearly shows an all-electronic system which is basically that used today. Of course in 1930 no electronic pick-up tube of any kind was in existence and the only cathode-ray tube available was the gas focused type used for oscillograph work.

At about this time, the Gramophone Company began to take a serious interest in Television. The Research Laboratory under Mr. G. E. Condliffe decided to produce a picture of greatly improved definition by effectively combining five separate channels, each using mechanical scanning. C. O.

mission was from film with a lens drum scanner, and reproduction was by mirror drum and multiple Kerr Cell. Naturally, the composite nature of the picture was apparent but the demonstration served to confirm the value of increased definition and also gave useful experience in basic television technique.

Shortly after this demonstration, The Gramophone Company and the Columbia Graphophone Company merged to form Electric and Musical Industries, Ltd. The research teams of the two companies were also merged. Mr. I. Shoenberg of Columbia became Director of Research and among the people he brought with him to Hayes were A. D. Blumlein, P. W. Willans and E. C. Cork, who were to play an important part in the development of television.

Decision to abandon Mechanical Scanning

One of the first questions which had to be answered was whether development should proceed on mechanical or electronic lines. Mechanical scanners had the advantage that they had been made successfully whereas the electronic scanner did not exist as a practical device. On the other hand, electronic scanning had many potential advantages. The receiver would not need any moving parts and definition would not be limited by mechanical considerations. It was therefore decided to venture into the unknown and to attempt to develop a receiver using a cathode-ray tube.

The first difficulty was that the only cathode-ray tube generally available at the time was a soft tube using gas focusing which is only satisfactory if the beam current does not vary much. As the beam had to vary a great deal to reproduce the light and shade of the picture, this was most unsatisfactory. The possibility of focusing electrons by electric or magnetic fields had been suggested in a few theoretical papers but practical knowledge was scanty. Again it was decided to take the bold course and to attempt to produce a hard tube in which the beam would be focused by electron optics and not be subject to the defocusing and variable behaviour of the soft tube. This may now seem to have been the obvious course to take but there were others who preferred to use the gas focused tube and showed great ingenuity in applying it to television receivers. You will no doubt recall the variable velocity methods used by the Cossor Company, which enabled the tube to be operated at fixed beam current. The principle was that the apparent brightness of a picture element could be varied by changing the scanning velocity. Thus, to make a picture element appear brighter the scanning velocity was reduced when that element was scanned so that the beam was active for a longer time.

W. F. Tedham and J. D. McGee were the pioneers of the hard cathode-ray tube development. Later, L. F. Broadway took a hand in this work and turned the experimental electro-statically focused hard tube into a form suitable for large scale production. The practical knowledge of electron beam focusing gained from this early work was also very useful in connection with the electronic pick-up tube which was constantly under consideration.

The development of the cathode-ray receiver could not proceed without a source of test signals, so a mechanical scanner developed from C. O. Browne's 5-channel equipment was at first used for this purpose. A radio link of some kind was also necessary and, as our hands were so full, we hired a shortwave transmitter working on 7 metres from the Marconi Company. This arrangement later developed into the Marconi-E.M.I. Company,

with Marconi's taking care of the radio frequency transmitter business and E.M.I. being responsible for the video side. The Marconi transmitter and an aerial on a tall mast were set up at Hayes. This transmitter was modulated by C. O. Browne's film scanner operating on 120 lines. Receivers using cathode-ray tubes were developed and field tested on signals from this transmitter.

The D.C. Component

Although the importance of the D.C. component of television signals was understood at this time, no practical method for dealing with it was in sight. Direct coupled amplifiers were known but were so unstable that high amplification was impossible. Transmission of the D.C. component by a separate carrier was a possibility, but this would have increased the bandwidth of the channel and would also have complicated the receiver. Consequently, it was general practice not to transmit the D.C. component but to leave it to the receiver operator to adjust the background brightness continuously so that the picture looked about right.

D.C. Reinsertion

The solution to this problem, namely, D.C. reinsertion, was invented by P. W. Willans in 1933[3] and this solution, together with subsequent refinements invented by A. D. Blumlein and C. O. Browne, led to the D.C. reinsertion technique which is standard practice today. The basic principle of D.C. reinsertion is that, providing care is taken to include datum signals which would, were the D.C. component retained, be at a constant level, the D.C. component may be lost and subsequently recovered by bringing the datum signals to a fixed level. Willans proposed to make the datum signal either a maximum or a minimum signal. The datum signals can then be brought to a fixed level by means of a peak rectifier. If the datum signals tend to rise, the peak rectifier charges its associated blocking capacitor and counteracts this rise; if the datum signals tend to fall the charge in this capacitor leaks away through its associated leak resistance and this fall is counteracted.

It is often inconvenient to make the datum signal a maximum or a minimum signal. For example, if the picture signals are positive and the sync. signals are negative, black will have an intermediate level. This limitation of Willans' proposal was overcome by the " clamp " method proposed by Browne, Blumlein and F. Blythen.[4] They

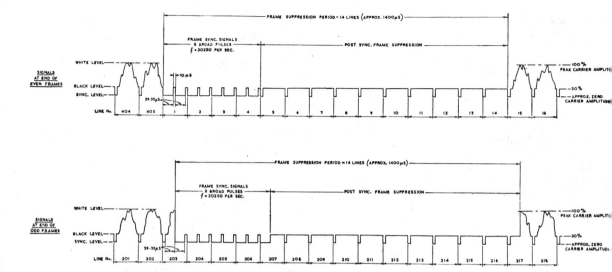

Fig. 2. Waveform of the British 405-line high-definition system

These ideas gave good hopes that it would be possible to transmit the D.C. component and they were followed up actively. They turned out to be quite practical and were adopted with very beneficial results. The transmitter operated much more efficiently because the signals no longer drifted about as the picture changed and the transmitter could be more fully modulated. Reception was greatly improved because constant readjustment of the black level by the viewer was no longer required.

Choice of Synchronizing Signal Waveforms

Another basic question which had to be decided was how the scanning at the receiver and transmitter was to be synchronized. Very many methods of synchronization had been proposed and it was quite a problem to select the most suitable method. The method eventually chosen was to transmit line and frame synchronizing signals along the same channel as the picture signals during intervals between the scanning of successive lines and frames. To prevent the picture signals from interfering with the synchronization, the picture signals and the synchronizing signals were confined to different amplitude ranges. The picture signals were arranged to extend on one side of the amplitude level corresponding to black, and the synchronizing signals extended in the opposite direction, i.e., the so-called blacker than

black direction. This selection of waveform gave rise to the difficulty that the line synchronizing signal was lost during the time when the frame synchronizing signal was being transmitted because the frame signal occupied several line periods. The consequence of this was that the line scanning oscillator of the receiver changed frequency at these times and the vertical sides of the picture on the receiver screen bent over at the top of the picture—an effect known as " hooking ". C. O. Browne, E. L. C. White and W. S. Percival[5] devised the way out of this difficulty. Their proposal was to replace the continuous frame synchronizing signal by a series of lengthened line pulses, each of these lengthened line pulses commencing at the appropriate timing to hold the line scanning oscillator in synchronism when the frame signal sequence was transmitted. (Fig. 2).

The Emitron

While the fundamentals of the system were being worked out with the signals provided by the mechanical film scanner, research on electronic pick-up tubes was continuing. W. F. Tedham and J. D. McGee built tubes with electrostatically focused guns and photo-electric targets of various kinds and, as early as 1932, recognizable signals were obtained. The great difficulty at this time was to understand the mechanism by which picture signals were derived from the electronic pick-up tube. McGee's researches eventually showed that secondary emission phenomena played a vital part in the signal generation. The early pick-up tubes were operated with scanning beams of comparatively high velocity which gave rise to copious secondary emission from the mosaic elements and almost anything could happen to the potential of the element. The element could rise in potential if it lost more secondary electrons than the primary electrons received from the scanning beam. Alternatively, it could fall in potential if it lost fewer secondary electrons than the primary electrons it gained. As the secondary emission factor was dependent upon the energy of the scanning beam, the same element could behave differently according to the voltages put on the tube. Another complicating factor was that the secondary emission from one element generally spread over neighbouring elements and gave rise to unpleasant shading effects in the television picture.

Cathode Potential Stabilization

When the part played by secondary emission was at last understood it became clear that many benefits would follow if secondary emission could be suppressed. This led to the proposal of cathode potential stabilization by J. D. McGee and A. D. Blumlein in 1934.[6] Secondary emission increases rapidly as the velocity of the bombarding electrons increases, and if a low velocity beam is used the secondary emission is slight and the mosaic element receives a net negative charge and is lowered in potential to the potential of the gun cathode. Under these conditions the photo-emission from the mosaic is efficiently collected, so that the tube is sensitive, and the secondary emission is also collected so that there is no spreading on to adjacent elements and consequently no spurious shading effects.

In spite of the attractive possibilities offered by cathode potential stabilization the urgent demand for an electron pick-up tube made it necessary to concentrate on the Emitron because this was giving picture signals of a sort and was nearer to being a practical proposition. Unfortunately, the secondary emission from the photo-electric mosaic of the Emitron was such that the spurious signal due to the spreading of the secondary electrons was much greater than the wanted picture signal. C. O. Browne was somewhat of a purist and the signals coming out of his mechanical film scanner were a very close copy of the text book waveforms. When he first saw on the monitor screen the

signals coming from the early electronic pick-up tubes he was greatly shocked and exclaimed " What do you expect me to do with signals like these?" It would have been very easy to decide that the electronic pick-up tube really would not do at all and to revert to the further development of the mechanical scanner. However, the great potentialities of the electronic pick-up tube were so attractive that it was decided to proceed with' the Emitron and to hope that all the practical difficulties would eventually be overcome. At the time in question this was a considerable act of faith and, although it is easy now when the faith has been justified to think that the choice was an easy one to make, it caused considerable heart-searching at the time.

Special Circuitry for the Emitron

Once it had been decided to back the Emitron, the circuit people got to work to devise circuits which would sift out the wanted signal from the great waves of spurious signals. Blumlein, Browne and White devised the compensating waveform generators which could be used to neutralize the main components of the spurious signals and gave them the names which have since become part of the technical language of this art, namely, " tilt " and " bend ". They also produced other circuits for eliminating the large amplitude spurious signals which were developed by the pick-up tube during flyback; these were the circuits now so well known as " suppression circuits ". Then they had to replace the mechanical synchronizing pulse generator used in the mechanical film scanner by an all-electronic pulse generator . This was done by means of new pulse circuits largely devised by Blumlein and White. At the time all this represented a major effort in circuit technique and without such advances, signals from the early pick-up tubes could never have been used.[7] Fig. 3 shows the signals before and after correction.

Further Development of the Emitron

Progress was also made by McGee and his people on the vacuum physics side, notably in regard to the sensitivity and colour response of the photo-electric mosaic of the Emitron. This work was full of disappointments and it often happened that the first tube made in accordance with some new formula would give very good results and then subsequently it would be found that apparently identical tubes would give very poor results. In this connection it is interesting to recall an accident which led to a useful advance.

One of the laboratory assistants made some mistake in processing which gave, to everyone's surprise, a tube of improved performance. It was only after lengthy cross-examination of the unfortunate assistant who had made the mistake that the significant change which he had made was at last identified. It was found that he had made a film of silver too thin and it was this reduction in thickness which gave the improvement. This chance mistake was subsequently incorporated as part of the standard processing for Emitrons. Another improvement came from the use of bias lighting of the mosiac of the Emitron by means of a small pea lamp which could be adjusted to reduce the spurious signals from the tube.

The choice of 405 lines

Towards the end of 1934 the progress of the work on the Emitron and on the new circuits required for its satisfactory operation made an all-electronic television system a practical possibility. Fortunately, this point was reached before orders for the Alexandra Palace equipment had been placed. As Director of Research, Mr. Shoenberg had to make many difficult and important decisions but perhaps one of the most difficult of all was the choice of the higher definition standard made possible by the all-electronic system. He finally decided early in 1935 to adopt 405-line definition. Mr. Shoenberg made this choice knowing that receivers available at the time could not be expected to deal with the full bandwidth which 405-line scanning would require but his view was that it was better that the early pictures should be somewhat lacking in definition along the line so that later developments in receiver design which he was sure would take place could be usefully employed without any change of standards. At the time many people thought that he was taking a most unnecessary risk. After all, the specification laid down by the Television Advisory Committee was satisfied by a definition of 240 lines, so why face all the difficulties involved in the much higher standard of 405 lines? It is perhaps some indication of the doubts felt on this point that the Committee decided that a 240-line equipment should also be installed at Alexandra Palace side by side with the 405-line equipment. Events have amply justified Mr. Shoenberg's courageous and far seeing decision. Even today the full potentialities of the 405-line standard are not always exploited but, when they are fully exploited, I think everyone will agree that a picture of excellent entertainment value can be produced.

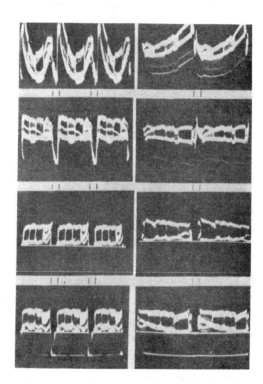

Fig. 3. Line and frame waveforms before and after correction by compensating waveforms, showing effect of spurious signals.

The 405-line standard required a considerable increase of bandwidth of the transmitter and this gave rise to unexpected aerial problems. It was necessary to have a high aerial to give good coverage and in consequence a long vertical feeder run had to be used. The aerial was not matched to the feeder over the full bandwidth in use and this gave rise to echoes which had a considerable time delay owing to the long feeder and could in consequence be seen in the received picture. Other echoes set up by the feeder insulators could also be seen. E. C. Cork and J. L. Pawsey overcame these troubles by devising new networks giving accurate matching between feeder and aerial over a wide band of frequencies and also by introducing further echoes to cancel the echoes set up by the feeder insulators.[8]

Adoption of E.M.I. System

The E.M.I. television system was officially adopted for television broadcasting in this country early in 1937 after a short trial run in conjunction with the alternative 240 line system. Apart from the interruption during the war years, the same system has been used by the B.B.C. ever since. After the war, careful consideration was given to possible changes in the system and it was decided to continue with it unchanged. Since that time many other systems have been proposed but basically they only differ from the old system in

research and development at E.M.I., I think the record of achievement is quite remarkable. In 1935, for example, only about 30 senior staff with about 120 assistants were working in the Laboratories. There was no time for mistakes. The correct decisions had to be made at each point to allow the practical results to be obtained in time

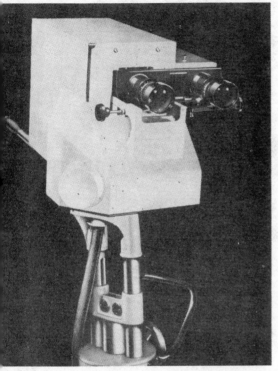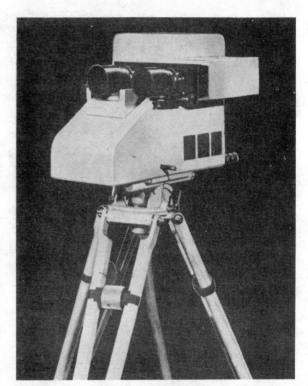

Fig. 4. The new Super-Emitron camera compared with the original Emitron camera

regard to number of lines and sense of modulation. No doubt there is a case for increasing the number of lines somewhat in countries in which television is being introduced for the first time. As to the sense of the modulation, positive modulation was chosen by E.M.I. after lengthy field trials in comparison with the negative modulation system for the reason that it is inherently better for the synchronization of the receiver. With synchronizing pulses going down to zero carrier noise cannot trip the scanning oscillators prematurely. If the synchronizing pulses are in the sense of maximum carrier, then noise pulses just preceding the synchronizing pulse can trip the scanning oscillator and give rise to all the troubles which have caused the Americans to adopt flywheel scanning.

Looking back on the exciting years of television

with the limited effort available. In addition to all the work involved in developing the all-electronic television system, the Laboratory staff designed and made with some factory assistance all the 405-line video equipment and the associated sound equipment required for the opening of Alexandra Palace in 1936. This equipment, which came straight off the drawing board, has continued in operation to this day. They also brought the cathode-ray tube receiver to the stage at which it could be handed over to the factory for large scale production.

The First Outside Broadcasts

After the Alexandra Palace equipment had been made and installed, effort was transferred to Outside Broadcast equipment. Three Emitron camera channels with all necessary control equipment built into a motor coach were first used for the Coronation broadcast in May, 1937.

To make possible, relays from London theatres and exhibitions a cable link was developed. This used a new type of wide band equalizer invented by J. Collard and M. B. Manifold and a new type

outcome of the cathode potential stabilization to which I have already referred. I expect you have all noticed the excellent quality of the pictures given by the C.P.S. equipment which is now used by the B.B.C. for the Childrens' programmes and for Drama. I should also mention our Flying Spot film scanners based on an invention by

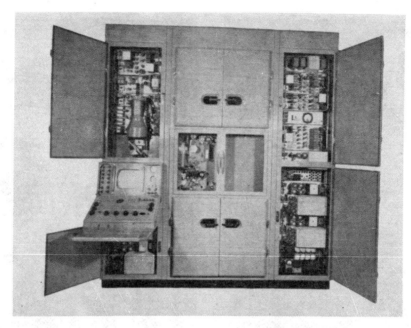

Fig. 5. The E.M.I. flying spot film scanner as used by the B.B.C. today

of cable invented by Blumlein and Cork.[9] It was installed between Alexandra Palace and Broadcasting House in 1938 and later extended to the West End. Camera sensitivity was always a problem, particularly in the case of outside broadcasts, and a significant step forward was the introduction of the Super-Emitron in 1938. (Fig. 4). This was the outcome of the proposal by G. Lubszynski and S. Rodda in 1934[10] to use secondary emission multiplication of the photoemission to increase sensitivity. Pick-up tubes operating on this principle are in wide use today.

An event of historical interest was the experimental use of television in an aircraft in 1936 which I think must have been the first case of airborne television. The experiment was so successful that equipment was later developed for the Services.

Post-War Advances

Among post-war developments I should mention the C.P.S. Emitron, which is the practical

Condliffe in 1936[11] which have raised the B.B.C. Film transmissions to such a high level and made them the envy of the American engineers. (Fig. 5).

With the post-war break-up of the Marconi-E.M.I. Company it became necessary for us to deal with the radio frequency transmitter business which we had previously left to the Marconi Company. The first stage in this new policy was the Sutton Coldfield transmitter which was designed by E.M.I. and also manufactured by E.M.I. with assistance from a sub-contractor because our manufacturing capacity was not sufficiently developed at the time. When this transmitter went on the air in 1949 it was the most powerful television transmitter in the world. The next stage was the Kirk o'Shotts transmitter which is an entirely different E.M.I. design with many important new features, and this time it was manufactured by E.M.I. without outside assistance. This transmitter, and the similar transmitter being installed at Wenvoe are now the most powerful in the world. I think this successful entry into an entirely new field reflects great credit on the E.M.I. engineers chiefly concerned, namely, H. A. M. Clark, E. L. C. White, E. Nind and E. McP. Leyton. (Fig. 6).

Another post-war E.M.I. venture is the mobile microwave link which was developed by Collard. This link has found favour with the B.B.C.,

formative period of television R.C.A. and E.M.I. had certain commercial relations but these relations did not involve ownership of E.M.I. or control of E.M.I. policy by R.C.A., nor was there any exchange of research or manufacturing information on radio or television between the two companies. It is true that technical publications

Fig. 6. Vision control desk at the B.B.C. Station at Kirk o'Shotts

particularly for its long range, and two such links played a valuable part in the recent relay from Paris. (Fig. 7).

E.M.I.-R.C.A. Relations

It has often been suggested that British television is merely a copy of American television and that everything of significance in the E.M.I. system came from R.C.A.; for example, the Emitron is nothing more than the Iconoscope and the C.P.S. Emitron is only the Orthicon. This is quite untrue and I should like to take this opportunity to clarify matters. Throughout the

and copies of the specifications of patent applications were exchanged, but " know how " was not exchanged because R.C.A. was at that time only prepared to agree to such an exchange for a payment, which E.M.I. was not prepared to make. It is significant that in 1937, when the E.M.I. system was a proved success, a free exchange of " know how " was agreed. Thus, the position was that R.C.A. and E.M.I. were developing television on a basis of friendly but quite spirited competition, with neither party giving away any secrets to the other. In the case of the Emitron, we were naturally interested in rumours that

Zworykin of R.C.A. had obtained picture signals from an electronic pick-up tube. As in the case of the Atomic Bomb, it is helpful to know that such a bomb can be made to work, but in an atmosphere of secrecy it is no mean task to repeat the achievement. The Emitron is basically the same as the Iconoscope, but it was separately

Fig. 7. An early form of mobile short-wave link

developed and did differ in many important details. I think, for example, it has higher sensitivity and better colour response and I believe that some of the advantages of the Emitron were subsequently incorporated in the Iconoscope.

As to the Orthicon, this tube first appeared in 1939, and our patent on cathode potential stabilization is dated 1934.

Perhaps the best answer to those who think British television owes everything to America is to quote a few facts: When Mr. Shoenberg introduced 405-line all-electronic television early in 1935, the highest definition used by the Americans was 343 lines. They did not raise their standard

of definition above 405 lines until 1937, when they began to transmit experimentally on 441 lines with no D.C. component and with a frame synchronizing signal waveform very different from the British waveform. At about this time two American engineers (H. M. Lewis and A. V. Loughren) came to this country to study British television and they reported in the well-known American technical periodical *Electronics*[12] that the British standards constituted a major improvement over American practice. Four years later, in 1941, when Public television broadcasting officially commenced in the U.S.A. the D.C. component was transmitted and the British frame synchronizing signal waveform was also adopted.

It is also interesting to note that the Image Orthicon, which is now the foundation of American television, employs secondary electron amplification of the photo-current as in the Super-Emitron and cathode potential stabilization as in the C.P.S. Emitron.

A. D. Blumlein

You would, I am sure, like me to say something about Alan Blumlein, who was such an outstanding member of the E.M.I. Research team. I first met him when he came to E.M.I. from Columbia in 1931. He was then an experienced telephone engineer and an expert in acoustics. He had completed the development of an entirely new sound recording system for Columbia and this system was soon used for all E.M.I. recordings. When he turned his mind to television at E.M.I. he produced so many inventions of importance that it is difficult to select from them. He invented resonant return scanning as early as 1932 and, together with E. L. C. White, he added the now well-known H.T. boost in 1936. He had a large share in the conception of the D.C. reinsertion circuit and the many pulse circuits which had to be worked out for the successful application of electronic television. His inventions were not confined to the circuit field and his name appears on many of our patents relating to vacuum tubes and acoustics. Although Blumlein made many inventions, I think he always regarded himself as being an engineer rather than an inventor. His view was that a good circuit engineer should be able to design circuits to meet a given specification without any need for extensive experimental adjustments after they were constructed. He was consequently a great advocate of negative feedback to overcome variations in performance which were normally caused by valves and other components having variable characteristics.

Fig. 8. A. D. Blumlein reading the paper on the E.M.I. High-definition system before the I.E.E., 1938

Although he had the gift of lucid exposition, unfortunately, he did not find time to write many papers. The few he did write are models of clarity and I think his paper on the E.M.I. waveform which he read to the I.E.E. in 1938 is a classic.[13]

He was a man of amazing energy and enthusiasm. He was not content simply to invent solutions to problems but always wanted to try out the solutions in a practical way and then to put them on a proper engineering basis if they showed any promise.

This same energy and enthusiasm were equally apparent in all his other activities. He was most stimulating in his conversation on any subject. He was also a great favourite with young children because he could enter into their fun especially if they happened to have toys of a mechanical or electrical nature.

When the war commenced in 1939, his talents were applied to matters of greater national importance than television. He was mainly concerned with the most difficult airborne radar problems and he met his death while flight testing experimental radar equipment in 1942. He was then only 38 years old. Two of his colleagues from E.M.I., C. O. Browne and F. Blythen, also died with him.

Had he lived he would have risen to the highest ranks of his profession. Even though his life was so short he will long be remembered as a man who made outstanding contributions to British television.

References

1. *Proc. I.E.E.* 1952, Vol. 99, Part 111A, No. 17, p. 25.
2. *Journal of the Röntgen Society* 1912, Vol. 8, p. 1.
3. British Patent No. 422,906. (13.4.33)
4. British Patent No. 449,242. (18.9.34)
5. British Patent No. 425,220. (8.8.33)
6. British Patent No. 446,661. (3.8.34)
7. British Patents Nos. 450,675 (19.11.34); 462,110 (24.6.35); and 471,731 (4.12.35).
8. British Patents Nos. 464,825 (22.10.35); and 469,245 (21.11.35).
9. British Patents Nos. 452,772 (25.2.35); and 476,799 (3.3.36).
10. British Patent No. 442,666 (12.5.34).
11. British Patent No. 475,032 (9.4.36).
12. *Electronics* 1937, Vol. 10, No. 10, p. 32.
13. *Journal I.E.E.* 1938, Vol. 83, p. 758.

DISTANT ELECTRIC VISION

J[ames] D. McGee

Distant Electric Vision*

J. D. McGEE†

Summary—An outline of the history of television is followed by a detailed description of the design and development of some English television pickup tubes.

JUST OVER FORTY YEARS AGO ·a short letter appeared in *Nature* under the title "Distant Electric Vision." It was signed A. A. Campbell-Swinton.[1] I have ventured to borrow that same title for my paper, since I would like you to regard what I have to write as a part of the story of the development of the brilliant idea first proposed in that letter before even the word "Television" had been coined. I shall return to Campbell-Swinton's proposal presently, but first I must review briefly what had gone before. The photoelectric effect, discovered in 1873, was very soon realized to be the key to "distant electric vision." The first practical device for generation of television signals was, in fact, proposed only eleven years later by Nipkow.[2] The apparatus proposed by Nipkow is shown in Fig. 1. An image *4* of the scene to be transmitted *5* is formed by the lens *6* on the

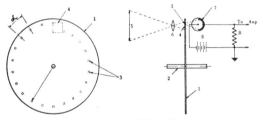

Fig. 1—The Nipkow disk.

surface of a disk *1* which can rotate about its axis *2*. Around the periphery of this disk is a spiral of small apertures *3* spaced at equal angular intervals and each spaced radially from the center of the disk at distances increasing successively by the height of an aperture. At any given time light can pass through one aperture only and enter the photocell *7*, where it releases an electron current proportional to the intensity of the light; that is, proportional to the brightness of the image at the point where the aperture happens to be at that instant. These electrons are collected on the anode of the photocell by the electric field maintained by the battery *8* and the

* Decimal classification: R095×R583.6. Paper received by the Institute, November 4, 1949. Presented, I.R.E. Radio Engineering Convention, Sydney, Australia, November, 1948. (Since Dr. McGee was unable to present this paper personally to the Convention, the entire manuscript was delivered by means of a magnetic tape recorder.) Reprinted from the *Proceedings of the Institution of Radio Engineers*, Australia, vol. 10, pp. 211–223; August, 1949.
† E.M.I. Research Laboratories, Ltd., Hayes, Middlesex, England.
[1] A. A. Campbell-Swinton, "Distant electric vision," *Nature*, (London), vol. 78, p. 151; June, 1908.
[2] P. Nipkow, Deutsches Reichs Patent No. 30,105; January 6, 1884.

current passing through the resistance *R* produces a voltage change across it. As the disk rotates, the aperture will scan across the optical image, and the light passing through it will fluctuate proportionally to the light and shade along that line of the image. The voltage fluctuations across the resistance *R* will be also proportional to the light fluctuations. Thus, a "picture signal" will have been generated for that line of the image. As one aperture completes its scan of one "line" of the image, the next aperture begins to scan the next line of the picture, and so on, until the whole image has been scanned in one complete revolution of the disk.

For modern television pictures of, say, 405 lines and 25 pictures per second, the disk must rotate at 1,500 rpm. The diameter of the disk must be as large as possible in order that the apertures may be made of a reasonable size; otherwise, the signal current will be very small and the signal-to-noise ratio very poor. However, a disk of, say, 3 feet in diameter rotating with high precision at 1,500 rpm, is not exactly a convenient or mobile piece of equipment. Moreover, an amplifier of sufficient bandwidth for a 405-line television picture signal (say, 3 Mc) requires a peak input signal of about 0.1 μa to override by an order of magnitude the noise of the first stage of the amplifier. If we assume a high photoelectric efficiency of 100 μa per lumen for the photocell it follows that 10^{-3} lumens must fall into the photocell to give this signal current in the peak whites of the image. Since there are 200,000 picture points in such a picture, this corresponds to a light flux of 200 lumens in the whole image, assuming it to be of uniform brightness. Again, assuming an area of the image on the disk of 4 × 5 inches, this corresponds to light flux of 1,440 lumens per square foot or 1,440 foot-candles illumination.

The illumination in the plane of the image I_i is related to the illumination on the scene I_s by the well-known formula

$$I_i = \frac{I_s \times \gamma}{4 A^2},$$

where γ is the reflection coefficient of the scene which may be taken as unity for a perfect white diffusing surface, and A is the numerical aperture of the lens, say $f/2$.
Then

$$I_s = \frac{I_i \times 4 A^2}{\gamma} = 23,000 \text{ foot-candles.}$$

Such illumination is about a hundred times greater than that used in television studios today and would be quite intolerable.

The advent of the electron multiplier made it possible to multiply the initial photoelectric current by a very large factor, 10^6 or more, without seriously changing the signal-to-noise ratio of the initial current which is due to probability fluctuations in the photoelectric emission. In this way the input signal to the amplifier can be made large compared with the noise of the first stage, and the limiting noise then becomes that of the primary photoelectric current. A simple calculation shows that the illumination required is then very much less but still about 1,000 foot-candles, even when the most favorable conditions are assumed.

Thus, the inconvenience of the mechanical scanner, together with its very low sensitivity, has doomed it to oblivion except possibly for film scanning, but, even here, all-electronic systems have found greatest favor.

CAMPBELL-SWINTON'S PROPOSAL

I never cease to wonder at the vision and imagination of that great television engineer, the late A. A. Campbell-Swinton, F.R.S., who as long ago as 1908 not only saw clearly the fundamental limitations of the mechanical systems, but actually proposed the all-electronic system which, with modern technical embellishments, has become the television system of today. I believe I need not apologize to you for showing you in Fig. 2 the actual schematic diagram of the system for "distant electric vi-

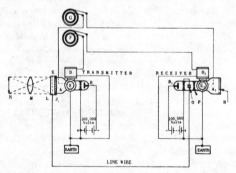

Fig. 2—Electronic television system proposed by Campbell-Swinton.

sion" which he proposed in 1908 and amplified in a paper in 1911.[3] In this figure, you see the transmitter on the left; the receiver on the right. The transmitter is a cathode-ray tube A, in which an electron beam from the cathode B is scanned by crossed magnetic fields D and E line by line over a mosaic of photoelectric cells J. An optical image of a scene N is focused by the lens M on to the other side of this mosaic of photocells which become charged by the loss of photoelectrons. These charges are discharged at each scan of the electron beam to the grid L. These fluctuating "picture signals" are conducted over a line co modulate an electron beam which scans the

fluorescent screen of the receiving cathode-ray tube in synchronism with that of the transmitting tube. Here are two ideas of capital importance; the use of inertia-less cathode-ray beams for scanning at very high speed, and the mosaic of photocells in the pickup tube. I think one may well claim that this was the seed from which present-day television engineering has grown.

FARNSWORTH DISSECTOR

One feature of Campbell-Swinton's scheme was realized in the Farnsworth dissector tube,[4] namely, that of all-electric operation. This tube was first described in 1934 and, because it still may find an application in film scanning, it is worthy of a brief description here.

The tube and associated equipment are shown in Fig. 3. An optical image of the scene is focused on a transparent conducting photoelectric layer *1* on the inside surface of the flat end wall *2* of the high-vacuum tube *3*. Electrons are emitted from this surface under the influence of, and proportional in number to, the

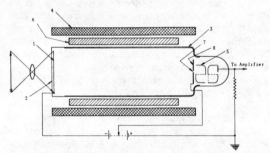

Fig. 3—Diagram of Farnsworth dissector tube.

incident light. These electrons are accelerated by an axial electrostatic field and focused by means of a uniform axial magnetic field produced by the solenoid *4* to form an electron image in the plane of the electrode *8*. A pencil of electrons passes through the small aperture *7* and falls on the first plate *5* of a multistage electron multiplier, the output of which is fed to a suitable amplifier. Two pairs of coils, one of which is shown at *6*, actuated by frame and line frequency currents of sawtooth wave form, produce two transverse magnetic fields at right angles to one another which scan the whole electron image across the aperture in a series of lines. Since the aperture *7* must allow only those electrons from one picture point to pass through to the multiplier, it follows that for a 400-line television picture the linear dimensions of the aperture must be 1/400th of the dimensions of the electron image, and, since there are 200,000 picture points in such a picture, only one part in 200,000 of the electron emission from the photocathode passes through the aperture. In practice, this current is so small (of the order of 10^{-11} amperes) that if it were fed directly to the input of an amplifier of the necessary bandwidth,

[3] A. A. Campbell-Swinton, "Presidential address," *Jour. Roentgen Soc.*, vol. 8, p. 7; January, 1912.

[4] P. T. Farnsworth, "Television by electron image scanning," *Jour. Frank. Inst.*, vol. 218, p. 411; October, 1934.

the signal produced would be far smaller than the amplifier noise. However, by multiplying the primary electron current in a multiplier by a factor of 10^6 or more the output can be increased until the signal is much greater than the amplifier noise. The signal-to-noise ratio of the signal is then determined by the random fluctuations of the primary photoelectric current which is not changed appreciably in the process of multiplication. Thus, the Farnsworth dissector is the electric analogue of the Nipkow disk and to a first approximation, the sensitivities of both signal-generating systems are the same. Minor advantages make the Farnsworth dissector slightly more efficient, but it still requires a scene illumination of about 1,000 foot-candles. This is, of course, a hundred times the illumination required by modern television cameras.

Charge Storage Principle

It will have been noticed that, in both the Nipkow and the Farnsworth methods, only a minute fraction of the light from the scene is effective in producing picture signal—in fact, the light from one picture element. Less than 10^{-5} of the light or photoelectric current is used. It is obvious that if all of the photoelectrons released by the light in the Farnsworth dissector could be stored during the frame-scanning period and discharged once per frame period, the gain in efficiency would be enormous—a factor of 10^5 or more.

The obvious line of attack on this problem was clearly suggested by Campbell-Swinton's mosaic of photoelectric cells. If an optical image is formed on a mosaic of minute photoelectric cathodes, each of which is associated with a minute condenser, and the photoelectrons liberated from the cathodes by the light are saturated to a common anode, charges will be built up on the condensers which are at all points proportional to the incident light. Thus, a reproduction of the optical image is built up in electric charges. If now this mosaic of condensers is scanned by a suitable commutator, e.g., a beam of electrons, which discharges those condensers in the area of one picture point simultaneously through a signal resistance, the potential fluctuations across this resistance will be the required picture signal.

A single cell of such a system is shown diagrammatically in Fig. 4. Light L falling on the cathode of

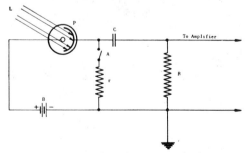

Fig. 4—Elementary charge storage circuit.

the photocell P liberates electrons in proportion to its intensity. These are collected by the anode which is maintained at a positive potential by the battery B. This current flowing through the resistance R charges the condenser C. If now, once per frame period, this condenser is discharged by the commutator A (which may be an electron beam with an impedance r) a pulse of charge will pass through the resistance R proportional to the integrated light flux on the photocell during the period of storage. The potential change across R will be proportional to the discharge current and may be applied to the grid of the first valve of an amplifier chain. The discharge time of the circuit consisting of C, r, and R must be not less than the time required to scan one picture element, i.e., 2×10^{-7} seconds. The potential drop across R during the charging period is the picture signal that would be obtained without storage; hence, the increase in signal strength due to storage is the ratio of the charging time (0.04 second) to the discharging time 2×10^{-7} seconds of the condenser C. That is, a gain of 2×10^5 is achieved. If in practice only a small part of this theoretical gain could be realized, say 1 per cent, it would still represent an increase in sensitivity by a factor of 2,000.

The first account of a television pickup tube employing charge storage was published by Zworykin in 1933.[5] This tube was named the "iconoscope" and it has been the main tube employed in American television cameras until quite recently. Work had been in progress in the Electric and Musical Industries Ltd. Laboratories and, in 1932, Tedham and the author succeeded in producing television signals with a tube which proved to be fundamentally the same as the iconoscope but which differed from it in many important practical details. Continuation of this work led to the development of the emitron[6,7] which was used from the beginning of the first regular public television service instituted by the BBC in London in 1936. It is still the only tube used for studio broadcasts by the BBC. Further development led to the super-emitron.[7,8]

I will now give a brief outline of the operation of the emitron and super-emitron before going on to describe the CPS emitron which has been developed since the end of the war and has only recently gone into service.

The Emitron

The tube with its immediate operating circuits is shown in Fig. 5, while a very much enlarged section of the target is shown in the inset. The highly evacuated envelope *1* carries a side tube in which a more or less conventional electron gun is mounted. The gun consists

[5] V. K. Zworykin, "Television with cathode-ray tubes," *Jour. IEE*, vol. 73, p. 437; 1933.
[6] J. D. McGee and H. G. Lubszynski, "E.M.I. cathode-ray television transmission tubes," *Jour. IEE*, vol. 84, p. 468; April, 1939.
[7] J. D. McGee, "The development of high definition electronic television in Great Britain," *Proc. I.R.E.* (Australia), *Proc. World Radio Convention*, Sidney, Australia; April, 1938.
[8] H. G. Lubszynski and S. Rodda, British Patent No. 442,666; May 12, 1934.

of a cathode *5*, modulator *6*, first anode *7*, and second anode *8*. The latter is held at earth potential and extends some distance into the spherical part of the tube where it acts as anode for the photoelectric mosaic also. The cathode of the electron gun is held at about −1,500 volts and the electron beam is focused by the electrostatic electron lens between the first and second anodes, *7* and *8*. The beam is scanned over the mosaic by two pairs of coils, one of which is shown at *9*. Since the beam falls obliquely onto the target, a conventional scanning raster would appear as a trapezium or keystone shape. This must be corrected by modulating the line-scan amplitude by the frame-scan amplitude. This is termed "keystone correction." Further, the frame-scan wave form must be slightly distorted from linear to obtain even spacing of the lines from top to bottom of the mosaic. The electron beam comes to a focus on the surface of a sphere, the center of which is the point of deflection of the beam. This sphere intersects the plane of the target in a circle and at all points off this circle the beam is more or less out of focus. The loss of definition due to this difficulty is minimized by restricting the beam to a very narrow pencil by limiting apertures in the gun so that its depth of focus is increased. This can be done because only a very small beam current of a few tenths of a microampere is required. All these inconveniences in design are accepted in order to be able to project the optical image normally onto the target. The lens *13* forms an image of the scene through the flat window *4* on the surface of the target, which is also scanned by the electron beam. The lens is usually $6\frac{1}{2}''$ focal length with a numerical aperture of $f/3$.

The target is 4×5 inches (if the agreed picture aspect ratio is to be 5 to 4). A small section of it, greatly magnified, is shown in the inset of Fig. 5. It consists of a sheet of highly insulating dielectric *D*—for example, mica—coated on the side which is not scanned by the electron beam with a conducting layer *S*, known as the signal plate. This signal plate is connected to earth through a signal resistance and to the grid of the first tube of the amplifier. On the side of the dielectric which is

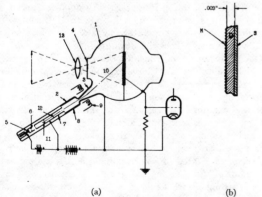

(a)   (b)

Fig. 5—(a) The emitron. (b) Cross section of emitron target.

scanned by the electron beam is formed the photoelectric mosaic *M*. This may be regarded as a vast number of minute islands of photosensitized metal each separated and highly insulated from its neighbors. Also, each mosaic element forms a small condenser with the signal plate.

The basic material of the mosaic elements is usually silver which can be formed into a mosaic by evaporating a very thin layer and aggregating this by heating. The mosaic may also be formed by evaporating the metal through a fine metal mesh which acts as a stencil and in several other ways. The silver mosaic is made photosensitive by oxidation and subsequent treatment with cesium. Here a difficulty is encountered. It is difficult to obtain high photosensitivity of the mosaic elements while preserving good insulation between them because the metallic cesium tends to form a slightly conducting film over the surface of the dielectric between them. A compromise has to be accepted by which we obtain about half the maximum possible sensitivity. Also, the mosaic elements cannot cover more than about 50 per cent of the target surface so that another factor of two is lost in photoelectric efficiency. We can obtain an efficiency of about 10 μa per lumen instead of over 30 μa per lumen that can be obtained from a normal photocell. These points are important and will be referred to when we consider the super-emitron.

We must next consider the mechanism of signal generation in the emitron. This is a complex problem and I ask your pardon if my treatment of it in this paper leaves many unsatisfactory loose ends.

In its simplest form we may be sure that the light of the image which falls on the mosaic liberates photoelectrons at all points in numbers proportional to the light intensity. If these negative electrons are removed from the mosaic, for example, to the anode of the tube, a distribution of positive charges will be built up continually on the minute condensers formed by the mosaic elements and the signal plate. If now the electron beam scans the mosaic line by line, neutralizing these positive charges point by point, equal and opposite (i.e., negative) charges will flow to earth through the signal resistance *R*. These charges and the voltages produced across *R* will be proportional to the integrated intensity at the corresponding points in the image over the previous frame period, i.e., 1/25th of a second. Thus a rapidly moving object should be reproduced with a loss of definition corresponding to that seen in a photograph taken with a time exposure of 1/25th of a second. In fact, this is not found to be the case. A television picture of a rapidly moving object produced by an emitron shows it as a series of sharply defined images. This means that uniform charge storage cannot be taking place during the whole frame period. Moreover spurious signals are produced which cannot be explained on this simple theory.

These experimental observations can only be explained when we take into account the fact that the scanning beam electrons, which reach the mosaic with an energy of 1,500 electron volts, each releases between 5 and 10 secondary electrons. These relatively slow secondary electrons stabilize the potential of the mosaic at approximately that of the nearest electrode of fixed potential, i.e., the second anode. There is therefore no electrical field of sufficient strength to saturate the photoelectrons away from the mosaic. If we now consider the mosaic being scanned by the beam but without light falling on it, it is clear that over a period of time it can neither gain nor lose charge; that is, only one secondary electron can leave the mosaic for each primary falling on it. The remaining secondary electrons must return to some part of the mosaic, but not necessarily the point from which they were released.

Now the point on the mosaic on which the beam is falling at any instant will initially lose between 5 and 10 secondary electrons for every incident primary electron. Thus, its potential will be driven positive until it can recapture all but one of the secondaries released by each primary electron. This charge and discharge of mosaic elements is illustrated for two successive elements A and B in Fig. 6. It is estimated that the poten-

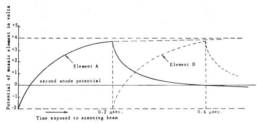

Fig. 6—Charge and discharge curves of mosaic elements.

tial rise of the mosaic during scanning is about $+6$ volts relative to the surrounding areas. As the scanning beam passes on to the next point on the mosaic the secondary electrons liberated from it tend to be deflected towards, and captured by, those areas which have just been scanned since they are the most positive areas in the neighborhood.

It will, I think, be clear that this exchange mechanism cannot be uniform over the whole scanning cycle. The charges on the surrounding areas of the mosaic must be very different at the beginning and end of line and frame scans and this will influence locally the sharing of secondary electrons between the second anode and the surrounding mosaic. Other factors such as the asymmetry of the tube, variations in the focus of the scanning beam, local charges due to the light image, etc., can also affect this sharing of electrons between the mosaic and the second anode. It is this variation in the sharing of the secondary electrons between the mosaic and the second anode that causes the spurious signals. These are illustrated in Fig. 7, in which line A shows the wave form of the uncorrected signals for three successive lines. There are large spurious pulses at the be-

ginning and end of each line, and a spurious low-frequency signal known as "shading" is superimposed on the true signal. In line B is shown the artificially generated correcting signals, "Tilt" and "Bend," which when mixed with the crude signals result in the train of signals shown in line C. The large spurious signals between lines are then suppressed, leaving the corrected signal as it appears in line D. Finally, synchronizing signals are inserted between the lines as shown in line E.

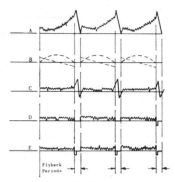

Fig. 7—Emitron signals.

If, when once corrected, the shading signals remained constant, they would be of little consequence. Unfortunately they do not, but change appreciably with distribution and intensity of light falling on the mosaic. Hence, the shading controls must be constantly adjusted as the scene changes in front of the camera and, at the lower limit of light levels when the amplifier gain must be increased, the satisfactory correction of shading becomes almost impossible. In fact, the lowest level of illumination at which an emitron camera will work satisfactorily is usually determined by uncorrected shading signals rather than by amplifier noise. Another important result of this mechanism of operation is the fact that at no part of the scanning cycle is a signal generated which can be regarded as corresponding to the true black level of the picture. This is a great disadvantage, since there is then no absolute method of re-establishing the true black level of a picture after it has passed through an alternating-current system. It must be left to the arbitrary adjustment by some operator.

We have seen how the unwanted signals are generated. We must now consider the true picture signals which are wanted. As I have explained above, each element of the mosaic is driven strongly positive by the scanning beam and then gradually falls in potential until the scanning beam reaches it on the next cycle. Thus, there is an area of the mosaic that has just been scanned which is about 6 volts positive relative to those elements which are about to be scanned. Hence, any photoelectrons liberated in a narrow band across the mosaic in front of the scanning beam will find themselves in a quite strong electric field attracting them to those areas of the mosaic which have just been scanned. This narrow band within which the photoemission is saturated is estimated to extend something like 1 centimeter beyond the last line to be scanned. Most of the photoelectrons liberated in this area leave the mosaic element from which they were liberated and consequently charge them positively. They are collected in a random distribution on the mosaic areas that have just been scanned which they drive more negative in potential. Beyond the narrow band of mosaic from which the photoelectrons are completely removed, the percentage of photoelectrons collected will decrease with distance. Over the greater part of the rest of the mosaic the photoelectrons which are liberated by the light will fall back onto the mosaic at random and so will not contribute appreciably to the stored charges. As the scanning beam passes over these elements on which varying positive charges have been stored, they will all be driven up to a fairly constant peak positive potential. Thus, at a point which has been exposed to light and on which positive charge has been stored, the potential change on being scanned will be less than for an element which has not been exposed to light. It is, in fact, the differences in these potential changes as the beam scans the mosaic that constitute the picture signal.

It will be clear from the above that the emitron operates in a manner very similar to a camera using a focal-plane shutter, and the charges that give rise to picture signal are built up in a very short period of time, about 1/250th of a second, before the scanning beam reaches each part of the mosaic. This explains why fast-moving objects appear as a succession of sharp images rather than as a continuous blurred image.

This same mechanism of operation explains the low sensitivity of an emitron which gives only about 5 per cent of the sensitivity to be expected from a full storage tube. It will be seen that the photoelectric emission is used to build up charges on the mosaic only during a small fraction of each frame period.

Another consequence of this mechanism is that the signal output is not proportional to the light input. This is to be expected since, as a mosaic element which is exposed to light becomes more positive, it tends to collect secondary electrons itself. If we express the signal S in terms of the illumination of the mosaic I by the formula

$$S = I^\gamma,$$

we find that γ is not greater than about 0.75. This was rather a lucky accident, since this reduction in contrast of the generated signal compensates the inevitable increase in contrast which occurs when such signals are displayed on a normal cathode-ray receiving tube. However, when the scene to be televised lacks contrast there is nothing that can be done to improve its appearance to the viewer.

One final curiosity which can frequently be noticed in an emitron picture can be explained on this theory. If the picture to be televised has horizontal lines that are emphasized, it is found that a white horizontal bar is followed by a black streak extending possibly half the width of the picture in the direction of scanning and vice versa. This is believed to be due to the charges on the mosaic that are about to be scanned influencing the level of the peak positive potential to which the mosaic elements are driven by the scanning beam. This may be regarded as a loss in response at the lower frequencies and it can be corrected to some extent by introducing a bass boost in the amplifier chain between 5 and 10 kc.

The emitron tube mounted on the base of the camera is shown in Fig. 8. The first few stages of the amplifier are built into the camera chassis in order to keep the input capacity of the amplifier as low as possible. Also, the line and frame scan amplifiers are built into this chassis. The complete camera is shown in Fig. 9, showing the objective lens and the view-finding and focusing lens. Such a camera will give a very satisfactory picture

Fig. 8—Emitron tube mounted on base.

Fig. 9—Emitron camera.

with about 200 foot-candles of illumination incident on a studio scene when the lens is at full aperture ($f/3$). However, under these conditions the depth of focus is very small, which imposes restrictions on the studio production. This can be improved only by increasing the light and stopping down the lens. A worthwhile increase in light becomes expensive and uncomfortable. The definition of the picture is excellent and the geometrical distortion of the image practically negligible. However, the pictures are frequently marred by uncorrected shading and the absence of a definite black level. Moving objects do not appear to lose definition and very rapidly moving objects may even appear as a series of well-defined separate images. It is a doubtful point whether this is preferable to a single blurred image which is more nearly what is seen by the eye.

The Super-Emitron

I have spent a considerable time over the details of the theory of the emitron because it is mostly applicable to the super-emitron also. This type of tube was proposed by Lubszynski and Rodda in 1934, was developed in the Electric and Musical Industries Ltd. Laboratories[8] and was first used by the BBC for an outside broadcast in November, 1937.

The tube and associated equipment are shown in Fig. 10. The electron gun *1*, scanning system *2*, and target, are similar to those of the emitron except that in this tube the mosaic is not photosensitive but is a good

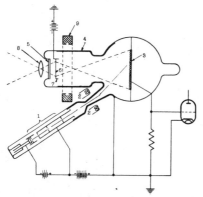

Fig. 10—The super-emitron.

emitter of secondary electrons. In place of the flat window of the emitron, a tubular extension *4*, is provided with a flat window *5* at its end. Inside this flat window a sheet of transparent mica or glass is mounted, and a transparent conducting photocathode *6* is formed on that side of it which faces the target. Contact is made to this photocathode from a solid metal border *7* and it is held at about 500 volts negative relative to the metal coating on the tube wall, which is an extension of the gun second anode. A light image is formed on the photocathode by the lens *8* and the photoelectrons liberated by the light are accelerated by the axial electric field

towards the target. A magnetic field produced by a current in the coil *9* brings these electrons to a focus on the surface of the target. Each of these photoelectrons impinging on the mosaic surface of the target with an energy of about 500 electrons volts will liberate between 5 and 10 secondary electrons. These relatively low-velocity secondary electrons play much the same role in the generation of picture signals as do the photoelectrons liberated from the mosaic of the emitron by the direct action of light.

Two significant advantages have been gained in this tube. First, the functions of photoelectric emission and insulating mosaic have been separated. The photoelectric sensitivity can be pushed up as high as possible without ruining the mosaic insulation. In fact, photosensitivity of 30 to 40 μa per lumen can be reached as compared with 10 μa per lumen for the emitron. Second, each photoelectron that impinges on the mosaic liberates at least 5 secondaries. Thus, for the same amount of light gathered by the objective lens the number of electrons liberated from the mosaic is between 15 and 20 times greater in the super-emitron than in the emitron. Since the average energy of these secondary electrons is about ten times that of photoelectrons, it is not to be expected that the mechanism of signal generation would be the same in both tubes. In fact, noticeable differences are observed. The super-emitron shows less evidence of the very short-term charge storage, which is so characteristic of the emitron, and more evidence of uniform charge storage during the whole frame period. This is shown by rather more blurring of moving objects in pictures transmitted by a super-emitron as compared with those transmitted by an emitron. The increase in efficiency of the super-emitron over the emitron does agree fairly well with the factor of 15 to 20 given above.

Another useful feature of the super-emitron is that the electron image can be magnified between the photocathode and the mosaic. Thus, we usually start with an optical image 1×0.8 inch and magnify it to cover a mosaic 5×4 inches. Then for normal work a lens of 2-inch focal length and $f/2$ or better can be used. This lens has a light gathering capacity of about 25 per cent of the normal lens of the emitron camera, but it gives a much greater depth of focus. Also, telephoto lenses may be used conveniently and a series of lenses can be mounted on a rotating turret.

Since, of the total increase in efficiency of the super-emitron over the emitron by a factor of 15 to 20, a factor of 4 has been used in improving the optical conditions, we are left with a factor of between 4 and 5, as the increase in the sensitivity of the super-emitron camera over the emitron camera. That is, a super-emitron camera will give the same quality of picture at 40 to 50 foot-candles as an emitron camera will give at 200 foot-candles incident illumination.

A practical advantage of the super-emitron is due to the fact that the mosaic is not photosensitive as in the emitron, where the action of the electron beam slowly reduces the photosensitivity and so shortens the life of the tube. The scanning beam has no noticeable effect on the super-emitron mosaic and the photocathode has a very long life.

In focusing the electron image onto the mosaic, the whole image is rotated through an angle of 20° to 30° about the axis of the electron lens. This is easily compensated by rotating the whole tube by the same amount about the same axis. More important is the geometrical distortion introduced by aberrations in the electron lens, which result in peripheral points of the electron image being rotated slightly more than points nearer the center. By careful design it has been found possible to reduce this distortion to a tolerable magnitude.

Very efficient magnetic screening between the focusing field and the scanning fields is necessary. Other-

wise, the former will distort the scanning raster and the latter will cause the electron image to oscillate slightly on the mosaic and definition will be impaired.

The super-emitron suffers from the same inconveniences due to oblique scanning and generates the same type of spurious signal, shading signal, and streaking. Also, it does not produce a signal corresponding to true black.

A super-emitron camera with its cover removed is shown in Fig. 11. The magnetic focusing coil is easily visible and also the rotation of the tube to compensate for the rotation of the electron image. The complete super-emitron camera is shown in Fig. 12. Its greater sensitivity and adaptability to telephoto shots are of great value in outside broadcasts.

A small-size super-emitron tube has recently been developed which has a very similar performance to that of the standard tube. Its smaller size facilitates camera design, especially in regard to the lens system, since shorter focal length lenses can be used to give the same angle of view.

THE CPS EMITRON

We had realized as early as 1934 that it is the uncontrolled secondary electrons that are responsible for the spurious signals and low sensitivities of the emitron. In that year the late A. D. Blumlein and the author proposed a method which, it was hoped, would eliminate these troubles.[9] The fundamental principle of this scheme was to scan the mosaic with a beam of electrons which have so little energy when they reach its surface that they liberate substantially no secondary electrons. Under these conditions the mosaic must be driven negative until it reaches the potential of the cathode from which the electron beam originates. Since there are no secondary electrons, we argued, there can be no spurious signals and this has, in fact, proved to be true. Furthermore, since a strong field gradient is established in front of the mosaic which decelerates the approaching beam electrons, this must also serve to saturate away from the mosaic all photoelectrons that are liberated by the light of the image. Thus, there should be full storage of the photoelectric charges during the whole scanning cycle, leading to a great improvement in efficiency, which also proved to be true.

Considerable work had been done in our Laboratories before the war on tubes of this type and it had been realized by my colleague, Lubszynski, that the essential condition for stable operation of tubes in this manner is that the electron beam should fall almost normally onto the mosaic.[10] That is to say, the mosaic must be scanned orthogonally. However it is not because of this feature that the name "Orthicon" has been coined, but

Fig. 11—Super-emitron camera without cover.

Fig. 12—Complete super-emitron camera.

[9] A. D. Blumlein and J. D. McGee, British Patent No. 446,661; August 3, 1934.
[10] H. G. Lubszynski, British Patent No. 468,965; January 15, 1936.

because the signals generated are strictly proportional to the incident light.[11] We prefer to call this type of tube by the initials "CPS" of the earlier and even more fundamental principle employed of "cathode potential stabilization" of the mosaic.

Pre-war tubes of this type did not prove good enough to displace the emitron or super-emitron, but since the end of the war we have returned to this line of work and now have a CPS emitron with characteristics which I believe will interest you. An experimental camera using this tube was first used by the BBC to televise the royal wedding procession in November, 1947, and a properly engineered three-camera mobile unit was used for the first time in August, 1948, to televise those events of the Olympic Games that took place in the Empire Swimming Pool at Wembley, near London.

The tube and its associated driving coils is shown diagrammatically in Fig. 13. The cylindrical glass tube *1*, about $2\frac{1}{2}$ inches in diameter, has a narrow neck at one end in which a simple electron gun *2* is mounted. This consists of a cathode held at earth potential, a modulator slightly negative, and a first anode held at about +300 volts. This latter electrode accelerates the electron beam and restricts it by an aperture at its center to a diameter

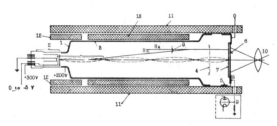

Fig. 13—Sectional diagram of CPS emitron.

of about 0.002 inch. The electrons are projected along the axis of the middle section of the tube which has a metal coating *3* on the walls extending from the first anode to slightly beyond a fine metal mesh *4*. This "wall anode" and the metal mesh are held at the same potential, about +200 volts. On the wall near the end of the large diameter section of the tube is a short cylindrical electrode *5* known as the decelerator. It is held at approximately earth potential. This end of the tube is closed with a flat polished window *6* and the target *7* is mounted on the flat inner surface of this window. This target is similar to the emitron target. It consists of a thin sheet of transparent dielectric which may be mica or glass between 0.002 and 0.004 inch thick. The surface of this dielectric facing towards the window is coated with a conducting but transparent metal layer, the signal plate. Contact is made to this layer through a metal seal at the edge of the window from a small metal plate *8* on the

[11] A. Rose and H. Iams, "Television pickup tubes using low-velocity electron-beam scanning," Proc. I.R.E., vol. 27, pp. 547–554; September, 1939.

front surface of the window. Here a spring makes contact to the first stage of the head amplifier *9*.

The photosensitive mosaic is formed on that side of the dielectric which is scanned by the beam. It is 35 mm ×44 mm in standard tubes and 20 mm×25 mm in special purpose tubes. The mosaic is framed by an opaque metal border which is connected electrically to the signal plate. The optical image is focused onto this mosaic by the lens system *10* through the glass window, the transparent metal signal plate, and the dielectric. The light passes through the transparent elements of the mosaic itself to liberate photoelectrons from the side on which the electron beam falls. In the past, two reasons for the poor performance of tubes of this type were (1) the strong absorption of light in the signal plate and (2) the very poor photoelectric efficiency of the mosaics then used. The former loss has been reduced from 60 per cent to about 30 per cent while the efficiency of the mosaic has been increased from about 3 μa per lumen to between 12 and 15 μa per lumen. It had been known for ten years that the most efficient photo surface for visible light was a layer of antimony treated with cesium which is in fact used in the conducting photocathode of the super-emitron. It had proved impossible, however, to make a mosaic of this material by the usual techniques. This problem was solved when my colleague, Holman, produced extremely fine metal meshes having 1,000 meshes per inch or one million apertures per square inch and a shadow ratio of as low as 30 per cent. Having been provided with such a mesh we are able to use it as a stencil to make the mosaic. It is held in close contact with the surface of the dielectric, while the basic antimony elements of the mosaic are evaporated through it. This gives a mosaic having about 1,750 elements in the line direction and 1,400 in the frame direction, and a total for the whole mosaic of 2.5 million elements. Thus, there are over ten mosaic elements for each picture point of a 405-line picture. Fig. 14 shows on the left a microphotograph of a small piece of this metal mesh and on the right a similar photograph of the mosaic elements formed by using it as a stencil.

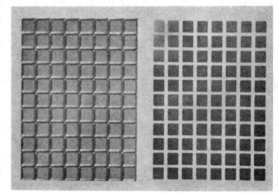

Fig. 14—Microphotographs of mesh (left) and mosaic (right) for the CPS emitron.

It is interesting to look back to the first emitron mosaic made in our laboratories in 1932 which we made by evaporation through a mesh having 2,500 apertures per square inch.

I must now refer back to Fig. 13 to explain the focusing and scanning systems 4. The tube is surrounded by a solenoid 11 which produces a fairly uniform axial magnetic field of about 50 gauss. As the electron beam is projected along the axis of the tube each electron traverses a spiral path and all return periodically to a focus in the well-known manner. The axial magnetic field and the electric accelerating fields are so adjusted as to bring the electrons to a focus on the surface of the mosaic. If the electron beam is not projected accurately parallel to the lines of force of the axial magnetic field, the whole beam will follow a spiral path and poor focus quality will result. It is impossible to guarantee mechanical accuracy of the electron gun, so two pairs of saddle coils are provided, one of which is shown at 12, to give a steady transverse magnetic field where the beam leaves the gun. By suitable adjustment of the resultant field from these coils the beam can be deflected at the beginning of its path to bring it into alignment with the axial focusing field. Hence these coils have become known as the "alignment coils." On the long medium diameter section of the tube two more pairs of saddle coils 13 are mounted, producing transverse magnetic fields at right angles to one another. They are actuated with currents of sawtooth wave form, one at line frequency, the other at frame frequency. The maximum transverse field H_T produced by these coils is small (about 5 gauss) compared with the axial field H_A (50 gauss). So the instantaneous effect of this transverse field is to distort or bend the resultant field to make a small angle θ with the axis, where $\tan \theta = H_T/H_A$. Outside the region enclosed by these coils the magnetic lines of force again become parallel to the axis. Provided that the distortion in the magnetic lines of force is not too great, the spiral paths of the electrons will be so modified that they will follow the resultant lines of force through the deflecting fields, and after leaving the deflecting fields they continue to follow the lines of force which are now parallel to the tube axis. So the electron beam arrives normal to the mosaic.

If, having excluded all light from the mosaic and biased off the electron beam, we gradually apply volts to the electrodes of the tube, the mosaic will remain substantially at earth potential. If now we turn on the electron beam, the electrons will enter a retarding electric field as soon as they pass the mesh 4. Between this mesh and the mosaic they will be slowed down to an energy of a few electron-volts and hence, on reaching the mosaic surface, each electron has a very small chance of liberating a secondary electron. It follows that the beam will drive the mosaic more and more negative in potential until it has reached the potential of the thermionic cathode from which the electrons originated. With the scanning fields operating, the whole mosaic is quickly established at cathode potential. Some corrections to this simple theory due to such effects as space charge, contact potential, and lateral energies of beam electrons, should be added here if time permitted, but they do not affect the final result much. When the beam electrons can no longer reach the mosaic they are reflected, accelerated towards the mesh 4, retrace their outward paths very approximately, and are captured on the first anode.

If now a light image is formed on the mosaic, the photoelectrons liberated by the light are accelerated towards the mesh 4. The electric field is sufficient to give full saturation of the photoelectric emission. Thus, a positive charge image is constantly being built up on the mosaic and at each passage of the beam the mosaic elements are driven back to cathode potential. The electron beam must be of sufficient intensity to discharge completely at each scan those mosaic elements exposed to the brightest parts of the image. Even in these peak white areas only about 10 per cent of the beam current is effective in discharging mosaic elements. In practice, a beam current of about 1 μa is used and the discharging current, which is the effective signal current, is about 0.1 μa. The residue of unused beam current returns to the electron gun. Obviously, in absolutely black areas of the image no beam electrons are accepted by the mosaic, all being returned to the first anode. That is to say, true black level in the picture signal corresponds to the state where the mosaic is receiving no electrons from the beam. It is obvious that this is the signal level produced during the return strokes of the scanning. Thus, there is provided automatically, when the beam is biased to zero, a signal which corresponds to true black, even though there may be no completely black area in the picture being transmitted. This gives a great advantage over the older tubes in that however extremely the light content of the picture may vary, the black level remains rock-steady. Consequently all the other gradations of picture brightness are truly reproduced.

A less desirable feature of this tube, which is common to all full-storage tubes, is the fact that movement of the image results in a blurring of the charge image and so of the reproduced picture. This loss of definition in individual pictures is equivalent to that observed in a photograph taken with 1/25th of a second exposure. However, just as in a cinematograph picture of moving objects, the eye is able to integrate the successive pictures to some extent and the total effect is not nearly so disturbing as would be expected from examining one single fame as a "still."

The capacity of the mosaic to the signal plate has an important bearing on this question of blurring of moving objects. If the dielectric is too thin, giving a capacity that is too great, the potential changes of the mosaic elements will be small. Hence a small proportion of the beam electrons will be able to reach the mosaic elements and the discharge will be inefficient. In these circumstances it is quite possible for charges to remain incompletely discharged for several frame periods which shows as serious blurring of moving objects and trailing after white objects. On the other hand, if the dielectric is too thick, the potentials of the mosaic elements may rise so high that the beam electrons have sufficient energy on reaching them to liberate almost one secondary electron per primary. Under these conditions the discharging is also inefficient with similar results. A compromise must be reached between these two extremes, and it is found possible to arrive at a fairly satisfactory result.

If a large amount of light falls on the mosaic it is possible for the photoemission to exceed the beam current. The mosaic elements will then rise rapidly in potential until the beam electrons can reach them with sufficient energy to liberate more than one secondary per primary. The beam then drives the mosaic elements still more positive, neighboring elements follow suit, and in a few seconds the mosaic will reach the potential of the wall anode. It then functions as a very bad emitron. This is corrected by first reducing the light to a reasonable level and then lowering the wall anode to earth for a few seconds and raising it again. Earlier tubes of this type were very unstable in this way and required very careful operation. This new tube is much more stable for the reason that the strips of highly insulating dielectric surface that separate the mosaic elements do not lose photoelectrons, and hence tend to remain at cathode potential. They therefore exert a stabilizing influence and may even reduce or stop altogether the escape of photoelectrons from those elements that have drifted dangerously positive.

In general, this control action of the "grid" of strips of insulating surface only comes into action when the potential of the mosaic element has risen well above the normal level for peak white. Over the normal working range of mosaic illumination it has little effect, and the charges built up are proportional to the image brightness; or the device has $\gamma = 1$. Since the normal cathode-ray tube increases the contrast of the signals which modulate its beam, this results in a picture with far too much contrast. The highlights are too bright and the darker tones are too dark. If the scene to be televised is very "flat" and lacking in contrast (imagine, for example, a grey winter's afternoon) this can be a good thing, since a much more interesting picture is presented to the viewer. But, on most occasions, a much more satisfactory result can be obtained by reducing the contrast of the picture. Thanks to the fact that this tube does give a definite black level to work on, my colleague White and his circuit experts have succeeded in doing this very satisfactorily. So now for the first time in the new CPS emitron cameras the contrast of the picture can be adjusted to give the most pleasing result.

You may have noticed a mesh 4 stretched across the path of the electron beam. This is to eliminate a white spot which may appear in the center of the picture which is produced by gas ions. The electrons pass along the tube, collide with molecules of residual gas, and produce positive ions. In the absence of this mesh 4, these ions drift towards the mosaic under the influence of the electric field which retards electrons but accelerates positive ions. The convergence of this electric field also tends to concentrate these ions in a diffuse spot in the center of the mosaic. There they build up positive charges which are discharged by the beam, giving a white signal. The mesh 4 held at the same potential as the wall anode cuts off the electric field which collects these ions from some distance back along the tube axis. Any ions produced between the mesh and the mosaic are uniformly distributed over the mosaic area and hence do not appear as a disturbing signal in the picture. The mesh 4 need not be of such fine pitch as that used to make the mosaic, and it is placed at an antinode where the beam is spread to its greatest diameter and covers some hundreds of mesh apertures. Thus the modulation of the beam by interception of electrons on the mesh is negligible.

Fig. 15 is a picture of a CPS emitron tube. This shows the general appearance, but unfortunately the details of construction are not shown. Fig. 16 gives a half-front view of the CPS emitron camera. I can only briefly outline its design. The tube with its driving coils and head amplifier is mounted rigidly on a platform which can be racked back and forth on a very accurate "tramway," so the lenses are fixed and focusing is done by moving

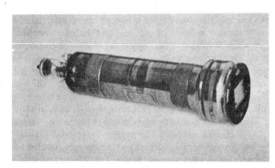

Fig. 15—The CPS emitron tube.

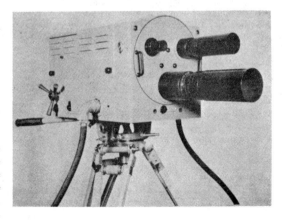

Fig. 16—CPS emitron camera (front view).

the tube. Two interchangeable lens turrets are provided, one having lenses 2.5-, 4-, and 6-inch focal length, all of maximum aperture $f/1.9$; the other having lenses of focal lengths 6-, 12-, and 17-inch and maximum aperture $f/4.5$.

Thus a range of angles of view from 5.5° to 40° is provided. A modified tube can now be provided with a smaller mosaic which gives an angle of view of 3 inches with the 17-inch telephoto lens. The iris diaphragms of the three lenses on each turret are ganged together and can be operated by a handle at the back of the camera. Thus all lenses are set at the same f number, so that when the turret is rotated the picture always comes up at the same signal strength. The turrets can be interchanged in a few minutes—if necessary, during a broadcast.

Fig. 17 shows a half-back view of the camera. At the side is the focusing handle and just below the viewing mask can be seen a large lever for rotating the turret and a smaller concentric lever for controlling the iris

Fig. 17—CPS emitron camera (back view).

Fig. 18—Interior of CPS emitron control van.

diaphragm. An electronic view finder is built into the camera and is observed by the operator through the mask which is seen on the back of the camera. This view finder is a small cathode-ray-tube television receiver on which the picture signals being generated by the camera are displayed. Thus it enables the operator to view-find and focus without an independent optical system.

In Fig. 18 the interior of the mobile control van is illustrated. Four similar units can be seen. Of the three on the right, each controls one of the three cameras that can be operated simultaneously. One operator controls each camera channel and is responsible for maintaining a satisfactory picture on his monitor. The fourth unit on the extreme left is the fading and mixing unit on which the picture from any camera can be selected for transmission. Above the four main units is the radio monitor where the picture as seen by the home viewer is displayed.

Conclusion

I would now like to summarize as far as I am able the relative merits of the television pickup tubes that have been described.

The quality of a television service depends on a large number of more or less independent features of the pickup tube that is used. The ten most important features are as follows:

1. Sensitivity and the closely related depth of focus.
2. Picture definition.
3. Picture geometry.
4. Prevalence of background blemishes.
5. Color response.
6. Spurious signals.
7. Stability to variations in light.
8. Picture tone gradation.
9. Effect on picture of movement.
10. Life of the tube.

I will now try to give an estimate of the performance of the three tubes, emitron, super-emitron, and CPS emitron in respect of each of these ten characteristics.

1. Sensitivity

The emitron has low sensitivity requiring about 200 foot-candles of illumination incident on the scene, and even then has very small depth of focus. The super-emitron has moderate sensitivity requiring about 50 foot-candles but has good depth of focus even with the lens at full aperture. The CPS emitron has still better sensitivity, requiring about 10 foot-candles, and also has good depth of focus at full aperture. Under reasonable illumination the lens can be stopped down and great depth of focus obtained. Even at much lower illumination a reasonable picture can be obtained, for though the signal-to-noise ratio becomes worse, the picture is not vitiated by uncontrollable and variable shading. It makes possible direct transmissions from locations where extra lighting is impractical or impossible, e.g., theatres, public functions, and the like.

2. Picture Definition

All three tubes are very good for 405-line pictures, but the super-emitron is somewhat better than the emitron, and the CPS emitron is the best of the three. The limiting resolutions obtainable are as follows:

Emitron	800 lines
Super-emitron	1,000 lines
CPS emitron	1,250 lines.

The definition of the latter is more uniform over the whole picture than either of the other tubes.

3. Picture Geometry

The emitron is excellent, the super-emitron is the worst, being prone to spiral distortion, while the CPS emitron is about as good as the emitron.

4. Background

All tubes can be made with very clean backgrounds and there is little to choose between them in this respect, although extreme care is necessary in manufacture.

5. Color Response

The response of the emitron to colors is very close to the response of the human eye with a little more response to the red and infrared than is desirable. However, very light make-up is all that is necessary. The super-emitron and CPS emitron have the same response and are rather less sensitive to red than the eye. They have no response to the infrared. There is little to choose between the tubes on this score.

6. Spurious Signals

Strong and variable spurious signals are produced by the emitron and super-emitron which require continual correction. The bad effects due to them are seldom completely eliminated and they become serious at low light levels. The CPS emitron has no spurious signals that affect the viewer.

7. Stability to Light Variations

The emitron and super-emitron are completely stable to light, though to get the best results the amount of light falling on the photosurface must be adjusted. The CPS emitron is fairly stable and will cope with most light variations met with in practice. To get the best results, the amount of light falling on the mosaic must be kept within a range of 4 to 1 by adjustment of the iris diaphragm. Excessive light can destabilize the mosaic and put the tube out of operation for a few seconds.

8. Picture Tone Gradation

The emitron and super-emitron give low contrast pictures which are a good average compromise for most cases. However, their contrast is fixed and nothing can be done to alter it, since there is no fixed black level in the picture signal. For this reason also, the general level of a transmitted picture depends on the manual adjustment of an operator, who may not even see the actual scene being televised. It is therefore quite arbitrary. Furthermore, the uncorrected shading results in some fogging of the dark areas of the picture. The CPS emitron gives very true blacks and, whatever the distribution of light and shade in the picture, the reproduction is accurate. The gradation between black and white is excellent and is not vitiated by uncorrected shading. The signal output is proportional to light input which gives too high a contrast law for general use. However, since the black level is fixed, control can be applied successfully, and the contrast of the transmitted picture can be adjusted to optimum.

9. Effect on Definition of Movement in the Picture

Emitron pictures of normal movement remain very sharp, equivalent to an exposure time of 1/250 second, but very rapidly moving objects appear as a succession of separate images which sometimes leads to curious results. The super-emitron gives pictures which are not so sharp, some blurring being noticeable in cases of rapid movement equivalent to an exposure time of, say, 1/100 second. The CPS emitron gives pictures at best corresponding to the equivalent of 1/25 second exposure, and so some loss of definition of rapidly moving objects is noticeable.

10. Life

The emitron has a relatively short life of around 100 hours. The super-emitron has a long life—probably between 500 and 1,000 hours. The CPS Emitron has not yet had sufficient use to be certain on this point but some tubes have already done 200 to 300 hours of operation without any serious deterioration occurring, so I believe the prospect for a long life is good.

You will see from this brief summary that there is as yet no perfect television pickup tube. I leave it to you to give marks and to add up the score. In the end the question of what is, and what is not, a good television picture is to some extent subjective, and people are influenced by what they have become accustomed to. The only final test is a long series of broadcasts ranging over a wide variety of subjects. The virtues and vices of the emitron and super-emitron are well known. The CPS emitron has had, as yet, only one serious test—the Olympic events from the Empire Pool at Wembley. It is generally agreed, I think, that it survived this severe test with flying colors.

Acknowledgments

I wish to thank Sir Ernest Fisk, I. Shoenberg and G. E. Condliffe, and the Directors of Electrical and Musical Industries, Ltd. Research Laboratories, Inc., for permission to present this paper to you; my colleagues of Electrical and Musical Industries, Ltd., who have done so much of the work, and last but not least my friend, J. Briton, who I am sure will have helped to make this cold meal considerably more palatable.

A SHORT HISTORY OF
TELEVISION RECORDING

Albert Abramson

A Short History of Television Recording

By ALBERT ABRAMSON

This paper describes the development of the three basic television recording processes since 1927. It also describes the film-recording processes used in both the United States and Great Britain. The introduction of television recordings made on magnetic material in both monochrome and color is noted. The paper concludes with a short résumé on the new art of electronic motion pictures produced with television cameras and recording facilities.

TELEVISION recording is an important part of the television industry today. Film recordings are made for a variety of useful purposes, some of which are: to compensate for time differentials, to delay presentation of a program to a more convenient time, and to provide network service to stations not connected by radio-relay or coaxial cable. In addition, programs may be recorded in advance to allow personnel to be elsewhere when the program is telecast, or even to provide a reserve program in case of emergency. As a result, the amount of film used to record programs by the major networks in the United States far surpasses that used by conventional film making means.

This recording of the television signal has been complicated lately with the adoption of a compatible color-television system. Since more and more network programs will be presented in color, a satisfactory commercial system of color recording must be developed that will have the same flexibility and speed as the monochrome recording. One of the answers to recording in color seems to have been solved with the introduction of magnetic television recording. This has been presented in both black-and-white and in full color, although only in a developmental stage. Magnetic television recording has many advantages such as: immediate playback of the picture, no development or chemical processing needed, and the saving if the magnetic material is used over and over.

Finally the possibilities of actually making high-quality motion pictures by means of television recording and electronic (television) cameras opens up a whole new field of endeavor. The advantages of this method have long been recognized but it has been only lately that equipment capable of the definition required has become available. It is to be expected that this process will eventually replace the more conventional film making methods.

Presented on October 19, 1954, at the Society's Convention at Los Angeles, by Albert Abramson, CBS Television, 1313 N. Vine St., Hollywood, Calif.
(This paper was received on December 22, 1954.)

Historical Development

Phonovision by Baird. Television recording is almost as old as television itself. To find the first efforts at recording the television image we must go back to the work of John L. Baird in England in 1927. Baird was a restless experimenter who covered the whole field of television. He tried long distance television, night-time television, stereoscopic television and color television.[1] It is little wonder that he even tried recording television. This he did on a machine called the Phonoscope.[2] Baird was experimenting with a process of "Phonovision" which was the recording of the television signal on phonograph records. At this time, Baird was transmitting a 30-line picture at a rate of $12\frac{1}{2}$ frames/sec. Thus the signal was actually of a very low frequency, so low, that it was easily carried on a regular telephone line or impressed on a wax record. The amplified signal was carried to an ordinary stylus head where the "picture" was converted into vibrations on the surface of the record. If desired, a synchronized record could be made of both the picture and the sound with either a double track being made, on one record, or else recording the sound on an accompanying record. To reproduce this record all that was necessary was a turntable synchronized with a scanning disc. The vibrations were converted back into electrical impulses which were fed to a neon light which illuminated the apertures in the scanning disc. Undoubtedly they were of poor quality for they were subject to the limitations of the crude mechanical system that produced them and there were other losses in the recording and reproduction processes. Baird soon tired of this facet of television and went on to more promising aspects of the field. However, we must credit him with making the first television recordings.

Rtcheouloff's Magnetic Recorder. While Baird was experimenting with his stylus recordings of the television signal another man filed a patent in England on January 4, 1927, for a process of recording the television signal on magnetic material. This was B. Rtcheouloff who in-

dicated apparatus "...adapted for the production of a magnetic record of the Poulsen telegraphone type." Poulsen was the Danish physicist who invented magnetic recording in 1898. Rtcheouloff's patent indicated that the accompanying sound was to be recorded on the opposite side of the magnetic material. At the receiving end the record was to feed several television receivers and the telegraphone receivers.[3] There is no indication that this apparatus was ever built.

Hartley and Ives. On September 14, 1927, Hartley and Ives of Electrical Research Products proposed a new method of "interposed" film at both the transmitting and receiving ends. The reasons given were as follows: "Television of background details is improved and increased illumination is obtained by taking a kinematographic film of the scene to be transmitted." Thus they proposed a method of television in which the scene to be transmitted would first be filmed by conventional methods and then the resultant film would be scanned for transmission. They also stated that "...preferably a photographic process is also interposed at the television receiving station."[4] This, of course, is the basis of the "intermediate film" process which later came into being as one means of producing large screen television.

Thus the period of 1926—27 saw the birth of the three basic television recording processes; however, of the three, only one, the film recording process or "intermediate film" method, was to be of any consequence for the next twenty-five years.

During the early 1930's there were many attempts to project a large-screen television image. Most of these used large scanning disks with powerful arc lights modulated by a Kerr cell. But many experimenters tried to take advantage of the regular motion-picture film projector and its greater light-throwing capacity.

Lee De Forest's Large-Screen Projector. De Forest and his associates filed a patent on April 24, 1931, for a method of recording pictures, film or events, "at the receiver by the etching action of an electrical discharge upon a suitable coating applied to a moving picture film or strip."[5] Their apparatus consisted of a revolving wheel with a series of needle-points. These needle-points were connected to the receiving apparatus which impressed the video signal upon them. These points passed over a strip of moving 35mm film which was coated with pure metallic silver. As the impulses varied, so did the etching action of the needle-points as they passed over the film. Thus the dark and light portions of the picture were to be reproduced as modulated lines on the film. This etched film was to be projected on a standard motion-picture machine.[6] However, due to many difficulties, this method was soon abandoned as impracticable.

The Intermediate-Film Process

The Intermediate-Film Transmitter. Another attempt at large-screen television was made by Fernseh A.G. in Germany. In 1932 they introduced their "intermediate film" transmitter at the Berlin Radio Exhibition. This was a television apparatus that first photographed the image to be transmitted by means of an ordinary motion-picture camera.[7] The scene was photographed on film prepared with a rapid and sensitive surface. This film was then passed through tanks where it was developed, fixed and washed, and while still wet (or in some cases after it had been partially dried) fed through a film gate in the last tank. Here it was fed to a scanning disk where it was dissected for transmission. After transmission, the film was either resensitized for immediate re-use or else saved for future transmission.

The Intermediate-Film Receiver. At the 1933 Berlin Radio Exhibition Fernseh again demonstrated this intermediate-film transmitter. Also in this year they demonstrated for the first time their "intermediate film" receiver for large-screen television.[8] Here the received signal was "recorded" on motion-picture film and then rapidly processed and projected by a standard motion-picture machine onto a full-size screen.

At the receiving end, the television signals were made to modulate a powerful beam of light by means of a Kerr cell. Between the cell and the film was a scanning disk with 90 hexagonal holes. This was rotated at a speed of 3000 rpm. Thus with 25 frames/sec a 180-line picture was obtained. The resultant light was focused on the film by a special optical system.

An image of the aperture in the disk was focused onto the sensitized film that was passing down a recording window. In this manner a series of adjacent lines of varying amounts of light and shade along each line, were imprinted on the film, so building up a picture. The film was rapidly developed and fixed. It was then fed into a theater projector of the usual intermittent type. The picture was 10 ft by 13 ft in size.

Thus the first television film recordings were made in 1933. This system was again demonstrated in 1934 using the same apparatus. The film was either saved

or resensitized as in the intermediate-film transmitting process. The results were often marred by blotches on the film. There was a delay of some 20 sec between the time the image was received and the time it was projected on the screen.

In 1935 the intermediate-film receiving system dispensed with the mechanical scanning disk and the Kerr cell in favor of a cathode-ray picture tube. A patent was taken out by Rolf Möller of Fernseh A.G. in Germany on December 12, 1934, for recording television images on film from a cathode-ray tube using continuously moving film. This apparatus was shown at the 1935 Berlin Radio Exhibition.[9] Thus the first cathode-ray film recordings were made during the period of late 1934 and shown publicly in 1935. However, this new intermediate-film receiving system was not successful and was not shown at the annual radio exhibition in 1936.

The Visiogram. In England there also were interesting attempts made to record television images. A novel machine called the "Visiogram" was developed by Edison Bell Ltd. in 1934. Motion-picture film was used, with the television signals being recorded thereon by the variable-density method familiar in sound techniques. The video signal was not converted into the usual light values of the scene itself but into a modulated "sound" track of the image. By means of a simple attachment the film signals were to be translated back into a visual image in an ordinary television receiver. In a demonstration given to the press the results were extremely poor.[10] Both intermediate-film transmitters and receivers were studied in the laboratories of Fernseh A.G. in 1937; the intermediate-film receiver method disappeared in Europe after 1937. It was revived after World War II when one of the major American motion-picture companies turned to it as one solution to the large-screen television problem. It was to be many years before it would be possible to project an image as large or as bright with a cathode-ray projector as with the intermediate-film receiving process.

Early Film Recording in the United States

In 1938 the first attempts were made in the United States to record on film the screen of a cathode-ray picture tube. These early efforts used standard silent, 16mm, spring-wound cameras operating at 16 frames/sec. With the low light intensity of the monitor screen, it was necessary to use the fastest film emulsion then available.

Since the cameras were nonsynchronous with the 30-frame rate of the television screen, the film recordings were marred by the appearance of

banding or horizontal lines (shutter bar) of over and under exposure caused by the uneven matching of the odd and even fields recorded on each frame of film.

The film was then recorded at 15 frames/sec which succeeded in eliminating banding but was successful in recording only every other frame of the television 30-frame picture.

It became obvious that if commercial use was to be made of television film recordings in the United States, the 30-frame television picture would have to be recorded on film at 24 frames/sec to conform with the speed of standard 16mm sound film, thereby permitting projection of the film either in a conventional sound projector for direct viewing, or by a standard projector for rebroadcasting by television. Development of a suitable commercial television recording camera was to continue for the next ten years before a practical system was perfected.

Recording of Airborne Television Transmissions. Although commercial television started in the United States in 1939 there was no further development in television film recording until the middle of World War II. Experiments with airborne television equipment such as "Project Ring" and the "Block" system were carried out. With the development of "Block" and "Ring" equipment, it became necessary to make permanent records of the transmissions of this apparatus. Motion-picture film cameras were used to record the television images sent by these developments from aircraft and guided missiles. Film cameras were installed on television receivers on the ground and in other aircraft. One of these early motion-picture cameras was a standard Air Force camera with a speed control to adjust the shutter to about 8 frames/sec. Speeds as low as 4 frames/sec were available. The recorded pictures were very poor due to the different standards of the transmissions, the low light intensities of the recording monitors, and to the many steps involved in the photographic processes. Shutter banding was noticeable in the film but did not destroy its value as a record. Further work was done with a Cine Special camera at 15 frames/sec with a 170° open shutter.[11]

During the immediate postwar period there was created a new interest in the recording of television images. The U.S. Navy started a series of experiments with its airborne television equipment. The first postwar black-and-white television film recordings were made on March 21, 1946, at the Naval Air Station at Anacostia, D.C. These were secured during a public demonstration of the Navy "Block" and "Ring" airborne television equipment.

Need for Commercial Film Recording.
The rapid growth of the television industry also necessitated the use of commercial film recordings. Television stations in the United States were opening faster than the telephone company could lay coaxial cable or erect radio-relay stations. With many new stations commencing operation it became necessary to serve them with program material from the two great centers of production, New York and Hollywood. Therefore the television film recording or transcription filled this need for program material.

Paramount's Large Screen Method. The film recording became one of the prime methods for large-screen television also. Paramount Pictures chose the "intermediate-film" method of recording for their large-screen television system. Paramount selected this system over the other immediate cathode-ray methods for a variety of reasons. Some of these were: the opportunity for cutting and editing the program before presentation; flexibility of programming around the regular film showing; and the use of regular projection equipment at the usual high light values.

With these advantages in mind, Paramount developed an intermediate-film system which used 35mm film exclusively. It used a special 35mm single-system (both sound and picture recorded on the same film), recording camera built by the Akeley Camera Co. This camera was unique at the time in that it used an electronic shutter. It could be loaded with 12,000 ft of film which permitted recording of over two hours. A Cooke $f/1.3$ coated lens was used at normal aperture, $f/2.3$. Du Pont Type 228, fine-grain, master positive film or Eastman 5302 film was used for recording either positive or negative pictures.

The film was processed by high-speed developing machines, in approximately 66 sec and was fed by a chute directly to a standard motion-picture projector.[12]

The Eastman Television Recorder. In January 1948 Eastman Kodak announced their new 16mm motion-picture camera for rerecording television programs on film. It had been developed in cooperation with NBC and the Allen B. Du Mont studios. It featured a 1200-ft magazine for continuous recording of a half-hour program, separate synchronous-motor drives for the shutter and film-transport mechanism an $f/1.6$ 2-in. lens, and a 72° closed shutter. The pulldown time was 57 degrees. Other features included a "bloop" light to provide registration with a sound-film recorder, a film loop-loss indicator, and appropriate footage indicators.[13]

First Color-Television Recordings. The first color film recording was made on August 18, 1949, in Washington, D.C. The film camera used was the Navy's

Berndt-Maurer with a 25mm $f/1.4$ Cine Ektar lens. Daylight Type Kodachrome was used. Exposures were made at 15 frames/sec synchronous, and at approximately 8 and 4 frames using the hand crank. The results were quite promising, in that the exposures at 8 and 4 frames were both adequate.

Other cameras were used and it was claimed that the "...first completely successful color recordings were made from a CBS Color Television receiver at the speed of 25 frames per second." On February 6, 7, and 8, 1950, sound was recorded with the picture to make the first sound and picture color film recordings.

The first color recording of the RCA "dot sequential" color television system was made at the RCA Silver Spring Laboratory on March 10, 1950. The initial recording was made at 15 frames/sec with a 180° open shutter, the exposure time being one-thirtieth of a second. It was claimed that all exposures were good and, "...strangely enough, it was the consensus of opinion that the film record was superior in quality to the image on the color television receiver as viewed with the naked eye. This phenomenon may be partially explained by the fact that the recording camera lens was located on the axis of, and normal to, the color television image, whereas the observers were forced to view the image from an 'off center' position."

The last of the three experimental systems, the "line sequential" system of Color Television, Inc., San Francisco, was recorded from an RCA Receiver on March 16, 1950, for the first time. This was made at 15 frames/sec.[14]

Early Film Recording in Great Britain

Similar progress in television film recording took place in England immediately after World War II. However, the need for recordings was different since there was only one television station in operation right after the war. The English realized that many topical events occurred when the majority of viewers were unable to see the direct transmission. Also they considered it a waste of time when recording topical or news events to have both newsreel cameramen and television cameras cover the same event. This is especially true when the televised event can be recorded and readied for broadcasting in such a short time compared with the regular filmed version. In addition the British had developed the habit of repeating dramatic programs a few days after the original performance. The use of television film recording allowed them to repeat a performance without added expense of cast and crew.

188½-Line Recorder. The 1947 efforts of the British to record television programs on film were done along the same pattern as the Americans. They tried to use

intermittent recording cameras with quick pulldown times. Here they faced a tremendous problem. They had to record a 25-frame television picture at 24 frames/sec and the amount of pulldown time was about 12°, so they compromised by recording only 50% of the television picture using the other 50% for pulldown time. These recordings were made on 35mm film but recorded only 188½ lines of the British 405-line picture.[15]

16⅔-Frame Recorder. Later in 1947 they used another method of intermittent recording. A special shutter was designed which recorded for 240° and was closed for 120°. This produced a film recording that was nonstandard, being recorded at 16⅔ frames/sec.[16] This recorder was also abandoned for a new one that used no intermittent mechanism at all.

The Mechau 35mm Recorder. Early in 1948 a 35mm continuous motion picture projector, the Mechau made by A.E.G. in Germany, was converted to a camera for continuous recording. It had a rotating mirror drum which for all practical purposes produced a stationary film frame. It used a form of optical compensation where as the drum rotated in sync with the film, the varying tilt of the mirrors made the reflected images of the television picture follow the film on its downward course. Thus the image was stationary in relation to the film. In this way a succession of images was formed on the film as it passed through the gate, the brilliance of each image rising from zero at the top of the gate, then increasing to a constant intensity over the central part of the gate and finally falling to zero at the bottom of the gate. This method eliminated any frame rate difference. This machine also eliminated the high rate of pulldown and the problem of the "picture splice" in the center of the frame common to the United States. There were no lines of the television picture lost in the recording process. It was claimed that this method could be used in the United States by blacking out a portion of the mirrors to avoid more than two fields being recorded on a single frame.[17]

Experiments with this equipment showed that the mirror drum was fully capable of providing correct optical compensation; however, the film transport mechanism did not attain the same high standard. Therefore the equipment was redesigned completely and three machines of this new type were to be installed in 1953 at the Lime Grove television studios.[18]

Application of "Spot Wobble." In addition, there was added an ingenious system of spot position modulation to eliminate the line structure that forms the television image. This was the application of a 10- to 15-mc modulation to the scanning beam. This caused the electron beam to oscillate vertically as it swept across a line and thus spread out. This increased the effective height of the scanning spot without increasing its width. It was claimed that a gain in light output in the order of two to one was made possible by the application of the "spot wobble."[19]

Double Gate 16mm Recorder. While this continuous recorder was more than satisfactory with 35mm film, more economical methods of recording became desirable by 1950 and development was concentrated on 16mm film. Due mainly to the fact that no continuous motion mechanism had been developed for 16mm film it was decided to develop a new recorder using an intermittent movement. This new 16mm film recorder had a double gate, that is, one gate above another, with an optical system capable of producing two images, identical in size, shape, and brightness at the normal 16mm frame spacing. A special pulldown mechanism was designed to work at 90°.

A full frame with two fields was recorded in the bottom aperture and then the second field of the first frame and the first field of the second frame recorded in the top aperture. The second field was lost and the film advanced two frames in this period. This cycle was then repeated. This gave an average film speed of 25 frames/sec. This recording camera gave good general definition with excellent interlacing. Movement was satisfactory although some jerkiness could be detected, especially on pan shots.[20]

Continuous Recording. In addition to the continuous mirror drum system and intermittent methods using either one or two gates there is a third method for recording television images on film. This was actually the oldest system of all as it was the basis of Fernseh's "intermediate-film receiver of the early 1930's. With the use of short decay and buildup phosphors, it was possible to make continuous recordings of television images from a cathode-ray picture tube, for example, merely by using the film motion as the vertical sweep and allowing it to spread a complete record of each frame of the television picture along the length of the film. There were no complications due to the difference in frame frequency and shutter frequency since there was no shutter used on the camera. This method is used in the present day "Ultrafax" facsimile recording apparatus.[21]

Magnetic Television Recording

Another advance in television recording was made on November 11, 1951, when the Electronic Division of Bing Crosby Enterprises gave its first demonstration of a video tape recorder in black-and-white.[22] The advantages of recording video signals on magnetic tape are many.

1. There is no lens system necessary, as there is in the film-recording camera. Thus all optical losses are eliminated.

2. The signals are recorded as electrical waveforms and not as visible images from the face of a cathode-ray monitor tube with its possible distortions and limitations.

3. There are no problems of pulldown time or frame-rate conversion.

4. There are no developing processes, and so losses due to chemical action and image transfer are avoided.

5. The video tape recording can be played back immediately.

6. Magnetic tape is cheaper than processed motion-picture film and can be used over many times if necessary.

7. It will eventually allow full color, stereoscopic pictures with stereophonic sound to be recorded on one strip of magnetic tape.

However, the recording of a video signal on magnetic tape presents many unique problems. Since the signal cannot be spread horizontally as it can in film recording, it must be spread along the length of the tape. It is possible to record the signal by use of special recording heads capable of responding to a 3- or 4-mc signal. Or the signal may be divided among a series of heads for recording. In either case the tape must be run many times faster than sound recording speed. The problem is also complicated by the fact that the tape must run at an absolute constant rate of speed. In addition, it must be remembered that the television signal consists of other necessary information such as synchronizing pulses. Finally, the tape must be played back on the same standard of definition (number of lines and frames/sec, etc.) as it was recorded.

A multiple-track method was chosen by the Electronics Division of Bing Crosby Enterprises for their first video tape recorder.[23] This apparatus used twelve recording heads. Ten were used to record the video signal, the eleventh was for a synchronizing track, and the twelfth was for recording audio. By combining an ingenious method of sampling each head in a stroboscopic manner with a unique switching device, an alternating signal was recorded on each track, with both positive and negative halves representing bits of picture information up to 1.69 mc for the whole group of ten heads. Early models of this apparatus used 1-in. brown oxide tape although later models use either one $\frac{1}{2}$- or $\frac{3}{8}$-in. tape. The tape ran at a speed of 100 ips. The recorder accommodated reels for 16 min of recording.

The October 2, 1952, demonstration of this video tape recorder proved that this process merited attention. The picture had the following good features:

1. The gray scale was outstandingly good.

2. The picture was sharp and clear. However, the following faults were apparent:

1. A diagonal pattern was always prominent.

2. Considerable flicker was noticeable.

3. Under certain conditions a series of ghosts was noticeable.[24]

Later developments by the Crosby Organization have resulted in highly improved definition of the picture and elimination of most of the previous deficiencies. The number of tracks required has been greatly reduced for black-and-white recording, thereby making the system more easily adaptable to color recording.

The RCA Video Tape Recorder. The Radio Corporation of America on December 1, 1953, demonstrated at the David Sarnoff Research Center a video tape recorder on both monochrome and color. It proved that the recording of images in color was as easily accomplished as in black-and-white. The recorder used paper thin plastic tape running at 30 fps. The reels were 17 in. in diameter and would record some four minutes of a program. RCA had achieved the recording of a 3-mc signal through the use of specially designed recording and reproducing heads which responded to frequencies much higher than the cutoff point for heads used in sound recording.[25]

The black-and-white programs were recorded on $\frac{1}{4}$-in. tape, using two tracks, one for the picture and synchronizing signal, and the other for the sound portion. The color program was recorded on $\frac{1}{2}$-in. tape, using five tracks. Three tracks were for the colors, red, blue and green, one was for the synchronizing signals and the last was for the audio portion. The playback of the color recording showed only a slight loss in definition, mostly in excess light values. There was a slight smearing, streaking and halo effect, as well as a high-frequency noise level hiss. Occasionally there was some jitter due to nonuniform speed control. However, the demonstration was considered to be an overwhelming success.

It is expected that these problems will be overcome and that video tape recording will emerge from the laboratory capable of reproducing pictures indistinguishable from the original "live" pickup. It is expected that this process will supplement if not supplant the film or visual recording.

The Electronic Motion Picture

A whole new field of television film recording is being introduced by the development of a completely new electronic picture recording system by High Definition Films Ltd., in London, which has a new concept of producing high-quality motion pictures, utilizing electronic (television) cameras and advanced film recording techniques.

High Definition Films Ltd. Norman Collins and T. C. Macnamara have carefully pointed out the limitations of ordinary film-making procedures while indicating the advantages of using the electronic camera.[26] Avoiding ordinary television broadcasting requirements, they are not bothered by such items as a restricted bandwidth, limited contrast range and tonal gradation, and the necessity for mixing in synchronizing signals. The whole apparatus is operated on closed-circuit under virtually laboratory conditions.

In the development of this equipment, it was decided to equal the standards of present 35mm motion-picture filming. This was accepted as some 30 to 40 lines/mm resolution. To equal this definition a 24-frame picture would have to have 992 lines with a bandwidth of 15.75 mc/sec. This may be increased to 1300 lines if necessary. However, it was felt that it would not be necessary to go much above a thousand lines to equal today's 35mm film standards.

The line scanning is sequential. It is known that interlacing is not needed for pure film recording purposes. Interlacing's main advantage of eliminating flicker while conserving available bandwidth does not overcome some of its more serious faults. These include "line-pairing," "line-crawl," and movement blur. Since in this system, the picture signals are kept separate from the synchronizing signal, the line frequency does not have to be related to the frame frequency. This simplifies the pulse generating apparatus and also enables the number of lines to be varied to suit the resolving power of the type of pickup tube selected.

When recording television images on film, any apparent flicker will be eliminated by the film projector due to its interruption of the light source two or three times during a frame. However, it was expected that monitoring would be difficult, but that the 24-frame flicker could be reduced by using cathode-ray screens having long delay times.

The recording unit is of the intermittent type. While it was felt that continuous motion was exceedingly attractive from many points of view the accuracy of registration which could be realized at the present state of development was insufficient for recording picture of the high definition required. In 1952 a standard film camera with a 70° pulldown was being used. The lens was also a standard motion-picture

75mm type operated at full aperture of $f/2$. Recording was on a slow, fine-grain sound recording stock.

Refinements of the High-Definition system were made and it was presented with the following features in 1954.[27]
1. The cameras in England were the Pye Radio type using the Pye Photicon image iconoscope pickup tube. Cameras for use in the United States and Canada were the General Precision Laboratory type using the 5826 image orthicon pickup tube.
2. The cameras used sequential scanning of either 625 or 834 lines per frame at 24 frames/sec.
3. The chain is essentially closed circuit with a bandwidth of 12 mc.
4. The picture and synchronizing signals are never mixed and a new method of signal control has been devised.
5. A special "staircase" signal is present on all picture monitors in the form of a step wedge. Its presence allows accurate adjustment of the signal amplitude and lift.
6. On the recording monitor two photocells are used for measuring brightness of the first and tenth suppression steps on the kinescope tube face. The monitor tube is a special HDF/Cintel tube with a 9-in. diameter flat screen, free from granularity, and aluminum backed. It presents a 3 × 4 in. picture.
7. The High Definition system employs a spot wobble oscillator to eliminate line structure.
8. There are two recording channels using HDF/Moy 35mm film cameras. These cameras have a specially designed 20° pulldown mechanism with a special film accelerator.
9. This system is designed to use any kind of sound recording method.

High Definition Films has acquired studios in London and was producing demonstration films by this process early in 1954.

Television recording has come a long way from Baird's crude grammophone recordings to today's high-definition film recordings. The television recording has an important place in the commercial television industry. It promises to play an even more important role in the motion-picture industry of tomorrow. The electronic motion picture using neither film cameras nor motion-picture film is an actuality. The television camera and the magnetic video recorder will allow the motion picture a perfection and flexibility that has never before been attained.

References

1. Sydney A. Moseley and H. J. Barton Chapple, *Television Today and Tomorrow*, Sir Isaac Pitman & Sons, Ltd., London, 4th ed., 1934.
2. A. Dinsdale, "Television sees in darkness and records its impressions," *Radio News*, 8: 1422–1423, June 1927.
3. B. Rtcheouloff: Brit. Pat. 288,680, June 7, 1928.
4. R. V. L. Hartley and H. E. Ives: Brit. Pat. 297,078, Mar. 19, 1928.
5. American Television Laboratories: Brit. Pat. 386,183, Apr. 14, 1932.
6. Lee de Forest, *Father of Radio*, Wilcox & Follett, Chicago, 1950, pp. 418–422.
7. A. T. Stoyanowsky, "A new process of television out of doors," *Jour. SMPE*, 20: 437, May 1933.
8. J. L. Baird, "The Kerr Cell and its use in television," *Jour. TV Soc.*, 11: 110–124, Jan. 1935 – Dec. 1938.
9. E. H. Traub, "Television at the Berlin Radio Exhibition," *Jour. TV Soc.*, 11: 56, Jan. 1935 – Dec. 1938, p. 56.
10. London Times, Nov. 29, 1934.
11. Robert M. Fraser, "Motion picture photography of television images," *RCA Review*, 9: 202–217, June 1948.
12. Richard Hodgson, "Theater television system," *Jour. SMPE*, 52: 540–548, May 1949.
13. J. L. Boon, W. Feldman and J. Stoiber, "Television recording camera," *Jour. SMPE*, 51: 117–126, Aug. 1948.
14. W. R. Fraser and G. J. Badgley, "Motion picture color photography of color television images," *Jour. SMPE*, 54: 736, June 1950.
15. H. W. Baker and W. D. Kemp, "The recording of television programmes," *B.B.C. Quarterly*, 6: 1, pp. 236–248, Winter, 1949–50.
16. D. A. Smith, "Television recording," *Wireless World*, 55: 305, 1949.
17. W. D. Kemp, "Video recordings improved by the use of continuously moving film," *Tele-Tech*, 9: 32–35, 62–63, Nov. 1949.
18. W. D. Kemp, "Television recording," an abstract (75% of the original paper, presented at the Convention on the British Contribution to Television, Apr. 28–May 3, 1952), *Jour. SMPTE*, 60: 367–384, Apr. 1953.
19. P. J. Herbst, R. O. Drew and J. M. Brunbaugh, "Factors affecting the quality of kinerecording," *Jour. SMPTE*, 58: 85–104, Feb. 1952.
20. W. D. Kemp, "A new television recording camera," *J. Brit. Kinemat. Soc.*, 21: 39–56, Aug. 1952.
21. D. S. Bond and V. J. Duke, "Ultrafax," *RCA Review*, 9: 99–115, Mar. 1948.
22. Frederick Foster, "Motion Pictures on Tape," *Am. Cinemat.*, 32: 500, Dec. 1951.
23. "Video magnetic tape recorder," *Tele-Tech*, 13: May 1954, 77, 127–129, May 1954.
24. John T. Mullen, "Video tape recording," Speech delivered at 7th Annual NARTB Conference, Thursday, Apr. 30, 1953. (A brief illustrated description appears in the New Products column, *Jour. SMPTE*, 62: 323–324, Apr. 1954.
25. H. F. Olson, W. D. Houghton, A. R. Morgan, J. Zenel, M. Artzt, J. G. Woodward, and J. T. Fischer, "A system for recording and reproducing television signals," *RCA Review*, 15: 3–17, Mar. 1954.
26. Norman Collins and T. C. Macnamara, "The electronic camera in film-making," *J. Inst. Elec. Engrs.* (London), 99: Part 111A, No. 20: 673–679, 1952. (Reprinted, *Jour. SMPTE*, 59: 445–461, Dec. 1952. Norman Collins and T. C. Macnamara, "High definition films," *J. Brit. Kinemat. Soc.*, 21: 32–38, Aug. 1952.
27. Information obtained from a brochure sent by High Definition Films, London, England, entitled *High Definition Electronic Picture Recording System*, Technical Description, May 1954.

A SHORT HISTORY OF
TELEVISION RECORDING

Part II

Albert Abramson

A Short History of Television Recording: Part II

By ALBERT ABRAMSON

This is a survey of television recording from 1955 to 1970. The history begins with the kinescope recording era, both in monochrome and color, and continues with the introduction of the first commercial magnetic videotape recorder by Ampex in April 1956. It describes the first helical-scan videotape recorder introduced by Toshiba in Japan in Sept. 1959, and it details progress made to improve videotape quality culminating in the adoption of new "high-band" recording standards in 1964. The field of stop-, slow- and reverse-motion magnetic disc recorders is surveyed. The history describes progress made in thermoplastic and electron-beam recording in a vacuum; high-speed magnetic contact printing is discussed. The resurgence of color kinescope recording as a means of producing electronic motion pictures is narrated. The cassette/cartridge revolution is covered from the introduction of the CBS EVR system to the Teldec method of video playback from a phonograph disc.

THE FIRST PART of this paper was presented in 1954. It reviewed the development of television recording from the earliest efforts of John Logie Baird to record a video signal on a gramophone disc, through the early efforts at magnetic recording, to the ill-fated efforts of High-Definition Films in London to produce electronic motion pictures using television film recording early in 1954. A complete survey of the three basic recording methods (phonograph disc, motion-picture film and magnetic tape) was made.[1]

TELEVISION FILM RECORDING

The Hot-Kine Era

In 1955 the American television industry continued to improve the established "kinescope recording" or "kine" process. The 3-hour East-West time differential required a rapid method of television recording since all three networks desired to maintain a uniform program schedule. Thus an era of "hot-kines" or "quick-kines" arose in which the recording and playback of a monochrome television program was made possible in 3 hours or less. In a typical recording operation, the program was recorded on the West Coast on two machines simultaneously, one 35mm and one 16mm. The 16mm recording had an optical track but the 35mm had no sound track, its sound being recorded on either magnetic film or magnetic tape. The program was recorded in 30-min segments and was pulled out of the recorder tails out and rushed to a developing laboratory where it was quickly developed, washed and dried. These recordings were then projected as negatives with the 35mm as the "Air Copy" and the 16mm as "pro-

Presented on 27 Apr. 1971 at the Society's Technical Conference in Los Angeles by Albert Abramson, 6643 Mammoth Ave., Van Nuys, CA 91405.
(This paper, first received on 5 April 1971, was received in final form on 22 Jan. 1973.)

tection." This "hot-kine" process had begun with the completion of the transcontinental microwave system in Sept. 1951.[2] In Dec. 1953 the NTSC color system was adopted in the USA and a color recording process became necessary. It was reported in 1954 that NBC was developing a color-kine system using a Triniscope (three separate 10-in kinescopes) optically combined by means of mirrors and recorded on 35mm color negative film. The results were only "reasonable" so a color hot-kine process needed another solution.[3]

Lenticular Film Color Recording

Early in 1955 Eastman Kodak announced that NBC would be using a lenticular color film process by the fall of 1956.[4] Since lenticular film lent itself to the same quick processing methods as monochrome film it could prove to be a solution to the color time-zone problem. The color process was made possible by the presence of many vertical lenticles or lenses which were embossed on the base of the film surface and acted as prisms to separate the color into color difference stripes. The special Eastman Type 5308 embossed kine-recording film had 25 lenticles/mm and approximately 390 across the width of a 35mm frame. The system used three kinescopes having P16 phosphor screens displaying negative polarity. A special prism was placed in the position of the color filter. Since the light from each kinescope could only reach that strip behind the lenticle reserved for one color, the recorded film could be considered a color separation negative. As in the monochrome kine process the lenticular film was quickly processed and threaded on a special projector. This consisted of a three-vidicon film chain with a mask in front of each camera tube in order to pass only the correct color information.[5] NBC inaugurated the lenticular color-kine process on a 3-h delay on 29 Sept.

1956. This continued on a daily basis until 19 Feb. 1958 when it was announced that all time-zone-delay would be done by means of color videotape recording. This began at NBC on 27 Apr. 1958 and signalled the end of the hot-kine era, but did not mean the end of the television film recording process. On the contrary, much effort was put into perfecting the process to take advantage of its flexibility and interchangeability.

NBC Color Kinescope

In 1957 NBC in New York was experimenting with photographing the color image from a 21-in (533-mm) shadow-mask kinescope onto 16mm Anscochrome. Picture quality was acceptable only for reference purposes. The soundtrack, obtained by single system direct positive recording, gave less acceptable results. However, a blue-sensitive photocell in the projector improved the sound quality.[6] In Oct. 1962 NBC reported further progress on color kinescope recording. The most recent system used specially designed 7-in (178-mm) color kinescopes with dichroic mirrors. Some successful attempts were made to photograph tricolor 21-in kinescopes. Another experiment concerned a "telecentric" lens system employing three 5-in (127-mm) color projection kinescopes. The three images were added together by means of front surface reflectors only. A demonstration of this system was made of (a) a 16mm film (Ektachrome Commercial) that was printed onto internegative and into positive color prints, (b) a 16mm print made from an original 35mm camera negative and (c) a 35mm print made from the same camera negative as the 16mm print.[7]

BRYG System

In Apr. 1958 a nonstandard color television system, the BRYG method was proposed to separate the color information on the basis of luminance, red and blue. It was planned to use half of the film (either 35mm or 16mm) for the luminance information and a quarter area each for the red and blue information. Both film travel and optics were to be nonstandard. It was proposed for reproduction that electronic matrixing be employed to produce green by subtracting both red and blue from the luminance signal.[8]

Vidtronic Process

In June 1966 the Technicolor Corp. formed the Vidtronics Division to concentrate on the transfer of color video tape to 35mm, 16mm and super-8 color film. In Apr. 1967 details of the color tape to film process were announced. Basically it was a method of separating the red, green and blue signals from the videotape. The signal was completely decoded; the red, green and blue information was taken out as black-and-white images which represented the information in the red, green and blue signals. These signals were electronically enhanced, displayed on a monitor and recorded on film separately. The tape was played three times to produce the three separation signals. The three separation elements in monochrome were recombined by the normal Technicolor laboratory process to produce a final color print in either 35mm or 16mm. The process required about 5 to 7 days from the time the tape arrived until the prints were ready.[9]

NHK Color Kinescope

In Nov. 1966 NHK in Japan began using a new color-kinescoping system with three separate color kinescope tubes that had adequate luminance. Since each of the primary colors was independent of the others, it was possible to obtain perfect matching between tubes and the high-sensitivity color reversal film used. Thus color pictures of excellent resolution were obtained.[10]

ACME-Chrome Film Transfer

In July 1967 Acme Film and Video-Tape Labs of Los Angeles, Calif., announced its new videotape-to-film process. The process consisted of using a composite display of RGB picture information. This display was recorded on Eastman 7254 high-speed color negative film. Acme had been using a similar process since 1963 using Ektachrome 7255. It was claimed that it was possible to get a finished color film print in some 8 hr.[11]

VIDEOTAPE RECORDING (TRANSVERSE SCAN)

The Ampex Revolution

In Feb. 1956 George I. Long, President of the Ampex Corp., Redwood City, Calif., announced in a letter to stockholders that "Ampex has constructed a laboratory version of what is believed to be a practical system for the recording and reproduction of TV pictures on magnetic tape."[12] On Saturday, 14 Apr. 1956 Ampex demonstrated their first prototype video recorders to some 200 CBS-TV affiliates gathered for the NARTB Convention in Chicago and simultaneously at its Redwood City Laboratory.[13] These prototypes had many ingenious features. They used 2 in (50.8 mm) wide magnetic tape that ran at a speed of 15 in/s (381 mm/s) past a magnetic head assembly which

rotated at a high rate of speed. The specifications revealed a horizontal resolution of 320 lines, a video bandwidth of 4-MHz, video peak-to-peak SNR of 30 dB or more and an audio track that went beyond 10 kHz. The recorder was

the result of a program led by Charles P. Ginsburg and a special team of engineers.[14] This work had begun in the fall of 1951. Three methods of utilizing a rotating head were proposed to avoid either high tape speed or time-division

Arcuate system using a three-head disc: left, the disc; right, scanning pattern. (Courtesy of Ampex Corp.)

The mark II Machine in 1955. (Courtesy of Ampex Corp.)

multiplexing. They were: (1) arcuate scan, (2) transverse scan and (3) helical scan. In Oct. 1952 the first almost recognizable pictures were shown using an arcuate system comprising three magnetic heads mounted on the flat surface of a spinning drum over 2-in magnetic tape driven at 30 in/s (762 mm/s).[15] This method was soon superseded by transverse scanning where four magnetic heads were mounted on the edge of a 2-in drum which rotated at 14,400 r/min. Much additional work was undertaken to perfect the vacuum guide system as well as a unique FM modulation/demodulation system as a means of controlling the rapidly fluctuating head signals. The FM system conveniently solved the problem of recording a wide bandpass signal from dc to 4 MHz. The FM carrier frequency was 5.0 MHz; sync was 4.28 MHz and peak white 6.8 MHz. Besides the transverse video track, the recorder had three more magnetic tracks, an audio track, a cue track and a 240-Hz control track which was used to control the relative position of the head disc and the capstan shaft during replay.[16] On 30 Nov. 1956 the first videotape recording was aired by CBS-Hollywood to the Pacific Coast Network. This first videotape playback was backed up by both 35mm and 16mm kinescope protection. It was reported on 14 Oct. 1957 that Ampex and RCA had reached an agreement for the exchange of patent licenses covering videotape recording and reproducing systems in both monochrome and color.[17]

RCA Color Videotape Recorder

On 25 Oct. 1957 RCA demonstrated their prototype color videotape recorder to the press from Camden, N.J. Although designed for color, it was noted that this machine could also be used in monochrome.[18] The color processing in the RCA Television Tape Recorder was to cancel phase drift in the chrominance signal by translating this signal to a higher frequency spectrum and then heterodyning this translated signal with a signal which also contained the phase drift and was of such a frequency that the difference signal frequencies fell back into the original frequency band. If this was derived from a signal recorded on the tape, it would contain the same phase-drift effects as those in the translated chrominance signal, but the difference signal obtained by heterodyning would be free of phase drift because errors had been cancelled by subtraction.[19]

Ampex Color

The Ampex color system was different. The encoded chrominance information was reduced back to its I and Q video components and then re-encoded with the luminance signal. Burst was added. The signal coming from the tape was

also divided by appropriate filter networks into a luminance channel of approximately 3 MHz in bandwidth and a chrominance channel of 2 MHz. The burst signal from the tape was used to drive the burst lock oscillator and the nominal 3.58-MHz signal thus generated was phase shifted and applied to a pair of diode clamp demodulators, providing quadrature demodulation of the color video information. Since the regenerated 3.58-MHz signal was being corrected in phase at the start of each horizontal line, its relation to the encoded chrominance signal on each line remained constant and therefore the recovery of the color signal was reasonably precise.[20]

Intersync

In Apr 1960 Ampex announced a new accessory called Intersync to be used with the Ampex videotape recorder.[21] This was a device which enabled the capstan servo and head drum servo of the videotape machine to be brought into positional coincidence with an external sync source. Thus the recorder could be locked line by line as well as vertically, field for field. Thus it was possible to cut to or from a videotape machine without getting any rolls or discontinuity. Also, dissolves and special effects were made possible.[22]

Amtec

Also in Apr. 1960 Ampex demonstrated a new automatic time base error compensator called Amtec. Amtec consisted basically of electrically controlled continuously variable delay lines, the delay of which varied in sympathy with the time variations present in the input voltage. The correction voltage was obtained by comparing the signal incoming to the equipment with an internally generated train of flywheel pulses. Thus a positive or negative delay in each line brought its starting time to within 0.03 μs of a stable external reference. This device, which was developed at CBS-Chicago by Charles Coleman, was first called Coltec.[23]

Colortec

In May 1961 Ampex demonstrated their new direct color recovery system called Colortec at the NAB Convention in Washington. In operation it was said to accomplish line-by-line compensation of timing errors in the composite color signal by sampling burst phase of the signal each horizontal interval, with respect to the external 3.58-MHz signal. The instantaneous phase difference between the sampled and reference signals was converted to a proportional voltage which adjusted the delay time of a voltage controlled delay line in the video signal path. Thus the resultant signal was within the required stability limits for direct color playback.[24]

VR2000

In Apr. 1964 Ampex announced their new VR2000 color compatible television tape recorder. This machine introduced a new "high-band" record/reproduce standard. This new concept in videotape recording was developed primarily to serve the more stringent requirements of the BBC's new 625-line system on BBC 2 which went into operation on 20 Apr. 1964.[25] It had long been realized that the original recording standards, while more than adequate for the USA 525-line and the BBC 405-line standards, were lacking in both frequency response and SNR for the more demanding 625-line systems.[26] As early as 1958 when the first Ampex machines were delivered to the German television authority it became apparent that much work remained to deliver the higher quality required.[27] Negotiations were instituted under the auspices of the EBU to allow a higher carrier frequency to become standard for European broadcasters to permit a higher video bandwidth than the figure of 4.5 MHz.[28] The new British standard was designed to operate on the third-order shelf with a 7.8-MHz carrier frequency; sync tip was 7.16 MHz and peak white was 9.3 MHz. A pre-emphasis boost of subcarrier was 8 dB and the -3dB point for the bandwidth was 6.0 MHz. The SNR, unweighted, was 43 dB.[29] New video heads were developed so that frequencies in excess of 11 MHz could be recovered from the tape. The differential gain was less than 5%, differential phase less than 5%, color phase error on 75% saturated bar $\neg 2°$. It had better than 2% K factor. During the development stage, Ampex delivered two modified VR1000C recorders (called VR-1000Ds); they had been specifically modified with the new high-band modulation/demodulation system and Nuvistor preamps to permit evaluation of this new recording technology. Then early in 1964 the BBC received the first VR2000s delivered to anyone in the world.[30] The new machines had such excellent quality that it was now possible to make up to fourth generation dubs in color with minimum loss of quality. The only drawback was that there were now three sets of recording standards.

Automatic Velocity Compensator

In Mar. 1966 the Ampex Corp. announced an Automatic Velocity Compensator to eliminate color banding due to differential velocity errors.[31] The compensator was designed around the fact that the mechanical conditions which produced this error were generally fixed and cyclic, repeating themselves with each rotation of the head. A memory bank containing 64 individual capacitor stores was provided with a series of incremental error voltages that were derived from the combined geometric and color correcting delay line errors. Before any given line was repeated on subsequent rotations of the head assembly, the capacitor store representing that line was read out and its error signal was applied to a ramp-forming capacitor which generated a waveform whose slope was proportional to the velocity error component. This ramp was applied to the geometric correcting delay line as a modulation voltage which caused the Amtec delay line to change its delay over the active period of the television line from the pre-set level established at the start of the line to the level predicted by measurement of the incremental error of the previous rotation.[32]

VR3000

In Apr. 1967 Ampex announced a new battery-powered portable videotape and camera combination, the model VR 3000. The recorder weighed 35 lb (16 kg) and could produce either high-band or low-band monochrome tape that could be played on any standard transverse recorder. It was also capable of recording high-band color from studio color cameras with no modification. The machine recorded some 20 min of picture and sound on an 8-in (203-mm) reel of 2-in (50.8-mm) tape. The battery permitted some 20 min of preview operations of the camera only or "live" telecasting. The accompanying camera which weighed some 13 lb (6 kg) used a Plumbicon tube.[33]

AVR-1

In Apr. 1970 the Ampex announced their new "third generation" videotape recorder Ampex Model AVR-1.[34] It had a new tape transport consisting of a capstan tape loop and a vacuum column which required only the capstan to be up

to speed. It featured an instant play feature: video was available in 200 ms. Threading was automatic with air-guides in the tape path. The vacuum tape guide automatically withdrew from the head for both loading and cleaning. The output was continuously synchronized and automatically adjusted for maximum picture quality. It could record nonsynchronous picture source material. It used a new Mark XX video head assembly equipped with air bearings and a rotary transformer for longer life, lower noise and high dependability. It offered a revolutionary time-base error correction system that eliminated pre-roll requirements. The AVR-1 was designed to be interfaced with computers used in automatic programers.[35]

VIDEOTAPE RECORDING (HELICAL SCAN)

TOSHIBA VTR-1

In Sept. 1959 a Toshiba VTR-1 prototype videotape recorder was demonstrated to the public at the Matsuda Research Laboratory, Kawasaki, Japan. This was the first helical-scan videotape machine. It had one video head and the tape was run in a helical loop around a cylinder containing the video head. It used 2-in (50.8-mm) tape running at 15 in/s (381 mm/s). An FM modulation system was used to conveniently record the signal on tape. The SNR was about 35 dB. Due to the single head, the video signal was interrupted for about 100–300 μs. A special inhibition gate was used. A processing amplifier was set up to reshape the synchronizing waveform, in which the horizontal sync pulses from this part of the tape were cleaned up, with new pulses being generated, re-inserted and a standard television waveform reproduced. It was claimed that

this new system could reproduce pictures at any speed, whether fast forward, slow forward, rewinding, or even stopped. This was because each television field was recorded on one long track, so that as the head rotated at 60 Hz it would repeat one field even when in the stopped position. It was claimed that work on the system had begun in 1953 and the first experimental machine completed in 1958. At the end of 1961 one unit had been delivered to NHK for evaluation and test. In addition to the one-head machine, Toshiba described a two-headed helical-scan mechanism in which the tape was only half-wrapped around the video head.[36]

Victor Color Videotape Recorder

In Jan. 1961 the Victor Company of Japan announced a new two-head helical-scan color videotape recorder.[37] Actually two sets of magnetic heads were used — two wide track heads being used for recording and two narrower heads for monitoring (while recording) and playback. The two recording heads were mounted 180° apart at the edge of the rotating drum and produced one complete television picture for each 360° of rotation. It was claimed that editing and splicing were made simple because this machine could reproduce a "single frame" while the tape was standing still. The picture could also be monitored while in reverse or fast forward. Tape speed was 15 in/s (381 mm/s) and the machine could record up to 90 min on 7,200 ft (2,200 m) of tape. It had an FM modulation system and a special color processor to produce standard NTSC color signals. This machine was marketed as the Telechrome JVC Model 770 in the USA.[38]

Machtronics MVR-10

In June 1962 a new American company, Machtronics Inc. of Mountain View, Calif., introduced a portable helical-scan videotape recorder built to high standards, the Model MVR-10. It used 1-in (25.4-mm) tape running at a speed of $7\frac{1}{2}$ in/s (190.5 mm/sec). The machine weighed 90 lb (41 kg) and occupied only 2.3 ft³ (0.065 m³). It was a two-head unit that contained an 8-in (203-mm) monitor. It had specifications of ±2 dB from 10 kHz to 3 MHz with respect to 100 kHz and down no more than 6 dB at 3.5 MHz. It had a horizontal resolution of 250 lines and the SNR was 40 dB or better rms to p-p video. It was all transistor and ran 90 min on a 10.5-in (267-mm) reel.[39] Late in 1962 a more advanced model MVR-11 was introduced. The self-contained monitor had been eliminated and the machine now weighed some 65 lb (29.5 kg). It measured some 24 × 10 × 13 in (61 × 25 × 33 cm). Late in 1963 the MVR-15 was intro-

Toshiba one-head helical-scan videotape recorder. (Courtesy of Toshiba Research and Development Center)

duced. It was claimed that this machine had broadcast capability. It was wired for remote control, had two audio tracks and used standard 60-Hz current. [40]

Precision Instrument Co

In Apr. 1963 another portable television tape recorder, Model PI-3V from Precision Instrument Co. of Palo Alto, Calif., was introduced. This was a 68-lb (31-kg) machine using helical scan with two long-life, video heads and a 180° tape wrap. This machine was both electrically and mechanically similar to the MVR-11 of Machtronics. [41] In June 1965 a newer model, the PI-4V, was introduced which had variable playback speed control to display a picture at normal speed, any degree of slowness or with the machine stopped. [42] In Oct. 1965 Precision Instrument announced the model PI-7100 portable videotape recorder. It included Variscan speed and directional control from 0 to 16 in (406 mm)/s. It included an all-electronic video head signal coupling and demodulation system which removed all mechanical couplings from the signal path. It featured stacked coaxial reels and helical-scan closed-loop drive recording on 1-in tape. [43]

Sony PV-100 Videocorder

In July 1964 the Sony Corp. announced a new videotape recorder using a 1.5-head system. This consisted of a main head which scanned the entire video and a small portion of the vertical sync pulse while the other head scanned the entire vertical sync pulse and a small portion of the video. Thus it was possible to record and play back one field picture without any switching. The video heads were made of S-alloy with claimed hardness and life five times that of conventional heads with 15 dB greater efficiency. It was claimed that this machine recorded and played back a picture of almost 300-line resolution with over 40-dB SNR. [44]

Westel WRC-150

In Apr. 1966 the Westel Co. of San Mateo, Calif., announced the first portable television camera and backpack recorder, the Model WRC-150. [45] The video recorder held a 30-min supply of 1-in magnetic tape. The recorder used the Coniscan principle which directed the tape at 10 in/s around an inverted three-piece, flat-top cone. The whole center section, which housed the single recording head, rotated against the tape. This created an air cushion to reduce tape friction. Westel used special motors of the printed-circuit type to achieve high torque and low inertia. The system contained a full-scale internal sync generator to provide a clean and stable source for all timing functions of the camera and recorder. The back pack recorder was a record-only module

weighing some 23 lb (10.4 kg) complete with tape and rechargeable nickel-cadmium batteries, and the cable-linked vidicon camera head weighed rather less than 7 lb (3 kg). Reels were coaxially mounted with a $6\frac{1}{2}$-in (165-mm) reel giving some 14 min of recording and the 8-in (203-mm) reel some 33 min. The camera head included a novel small CRT which could be switched to operate as an "A" scope. In the interest of low weight and power consumption, rewind and fast forward functions were not included and there was only one audio channel. Westel claimed specifications of 10 Hz to 4.2 MHz, ± 2 dB bandwidth and an SNR of 42 dB peak to peak. A console model, the Model WTR-100 was a professional videotape recorder weighing some 75 lb (34 kg). Electronic editing as well as an additional color module, the WCM-200, provided final electronic tape base correction and allowed direct recovery of NTSC color signals. [46]

IVC Model 800

In Nov. 1967 the International Video Corp. of Mountain View, Calif., introduced a new portable color video recorder, the IVC Model 800. The machine was a full 5-MHz NTSC color recorder weighing some 52 lb (23.6 kg) and recorded 60 min of programming on 2,150 ft (655 m) of 1-in (25.4 mm) videotape on an 8-in (203-mm) NAB reel. The machine ran at a speed of 6.9 in/s (175.26 mm/s) and was a one-head machine using a 360° alpha wrap. The modulation was pulse interval modulation, not FM. [47] In Nov. 1968 IVC introduced two other models, the IVC 801 for recording and playing SECAM color, and the IVC-811 designed for PAL color standards. [48]

VIDEOTAPE RECORDING (LONGITUDINAL SCAN)

RCA

On Mar. 26 1956 RCA announced that they "may be in position by late 1957 or early 1958 to have designed for NBC a magnetic tape recording system for actual broadcast use." The RCA magnetic tape recorder used $\frac{1}{2}$-in (12.7-mm) tape running at a speed of 20 ft/s (6.1 m/s) using seven tracks. Three tracks were for color, one for mixed high video components, one for sync and two for audio. [49]

VERA

In Apr. 1958 the BBC unveiled their new magnetic video recorder called Vision Electronic Recording Apparatus (VERA). This was a rather large machine which used $\frac{1}{2}$-in (12.7 mm) magnetic tape running at a speed of 200 in/s (5.08 m/s). It used a $20\frac{1}{2}$-in (52-cm) reel for 15 min of operation. It

made use of three tracks, one for direct video and a second for low frequency (below 100 kHz) using FM modulation. The third track carried the sound which was FM-modulated onto a carrier. Separate recording and reproducing head stacks were used so that the recorded signal could be monitored. A separate cue head was incorporated for putting a 30-kHz tone on the tape for editing purposes. BBC research engineers had spent some five years developing the recorder and two prototypes were in use. [50]

Telcan

In June 1963 Telcan Ltd. gave the first demonstration of the Telcan videotape machine. It was claimed that this demonstration gave very fair quality on 17-in (43-cm) and 21-in (53-cm) home receivers. This machine used $\frac{1}{4}$-in (6.35-mm) tape running at a speed of 120 in/s (3.048 m/s). A double-head system was used giving a playing time of 30 min on a standard 11-in (28-cm) spool by reversal of the tape after 15 min. The video bandwidth recorded and reproduced was 2.2 MHz peak white on 405 lines and the SNR was 28 dB. Telcan was planned to be a low-cost recorder for home taping of television programs. [51]

Fairchild/Winston

In Apr. 1964 a new low cost television tape recorder was introduced by the Winston Research Corp., Division of Fairchild Camera and Instrument Corp. of Los Angeles. The Fairchild/Winston Television Tape Recorder was a four-track machine that recorded on $\frac{1}{4}$-in (6.35-mm) tape running past a fixed head at 120 in/s (3.048 m/s). All recorded signals were combined on a single track through multiplexing. Quality was claimed to be comparable to normal home viewing. [52]

Ampex Model VR-303

In Mar. 1965 Ampex announced a 95-lb (43 kg) television tape recorder, the Model VR-303. This was a fixed-head machine using $\frac{1}{4}$-in (6.35-mm) tape. It could record up to 50 min of program material on a $12\frac{1}{2}$-in (32-cm) reel. The tape moved past fixed recording and playback heads at a speed of 100 in/s (2.54 m/s). This unit was designed to be used by nontechnical personnel with minimal instruction. [53]

Newell Associates

In July 1967 Newell Associates of Sunnyvale, Calif., announced a unique high-speed videotape recorder. This machine with only three basic moving parts ran at tape speeds of up to 1,000 in/s (25.4 m/s) with bandwidths of over 10 MHz. A speed of 4,000 in/s (101.6 m/s) with a bandwidth of more than 50 MHz was said to be possible using the principle of the new transport. This was accompanied by using a single power source at the rims of the tape rolls rather than through the tape itself. As a result, the tape never existed as an open filament but only in a solid packed condition. Since no power was transmitted through the tape, and the tape existed nowhere in an unsupported conditon in the system, rapid acceleration and deceleration could be accomplished without tension effects on the tape.[54]

VIDEO DISC RECORDING

Philips Storage Device

In Nov. 1960 Philips Gloeilampenfabrieken, Eindhoven, The Netherlands, described a storage device for still pictures in which single frames could be displayed. This device consisted of a rotating magnetic wheel with a diameter of 40 cm (15.75 in). It had a rim 30 mm (1.18 in) wide on which the picture was stored. The picture was recorded and played back from a single video head positioned just above the rim. The wheel turned at 3,000 r/min and the signal was recorded by an FM modulation system. It was claimed that this system was first displayed at the 1958 Photokina Exhibition in Cologne, Germany, where it used a direct recording method. Here it was coupled to a TV camera in a booth. Philips was demonstrating flashbulbs and by means of this device were able to produce a still frame of the subject as if taken by a flashbulb. Another proposed use was for the display of x-ray photographs.[55]

Siemens & Halske

In the fall of 1961 Siemens & Halske AG, Karlsruhe/Munich, developed a magnetic single-frame storage device. A single field or a single frame was recorded on one or more tracks on a rotating storage foil. This information could be played back at once and continuously reproduced as a still image. The distance between the foil and base plate carrying the video heads was stabilized by an adjustable air current. Head distances of approximately 1 μm were attained and frequencies up to 10 MHz could be recorded. FM modulation was used in signal recording applying the double-heterodyne system. The recorded video band ranged up to 5 MHz. Additional electronic devices control the time of the recording, playback and erasure processes.[56]

MVR Videodisc Recorder

In July 1965 the MVR Corp. of Palo Alto, Calif., demonstrated a single-frame Videodisc recorder at a meeting of the SMPTE in San Francisco. This machine was capable of recording video on a magnetic disc up to 600 single frames, or 20 s continuous. Any frame

Magnetic wheel assembly. (Courtesy of NV Phillips Research Laboratories)

could be replayed or erased, and was capable of being replayed instantly.[57] In Aug. 1965 a television disc recorder, the Model VDR-210CF Videodisc built by the MVR Corp. to CBS specifications was used to record and play back action highlights of a football game on the CBS-Television Network. The 40-lb (18-kg) unit permitted recording of complete 20-s segments of the action which could then be replayed in regular motion or stopped to provide "freeze action" shots. It could be reset in $\frac{1}{5}$ s and incorporated appropriate video switching logic to provide smooth transitions between modes of operation and to insure transmission of a standard composite video and synchronizing signal.[58]

Sony Videomat and Video Color Demonstrator

In Mar. 1966 the Sony Corp. of America announced two machines using thin plastic discs to record and play back monochrome motion pictures and still pictures in color. The first machine was the Videomat which recorded monochrome motion pictures for immediate playback. This machine contained a camera, lights, discs and a 19-in (48-cm) television set. The disc had an aluminum rim with an "inside surface of material identified with videotape." The Video Color Demonstrator was reported to record up to 40 color still pictures on the disc for playback over a standard television monitor. The pictures could be repeated or erased from the disc if desired.[59]

MVR Slow-Motion Recorder

In Jan. 1967 the MVR Corp. of Palo Alto, Calif., announced the MVR Videodisc Slow Motion VDR 250 Recorder. This was an advanced model to be used for sports events, and it

featured a "Slow Motion Instant Replay." The slow-motion replay could be accomplished at slow, medium or fast speeds. The recorder also had the stop-action or "freeze" capability of the earlier model.[60]

Ampex HS:100

In Apr. 1967 Ampex announced a new high-band color disc recording system designed to provide color instant replays in slow- and stop-motion. The system, designated the HS-100, recorded on metal discs and could record and playback 30 s of action in high-band color. The system's capabilities included reverse action playback at either normal speed or slow-motion and frame-by-frame advance for animation or analysis of highlights. Any part of the 30-s recording could be re-cued within 4 s.[61]

Visual Electronics Disc Recorder, Model VM-90

In Apr. 1968 Visual Electronics Labs of Sunnyvale, Calif., described their new slow-motion video disc recorder. This was a joint effort of MVR and Visual Electronics which used a single aluminum disc which was first coated with nickel-cobalt and then a rhodium flash coating. The disc was addressed by two record/ playback heads, one on top of the disc and the other under it. They were driven by reversible stepping motors and recorded odd fields on top and even fields underneath. Recording and playback started at the inside edge of the disc. It would record 30 s of video using a stepper mode of operation. It could be played in either forward or reverse. In the slow- or stop-motion mode one field was used many times over. Reconstructing the signal, the system provided monochrome interlace or chroma dot interlace in color.[62]

VIDEO RECORDING (MISCELLANEOUS)

Thermoplastic Recording

In Dec. 1959 the General Electric Co. announced a new system for video recording called "thermoplastic recording."[63] This process employed a high-melting film base coated with a transparent conducting layer and then with a thin layer of low-melting thermoplastic. An electron beam in an evacuated chamber laid down a charge pattern on the surface of the thermoplastic layer in accordance with the information to be stored. The film was then heated to the melting point of the thermoplastic. Within milliseconds forces between the charges on the film and conducting layer depressed the surface where the charges occurred. The film was then cooled and the deformations were fixed on the surface. The recording could be reproduced by means of diffraction optics. It could be erased by heating the film above the melting point of the thermoplastic. This increased its conductivity, surface tension smoothed out the deformation and the film was ready for reuse.[64] In Aug. 1965 GE reported on current thermoplastic recorders, including a 16mm tape, an 8mm tape, and a continuous loop recorder. Recording in color and a wide-bandwidth 1029-line video recording were described.[65]

Electron Beam Recording

In Sept. 1961 the Eastman Kodak Co. made a brief announcement of a new process for direct electron-beam recording on film. This process differed from thermoplastic recording in that the electron beam exposed the silver halide emulsion directly. It was claimed that this method offered several advantages over normal kine-recording particularly in speed, photographic resolution and the absence of phosphor grain.[66] In May 1964 RCA announced the development of a feasibility model of an electron beam recorder. RCA claimed higher resolution capabilities due to the use of lower beam current than used in a conventional kine-recording.[67] In Nov. 1965 the 3M Co. described a laboratory model of an electron beam recorder. It used 16mm film and an intermittent shuttle mechanism phase-locked to the television signal together with a time deflection gate of the electron beam to convert from 30 frames/s for TV to 24 frames/s for film. Exposure control was obtained with a beam sampling servo system. A two-vacuum system providing its own seal minimized start-up time and prolonged the electron-gun filament life. It was claimed that this recorder made possible 16mm television recordings free from the fixed noise patterns of phosphor faceplates, a resolution in excess of 600 lines, and pictures

free from shutter bar.[68] In Nov. 1966 3M Co. announced a commercial electron beam recorder, the Model EBR100. This machine had a magazine capacity of 1200 ft (366 m) of 16mm film. The recorder weighed about 1000 lb (450 kg).[69] In Apr. 1970 the Mincom Division of 3M, Camarillo, Calif., introduced a new electron beam recorder for producing 16mm color film. This was the Chromabeam system of transferring NTSC color to color film. The Chromabeam recorder decoded the color video signal of 60 field/s into separate red, green and blue components. These color components were recorded on normal monochrome low-cost film at 60 frames/s in a specially determined sequence such as red, green, blue, red, green, etc. After processing, the monochrome separation film with the decoded colors was put into the Chromabeam printer. Here, it was printed on a standard 16mm color film. Using a rotating color filter, each frame of the color film was exposed in sequence to the red, green and blue separation images. It was claimed that this system provided a high-quality, low-noise picture with true color reproduction.[70]

Electron Beam Scanning

An interesting experiment was conducted at the Armour Research Foundation in May 1963 to use electron beam scanning to eliminate mechanical problems of magnetic recording and playback. The system used a high density sheet beam to energize a simplified 500-element magnetic head. For playback a tube was built with a line of fine high-permeability wires sealed into the envelope in the beam path. Results indicated that the electronic method was feasible, but needed refinement.[71]

Phonograph Disc Playback

In Sept. 1965 the Westinghouse Electric Co. of Pittsburgh, Pa., announced a new system called Phonovid for the recording and playing back of still pictures and sound on a television screen from an ordinary home tape machine or phonograph. The pictures could be line drawings, charts, printed text or photographs. The system displayed the tape pictures on a conventional home receiver or the tape pictures could be transferred to the grooves of a record and played back. Some 1,200 pictures and accompanying sound could be stored on a standard 7-in, 1,200-ft (18-cm, 366-m) reel of narrow slow-speed audiotape or some 400 still pictures and sound could be put on the two sides of a standard 12-in (30.5-cm) $33\frac{1}{3}$-r/min record called a Videodisc. The system consisted of a slow-scan camera (a special vidicon which could store an image for several minutes), a scan converter to convert the taped pictures to standard television signals,

an audiotape recorder and a standard TV receiver.[72]

Analog Photographic Playback

In Oct. 1968 an interesting television recording and playback system using photographic discs was described by members of the Stanford Research Institute. The video and sync were recorded as analog signals on a spiral track. A 12-in (30.5-cm) diameter disc could store a one-half hour television program. Pictorial information from a 16mm motion-picture film were recorded on a plate at 1/16 real time using high-pressure mercury arc sources and a Kerr-cell modulator. Playback was in real time and the results were encouraging.[73]

Laser Beam Recording

In Sept. 1970 CBS Laboratories announced a new laser beam color film recorder.[74] It was based on a special color combiner and beam deflector that enabled perfect registration of the recording beam. Three gas lasers provided the light source. The three laser beams were combined precisely on a single optical axis before deflection. The unit recorded at the standard rate of 24 frames/s and could record either 35mm or 16mm film. The prototype will accept any NTSC color signal.

HIGH SPEED CONTACT PRINTING

Panasonic Bifilar Tape Duplication

In Apr. 1969 the Panasonic Co. (Matsushita Electric Industrial Co. Ltd., Osaka, Japan) announced details of a new high-speed videotape contact printer. This was a "bifilar type duplication system" using a specially developed high-coercivity master dupe. This master tape had to be recorded on a special videotape recorder in order to produce a mirror-image master. This master and a slave tape were wound together very tightly because of a special pressure roller onto one reel, after which the transfer field was applied. Contact of the two surfaces was complete at the critical time of field application. It was claimed that even after more than 100 duplications, the duplicated slave tape remained at a constant relative output, without varying quality. It was possible to duplicate a 2,400-ft (730-m) reel of tape in less than 2 min.[75]

IBM High-Speed System

In Apr. 1969 International Business Machines Corp., Los Gatos, Calif., announced a method of contact printing videotape at high speeds. A mirror-image video master was recorded on high coercivity tape (700 oersteds) which was then held in contact with an unmagnetized tape of low coercivity (300 Oe) while both tapes were passed through a magnetic bias field. Since there must be absolutely no slippage between the master and copy tapes, a method of clamping the tapes by means of compressed air was developed. An experimental machine was developed which allowed high-density video information to be transferred at high speeds. Copies of uniformly good quality with an SNR of 42 dB had been made constantly at speeds up to 150 in/s (381 cm/s) which made it possible to copy a one-hour videotape program in 3 minutes. Experiments were made on 1-in (25.4-mm) tape recorded on a helical-scan machine running at 7.5 in/s (190.5 mm/s). The master tape declined slightly during the first 10 transfer cycles. After the tenth transfer, the signal output of the master tape had dropped off to a maximum of 2 to 3 dB below its original output. Experiments proved that even when copied some 2,000 times the replay performance was still excellent.[76]

Ampex Model ADR-150

In Apr. 1970 the Ampex displayed a model of a new high-speed contact videotape printer. All broadcast formats, color or monochrome, high- or low-band, foreign or domestic could be duplicated. The machine was designed for modular expandibility to accomodate up to five slave reel systems, permitting duplication of from one to five copies in a single operation. The machine used a dynamic transfer system whereby the duplication process occurred in a magnetic transfer chamber located in the center of the tape paths. After the magnetic transfer process the audio, cue and control tracks were played back and rerecorded on the slave copy. At tape transfer speeds of 150 in/s (381 cm/s), a rewind speed of 300 in/s (762 cm/s) and with allowance for threading, loading and unloading, approximately 25–30 copies of a 1-hr program could be made in an hour.[77]

Memorex Thermal Contact Method

In May 1970 the Memorex Corp. of Santa Clara, Calif., announced a thermal contact method for duplicating videotapes. Copy tapes were made on chromium dioxide tape while the master could be of any type. The thermal transfer process was dependent on the unique Curie point properties of CrO_2. Above the Curie temperature CrO_2 becomes paramagnetic and below the Curie point (down to about 105°C) chromium dioxide tape is extremely susceptible to external magnetic fields. Therefore, when a pre-recorded conventional iron oxide tape (master) was placed in contact (coating-to-coating) with a CrO_2 tape (slave) in a thermal environment which passed very rapidly down through the Curie temperature of the CrO_2 tape the magnetic field from the master tape coerced the CrO_2 tape

into magnetization under ideal magnetic conditions. As the chromium dioxide duplicate returned to room temperature, it regained its full magnetic properties and could actually have a higher output than the original iron oxide master.[78]

Du Pont Thermoremanent Process

In Oct. 1970 the Experimental Station, Photo Products Dept. of the Du Pont Co., described their "thermoremanent" process for the duplication of magnetic tape recording. This was a thermal process in which the unique Curie properties of chromium dioxide tape were used to make possible high-speed copies of magnetic tape programs. In the process the copy cools through the Curie point in the magnetic field of the Master. The coating of the Master tape remains 20° or more below its Curie point. It was reported that copies of $\frac{1}{2}$-in (12.7 mm) color video recording had been made with satisfactory SNRs and good audio and control track transfers.[79]

THE CASSETTE/CARTRIDGE REVOLUTION

Electronic Video Recording (EVR)

In Oct. 1967 the Columbia Broadcasting System introduced a new system called EVR.[80] Essentially this was a device to play back on conventional television receivers prerecorded programing from motion-picture film and television tape. The program, contained in a 7-in (18-cm) cartridge, was simply inserted in the automatic player. An unusual engineering aspect of EVR was that color programing was to be recorded in monochrome and yet reproduced in color on a color set, so that the cartridges were capable of carrying up to 1 hr in monochrome or a half-hour in color. In Dec. 1968 the first public demonstration of EVR was given and the results were excellent.[81] The system's ability to play single-frame as well as stop-motion, reverse and fast-forward was demonstrated. On 24 Mar. 1970 the first public demonstration of Color EVR was given. The color reception was good, with resolution comparable to 35mm film on television. The old monochrome version was obsolete and superseded by a new color-compatible version.[82] Details revealed that a special high-resolution silver halide film provided an essential interface between the electron beam recording, the printing and the playback processes. The electron beam recorder consisted of three sections all under a vacuum: the film magazine, the transport and the electron gun. The process started with a standard NTSC color signal being translated into an EVR format, processed and sent to the electron beam recorder. The signal was recorded on both sides of the 8.75-mm (0.344-in) film, monochrome on one

half and encoded color information on the other. Each EVR frame which was 1/60 s long held all the information as the regular 1/30-s television frame. A synchronization signal in the form of a small clear window in the middle of the frame was added to the luminance signal. This provided synchronization in the player between the film transport and the CRT. In the player, the imaging optics consisted of two lenses, two rhomboidal prisms, a lens mount and a film gate which held the film in a cylindrically curved image plane. The playback system used a special flying spot scanner which was scanned through the dual optical system. Light passing through the film was sent to photomultipliers which generated luminance and EVR color signals. The color signals were converted to NTSC form; sound and picture were put onto an RF carrier and this modulated carrier was transmitted to TV antennas for playing.[83]

Arvin Industries, Model CVRXXI

In Feb. 1968 Arvin Industries Inc., Electronic System Div., introduced a prototype color videotape recorder, the Model CVRXXI. It was a fixed-head machine using the high-speed indexing transport developed by Newell Associates. It could record in both color and monochrome. The machine used 4,800 ft (1,220 m) of $\frac{1}{2}$-in (12.7-mm) tape compressed in a self-threading cartridge. The tape ran at a speed of 160 in/s (4.064 m/s) so that each track ran for 6 min. As each track approached the end of the reel, the transport mechanism stopped and reversed direction in less than 1 s. It was claimed that its frequency response was flat from dc to 2 MHz. Color response was from dc to 500 kHz. Audio response was from dc to 20 kHz. Horizontal response was some 200 lines.[84]

Sony Color Cassette Recorder

In Apr. 1969 the Sony Corp. of America announced their first color cassette videotape recorder. The cassette used 1-in (25.4-mm) tape running at $3\frac{1}{4}$ in/s (82.55 mm/s) with two tracks. It measured about 6 × 10 × 3$\frac{3}{4}$ in (152 × 254 × 95 mm) and the two reels were mounted coaxially. The tape came off one reel, passed around a hub that was slightly larger than the head drum and wound onto the second reel. When the cassette was slipped into the machine the hub slipped over the drum; locking the cassette in place rotated the hub so that a window in it let the tape contact the drum. The recording of the color was a modification of the NTSC signal. The luminance signal was recorded by FM modulation of a subcarrier. Its frequency was about 3–4.5 MHz which left room for the NTSC color signal which was shifted to a center frequency of 900 kHz and recorded with a bandwidth of ± 700 kHz around the center frequency. The color signal was both amplitude and phase modulated like a regular NTSC color signal so it needed only to be shifted back into its rightful space in the signal spectrum for playback.[85]

JVC Video Cartridge Recorder

In June 1969 the Victor Company of Japan, in Yokohama, announced plans to market a video cartridge recorder. The recorder was to use $\frac{1}{2}$-in (12.7-mm) tape running at 7.5 in/s (190.5 mm/s) for a maximum playing time of 30 min. It was to be enclosed in a cartridge measuring $5.51 \times 5.51 \times 0.91$ in (140 $\times$ 140 $\times$ 23 mm). The tape to be recorded was inserted into a slot and the mechanism inside automatically threaded it. It was to use the direct FM combined system which would permit it to record the 4.5-MHz NTSC color signal at slow tape speed. Audio was to be on dual stereo tracks.[86]

RCA SelectaVision

On 30 Sept. 1969 RCA announced a new process called SelectaVision. This was a color television tape player built around lasers and holography and designed for home use. SelectaVision tapes were made from motion pictures, videotapes, slides or photographs. The program was recorded on conventional film by means of an electron-beam recorder. Through a complicated series of technological steps, a hologram master was produced. This master was fed through a set of pressure rollers with a transparent vinyl tape which impressed the holographic relief on its surface and was ready for home use. To play back, the clear tape, packed in a cartridge for easy handling was run smoothly through the light from a 2-mW helium-neon laser. The images produced were recovered by an inexpensive vidicon camera. The playback mechanism, the laser, and the TV camera were housed in the SelectaVision player which was connected to the TV antenna terminals of the home receiver.[87] A demonstration revealed a good monochrome image considering what the new technique involved although there was a slight speckling and moiré effect. The color flickered a bit, faded out every few seconds and did not look natural. There was no soundtrack, but it was announced that these problems would soon be overcome.[88]

Panasonic Magazine-Type Videotape Recorder

In Nov. 1969 Panasonic introduced a magazine-type videotape recorder. This was a prototype measuring $13.9 \times 14 \times 5.1$ in (353 $\times$ 356 $\times$ 130 mm). It used $\frac{1}{2}$-in (12.7-mm) tape running at 7.5 in/s (190.5 mm/s) enclosed in a "Philips" type cassette measuring $10.6 \times 6.4 \times$ 0.9 in (269 $\times$ 163 $\times$ 23 mm). Two developments by Panasonic allowed them to record NTSC color direct. Improved hot-press ferrite magnetic recording heads made it possible to record frequencies up to 10 MHz on $\frac{1}{2}$-in tape running at 7.5 in/s. Also an improved double heterodyne automatic phase control system removed jitter from the playback signal.[89]

Sony Color Videoplayer

In Nov. 1969 the Sony Corp. of America announced a new home color cassette recorder, different from the one introduced in Apr. 1969. The Sony Color Videoplayer used $\frac{3}{4}$-in (19.05-mm) tape running at 8 cm/s (3.15 in/s) enclosed in a cassette measuring $8 \times 5 \times 1\frac{1}{4}$ in (203 $\times$ 127 $\times$ 32 mm). It could record or playback up to 90 min. The recorded signal was processed NTSC, and 250-line color resolution was claimed. In addition to the video there were two audiotracks for either stereo recording or dual languages.[90] A still newer version of this machine was introduced in Mar. 1970.[91]

Philips Video Cassette Recorder (VCR)

In June 1970 details of the new Philips Video Cassette Recorder (VCR) were announced. It was a deck which could record and playback in both monochrome and color. It measured $22 \times 13 \times 6\frac{1}{2}$ in (56 $\times$ 33 $\times$ 16$\frac{1}{2}$ cm). The cassettes measured $5.8 \times 5.0 \times 1.4$ in (147 $\times$ 127 $\times$ 36 mm) and could play up to an hour. Each cassette had two coaxially mounted reels of chromium dioxide tape which ran at 5.6 in/s (142.2 mm/s). The system incorporated sound dubbing, independent recording of the audiotracks and "stop motion." A variety of models from monochrome players only to complete color recorders complete with tuners were to be made available.[92]

Teldec

In June 1970 the Teldec television disc recorder was first demonstrated in Berlin. A joint venture of AEG-Telefunken and English Decca, the demonstration of monochrome television was considered of excellent quality.[93] The disc was a sheet of PVC only 1 mm (0.03937 in) thick that revolved at 1,800 r/min (for American standards). The spacing between adjacent grooves was about 280 μin (0.0071 mm) with a density of about 3,500 grooves/in (140 grooves/mm). The pickup device was a pressure sensitive piezo-ceramic unit which was moved across the groove by a positive mechanical drive. The groove modulation was hill-and-dale and the sound was carried by pulse. position modulation which recorded the audio during the horizontal blanking portion of the video signal. The video bandwidth was 3 MHz for monochrome with an SNR of 40 dB. The playing time was 5 min for a 9-in (23 cm) disc and 12 min for a 12-in (30-cm) disc. The original recording made use of an FM system. The player could be either single-play or automatic.[94]

Avco CTV Cartrivision

In July 1970 the Cartridge Television Inc. division of Avco Corp., New York, announced the Avco CTV Cartrivision system. The principal component of the Cartrivision system was a solid-state combination receiver-recorder-playback device built into a single self-contained unit. The system was designed to play prerecorded cartridges in color and to record television programs off the air with instant playback. A special camera was to be available in order to make "home movies."[95]

Instavision

In Sept. 1970 Ampex announced their cartridge system called "Instavision." The system (described as the "smallest cartridge loading video recorder and/or player to date") was manufactured by Toamco, the company's joint venture with Toshiba in Japan. The recorder/player used $\frac{1}{2}$-in (12.7-mm) videotape enclosed in a circular plastic cartridge 4.6 in (117 mm) in diameter and 0.7 in. (18 mm) thick. The basic instrument weighed less than 16 lb (7$\frac{1}{4}$ kg) including batteries and measured $11 \times 13 \times 4.5$ in (279 $\times$ 330 $\times$ 114 mm). It permitted slow- and stop-motion recording and elementary editing. Two independent audio channels permitted flexibility in audio recording, including stereo playback. Video resolution was 300 lines in monochrome. Color resolution was compatible with standard color television receivers. The SNR was 42 dB. A lightweight monochrome camera weighing some 5 lb including a zoom lens and an electronic viewfinder were to be available.[96]

ELECTRONIC MOTION PICTURES

Harlow

After the failure of High Definition Films, Ltd., in London to produce motion pictures using electronic cameras in 1955, many advances were made in television film recorders and associated equipment. Many of the variables in the film recording process were either minimized or eliminated. Such items as exposure control, shutter-bar, equalization of the CRT, etc., were no longer a problem. Literally thousands of television film recordings (in 16mm) were made for industrial, commercial and military use. By 1965 the first true electronic motion picture was made. This was *Harlow* made in Apr. 1965 by Electronovision. *Harlow* cost some $1.5 million and was shot in 5 days.[97] Electronovision had previously recorded two other black-and-white shows, *Hamlet* and *The Tami Show* using TK60 RCA 4½-in (114.3-mm) image orthicon cameras and RCA TFR-1 kine-recorders.[98] But *Harlow* was the *first* effort to actually stage the production for electronic cameras using television techniques. The finished picture was released on 35mm motion-picture film and to all intents and purposes was a regular motion picture. Aesthetically a poor picture, it was nevertheless a milestone in motion-picture production because it was the first electronic motion picture ever made for theater release.[99]

Technivision

In Nov. 1969 the Technicolor Corp. announced plans for a new method of using videotape for making motion-picture films. This new method was to be called Technivision, and was to be a method for shooting theatrical film which would revolutionize the economics, production and post-production techniques of motion pictures. It was to use 2,000 scan lines which would provide fine enough definition for transfer of videotape to 35mm or 70mm film for large-screen projection. It was claimed that direct shooting on tape had advantages of speed, post-production flexibility, elimination of laboratory work and instant viewing. A three-color laser was to be used to convert the tape signal to film for theater projection.[100]

CONCLUSION

This survey has covered the history of television recording from 1955 to 1970. Many tremendous advances have been made. From a relatively simple process of film recording, we now have an exotic array of sophisticated recording devices encompassing the range from magnetic discs to color laser beam recording. The magnetic recording of picture and sound is as common as home movies. The promise of television recording as a means of reviving Hollywood feature film production seems imminent. The grammophone disc, which was the first video recording medium, is now back in the form of an inexpensive disc which can play back high-quality programs.

The most important advance was the practical method of magnetic recording introduced by Ampex in 1956. This has grown from a relatively simple (but ingenious) machine to a complicated production tool which can not only record but turn out intricate programming with ease and economy of operation. There is literally no effect, optical or otherwise, that can not be accomplished by electronic means. As yet, there has been limited progress made to produce major features by utilizing the new techniques of television recording and its

unique economical editing and post-production techniques.

For if Hollywood is to survive, she must take all the tried and true production methods learned over the past 60 years and discard them. She must completely break with the past: no half-hearted methods such as using television cameras only for viewfinders, no old-fashioned practices such as viewing rushes, editing and cutting of film, etc. She must go the whole route to electronic production, from original shooting, to electronic editing, opticals, special effects, audio dubbing and re-recording — so that the finished product is ready for a transfer to a high-quality color film for theater release. In an industry where time is literally money, these methods can certainly bring back most of the production that has gone elsewhere due to high costs.

REFERENCES

1. Albert Abramson, "A short history of television recording [Part I]," *Jour. SMPTE*, *64:* 72–76, Feb. 1955.
2. Charles W. Handley, "Progress Committee Report," *Jour. SMPTE*, *58:* 408, May 1952.
3. Charles R. Daily, "Progress Committee Report," *Jour. SMPTE*, *64:* 240, May 1955.
4. Lloyd Thompson, "Progress Committee Report," *Jour. SMPTE*, *65:* 258, May 1956.
5. C. H. Evans and R. B. Smith, "Color kinescope recording on embossed film," *Jour. SMPTE*, *65:* 365–372, July 1956;
   J. M. Brumbaugh, E. D. Goodale and R. D. Kell, "Color TV recording on black-and-white lenticular film," *IRE Trans. on Broadcast and Television Receivers*, BTR-3: No. 2, 71–75, Oct. 1957.
6. Lloyd Thompson, "Progress Committee Report for 1957," *Jour. SMPTE*, *67:* 304, May 1958.
7. Vernon J. Duke, "A status report on current experimentation in color kinescope recording," *Jour. SMPTE*, *72:* 711, Sept. 1963.
8. Ref. No. 6, p. 304.
9. "The Vidtronics color tape-to-film transfer system," *American Cinemat.*, 262–263, 294–296, Apr. 1967.
10. R. E. Putman, "Progress Committee Report for 1966," *Jour. SMTPE*, *76:* 445–446, May 1967.
11. "Electronic process: color tape to film transfer," *Broadcast Management/Engineering*, *3:* 6, July 1967.
12. "Ampex TV in experimental stage," *Broadcasting*, p. 109, Feb. 13, 1956.
13. "Revolutionary new TV tape process unveiled by Ampex," *Broadcasting*, pp. 72–74, Apr. 16, 1956.
14. The team was made up of the following men, Charles Anderson, Ray Dolby, Fred Pfost, Alex Maxey and Shelby Henderson. Walter Selsted and Myron Stolaroff made invaluable contributions.
15. M. S. E. Downing, "The world of helical scan," *BKS&T Jour.*, *52:* 344–354, Nov. 1970.
   Personal interview with Charles P. Ginsburg, Redwood City, Calif.,
16. Charles P. Ginsburg, "A new magnetic video recording system," New Products, *Jour. SMPTE*, *65:* 302–304, May 1956.
   Charles P. Ginsburg, "Comprehensive description of the Ampex video tape recorder," *Jour. SMPTE*, *66:* 177–182, April 1957.
   Charles E. Anderson, "The modulation system of the Ampex video tape recorder," *Jour. SMPTE*, *66:* 182–184, Apr. 1957.

Ray M. Dolby, "Rotary-head switching in the Ampex video tape recorder," *Jour. SMPTE*, *66:* 184–188, Apr. 1957.
Charles E. Anderson, "Signal translation through the Ampex Videotape Recorder," *Jour. SMPTE*, *67:* 721–725, Nov. 1958.
Ray M. Dolby, "The video processing amplifier in the Ampex Videotape Recorder," *Jour. SMPTE*, *67:* 726–279, Nov. 1958.
17. New Products, *Jour. SMPTE*, *66:* 718, Nov. 1957.
18. "RCA shows color VTR model," *Broadcasting*, p. 88, Oct. 28, 1957.
19. A. H. Lind, "Color processing in RCA video tape recorder," *Broadcast News*, *99:* 6–7, Feb. 1958.
20. Charles E. Anderson and Joseph Roizen, "A color videotape recorder," *Jour. SMPTE*, *68:* 667–671, Oct. 1959.
21. New Products, *Jour. SMPTE*, *69:* 295, Apr. 1960.
22. Aubrey Harris, "Time-base errors and their correction in magnetic television recorders," *Jour. SMPTE*, *70:* 489–494, July 1961.
23. K. B. Benson, "Video-tape recording interchangeability requirements," *Jour. SMPTE*, *69:* 861–867, Dec. 1960.
24. New Products, *Jour. SMPTE*, *70:* 575, July 1961.
   J. Roizen, "The use of videotape for colour television recording," Proceedings of the Convention on Television and Film Techniques, Sponsored by the Brit. Kinematograph Sound and Television Society, pp. 11–19, 20 and 21, Apr. 1961.
25. J. Roizen, "High-band video tape recording," *Broadcast Management/Engineering*, *1:* 57–59, Mar. 1965.
26. Kurt R. Machein, "The Ampex Videotape Recorder and its performance on foreign TV standards," *Jour. SMPTE*, *68:* 652–656, Sept. 1959.
27. Lloyd Thompson, "Progress committee report for 1958," *Jour. SMPTE*, *68:* 307, May 1959.
28. H. J. v. Braunmuhl, "The Ampex Videotape recording system, its performance and utilization," *E.B.U. Review*, Part A—Technical No. 51, pp. 9–13, Apr. 1959.
   Charles Akrich, "Improving the picture quality of television tape recordings," *E.B.U. Review*, Part A—Technical No. 76, pp. 266–271, Dec. 1962.
   Robert M. Morris, "Progress in video-tape standards, a committee report," *Jour. SMPTE*, *72:* 488–490, June 1963.
   C. Coleman and E. Jarl-Hansen, "A comparison of two high-band standards for television tape-recording," (In German). *Intern. Elektron. Rund.* No. 1, pp. 23–24, and 29–31, 1964.
   H. D. Felix, C. H. Coleman and P. W. Jensen, "The theory and design of FM systems for use in colour television tape recorders," Int. Conf. on Magnetic Recording, London 1964.
29. The American standard was different. It was designed to operate on the fourth-order shelf with a 7.9-MHz carrier frequency; sync tip was 7.05 MHz and peak white was 10 MHz. Pre-emphasis boost of subcarrier was 8 dB and the −3 dB point was 4.5 MHz. Signal-to-noise ratio, unweighted, was 46 dB.
   "Colour TV recorders," *Int. TV Tech. Review*, *5:* 327, Sept. 1964.
30. Ampex Readout, Dec. 1967.
31. New Products, *Jour. SMPTE*, *75:* 442, Apr. 1966.
32. C. H. Coleman, "Techniques for multiple generation colour video tapes—today and tomorrow," *Jour. Royal Television Society*, *11:* 184–189, Winter 1966/7.
33. New Products, *Jour. SMPTE*, *76:* 726, July 1967.
34. New Products, *Jour. SMPTE*, *79:* 564, June 1970.
35. "New concept in VTRs," *Broadcast Management/Engineering*, *6:* 26, May 1970.

36. N. Sawazaki, M. Yagi, M. Iwasaki, G. Inada and T. Tamaoki, "A new video-tape recording system," *Jour. SMPTE*, *69*: 868–871, Dec. 1960.

37. "Two head color VTR," *Japan Electronics*, *2*: 21, Jan. 1961.

38. "Advantages of the two head videotape system," *Int. TV Tech. Review*, *2*: 19–21, July 1961.

39. "Portable video tape recorder," *Electronic Products Magazine*, p. 43, June 1962.
"Best designs of 1962," *Product Engineering*, *34*: 84, May 13, 1963.

40. "Portable TV tape recorder," *Broadcast Management/Engineering*, *1*: 38, Feb. 1965.

41. New Products, *Jour. SMPTE*, *72*: 354, Apr. 1963.

42. New Products, *Jour. SMPTE*, *74*: 571–572, June 1965.

43. New Products, *Jour. SMPTE*, *74*: 986, Oct. 1965.

44. "Here is a videotape breakthrough," *Int. TV Tech. Review*, *5*: 259–260, July 1964.

45. New Products, *Jour. SMPTE*, *75*: 439, Apr. 1966.

46. Chauncey Jerome, "Coniscan: portable camera and video recorder," *Int. Broadcast Eng.*, No. 21, pp. 246–250, June 1966.

47. E. R. P. Leman and D. F. Eldridge, "The IVC one-inch helical scan VTR format," *Broadcast Management/Engineering*, *4*: 56–57, Mar. 1968.

48. New Products, *Jour. SMPTE*, *77*: 1273, Nov. 1968.

49. "Spectacular gains in TV reported at IRE meeting," *Broadcasting*, p. 96, March 26, 1956.

50. "VERA" the B.B.C. vision electronic recording apparatus," *Jour. Television Society*, *8*: 399–400, Apr.-June 1958.

51. E. P. L. Fisher, "Telcan," *Int. TV Tech. Review*, *4*: 238–239, July 1963.

52. New Products, *Jour. SMPTE*, *73*: 363, Apr. 1964.

53. New Products, *Jour. SMPTE*, *74*: 294, Mar. 1965.

54. New Products, *Jour. SMPTE*, *76*: 726, July 1967.

55. J. H. Wessels, "A magnetic wheel store for recording television signals," *Philips Tech. Review*, *22*: 1–10, 1960/61.

56. John M. Calhoun, "Progress committee report for 1961," *Jour. SMPTE*, *71*: 351, May 1962.

57. Section Meetings, *Jour. SMPTE*, *74*: 978, Oct. 1965.

58. Education/Industry News, *Jour. SMPTE*, *74*: 954, Oct. 1965.
Adrian B. Ettlinger and Price E. Fish, "A stop-action magnetic video disc recorder," *Jour. SMPTE*, *75*: 1086–1088, Nov. 1966.

59. New Products, *Jour. SMPTE*, *75*: 314, Mar. 1966.

60. New Products, *Jour. SMPTE*, *76*: 75, Jan. 1967.

61. New Products, *Jour. SMPTE*, *76*: 626, June 1967.

62. Clarence Boice, "New video disc slows and stops the action," *Broadcast Management/Engineering*, *4*: 48–49, June 1968.

63. "TPR recording," *Electronic Industries*, *19*: 76–79, Feb. 1960.

64. W. E. Glenn, "Thermoplastic recording," *Jour. SMPTE*, *69*: 577–580, Sept. 1960.

65. W. E. Glenn, "Thermoplastic recording: a progress report," *Jour. SMPTE*, *74*: 663–665, Aug. 1965.

66. John M. Calhoun, "Progress committee report for 1961," *Jour. SMPTE*, *71*: 334, May 1962.
A. A. Tarnowski and C. H. Evans, "Photographic data recording by direct exposure with electrons," *Jour. SMPTE*, *71*: 765–768, Oct. 1962.

67. W. J. Poch, "The development of a feasibility model of an electron beam film recorder," *Jour. SMPTE*, *73*: 778–782, Sept. 1964.

68. Edward W. Reed, Jr., "An electron-beam recorder," *Jour. SMPTE*, *75*: 195–197, Mar. 1966.

69. New Products, *Jour. SMPTE*, *75*: 1144–1146, Nov. 1966.

70. "New generation broadcast gear," *Broadcast Management/Engineering*, *6*: 26, May 1970.
Technical information from the 3M Information Manual "Chromabeam."

71. Marvin Camras, "Experiments with electron scanning for magnetic recording and playback of video," *IEEE Trans. on Broadcast & TV Receivers*, *BTR-9*: No. 1, 63–67, May 1963.

72. New Products, *Jour. SMPTE*, *74*: 880, Sept. 1965.

73. P. Rice, A. Macovski, E. D. Jones, H. Frohbach, R. Wayne Crews and A. W. Noon, "An experimental television recording and playback system using photographic discs," *Jour. SMPTE*, *79*: 997–1002, Nov. 1970.

74. "Color TV recorder takes forward step back to film," *Electronics*, *43*: 38–39, Sept. 28, 1970.

75. H. Sugaya, F. Kobayashi and M. Ono, "Magnetic tape duplication by contact printing at short wavelengths," *IEEE Trans. on Magnetics*, *MAG-5*: No. 3, 437–441, Sept. 1969.

76. R. J. van den Berg, "The design of a machine for high-speed duplication of video records," *Jour. SMPTE*, *78*: 709–711, Sept. 1969.

77. New Products, *Jour. SMPTE*, *79*: 562, June 1970.
D. D. Esterly, "Contact duplication of transverse videotape recordings," *Jour. SMPTE*, *79*: 903–907, Oct. 1970.

78. "Memorex develops new thermal process," *Broadcasting*, *78*: 56, May 4, 1970.
W. B. Hendershot III, "Thermal contact duplication of videotape," *Jour. SMPTE*, *80*: 175–176, Mar. 1971.

79. J. E. Dickens and L. K. Jordan, "Thermoremanent duplication of magnetic tape recordings," *Jour. SMPTE*, *80*: 177–178, Mar. 1971.

80. "New video process," *Broadcast Management/Engineering*, *3*: 8, Oct. 1967.
New Products, *Jour. SMPTE*, *76*: 1165–1166, Nov. 1967.

81. "EVR—newest visual medium," *Broadcast Management/Engineering*, *5*: 41, Jan. 1969.

82. "Color EVR debuts," *Broadcast Management/Engineering*, *6*: 8, May 1970.

83. P. C. Goldmark, "Taking a look at color EVR from the inside out," *Electronics*, *43*: 94–101, Apr. 27, 1970.
P. C. Goldmark, et al., "Colour electronic video recording," *Jour. SMPTE*, *79*: 677–686, Aug. 1970.

84. "Home color," *Electronics*, *41*: 47–48, Feb. 19, 1968.

85. "Color by cassette," *Electronics*, *42*: 239, May 12, 1969.

86. "Cartridge-type VTR to be marketed by Victor of Japan," *Electronics*, *42*: 213, June 23, 1969.

87. "RCA color-TV tape player by 1972," *Broadcasting*, *77*: 57–59, Oct. 6, 1969.
O. Doyce, "Pressing pictures on holographic tape is fast, inexpensive," *Electronics*, *42*: 109–114, Nov. 10, 1969.

88. "Selectavision, willing, able, not yet ready," *Electronics*, *42*: 43–44, Oct. 13, 1969.

89. "Toward compatible VTR's," *Electronics*, *42*: 236–237, Nov. 10, 1969.

90. "And another VTR from Sony," *Electronics*, *42*: 236, Nov. 10, 1969.

91. "Videotape player updated," *Broadcast Management/Engineering*, *6*: 8, May 1970.

92. *Tape Recording Magazine*, *14*: 189, June 1970.
"A European standard," *Radio Electronics*, *41*: 4, Dec. 1970.

93. Industrial News, *Jour. Royal TV Society*, *13*: 125, Sept./Oct. 1970.

94. Aubrey Harris, "The Teldec television disc," *Electronics World*, *85*: 36–37, Feb. 1971.

95. New Products, *Jour. SMPTE*, *79*: 894, Sept. 1970.

96. "Portable VTR reproduces color," *Electronics*, *43*: 155–156–158, Sept. 14, 1970. The name was later changed to "Instavideo."

97. "Using 4 TV cameras, Electronovision hopes to shoot Harlow in 5 days," *Daily Variety*, p. 3, Mar. 30, 1965.

98. "A new system for theater TV," *Education/Industry News*, *Jour. SMPTE*, *73*: 900, Oct. 1964.

99. Albert Abramson, "Picture quality: film vs. television," *Jour. SMPTE*, *77*: 614, June 1968.

100. "Technicolor devising vidtape system for feature film use," *Daily Variety*, p. 1, Nov. 20, 1969.

Edit. Note: No reviewers for the Board of Editors have been available to consider in detail this comprehensive paper; therefore readers are particularly urged by the Author and the Editor to send advice promptly about any inadvertent oversights or errors.

PART IV: COLOR TELEVISION

TELEVISION IN COLORS

Herbert E. Ives

Television in Colors

By HERBERT E. IVES
Research Department

OVER two years ago Bell Telephone Laboratories demonstrated a practical system of television. For the first time successful representations of objects at rest or in motion were transmitted electrically — over wires or through the ether — for considerable distances. The reproduction of the scene then transmitted was in monochrome — the orange-red color of the neon lamp. Recent developments of the Laboratories, however, have made it possible to reproduce scenes with their true color values. The appearance of reality in the reproduced scene is thus greatly enhanced.

One of the most significant features of this new achievement is that it does not require completely new apparatus. The same light sources, driving motors, scanning discs, synchronizing systems, and the same type of circuit and method of amplification are used as in the monochromatic system. The only new features are the type and arrangements of the photo-electric cells at the sending end, and the type and arrangements of the neon and argon lamps at the receiving end. The out-

Side view of sending apparatus with doors of cabinets opened. With the exception of the photoelectric cabinet at the left, the apparatus is identical with that used for the original demonstration of monochromatic television

standing contributions that have made the present achievement possible are a new photo-electric cell, new gas cells for reproducing the image, and the ing landscapes in order to make the blue sky appear properly dark — this defect is corrected and the images assume their correct values of light and

One of the colored gelatin filters partly pulled back to reveal the color-sensitive photoelectric cell in the double container

equipment which is associated directly with them.

To render the correct tone of colored objects, it was necessary to obtain photo-electric cells which — like the modern orthochromatic or panchromatic plate — would be sensitive throughout the visible spectrum. This requirement has been satisfactorily met. Through the work of A. R. Olpin and G. R. Stilwell a new kind of photoelectric cell has been developed, which uses sodium in place of potassium. Its active surface is sensitized by a complicated process using sulphur vapor and oxygen instead of by a glow discharge of hydrogen as with the former type of cell.

The response of the new cell to color, instead of stopping in the blue-green region, continues all the way to the deep red. Because the former potassium cells were responsive only to the blue end of the spectrum, objects of a yellowish color appeared darker than they should have and the tone of the reproduced scene was not quite correct. This disadvantage applied particularly to persons of dark or tanned complexion. When the new cells are used in the original television apparatus and with yellow filters — similar to those used in photograph-

shade no matter what the color of the object or the complexion of the sitter. It is the availability of the new photo-electric cells which makes color television possible by their use.

The development of color television has been greatly simplified by the fact that as far as the eye is concerned any color may be represented by the proper mixture of just three fundamental colors — red, green, and blue. This fact was utilized in the development of color photography, and all the research that had been done in that field was available as background for color television. A host of methods of combining the three basic colors to form the reproduced image was available but, insofar as the sending or scanning end is concerned, a method was developed which has no counterpart in color photography. The method of "beam scanning" — used in the first television demonstration* — has been employed.

To apply this method to color television, three sets of photo-electric cells are employed in place of the one set used before. Each of these sets is provided with color filters made up of sheets of colored gelatine. One set

* See the RECORD, June, 1928, page 325.

has filters of an orange-red color which make the cells see things as the hypothetical red sensitive nerves of the retina see them; another set has yellow-green filters to give the green signal, and the third set has greenish-blue filters which perform a corresponding function for the blue constituent of vision. The scanning disc and the light source are the same as with the beam scanning arrangement used in monochromatic television. The only difference is in the photo-electric cells, and thanks to the tri-chromatic nature of color vision, it is only necessary to have three times the number of cells used previously to reproduce all colors. Three series of television signals, one for each set of cells, are generated instead of one and three channels are used for the transmission of the television signals.

The photo-electric cell container, or "cage," has been built in a somewhat different form from that used in our first demonstration. There three cells were used arranged in an inverted "U" in a plane in front of the object. In the new photo-cell cage twenty-four cells are employed, two with "blue" filters, eight with "green" filters, and fourteen with "red" filters. These numbers are so chosen with respect to the relative sensitiveness of the cells to different colors that the photo-electric signals are of about equal value for the three colors. The cells are placed in three banks, one bank in front of and above the position of the scanned object, one bank diagonally to the right, and another bank diagonally to the left, so that the cells receive light from both sides of the object and above. In placing the cells they are so distributed by color as to give no predominance in any direction to any color. In addition large sheets

of rough pressed glass are set up some distance in front of the cell containers so that the light reflected from the object to the cells is well diffused.

The television signals produced in the color sensitive photo-electric cells through the color filters are no dif-

During transmission one sits in front of the apparatus and sees the large sheets of diffusing glass behind which are the color-sensitive photoelectric cells

ferent electrically from those used in monochromatic television. Three sets of amplifiers are required, one for each color, and three communication channels in place of one, but the com-

Herbert E. Ives and A. L. Johnsrud adjusting the receiving apparatus of the color television apparatus

grid* could be employed similar to that used for the earlier demonstration but it would consist of three parallel tubes instead of a single one.

Thus far the television images have been received in a manner similar essentially to our method for monochromatic television. The surface of a disc similar to that used at the sending end is viewed, and the light from the receiving lamp is focussed on the pupil of the observer's eye by suitable lenses. To combine the light of the three lamps, they are placed at some distance behind the scanning disc and two semi-transparent mirrors are set up at right angles to each other but each at 45° to the line of sight.

The disc and motor drive for the color television apparatus are the same as for monochromatic television. The mirror and colored filters are in the small box behind the disc. One of the water-cooled argon lamps appears above the motor

munication channels are exactly similar to those which were used with the same scanning disc before.

For color television the three images must be received in their appropriate colors, and viewed simultaneously and in superposition. The first problem was to find light sources which, like the neon lamp previously used, would respond with the requisite fidelity to the high-frequency signals of television, and at the same time give red, green, and blue light. With such lamps available a decision would have to be made as to how the three colors could best be combined to form a single image.

Several methods of reception are possible. For displaying the transmitted image to a large audience a

One lamp is then viewed directly through both mirrors and one lamp is

*See the RECORD, May, 1927, page 329.

{442}

seen by reflection from each, as illustrated by the accompanying diagram.

The matter of suitable lamps to provide the red, green, and blue light has required a great deal of study. There is no difficulty about the red light because the neon glow lamp which has been used previously in television can be transformed into a suitable red light by interposing a red filter. For the sources of green and blue light nothing nearly so efficient as the neon lamp was available. The decision finally made was to use another one of the noble gases—argon—which has a very considerable number of emission lines in the blue and green region of the spectrum. Two argon lamps are employed, one with a blue filter to transmit the blue lines and one with a green filter trans-parent to the green lines of its spectrum.

These argon lamps unfortunately are not nearly so bright as neon lamps and it was, therefore, necessary to use various expedients to increase their

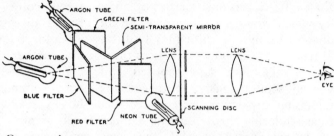

One semi-transparent mirror reflects red light from the neon tube; one reflects green light from one argon tube; and through both mirrors passes blue light from the other argon tube

effective brilliancy. Special lamps to work at high current densities were constructed with long narrow and hollow cathodes so that streams of cold water could cool them. The cathode is viewed end-on. This greatly foreshortens the thin glowing layer of gas and thus increases its apparent brightness. Even so it is necessary to operate these lamps from a special "I" tube amplifier to obtain currents as high as 200 milliamperes.

The receiving apparatus at present consists of one of the 16 inch television discs used in our earlier experimental work. Behind it are the three special lamps and a lens system which focusses the light into a small aperture in front of the disc. The observer looking into this aperture receives through each hole of the disc as it passes by, light from the three lamps— each controlled by its appropriate signal from the sending end. When the intensities of

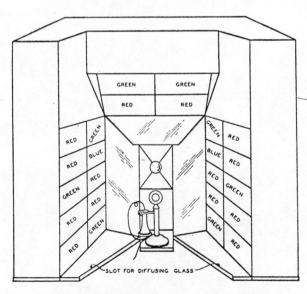

The grouping of the colored filters in front of the color-sensitive photoelectric cells is shown in this perspective sketch of the television transmitter

the three images are properly adjusted he therefore sees an image in its true colors, and with the general appearance of a small colored motion picture.

Satisfactory television in colors is a far more difficult task than is monochromatic television. Errors of quality which would pass unnoticed in an image of only one color may be fatal to true color reproduction where three such images are superimposed and viewed simultaneously. In three-color television any deviations from correct tone rendering throw out the balance of the colors so that while the three images might be adjusted to give certain colors properly, others would suffer from excess or deficiency of certain of the constituents. A further source of erroneous color exists at the scanning end. If the light from the object were not distributed equally to all the cells, the object would appear as if illuminated by lights of different colors shining on it from different directions.

Color television constitutes a definite further step in the solution of the many problems presented in the electrical communication of images. It is, however, obviously more expensive as well as more difficult than the earlier monochromatic form, involving extra communication channels as well as additional apparatus.

COLOR TELEVISION
DEMONSTRATED BY CBS ENGINEERS

Color Television
Demonstrated by CBS Engineers

Old color-movie filter disc principle modified and adapted to television by P. C. Goldmark and his staff. Excellent results obtained from color-film pick-up, producing 343-line images at 120 fields per second on standard channel width

ON September 4th, television in natural colors was demonstrated to members of the technical press by engineers of the Columbia Broadcasting System in New York. The results were impressive, even startling to the uninitiated, although certain limitations were pointed out by Dr. Peter C. Goldmark, in charge of television engineering. These limitations were in his opinion either counterbalanced by the addition of color to the picture or else were removable by employing a more sensitive pick-up device in the camera. The system employed is based on a technique old in the modern picture art, but refined and adapted to television in several important particulars by Dr. Goldmark and his staff. The reproductions had all the color values, so far as could be determined by visual inspection, of the Kodachrome color film from which they were transmitted.

Briefly the color television system employs a standard camera tube (in the demonstration a Farnsworth image dissector tube was used), a standard television channel, and a standard white-screen picture tube having higher than usual brilliance. Color is introduced by employing rotating discs containing red, green, and blue color filter segments, one disc placed in front of the camera and one in front of the picture tube. These discs are rotated synchronously at such a speed (actually 1200 rpm) that the light entering the camera during successive scanning fields passes through successive filter segments. Thus during a given field, only red light enters the camera, during the next field only green light, and during the third field only blue light. At the receiver, the light emerging from the picture tube screen passes through filter segments of corresponding color during the corresponding scanning fields. The rate at which the fields

follow one another is fast enough (actually 120 per second) so that the color impressions blend in the mind of the observer and an accurate tri-chromatic reproduction is obtained.

Since three separate color impressions are sent, it might appear that three times as many pictures must be transmitted in a given time as would be necessary for black and white pictures. But such a three-to-one ratio was found unnecessary. Actually the rate of sending pictures was increased by a ratio of only two to one. The progression of scanning and color impressions is shown in the accompanying diagram. The scanning fields are sent at a rate of 120 per second, which is twice the rate used in the R.M.A. standard for black and white pictures. The interlacing is two-to-one, so that 60 complete frames are sent per second. In scanning one complete frame, however, only two

Dr. P. C. Goldmark, left, and J. N. Dyer of the C.B.S. television staff at the film-scanning equipment. The filter disc may be seen in front of Mr. Dyer's right hand

Dr. Goldmark with equipment for projecting images from colored slides. The image dissector is at the right. Kodachrome transparencies were used as subject matter

	Scanning Field A (1/120th Second)	Scanning Field B	Scanning Frame (1/60th Second)			Color Progression (1/40th Second)			
Time Interval	$1/120$ Sec.	$1/120$	$1/120$	$1/120$	$1/120$	$1/120$	$1/120$	$1/120$	$1/120$	$1/120$
Scanning Field	A	B	A	B	A	B	A	B	A	B
Color of Filter	Red	Green	Blue	Red	Green	Blue	Red	Green	Blue	Red

Color-Scanning Coincidence — 1/20 Second

Table showing the succession of scanning fields and filter-disc segments. A cycle is completed in six fields, while each color is scanned every third field

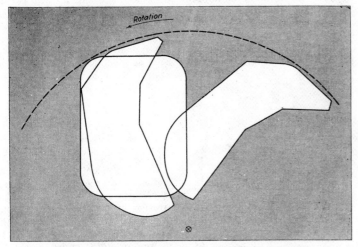

The shape of the colored segments on the filter disc at the receiver is chosen so that the light from the scanning spot has ample opportunity to decay before the segment passes by

of the color segments are employed. The third filter segment is employed on the next field. Then the progression of color segments repeats itself as shown. At the end of six scanning fields, two complete color progressions have occured, and the process then repeats itself. Note that a given scanning field, for example that containing the even-numbered lines, is scanned through a given filter segment, say red, 20 times per second, but that the effective frame repetition rate, so far as scanning is concerned, is 60 per second. Since this frame-repetition rate is higher than that of the R.M.A. standard, the representation of motion is smoother.

The over-all flicker is also correspondingly reduced. However, if a portion of the picture is a solid color corresponding to that of one of the filter segments, the other two segments do not transmit light to the camera, and the flicker rate for that particular color is 40 per second since the given filter segment passes in front of the camera that many times per second. This rate is high enough to avoid visible flicker even at a brilliance higher than would be ordinarily used in the home. On blended colors the effective flicker rates are higher, 80 or 120 per second, which are well above the critical range.

No flicker effects were noticeable at the demonstration, although a wide variety of pure and blended colors was shown, and the brilliance was adequate for viewing in the presence of illumination from overhead fixtures. Incidentally, the colored pictures show a remarkable resistance to the effect of external illumination, due largely to the fact that color contrasts are not effected so greatly as are simple black and white contrasts, when the picture is flooded with external light.

The immediate effect of increas-ing the scanning rate by a ratio of two-to-one is to reduce the number of lines in the picture, for a given bandwidth, by a factor equal to the square root of two. Thus if a 441-line image is assumed for black and white, the number of lines for the corresponding colored picture is 441/1.41 or 313. If 507 lines is assumed for black and white (this figure being closer to the optimum for the standard 6-Mc television chanel), the color pictures have 507/1.41 or 360 lines. For convenience, the CBS engineers chose the intermediate value of 343 lines. While any number of lines may be used, of course, depending on the bandwidth, the frame rate and on the relative degree of horizontal and vertical definition desired, it still remains that the colored pictures must have 41 per cent less resolution than the black and white pictures which can be sent over the same channel. It is the opinion of the C.B.S. workers, backed up by most of those who have viewed the demonstrations, that the addition of color more than counterbalances the loss of detail. In the first place the contrast of the color image is very greatly enhanced, since color contrasts are effective. But the principal difference is one of quality, since information is conveyed by color that cannot be conveyed in any other way. Hence the effective realism is very much improved.

A scheme developed by Dr. Goldmark to improve the detail for a given bandwidth involves quadruple interlacing. If a field rate of 180 per second is employed and the interlacing is four-to-one, the scanning frame rate is 45 per second. The reduction in detail relative to a black and white picture on the same channel is then only about 22 per cent. Such quadruple interlacing permits adequate rendition of the color progressions, excepting possibly the effect of interline flicker when a large patch of a solid pure color is to be reproduced. This might conceivably be a limitation, but one which could be taken care of by care in program technique and camera operation.

The Loss of Light in the Filters

Another important, and as yet inescapable, limitation is the loss of light due to absorption in the color filter segments. Measurements made at C.B.S. show that the average light transmission efficiency of the filter disc is of the order of 30 per cent. This figure applies to the standard Wratten gelatin films used, number 25 for the red, number 47 for the blue, and number 58 for the green. To produce a given output signal from the camera tube, therefore, roughly 3 times as much light is required for color as for black and white transmission. Unless the transmission efficiencies of the filters can be improved, the color system will always work under this disadvantage, but with the advent of greater sensitivity in pick-up tubes,

previously stated. For the time being, this fact limits the transmission to images picked up from colored motion picture film, through which sufficiently concentrated light can be directed to produce a satisfactory signal. Before direct pickup of live talent subjects can be accomplished, the system must be used with storage-type camera tubes. Practical methods for so doing have been devised, and the required equipment nears completion.

One such scheme involves the use of an optical method of interlacing. The problem in using a storage-type camera tube is that no stored

sage through optically flat inclined glass plates which are attached to the filter disc, one plate after every second filter segment. This optical method of producing interlacing avoids the storage of charge between successive fields, while retaining the high sensitivity of the storage type tube. With the more sensitive type of storage camera tube, perfectly adequate pick-up of studio scenes under the light levels now used should be possible. For pickups under adverse conditions of illumination, the color filter could be dispensed with at the transmitter and a black and white image transmitted without any other change being required in any other part of the system. If flexible scanning circuits are adopted as standard in the system, it would be possible, if desired, to increase the detail of the image during such black-and-white transmissions.

Mechanical Details and Synchronization

At the transmitter a continuous motion-picture projector is used, which was operated in the demonstration at 60 frames per second, purely for convenience, although the standard rate of 24 per second can be used readily if minor mechanical and optical alterations are made. The filter disc, about 7.5 inches in diameter, is placed directly before the face of the image dissector. The disc itself has six filter segments, or two complete sets of red, green and blue. The shape of the segments is similar to that shown in the accompanying illustration. When a non-storage type of camera tube is used it is necessary only that the filter cover at any instant that portion of the image in the tube which is being scanned at that instant. The disc is constructed simply of an aluminum frame, which grips the gelatine filters. A protective plastic cover used in earlier work has been found unnecessary. The disc is driven so that the disc makes one revolution in 1/20th second, (1200 rpm) since this produces two color progressions of 1/40th second each, covering three fields of 1/120 second each, per revolution. The motor used in the demonstration was a 1800-rpm synchronous motor with a 6-to-4 gear reduction unit. The vertical scan-

At the press demonstration two receivers were shown, one producing images in black and white (at the left above), the other in color. Each receiver employed a nine-inch picture tube

the problem may not have any great practical importance.

The camera used in the demonstration was a Farnsworth image dissector. When infra-red light was removed by filters, the sensitivity of the pick-up was lowered to the point that noise was visible in the picture. At present the infra-red light is removed from the incident light because of the extremely high infra-red sensitivity of the dissector tube. Dissectors with improved color sensitivity characteristics are being produced. Experiments carried out with panchromatic storage-type pickup tubes, such as iconoscopes or orthicons, indicate that the light losses do not exceed two-thirds, as

charge can be allowed to "carry over" from one field to the next, otherwise part of the light received by the camera through, say, the red filter will be shown to the eye through the green filter at the receiver and the color values will be distorted. To avoid this effect, artificial pairing of the interlacing is introduced by any of the known methods (such as introducing a 60-cps square wave into the vertical deflection circuits). The scanning beam thus always passes over a line in the mosaic which has been discharged during the preceding scanning field. The image is displaced vertically on alternate fields by the width of one scanning line, by pas-

(Continued on page 73)

Color Television Demonstrated By CBS

(Continued from page 34)

ning circuits were driven at 120 cps, the horizontal at 20,580 cps (343 x 60).

At the receiver a similar disc about 20 inches in diameter is used, mounted within a standard cabinet so that the shaft driving the disc lies parallel to and just adjacent to the cathode ray tube. The cathode-ray tube itself, which was built in the C.B.S. laboratories, is nine inches in diameter. It employs magnetic focus and deflection, and a sulphide screen made especially for projection tubes. The electron gun produces a beam current several times as great as that usual in receiver tubes, without increase in spot size. A brighter than ordinary image is thereby produced. The second anode voltage is 7000 volts, the first anode voltage about 5000 volts. The power supply for the tube is a standard unit taken from a commercial receiver. The cabinet is of the direct-viewing type. An 1800 rpm motor and 6-to-4 gear reduction unit were used to drive this disc at 1200 rpm. Synchronism between the two discs was maintained by the use of a common alternating-current supply.

The shape of the segments in the receiving disc is dictated by two factors. First, when a segment is in front of the cathode-ray tube screen, its leading edge must cover at any instant the position of the scanning spot at the same instant. Moreover as the disc rotates, the filter must continue to cover that spot until the fluorescent decay of light has been substantially completed. These two considerations have led to the shape of filter segment shown in the accompanying illustration. Sufficient safety factor has been included in the area of the segment to allow for phase variations in the synchronization up to 10 or 15 electrical degrees. A disc large enough for a 12-inch tube is now being constructed.

In commercial practice, synchronization by the use of a common alternating current supply is not always feasible, nor desirable even when a common a-c system is available. Independent synchronization of the receiver motor can be obtained from the television signal sync pulses, which are used after suitable amplification to drive a small phonic motor which operates in tandem with the main drive motor. Calculations indicate that a single type 6L6 tube is sufficient to drive the phonic motor.

One of the mechanical design problems is that of reducing the noise associated with the motor, gear-reduction unit, and the windage caused by the motion of the disc through the air. While the receiver was quite noisy during the demonstration, the noise has been greatly reduced since that time, and there is no reason why this problem should not yield to straightforward mechanical design and sound-insulation technique.—D.G.F.

COMMUNICATES SIX WAYS AT ONCE

George A. Mead, New York State Commander of the American Legion in a recent broadcast from G. E. Co.'s television studios, during which his voice was carried over every known scientific means of voice communication, according to General Electric engineers. His voice, in addition to going out on the ultra-short wave band accompanying the television picture, was simultaneously carried by WGY on 790 kc., by WGEO on 9530 kc., by W2XB (television) on 77.75 Mc., by frequency modulation over W2XOY on 43.2 Mc., and over a light beam on 430 to 750 million megacycles, in addition to the wire telephone operating at frequencies of 150 to 2500 cps.

THE PRESENT STATUS OF COLOR TELEVISION

REPORT

OF THE

ADVISORY COMMITTEE ON COLOR TELEVISION

TO THE

COMMITTEE ON
INTERSTATE AND FOREIGN COMMERCE
UNITED STATES SENATE

PRESENTED BY MR. JOHNSON, OF COLORADO

JULY 14 (legislative day, JULY 1), 1950.—Ordered to be printed, with illustrations

UNITED STATES

GOVERNMENT PRINTING OFFICE

69765

WASHINGTON : 1950

UNITED STATES SENATE,
COMMITTEE ON INTERSTATE AND FOREIGN COMMERCE,
May 20, 1949.

Hon. E. U. CONDON,
*Director, National Bureau of Standards,
Washington, D. C.*

MY DEAR DR. CONDON: The question of the present-day commercial use of color television has been a matter of raging controversy within the radio world for many months. There is a woeful lack of authentic and dependable information on this subject.

Hundreds of applicants for television licenses as well as those now operating television stations are vitally affected by its settlement. The capital investment involved in the installation of a television station runs into a tremendous sum. The operational costs of such a station are extremely high also. All of these expenses must be recovered through advertising. Those who are experienced in advertising believe that if color television were available now, attractive local advertising revenues could be obtained due to the strong consumer demand for it.

The Federal Communications Commission has declined to authorize commercial licensing of color television. It seems reluctant to indicate when and if it will act with respect to authorizing commercial licensing of color. As we understand it, the Commission must first fix minimum standards for color television before licensing can be undertaken, but it refuses to attempt to do so on the premise that color television has not been developed sufficiently for standards to be determined.

Accordingly, it is greatly in the public interest that a sound, factual ascertainment be had now whether or not minimum standards can be fixed today, or in the very near future, so that color television might develop and progress with complete freedom under the stimulus of commercial competition.

One unit in the industry has demonstrated color television 6 megacycles wide and asserts that if the Commission would allocate frequencies and license commercial operation, it could go ahead "tomorrow." Another large unit in the industry also, has demonstrated color television of varying width from 6 to 18 megacycles but believes that color is not yet ready for commercial operation; that much more experimental work must be done and field tests made before commercial licensing should be undertaken. Still another unit in the industry is said to be of the opinion that color television is several years away.

My objective, and the objective of the Senate Committee on Interstate and Foreign Commerce, is to encourage development of the radio art and to press for a Nation-wide, competitive television service in the public interest. Our committee sees television as a great new industry, not only providing new jobs and a new source of wealth but as the greatest medium of entertainment and diffusion of knowledge yet known to man. We believe that it has made great advances but we are concerned that through delay in opening up the ultra-high frequencies and holding up color until such time as some electronic experts believe that color has reached a state of perfection, a chain of circumstances will have been created which will tend toward monopoly control of the entire television art.

We are anxious, also, to reduce as much as possible any sharp impact on both station licensees and the general public, who already have invested one-half billion dollars in receiver sets, of any sudden but eventually necessary conversion to color. It is our belief that if both potential licensees and the set-buying public are given all of the facts now with respect to color television, less exploitation will ensue and less wasteful expenditures will occur.

Frankly, it seems to us that this is the time to obtain these facts and make them public. The Commission has in effect a "freeze" on further television allocations in the VHF band. It faces the problem of opening up the UHF band in order to provide sufficient channel space for a competitive Nation-wide television system. Now, when there is at least the probability that both bands may be opened simultaneously for allocation, is the time to make certain regarding the color television situation so that, if it is technically feasible, the Commission might also simultaneously open color to commercial licensing in either or both bands.

It has occurred to me, therefore, that at this juncture you could be most helpful in giving this committee sound, impartial, scientific advice. I am anxious that you individually, or in association with a small group of scientific persons of repute, none of whom are employed by or have any connection directly or indirectly with any radio licensee or radio-equipment manufacturer, shall investigate officially this matter for the committee.

Specifically, I would like you and your group to visit the laboratories of the Radio Corp. of America, Columbia Broadcasting System, Du Mont, and any others engaged in color television research and development; confer with their engineers; witness demonstrations; ask questions, all with the purpose of coming to a definite opinion as to the present stage of development of color television. Your inquiries will necessitate an evaluation of present-day practicability of color television: in short, can a satisfactory color television picture be broadcast today in the VHF and UHF frequency bands?

We are aware, of course, that both transmission and reception equipment is not now available on a commercial scale but that is not a controlling factor in whether color television should or should not be licensed, or stations allocated. We are also aware that undoubtedly experience and further experiment will result in the development of a better color picture but that, also, is not a factor in the evaluation we seek. We realize, as you, that color television today is as different from what it will be in perhaps 5 years as were the old crystal radio sets as compared with present-day radio receivers. It is not necessary that the art be fully developed for minimum standards to be outlined.

I am particularly concerned with resolving once and for all the charges that have been made that the advance of color television has been held up by the Commission for reasons difficult for us to understand, and I feel certain that a committee headed by so eminent a scientist as you will help resolve these doubts and questions which have been tossed about

You will want, I assume, to confer with the engineers and laboratory personnel of the Federal Communications Commission as well as with the people in the industry. I feel certain that you will have the cooperation and willing assistance of the responsible officials of the industry in such a study, and I shall be pleased to ask them and any Government agencies who may be concerned to accord you and your group every assistance and cooperation.

I sincerely hope in the public interest that you will assume this difficult assignment. I shall be pleased to confer with you at your convenience.

Sincerely yours,

ED. C. JOHNSON, *Chairman.*

THE PRESENT STATUS OF COLOR TELEVISION

CHAPTER 1

SCOPE OF THE INVESTIGATION AND SOME BASIC CONCLUSIONS

1. Introduction

This report has been prepared at the request of the chairman of the Senate Committee on Interstate and Foreign Commerce. It represents an independent appraisal of the present status of color television in the United States and takes into account observations of the black-and-white television service now offered to the public as well as demonstrations of three color-television systems proposed for public use by Color Television, Inc., the Columbia Broadcasting System, and the Radio Corp. of America. The report is confined to technical factors, expressed so far as possible in nontechnical terms.

The report is organized as follows: Chapter 1 outlines the activity of the committee, describes the approach of the committee to its assignment, and sets forth some basic conclusions. Chapter 2 analyzes color-television service in general and lists the apparatus and performance characteristics by which competing color systems should be judged. Chapters 3, 4, and 5 describe, respectively, the three proposed color systems, in alphabetical order, viz, those of Color Television, Inc. the Columbia Broadcasting System, and the Radio Corp. of America. These chapters state the actual and potential performance of the systems, in terms of the characteristics listed in chapter 2. Chapter 6 consists of a comparison of the three color systems and the black-and-white system, and includes a tabular tally sheet on which the systems may be judged.

No recommendation for the adoption of a specific system is given, since the committee believes that the decision to adopt a system must include consideration of many social and economic factors not properly the concern of the technical analyst. It is hoped that the report will provide a comprehensive and understandable basis on which the technical factors may be considered in arriving at a decision.

2. Narrative of the committee activity

The Senate Advisory Committee on Color Television was appointed in June 1949 by its chairman, Dr. E. U. Condon, the Director of the National Bureau of Standards, in response to a request by Senator Edwin C. Johnson of Colorado, chairman of the Senate Committee on Interstate and Foreign Commerce. The letter from Senator Johnson to Dr. Condon requesting the investigation, dated May 20, 1949, precedes this report.

On May 26, 1949, the Federal Communications Commission announced that, at a hearing to be convened to consider expansion of the commercial television service, evidence would be taken concern-

1

ing the possibility of instituting a public color-television service. Excerpts from the FCC Public Notice No. 49–948 relating to the color-television aspects of this hearing are appended as annex A.

Meetings of the Senate advisory committee (hereinafter referred to as "the committee") were held August 3, 17–19, October 7–10, November 21–22, 1949, January 19, 20, February 1, 20, 23, March 11, 14, April 26, May 22, July 5, 6, 1950. During these meetings, demonstrations of color television were attended by two or more members of the committee as follows: CTI system, February 20, 23, March 14, 1950; CBS system October 6–10, November 21–22, 1949, January 20, February 1, 23, April 26, 1950; RCA system, October 6–10, November 21–22, 1949, January 19, 23, 1950; Hazeltine demonstration, May 22, 1950. These demonstrations included the comparative demonstrations of the color systems before the FCC held November 21–22, 1949 and February 23, 1950, at which all members of the committee, or designated alternates, were present.

At its meeting of November 21, 1949, the committee discussed the question of the basic terms of reference of the report, particularly regarding the availability of additional channels not then contemplated by the FCC proposals. As a result of this discussion, an inquiry was prepared and forwarded to Senator Johnson as of February 2, 1950. A copy of this inquiry is appended hereto as annex B.

Shortly thereafter, the formation of the President's Communications Policy Board was announced. In view of the contemplated activity of this Board, Senator Johnson advised the committee to proceed within the terms of reference proposed by the FCC, namely, to consider channels in the very-high-frequency (VHF) band from 54–88 and 174–216 megacycles, and channels in the ultra-high-frequency (UHF) band from 475 to 890 megacycles. Senator Johnson's reply is appended as annex C.

At its meeting of March 11, 1950, the committee met with Senator Johnson and discussed matters pertinent to the report. The final report, approved unanimously at the meeting of the committee, July 5–6, 1950, is presented herewith.

3. Terms of reference of the report

(a) *The 6-megacycle radio-frequency channel.*—This report is concerned only with color-television systems intended for a 6-megacycle radio-frequency channel; that is, a channel equal in width to that now assigned to black-and-white stations. Since color systems of superior performance have been demonstrated using channels wider than 6 megacycles, the justification for confining this report to the 6-megacycle channel is stated at the outset.

As shown in greater detail in chapter 2, the choice of the channel width in a television system is necessarily a compromise between quality and quantity; quality of the reproduced television image on the one hand and quantity of television service on the other.

If the radio channel width were doubled, a clearly perceptible improvement in the quality of the image should be apparent, but the number of channels available would be halved, thereby greatly reducing the possible number of stations.

Moreover, as the width of the channel is progressively increased, the corresponding improvement in picture quality apparent to the observer under normal viewing conditions becomes less pronounced. There is, in other words, a law of diminishing returns that ultimately affects the attempt to improve image quality by increasing the width of the channel. On the other hand, no such diminishing law affects the relation between channel width and the number of channels. Each time the channel width is doubled, the number of channels is halved, and this law holds without diminution as the channel width is increased.

Evidently a point is reached, as wider channels are considered, at which the slight improvement in image quality afforded by a substantial increase in channel width is not worth the reduction of service that would be entailed. The optimum channel width must be chosen, therefore, by a body qualified to judge that combination of quality of image and quantity of service which best serves the public interest.

This judgment has been entrusted by statute to the Federal Communications Commission, which established the 6-megacycle channel for black-and-white television as early as 1937. This channel width provides an image quality roughly comparable to that of 16-millimeter home motion pictures, and allows 12 channels to be assigned in the very-high-frequency spectrum, due account having been taken of the needs of other services.

When a color-television service is considered, the optimum compromise between quality and quantity, similarly determined as meeting the public interest, does not necessarily lead to the same value of channel width. In fact, the addition of color to the image brings about a degradation of certain other qualities in the image (particularly pictorial detail and freedom from flicker, see ch. 2) when the channel width is unchanged. To avoid degradation of these qualities in the color image, a wider channel must be assigned.

In the face of this fact, a mitigating circumstance has appeared, in the form of a new development (known as dot interlace, explained in ch. 2) which is capable of substantially improving the pictorial detail of the television image, without requiring any increase in the width of the channel.

Specifically, when dot interlace is adopted in a color-television system, the technique can provide a color image whose pictorial detail is substantially equal to that of the black-and-white images currently rendered to the public. This fact implies that color service, capable of being rendered on a 6-megacycle channel, may achieve a quality generally as satisfactory as that of current black-and-white broadcasts.

Another factor affecting the choice of channel widths is an economic one, relating to the investment by the public in black-and-white television receivers when color television is first offered as a regular public service. If the investment is substantial, when compared to the ultimate per capita investment to be expected in the then foreseeable future, it is desirable that after any change or extension of the television service, the service can be used with then-existing receivers with a minimum of expense, inconvenience, and/or degradation of the quality or quantity of the service. If the new service operates on channels wider than 6 megacycles, existing receivers cannot use the new service.

> Based on the foregoing analysis, the committee concludes that the allocation of 6-megacycle radio-frequency channels for color television is the proper compromise between the quality and the quantity of the color service.

(b) *Comprehensive nature of systems considered.*—In restricting its consideration to three color-television systems, the committee is aware that certain other systems, known to the members, might have been considered. The report is confined to these three systems, not merely because they are the ones actively proposed at present, but rather because they comprise, as a group, all of the basic types of sequential color systems.

Television images, as outlined in greater detail in chapter 2, consist of picture elements (dots) arranged along lines, the lines being assembled to cover the field (the picture area). A succession of fields is transmitted to create the illusion of continuity and motion in the image. The dot, the line, and the field are then the three basic elements of a television picture. No matter how the picture is analyzed in the television camera or synthesized at the receiver screen, the process of transmission can always be described in terms of these three elements.

It is most fortunate, therefore, that the systems of color television actively proposed are based respectively on these three attributes of the image. The RCA system is a dot system, since the color is assigned to successive picture elements, or dots, of the image. In the CTI system, a line system, the color is assigned to successive lines of the image. The CBS system is a field system, the color values being assigned to successive fields of the image. Other color systems (notably the simultaneous system developed in 1946 by RCA but discontinued in favor of the dot system) are known, but they are difficult, if not impossible, to adapt to a six-megacycle channel.

If, therefore, only six-megacycle systems are to be considered, the committee concludes that the color television system ultimately adopted must be either a dot-sequential system, a line-sequential system, or a field-sequential system. No other methods need be considered, in the light of present or foreseeable technical developments.

(c) *Mutually exclusive nature of the color systems.*—Because the three color systems herein discussed are based on fundamentally different aspects of the television image, they are to a very large extent mutually exclusive, so far as public service is concerned. All use the 6-megacycle channels, and in many other respects are similar (each uses the same type of sound system, for example). Each, however, is fundamentally different from the others in the way in which the color values are distributed among the dots, lines, and fields of the image, and this difference is so profound that the receivers for one system cannot be converted to another except at considerable expense.

At the present stage of the art, a universal receiver, capable of receiving transmissions of all three types, would represent three separate receivers in a single cabinet, with certain elements in common. Changing the connections of the common elements, to convert from the dot-system to the line-system or to the field-system of reception, would involve a highly complicated and vulnerable mechanism. Moreover, the compromises inherent in the design of such a universal receiver would most certainly impair the performance of at least one system and perhaps of all three systems.

Past experience, notably in Great Britain in 1936, with multiple standards of television transmission has proved that such action encourages a portion of the viewing public to purchase equipment which loses its value when the final decision is made among the multiple standards. The decision can be made, and should be made, on the basis of analyses and tests conducted prior to the inauguration of the public service. Moreover, these analyses and tests are well under way, and the final decision can be made without unwarranted delay. But any authorization of color-television transmission on a multiple-standards basis is a guaranty of confusion that may well impose a much greater delay in the development of the color-television service.

The committee concludes that one and only one of these systems should be licensed for service to the public and that therefore the decision among the dot- line-, and field-systems must be made in advance of the introduction of a color television service.

4. *Summary*

In summary, the committee bases this report on the following basic conclusion:

1. A 6-megacycle radio-frequency channel is adequate for color-television service and represents a proper compromise between quality and quantity of service.

2. The three systems of color television herein described comprise all of the basic systems of color television which need be considered for a 6-megacycle channel.

3. The three systems are mutually exclusive. One, and only one, of these systems must be chosen in advance of the inauguration of a public color television service.

CHAPTER 2

COLOR TELEVISION PRINCIPLES

5. *Natural vision versus television*

In natural vision, the scene before the observer is focused on the retina of the eye, which contains millions of tiny light sensitive elements, all of which are continuously exposed. These retinal elements are connected to the optic nerve, which comprises hundreds of thousands of separate fibers, each of which is capable of carrying a part of the visual impression to the brain. In this manner all parts of the scene are apprehended simultaneously.

When an artificial medium, such as motion pictures or television, is interposed between the scene and the eye, it is not practicable to imitate the continuous nature of the actual scene. Rather, it is necessary to present to the viewer a rapid succession of still pictures, each differing very slightly from the ones preceding and following it. In motion pictures, this is done by printing a succession of pictures on a strip of film and passing the film through a projector. Each picture on a movie film is a comprehensive still picture of the whole scene. In motion pictures, therefore, the camera apprehends the whole scene at once, but it does so in discontinuous fashion, one still picture at a time.

In television, a similar succession of still pictures is transmitted, but it is not practicable to transmit the whole area of each still picture at once. This would require hundreds of thousands of separate cable or radio circuits, corresponding to the hundreds of thousands of separate fibers in the optic nerve. Any channel of an electrical communication system using telephone lines or radio waves can transmit only one thing at a time. Therefore the picture must be analyzed into a finite number of picture elements whose light intensity must be converted into signals one after the other, sent over the communication channel, and reassembled on the viewing screen in the proper position, all within the time normally used in a motion picture for showing one frame. The number of picture elements which must be distinguished in television is dependent upon the detail desired.

If the processes of dissecting and reassembling are carried out very rapidly the whole receiver screen appears to be illuminated simultaneously.

6. Television scanning: "Reading" the content of the scene

Television images are dissected and reassembled by a process known as "scanning," a term which arises from the similarity of the process to the action of the eye in scanning pages of printed matter. The eye reads from left to right along the first line of type, then returns rapidly to the beginning of the next line, scans it, and so on until the bottom of the page is reached. The page is then turned, and the process is repeated on the next page.

In a televised scene, corresponding to the individual letters of print are rapid or gradual variations in light and shade, depending upon the nature of the scene. These variations are arranged in horizontal lines, like lines of print, and the lines are arranged parallel to one another, filling the picture area. In the television camera the variations along the uppermost line are explored rapidly from left to right, and signals are generated which correspond to the degree of light or shade along the line. When the first line is thus scanned, another line below the first is similarly covered, and so on, until the bottom of the picture is reached. In a communication system the signal cannot change abruptly from one value to another, and the shortest distance along the line which can be made to change from white to black is called a picture element. The shorter this distance the greater is the detail which can be transmitted, but the greater are the requirements on the system, particularly the wider is the band width which must be allocated to the communication channel. In some cases these picture elements may be made up of dots definitely located along the lines, but in the present black and white system variations may start or stop at any point and no dot structure is observable, although the line structure always is. However, for purposes of explanation these picture elements will be generally referred to as dots.

Like the lines of print in a double-spaced typewritten page, each line of elements is separated from the line above it and the line below it by a blank space of the same height as the line. The blank spaces are filled in during the next successive scanning of the picture area. The two sets of lines are thus "interlaced" one within the other. As these two sets of lines are scanned, all of the light and shade values, over the whole area of the picture, are translated into a succession of electrical counterparts. The picture is scanned in two sets of interlaced lines, rather than in one set of consecutive lines, in order to minimize flicker in the image. This is explained in section 8.

At the receiver, a spot of light moves across the picture screen in the same scanning motion. Acting under the control of the broadcast station, the spot changes in brightness as it moves along each line and thus re-creates the light and shadows of the original scene. Since the light spot on the viewing screen moves in precise step with the scanning process at the television camera, each dot of light falls in its proper place, and has its proper value of light or shade.

This description of scanning shows how important it is that the operation of the television system be standardized. Once the number of lines and the number of pictures per second have been established it is essential that all receivers be capable of operating with that number of lines and that number of pictures per second. Any change in the scanning standards adopted for the transmitter thus requires an exactly equivalent change in the scanning process at each of the many millions of receivers in the hands of the public.

7. Pictorial detail: How many dots in the picture?

The choice of scanning standards starts with this basic question: How many dots are required in the whole picture area to reproduce a picture of acceptable quality?

If there were no economic limitations, and if the radio spectrum was limitless, it might be desirable to transmit a picture containing many millions of dots. Thus an 8-by-10-inch printed photoengraving of the highest quality (150-line-per-inch halftone, printed on high-gloss paper), contains about 2,000,000 dots. Such a picture can be examined closely by the unaided eye, without the dots themselves becoming separately visible.

In television and motion pictures, it is not necessary to examine the picture minutely. When a performance is to be viewed continuously for many minutes or hours, in fact, it is necessary that the whole picture area be contained within such a field of view as to avoid excessive movement of the neck or eyes. For example, most people find it uncomfortable and fatiguing to view continuously a picture 1 foot high from a distance less than 3 feet. This ratio of viewing distance to picture height applies equally well with other picture sizes; i. e., the minimum viewing distance, to avoid excessive fatigue, is generally taken to be three times the height of the picture. Many individuals cannot look for long periods at a picture unless it is viewed from a considerably greater distance than this, say five to eight times the picture height. These points are indicated by the location of the seats chosen in a motion-picture theater by patrons who have a free choice.

When the image is to be viewed at a distance greater than three times the picture height a pictorial detail of several hundred thousand dots suffices, as against the millions of dots that would be required for closer inspection. If a larger number of dots were used, the excess would be wasted since the eye cannot perceive the additional detail from a distance.

This limit on required detail has led to the choice of various sizes of motion-picture film. Professional 35-millimeter film, as commonly projected in motion-picture theaters, has a pictorial detail equivalent to about 1,000,000 halftone dots. The 16-millimeter movie film, used by the advanced amateur, has the equivalent of 250,000 halftone dots in the picture area, when film and projector are in first-class shape. The average performance of 16-millimeter home-movie film and projectors is such, however, that the effective pictorial detail seldom exceeds the equivalent of 200,000 dots. The smallest movie film currently used is the 8-millimeter size.   This film has the equivalent of about 50,000 halftone dots in the picture area.

The pictorial detail offered by various motion-picture systems, professional and amateur, is a compromise.   The upper limit is set by the cost of film and processing, cameras, and projectors.   The lower limit is set by the reactions of the viewer, who objects to an image having so little detail that it is incapable of portraying a wide variety of subjects satisfactorily.   All those who have viewed 16-millimeter and 8-millimeter movies of the same subject matter are well aware of the greater sharpness of the larger film.   In payment for the superior performance of the 16-millimeter system, approximately four times as much money must be paid for film and processing for a given period of viewing time, relative to the 8-millimeter type.   Accordingly, economic factors have given the 8-millimeter film a commanding position in the amateur-film market.   At the other end of the scale, movie theaters employ virtually nothing but 35-millimeter film to meet the high standard required for elaborate and expensive productions.

In television, a similar compromise must be found, since it is expensive to set up a television system having too much detail in the image.   The expense resides not only in the extra cost of transmitting and receiving equipment but also in the extra space occupied by the television channels in the radio spectrum.   In a given portion of the spectrum, for example, the number of channels which can be accommodated varies in inverse proportion to the number of picture elements in the image, all other factors remaining unchanged.   Thus a change from a television system approximately equivalent to 16-millimeter home movies (200,000 dots) to one equivalent to 35-millimeter professional movies (1,000,000 dots), would force a reduction in the number of channels in the ratio of 5 to 1.   For the sake of completeness, it should be mentioned that, on the basis of geometrical resolution alone, a 200,000-dot motion-picture system would be superior to a 200,000-dot television system, because the line structure is not present in the motion picture.

Faced by this conflict between quality (pictorial detail) on the one hand, and quantity (number of stations and choice of programs) on the other, the Federal Communications Commission in 1941 adopted for public television broadcasting a black-and-white system having about the equivalent of 200,000 halftone dots in the picture area.   This choice appears to have merit, because it follows the standard of the best visual medium of entertainment hitherto used in homes, the 16-millimeter home-movie system.   More fundamentally the 200,000-dot television system permits the picture to be viewed at a distance as close as four times the picture height, without the picture structure's becoming too evident.   This viewing distance is close to the minimum value of three times the picture height, set by the fatigue factors previously discussed.

When it is decided that the television picture should be equivalent to 200,000 halftone dots, it is necessary to select the number of lines and the number of dots per line.   This is not a critical matter.   For example, a picture of 400 lines, each having 500 dots, would provide a 200,000-dot picture ($200,000 = 400 \times 500$).   A picture having 500 lines, each containing 400 dots, would serve equally well.   The present black-and-white system employs 525 lines, about 490 of which are actually visible on the screen, and each line has the equivalent of about 420 dots along its length. As previously stated, the 490 visible lines are actually scanned in two sets of 245 lines each, one set interlaced within the other.

Experience with the 525-line black-and-white system since 1941 has shown that it provides an adequate basis for a public television service, so far as pictorial detail is concerned. But this is not to imply that additional detail would not be desirable if it were available without excessively reducing the quantity of service. For this reason, the introduction of dot interlace to the black-and-white system is being considered. This recently developed technique would increase the pictorial detail of the black-and-white image from 200,000 dots to something over 350,000 dots, without any increase in channel width.

Before leaving the question of pictorial detail, it must be emphasized that this aspect of television-system performance is capable of a considerable degree of misinterpretation in comparing the merits of different proposals. The difficulty arises from the various types of subject matter which may be portrayed by television.

When a scene is viewed in a close-up shot, as for example when the face of a performer fills the whole screen, not much pictorial detail is required. To show the essential features and details of a face it is not necessary to use more than 50,000 dots, as experience with the 8-millimeter movie system has amply demonstrated. When, however, it is desired to show the whole area of a baseball diamond, or some other equally extensive subject, the requirement for pictorial detail is very much larger. In fact 200,000 picture elements may then be insufficient to show more than the bare outline of the individual players.

Since a television system is called upon to depict both close-ups and long shots, sufficient pictorial detail must be provided to take care of the long shots, despite the fact that a large part of the detail is wasted when close-up shots are being transmitted. A test of a television system which comprises only close-up shots does not reveal the pictorial-detail limit of the system. Such tests must show the whole range of subject matter for which the television system is intended.

Since the appreciation of pictorial detail is a highly individual reaction of the viewer, it is unlikely that complete agreement on this aspect of system performance will be reached by all participants in a test. But it is possible to state categorically the effect of pictorial detail in the following terms:

Consider a subject viewed in a close-up shot, and suppose that the camera moves back from the subject so that the close-up shot gradually becomes a medium-length shot and finally a long shot. At some point, as the camera recedes from the subject, a given viewer will find that the pictorial detail becomes inadequate and the portrayal is unsatisfactory. This is the point at which the pictorial detail of the image becomes the limiting factor, for that particular observer.

If now the picture detail in the television image is increased, the area viewed by the camera can be increased in the same proportion, without exceeding the critical limit set by that observer. Suppose, for example, that the number of dots is increased four times, from 50,000 to 200,000. Then the camera can take into view an area four times as great, with the same degree of visual satisfaction. In concrete terms, if the face of one actor can be shown, with a given degree of satisfaction, on a screen of 50,000 dots, four actors can be shown with the same degree of satisfaction with a screen of 200,000 dots; if the action covering 1,000 square feet of a basketball court is portrayed satisfactorily with a 50,000-dot image, action covering 4,000 square feet may be portrayed with the same satisfaction with 200,000 dots.

Any limitation in the detail of the television image constitutes, therefore, a limitation on the program director with respect to the area which he can pick up with a given degree of satisfaction. If the pictorial detail is low, say, 50,000 dots, the cameraman must use close-up shots almost exclusively; whereas if 200,000 dots were available, medium shots could be used with the same degree of satisfaction. Finally, if very high detail were available, say a million or more dots, long shots would display the same degree of visual distinction as medium shots and close-ups.

It follows that the flexibility with which the program director can use lenses and cameras is intimately tied up with the detail provided in the image, and any restriction on pictorial detail implies a restriction on the use of the camera. It is true that this restriction can be circumvented in many types of programs by rapid switching from camera to camera, each showing a close-up shot. In athletic contests and other large-scale presentations, however, the restriction on viewing angles may prevent the viewer from following the over-all aspect of the action. This limitation is clearly evident in telecasts of football and hockey, but is much less noticeable in the confined arena of a boxing or wrestling match.

The technical term for pictorial detail is "resolution," because this quantity represents the ability of the television image to resolve the fine details of the scenes it depicts. As we have seen, resolution is measured by the total number of equivalent halftone dots in the image. The number of equivalent dots along each line (conventionally measured as the number of dots in a distance equal to the picture height) is the "horizontal resolution." The number of dots resolved at right angles to the lines is known as the "vertical resolution." As outlined in the following chapters of this report, resolution, measured in the horizontal and vertical directions, is one of the basic criteria by which the proposed color-television systems must be compared.

8. Image continuity: How many pictures must be transmitted per second?

The second question in the choice of scanning standards is the number of complete pictures to be sent per second. In considering this question it is necessary to have clearly in mind the meaning of the terms "field" and "frame." In section 6 it was pointed out that the television image is scanned in two sets of lines, one interlaced within the other. One set of these lines, having blank spaces between lines, is known as a field. The lines of one field cover only one-half the area of the picture. The other half of the area (the space between the lines) is filled in by the lines of the next successively scanned field. Hence all points in the picture have been covered when two successive fields have been scanned. Two successive fields, comprising all the lines in the image, are known as a frame.

To insure continuity in the motion of the image, it is necessary that the fields succeed one another at a rapid rate. If the fields are presented at a rate slower than about 15 per second, the apparent motion in the image will be disjointed or "jerky." This corresponds to running a motion picture film through a projector at too slow a speed.

In practice the rate of scanning the successive fields must be much higher than this minimum value of 15 per second, because of another effect known as flicker. Flicker appears because the light on the screen is cut off between the successive pictures. If the rate of scanning successive fields is too low, the light on the screen will appear to blink on and off in a manner which is annoying to watch and induces severe visual fatigue. If the successive fields are scanned at a sufficiently rapid rate, however, the sensation from one picture persists throughout the dark interval between fields and the screen appears as if it were continuously illuminated.

The brighter the television image, the more perceptible is the flicker. Hence, in deciding how many fields must be scanned each second, it is necessary to decide how bright the picture must be, and then choose a field rate high enough to avoid flicker at that level of brightness.

Different compromises have been adopted in this respect in different countries. In Great Britain, the pictures are scanned at a rate of 50 fields per second, whereas in the United States, in the black-and-white system, they are scanned at 60 fields per second. The brightness at which flicker is perceptible goes up very much faster than the increase in field rates, with the result that the American rate of 60 per second permits pictures to be about 6 times as bright as the British pictures. In consequence, British receivers must be viewed in a darkened room, whereas most American receivers can be viewed satisfactorily in rooms illuminated by direct daylight.

Two types of flicker must be distinguished in comparing the performance of color television systems. The first is "large-area flicker", which applies to the whole area of the image, or to any bright part of the image occupying a substantial portion of the field of view. The more closely the image is viewed, the larger is the portion of the field of view occupied by the bright portions of the image, and the more noticeable is the large-area flicker effect.

The second type of flicker, known as small-area flicker, appears in areas having the size of a few picture elements or the width of a few scanning lines. This type of flicker is most noticeable on close inspection of the image, but it may be apparent at normal viewing distances under certain conditions.

One form of small-area flicker applies to individual scanning lines. We have noted that each picture is scanned successively in two sets of lines, one set interlaced within the other. Hence any one line in the image is illuminated only half of the time, and the flicker rate which applies to a single line is accordingly half that applying to the image as a whole. This low flicker rate gives rise to the so-called "interline flicker" and "line crawl". Interline flicker manifests itself as a blinking of thin horizontal lines in the image, such as the roof line of a house. Line crawl is an apparent motion of the lines upward or downward through the image, due to the successive illumination of adjacent lines in the picture, particularly in bright parts of the image.

Flicker can be controlled either by dimming the image or by increasing the rate at which the successive fields are scanned. Both methods of flicker control are subject to severe limitations. If the image is dimmed too far, the fine detail in the image cannot be perceived by the eye, and eyestrain results. Also, even at considerably brighter levels than the eystrain limit, it is necessary to darken the room to secure accurate rendition of all the shades of gray (in a black-and-white picture) or all the saturations of color (in a color picture). On the other hand, if the picture-scanning rate is raised too high, the spectrum space required by the signal becomes exorbitant, as explained in section 10, below. The choice is essentially a compromise. Experience has shown that rates between 48 and 60 fields per second are required to produce flicker-free pictures of adequate brightness.

In passing, it should be noted that the picture-scanning rate is sometimes expressed in frames per second. Since a frame contains two scanning fields, the frame rate is one-half the field rate. Thus the 60-field-per-second rate of the American black-and-white system may also be stated as 30 frames per second. The most useful term, applicable to both black-and-white and color systems, is the rate in fields per second, and this term is used in this report unless specifically stated otherwise.

9. Channel width: How many megacycles for a television station?

The channel width required by a television station is determined directly by the scanning specifications described in the preceding paragraphs; namely, the number of equivalent dots per picture and the number of fields transmitted per second.

The relationship between these quantities can be traced as follows: From section 7 we recall that the standard black-and-white picture corresponds to about 200,000 dots, and that these dots are distributed in two sets of interlaced lines. One set of the interlaced lines (one field) thus encompasses about 100,000 dots. From section 8, we recall that the fields are transmitted one after the other at a rate of 60 per second. Nominally, then, 100,000 dots must be transmitted in one-sixtieth of a second. Actually, since a portion of the lines is not visible in the picture the time available is about one-eightieth of a second. Consequently, the rate of transmitting dots (100,000 of them in one-eightieth of a second) is about 80 times 100,000 or 8,000,000 dots per second.

To transmit picture dots at a rate of 8,000,000 a second, it is necessary to employ a channel width of at least 4,000,000 cycles per second (4 megacycles). This band width of 4 megacycles is required for the picture alone and is referred to as the video channel. In addition to this 4-megacycle minimum requirement, channel space of about 0.2 megacycle must be allowed for the sound transmission, and additional space must be allowed to prevent mutual interference between the picture and sound signals of the station. Finally, a substantial amount of additional space (about 25 percent) must be allowed to permit proper operation of the television transmitter and receiver (to permit "vestigial side-band" operation). When all these requirements are added, the radio-frequency channel width required for transmitting picture dots at a rate of 8,000,000 per second, plus associated sound, is 6 megacycles.

The foregoing discussion shows that the channel width is determined fundamentally by the number of picture elements (dots) in each field multiplied by the number of fields transmitted per second. If the number of dots were increased from 200,000 to 400,000 per picture, the channel width would have to be doubled. Similarly if the number of fields per second were increased from 60 per second to 120 per second, the channel width would have to be doubled. If both the number of dots and the number of fields per second were doubled, the channel width would have to be quadrupled.

When the channel width is fixed at 6 megacycles, as is assumed throughout this report, and in the absence of dot interlace, the number of dots can be increased above the 200,000-dot limit only if the number of fields per second is correspondingly reduced below the 60-per-second figure. Alternatively, the number of fields per second can be increased only if the number of dots is proportionately decreased.

The dots represent pictorial detail, and the field rate determines the brightness at which flicker becomes apparent. Hence pictorial detail can be increased only at the risk of incurring flicker, and flicker can be controlled only by incurring a loss in pictorial detail, once the channel width and picture brightness have been decided upon.

The conflict between pictorial detail and flicker has occupied the center of the stage in television development for many years. One result of this conflict is the division of the lines in a television picture into two groups, one interlaced within the other. This technique of "line interlace" was developed as early as 1934 to reduce flicker while maintaining the pictorial detail at a satisfactory level. In line interlacing, the area of the image is illuminated twice while the pictorial detail (200,000 dots) is laid down only once. While interlacing introduces interline flicker and similar small-area defects, these faults are worth accepting in favor of the general reduction of flicker, and the permissible brightening of the picture.

Much more recently (first announced publicly in 1949), an extension of this principle known as "dot interlace" was developed. In dot interlace, the picture elements along each line are arranged with blank spaces between them. In other words, they are actual dots, and the blanks are filled with dots on the next scanning of that line. When added to the line interlace just discussed, the dot-interlace system permits the area of the picture to be illuminated four times while the pictorial detail is laid down once. When the frequency of illumination is maintained at 60 fields per second, dot interlace plus line interlace thus permits all the pictorial detail to be laid down in one-fifteenth second (actually about one-twentieth second when the blanked-off portions of the lines are taken into account). Dot interlace therefore permits 400,000 dots to be accommodated in the picture as contrasted with 200,000 dots when only line interlace is used.

10. Color reproduction: The role of primary colors

The addition of color values to a television picture involves the reproduction of the thousands of different colors which the eye can distinguish. This seemingly formidable task is vastly simplified by the fact, established in Newton's time, that all colors can be very closely represented by combining just three colors, known as primary colors.

There are two types of primary colors. When the reproduction is effected with layers of colored material, one on top of another, through which light must pass in succession, the so-called "subtractive primaries" must be used to obtain a satisfactory range of mixture colors. The subtractive primaries are red, blue, and yellow. These are the familiar primary paint colors known to students in elementary school. Subtractive primaries are used in oil and water-color paintings, in color printing, and in color photography (prints and transparencies). In color printing and photography, the primary colors used are a bluish red ("magenta"), a greenish blue ("cyan") and a greenish yellow. These subtractive primaries are the ones most commonly known to the public.

In color television, the reproduction is not effected with layers of colored material one over the other, but rather consists of individual lights of the primary colors presented one after the other in time sequence. For this type of color reproduction, the so-called "additive primaries" must be used. The additive primaries are red, blue, and green. If pieces of red glass and green glass are placed one beside the other (not one on top of the other) and white light is passed through them in such a manner that the red and green light thus formed falls on the same area of a viewing screen, the combined light will have a yellow color. If red, green, and blue glasses are similarly employed, the combined light on the screen will appear white, or near white.

With these primaries combined in proper proportions it is possible to reproduce any of the hues of the visible spectrum, plus purples which do not appear in the spectrum, plus all the shades of gray from white to black, as well as mixtures of the above. With only three primary colors it is not possible to reproduce all the spectrum colors exactly, but the color match can be made so close that only simultaneous inspection of the original color and the reproduction will reveal the difference. Experience with various types of color photography has shown, in fact, that a highly realistic rendition of natural colors can be achieved with three properly chosen primaries.

When only two primary colors are used, the rendition is very much less realistic. The primaries customarily used in the two-color process are a red-orange and a green-blue. With these, it is not possible to reproduce a pure (saturated) red, a pure blue, a pure yellow, or a pure green. Some improvements have been obtained by making the two primary colors change with brightness, but even so a system of such limitations cannot properly reproduce the colors in nature. For this reason, two-color processes have not been widely employed in motion pictures, nor have they been proposed for public color-television service. All of the color-television systems discussed in this report use the three additive primary colors, red, green, and blue.

Since at least three primary colors must be used to achieve realistic color reproduction, it follows that three color images must be transmitted by a color-television system. The three color images are transmitted in sequence, hence the name "sequential color television system." In the dot-sequential system, the primary colors are assigned to successive dots of the image. In the line-sequential system, the primary colors are assigned to successive lines of the image. In the field-sequential system, the colors are assigned to successive fields of the image.

The manner in which the colors are interspersed is discussed in detail in the following chapters relating to the three systems. Here it suffices to say that three separate images, one in each of the primary colors, must be dissected in a particular sequence at the transmitter and reassembled in the same sequence at the receiver.   The dissecting and reassembling processes are performed so rapidly that the primary colors are not separately perceived one after the other, but appear to the observer to blend or "fuse," as though they existed simultaneously.

Thus, while it is true that only one primary color is actually present on the receiver screen at any one instant in each of the three sequential systems here described, persistence of vision causes the picture screen to appear as if all three primary colors were present simultaneously throughout the area of the screen.   We may then conclude that a color television image is equivalent to three images superimposed one on top of the other, each image being made up of light of one of the primary colors.   As we shall see later, in each of the proposed systems the color images may be somewhat less detailed than the equivalent black-and-white image.   But this is a difference merely of degree. In principle, a three-color television system employs the equivalent of three images, each depicted in light of one of the primary colors.

An important implication of this principle is this: All other factors being equal, the video channel occupied by a three-color television system must be three times as wide as that required for an equivalent black-and-white system.   A color system equivalent in pictorial detail to the black-and-white system must transmit three images, each containing 200,000 dots, and all these images must be transmitted in the same time as that of one image in the black-and-white system. Hence the rate of transmitting picture elements (dots) in color is three times the rate in black and white, and the video channel width must be trebled to accommodate the transmission.

If a color-television system is to be fitted into a 6-megacycle channel, something must be sacrificed.   Either the dots must be reduced in number, thereby reducing the pictorial detail, or the number of scanning lines must be reduced, thereby again reducing the pictorial detail, or the rate of scanning the fields must be reduced, thereby incurring flicker unless the picture brightness is dimmed by a substantial amount, or the number of fields per color picture must be increased, thereby increasing the blurring of motion.   Some such compromise has been adopted in all three of the proposed color systems. The nature of the compromise, and its effect on the over-all image quality, is an important basis on which the systems must be compared.

The fact that three separate, apparently superimposed, images are involved in a color-television system gives rise to several potential sources of trouble.   The first is "improper registration".   The three primary-color images must be precisely the same size, have precisely the same shape, and must appear to lie one directly over the other if the color reproduction is to be accurate.   Lack of registration is familiar in color printing.   It occurs when the impression of one printing plate is out of position with respect to the other impressions.   The outlines of objects are thereby blurred, the fine detail of the image is obliterated, and objects are outlined with color fringes.

Considerable care is required in the design and operation of a color-television system to secure proper registration.   The three types of systems described herein employ different methods to secure registration, and their performance in this respect differs, as outlined in chapter 6.   In particular, the field-sequential camera has at present better performance in this respect than the dot- and line-sequential cameras.

The second source of difficulty, rooted in the sequential nature of a color-television image, is known as "color breakup". When the eye moves while viewing a color-television image, either casually or in following the motion of the image, the successive fields laid down on the screen occupy slightly different positions on the retina of the eye. If each successive field is displayed in one primary color, as in the field-sequential system, the separate primary colors are then visible in the form of fringes around the outlines of objects. This effect is present only in the field-sequential system, since in the dot- and line-sequential systems the color-switching rate is many times greater. Fortunately, the majority of observers possess, or soon acquire, a substantial tolerance for the color-breakup effect, under normal conditions.

The third effect is "color fringing". This occurs when a rapidly moving object is televised in color. If the object has color components in more than one primary (as do the vast majority of objects), and if the object is scanned in successive fields of different color (as in the field-sequential system), the object will be scanned in one color on one field and in another color on the next successive field. If the object is moving rapidly, its position on the screen will have changed between the successive scannings, and the object will appear fringed with color or, if the motion is very rapid, as several objects in different colors. Like color breakup, color fringing does not occur to a noticeable degree in the dot-sequential and line-sequential systems, since the color-switching rate is many times greater.

11. Color fidelity: How true is the color reproduction?

It is evidently of paramount importance that the reproduced colors be sufficiently faithful copies of the original colors to induce a sense of realism in the observer. A first requirement of faithful color rendition is that the primary colors employed at the receiver are chosen to cover a suitably wide gamut of colors. Since a free choice of phosphors and color filters is open to all, this factor does not necessarily operate to the advantage of one system over the others.

A second requirement for proper color rendition is color balance, a term that indicates that the brightnesses of the three primary-color images are in proper proportion. This is especially important when all three primaries are combined to produce a white (or nearly white) element in the picture. Unless the three primary images are in precise balance, the element intended to be white will exhibit a greenish, reddish, or bluish tinge (or other off-color tinge), depending on what primary (or what pair of them) is present in disproportionate brightness. Color balance is particularly important in reproducing the delicate tones of flesh color. A slight excess of green, for example, can transform a ruddy glow into a sickly pallor. Color balance requires the correct choice of camera color filters to accord with the receiver color primaries and with the lights used in the studio. It also depends upon the correct operation of the transmitter, the correct functioning of the receiver, and the correct adjustment of contrast. Here again these basic techniques of maintaining color balance can be used by all three systems, although there are differences in the ease with which it can be accomplished.

So long as the dots, lines, and fields occur in their proper places and in the proper sequence, and so long as the proper color balance is maintained, a high standard of color reproduction is possible in each of the proposed systems of color television.   The observed differences in color fidelity are ascribable partly to poor color balance and partly to lack of registration (of dots, lines, or fields with minor effects due to color breakup and color fringing as described in section 10.

12. The addition of black-and-white detail to a color image

In color printing it is customary to employ four impressions, one in each of the primary colors, and the fourth in black (or dark brown). The black plate impresses shades of gray over the colors.   One purpose of the black impression is to overcome an inherent shortcoming of the primary-color printing inks which, by themselves, are not able to represent as dark shades of gray as may be desired.   Another purpose is to provide one impression (the black impression) which carries the basic pictorial detail of the subject, and thus relax to some extent the need for precise register among the three primary-color impressions.

This printing technique suggests that a similar method might be used in color television.   If all the fine pictorial detail of a color-television image is presented in shades of gray, the detail of the primary colors may be allowed to be somewhat coarser without adverse effect on the over-all sharpness of the color image.   This would allow an image of given sharpness to be sent over a narrower channel than would be required if the primary colors were sent in full detail and no gray image was employed.

Suppose then that the color image is to have a pictorial detail corresponding to 200,000 dots, equal to the detail of the present standard black-and-white image, and is to be sent at a field rate of 60 per second with conventional line interlace.   Suppose further that all fine details having a width not greater than corresponding to two dots are transmitted only in shades of gray, wheras all details of width greater than two dots are transmitted in three colors.   Then the fine detail in the gray image corresponds to frequencies from 2 to 4 megacycles and thus requires a video channel width of 2 megacycles, while each of the three primary-color images correspond to frequencies from zero to 2 megacycles and thus require three more video channels of 2 megacycles.   The total video channel width is therefore 8 megacycles. By the method of dot interlace the three 2-megacycle color channels may now be interpersed and compressed.   In the RCA form of the dot-interlace system the color channels are compressed into a single video channel from zero to 4 megacycles, or two-thirds of the sum of the three channels.   This color dot signal is finally mixed with the fine detail gray signal and we have the entire picture signal occupying a video channel width of 4 megacycles, permitting it to be transmitted on a 6-megacycle radio channel.   (A more detailed description of dot interlacing the color images and mixing them with the fine detail gray image is given in chapter 5.)

Hence, by confining the finest detail to shades of gray, and by using dot interlace, it is possible to compress a color transmission into the same channel now occupied by the black-and-white transmission, and to retain substantially the full detail of the image (200,000 dots) and the full flicker-brightness performance (60 fields per second).

The technique of transmitting fine detail in shades of gray only is known as the "mixed highs" system, from the fact that the highest frequencies in the three color signals are mixed together before transmission to the receiver.

In the example given above, the dividing line between full-color transmission and gray-tone transmission was taken at a detail size equal to the width of two dots. The dividing line can be set at details considerably larger than this. In fact, in the RCA system, as described in chapter 5, certain practical shortcomings of the dot-interlace transmission process, currently embodied in the apparatus, reduce the detail transmitted in true color to items having the width of eight dots. Since there are some 420 dots or more to the line this still represents a very good color detail and no adverse effects are noted. Moreover, the shortcomings of the present apparatus in this respect are not fundamental and can be compensated rather exactly should the need arise.

The technique of transmitting fine detail in tones of gray is applicable only to the dot-sequential system of color television. It is not applicable to the line-sequential and field-sequential systems because these systems make no color distinction between the dots along any one line of the image. Hence, whatever detail is provided, as each line is scanned, must necessarily be provided in full to the particular color present in that line. Thus all three color images contain the fine detail, and there is no opportunity to confine the fine detail to a single (gray) image.

13. Relation of color service to existing black-and-white service

The principles discussed in earlier sections of this report refer to the intrinsic properties of sequential color systems which are rooted in the choice of scanning method. These properties determine the long-time utility of each system, since they are based on the fundamental attributes of human vision.

There are several additional properties of a less fundamental nature, but of great economic importance, which refer to the transition from the existing black-and-white service to the future color service. The problems of this transitional period will endure so long as both black-and-white and color transmissions are available in a given locality, and this situation may continue in many populous areas for an indefinite period. Accordingly the committee believes that the relative suitability of the color systems for public use must be judged, in part, in terms of their relation to the existing black-and-white service.

The transitional properties of the color systems are described by three terms—compatibility, adaptability, and convertibility—defined as follows:

Compatible color system.—A compatible color system is one capable of producing black-and-white images on existing black-and-white receivers without any modification of the receivers.[1]

Adaptable color system.—An adaptable color system is one in which existing black-and-white receivers can be modified to receive color transmissions in black-and-white.

---

[1] This definition was first advanced by the Joint Technical Advisory Committee in testifying before the FCC at the color hearing and adopted by the majority of those testifying thereafter. It is restricted to the rendition of color transmission on black-and-white sets, as defined. Another form of compatibility, sometimes called reverse compatibility, relates to the reception of black-and-white transmissions on color receivers. Since the latter type of compatibility can be possessed by color receivers of all three systems to a nearly equal degree, and will undoubtedly be possessed by all color receivers manufactured during the transitional phase, the committee believes that reverse compatibility is not an important distinction between systems.

Convertible color system.—A convertible color system is one in which existing black-and-white receivers can be modified to receive color transmission in color.

In comparing systems on the basis of adaptability or convertibility, the cost, inconvenience, and technical complexity associated with the modifications are evidently important considerations. Comparative quantitative data on these aspects are at present inconclusive, in view of the rapid state of development of the systems, but it is possible to give a qualitative estimate of the relative adaptability or convertibility of each system.

The transitional properties of each system are stated in chapters 3, 4 and 5, and compared in chapter 6.

14. System characteristics

Those performance characteristics which are of paramount importance in comparing color television systems are—

(a) *Resolution.*—The amount of pictorial detail or the number of picture elements (dots) contained within the picture area. The greater the number of dots, the more copious the pictorial detail in the reproduced image.

(b) *The flicker-brightness relationship.*—The rate at which the successive fields are scanned determines the maximum brightness of the reproduced picture, above which flicker becomes objectionably apparent.

(c) *Continuity of motion.*—The number of fields presented per second must be high enough to permit motion in the image to be rendered in apparently continuous fashion.

(d) *Effectiveness of channel utilization.*—Since the space in the radio spectrum for television channels is severely limited by the needs of other services, it is of paramount importance to determine the relative effectiveness of the color systems in utilizing the 6-megacycle channel. The preceding sections have shown that the channel width is devoted to the performance characteristics above named; that is, adequate resolution, adequate brightness without flicker, and adequate continuity of motion. A system whose performance is inadequate in any of these aspects makes relatively ineffective use of the channel. In comparing two systems having equally adequate performance in one or two of these aspects, the system having superior performance in the remaining aspect or aspects is defined as making the most effective use of the channel. On this basis it is possible to compare the systems, on a qualitative basis, with respect to channel utilization.

The techniques for improving channel utilization include line interlace, dot interlace, the mixed-high method, and the use of long-persistence receiver screen materials to reduce flicker.

(e) *Color fidelity.*—Color fidelity is the degree to which the television receiver reproduces the colors of the original scene. It is particularly important that the system be capable of maintaining color fidelity over extended periods of time.

(f) *Defects associated with superposition of primary-color images.*— These defects include improper registration, color breakup, and color fringing.

(g) *Cost of color receivers.*—A final basis of comparison is the cost of a color receiver having adequate performance in each of the respects listed above.   While it is manifestly necessary to take this factor into account in arriving at a decision between the systems, the presently available cost figures are, in the opinion of the committee, not indicative of the situation to be expected when manufacture of receivers actually commences on a large scale.   If, as seems probable, a tricolor tube is to be used in future receivers, no matter which system is adopted, the costs will be more nearly equal than if a rotating filter disk is used in one system (CBS), a three-tube dichroic-mirror receiver in another (RCA) and a triple-projection receiver in the third (CTI). In view of the fact that a definitive answer to the question of receiver costs cannot be available until the color service is actually instituted and large-scale production is under way, the committee believes that it will not be possible to take the relative receiver cost factors into consideration in arriving at the necessary policy decisions affecting color television.

In the following chapters these factors are related explicitly to the three proposed systems, and the apparatus used in each system is described as it relates to performance, complexity, and cost.

CHAPTER 3

THE CTI LINE-SEQUENTIAL SYSTEM

15. Introduction

The information on the CTI system, contained in this chapter, is based in part on the document "Written Comments of Color Television, Incorporated" dated August 25, 1949, submitted in evidence before the FCC hearing, and in part on verbal comments offered by representatives of CTI at the demonstrations of the system.   The description is based on the system as demonstrated by CTI on May 17, 1950, namely, that using the so-called "interlaced color shift."

16. The CTI scanning pattern: How the picture is put together

Figure 1 illustrates the manner in which the CTI line-sequential color television image is scanned.   The figure shows the scanning lines in the six fields required to make up a complete color picture.   The lines are separated by blank spaces of equal depth.   In the first field the topmost line (line 1) is scanned wholly in green, the next line below (line 3) wholly in blue, and the next line (line 5) wholly in red. As successive lines are scanned (lines 7, 9, 11, etc.) the sequence of lines in green, blue, and red is repeated until the bottom of the image is reached.   This completes the scanning of the first field.

Thereafter the second field is scanned in the same manner, also covering lines numbered 1, 3, 5, 7, 9, etc.   Line 1, this time is scanned in red, line 3 in green, line 5 in blue, and so on until the bottom of the second field is reached.

The third field is scanned next, again covering only the odd lines. This time line 1 is scanned in blue, line 3 in red, and line 5 in green, and so on until the last odd line of the field is reached.

The image has now been scanned in all three colors covering the odd lines only, and this process is then repeated for the even lines, which lie midway between the odd lines scanned in the first three fields.

In the fourth field the color sequence is as follows: line 2 is scanned in green, line 4 in blue, line 6 in red, and so on, down to the bottom of the field.   For the fifth field the color sequence is line 2 blue, line 4 red, line 6 green, and so on, and finally in the sixth field the color sequence is line 2 red, line 4 green, line 6 blue, and so on.

All the lines have now been scanned in all three colors and a complete color picture has been produced.

The image consists of 525 lines, about 490 of which are visible on the viewing screen, and the fields are scanned at a rate of 60 per second. The radio channel width used is 6 megacycles, corresponding to a video band width of about 4 megacycles. These numbers are identical to those employed in the standard black-and-white system. Consequently the number of picture elements per line is the same as in the black-and-white system, about 420 picture elements per line. The maximum number of picture elements in the image, comprising 490 visible lines each with 420 dots, is about 200,000.

The whole sequence of color scanning is completed after six fields have been scanned, and the sequence then repeats. Since the field scanning rate is 60 per second, there are one-sixth as many, or 10, complete color pictures per second.

17. *Essential equipment of the CTI system*

Before discussing the performance of the CTI system, it is necessary to describe briefly certain essential items of equipment, unique to this system. These include the camera at the transmitter and the viewing apparatus at the receiver (picture tubes and viewing screen).

The CTI camera employs one image orthicon camera tube, of the type commonly used in black-and-white broadcasts. When used for black-and-white transmissions, one lens focuses the image on the sensitive plate of the camera tube. When used for color transmission a system of color-selective filters is used for producing three images side by side on the sensitive plate, one for each of the three primary colors.

The lenses are so positioned that they form three images on the sensitive plate, one beside the other in a horizontal row. These images appear in the three filter colors, and are arranged in the order red, green, blue, from left to right.

The three images are scanned, from left to right as a group, by sweeping a beam of electrons across the sensitive plate. As the beam sweeps it creates an electrical signal proportional to the values of light and shade along a particular line in each of the images. Consequently as the beam sweeps once across the group of images, it scans first a line in red, then a line in green, and finally a line in blue.

The beam then scans across the group of three images, along an appropriate path parallel to the first, and thereby produces three more lines in red, green, blue, and so on. This scanning process continues, each passage of the beam across the group creating three lines in the three primary colors, until the bottom of the group of images is reached. The beam has now scanned a complete field, corresponding to one of the scanning patterns shown in figure 1 and described in section 16.

The beam then returns to the top and the scanning process is repeated across the group of images. By properly adjusting the starting point of the scanning process in each successive field the color sequence is arranged to conform with the scanning sequence described in section 16 and illustrated in figure 1.

The signal created by the camera is transmitted to the receiver. Here the images are reproduced on the screen of a picture tube. The screen is composed of three different types of fluorescent material, arranged side by side, one material producing red light, another green light, and the third, blue light. The scanning beam in the picture tube moves over this three-part screen in exactly the same pattern as the beam in the camera and thereby recreates on the screen three images side by side, in red, green, and blue. These images are, therefore, replicas of the optical image focused on the sensitive plate of the camera tube.

The three primary-color images are combined by projecting them through three lenses onto a common viewing screen. Care must be taken, in the scanning of the camera and picture tube and in the positioning of the camera and projection lenses, to insure that these three images are precisely in register on the viewing screen. A reproduction of the original scene in color thereby appears on the projection screen.

If a black-and-white receiver, of the type commercially available in the United States, is tuned to a color transmission from a CTI color camera, a black-and-white image results. This follows from the fact that the CTI system operates with 525 lines, 60 fields per second, which are identical to the scanning rates of the standard black-and-white system. For this reason, the CTI system is known as a compatible system, i. e., a color-television system which will provide a black-and-white version of the color transmission on present-day black-and-white receivers, without requiring any change in the receiver.

18. Performance characteristics of the CTI system

On the basis of the foregoing description of the CTI system, we can examine its performance characteristics in accordance with the outline presented in chapter 2, section 14.

The first of these characteristics is resolution, section 14 (*a*). Since the scanning rates and channel width of the CTI system are identical to those of the standard black-and-white system, the number of picture elements per line is, in theory, the same in the two systems, namely about 420, and the number of lines visible in the image is also the same, about 490. Therefore the over-all resolution of the CTI and black-and-white systems are the same, about 200,000 picture elements.

These resolution figures are based wholly on the geometry of the scanning pattern, and take no account of other effects, such as improper registration and line crawl, which may reduce the effective resolution available in the CTI system, as discussed below.

The second performance characteristic is the flicker-brightness relationship, section 14 (*b*). Since the number of fields per second in the CTI system is 60, the same number as in the black-and-white system, the large-area flicker performance is approximately the same. The small-area flicker effects are accentuated by the fact that each line is scanned in any one primary color only 10 times per second. Moreover, two lines of the same color in any one field are separated by two other lines in different colors plus three blank spaces. These two effects together cause an apparent motion of the lines upward or downward in the picture, known as line crawl. In common with all flicker effects, line crawl becomes more pronounced as the image becomes brighter. Consequently this effect may in fact set the upper limit on the acceptable brightness of a line-sequential color image.

The phenomenon of line crawl is accompanied by an apparent grouping of the lines and this effect reduces the apparent vertical resolution below the value set by the scanning pattern geometry.

Interline flicker is also pronounced in the image produced in this system, particularly when a primary color is being transmitted, because each line is then illuminated only 10 times per second. If the image is bright, sharply defined horizontal edges exhibit a marked blinking effect.

The third performance characteristic, section 14 (c), is continuity of motion. So far as large-area portions of the image are concerned the continuity is determined by the field rate, so the performance is not noticeably different from that of the black-and-white system. On the other hand, the sharpness of edges of colored objects in motion is noticeably affected by the fact that the complete color sequence occurs at a rate of 10 per second, whereas the complete sequence in black-and-white images occurs at 30 per second.

The fourth characteristic, section 14 (d), is effectiveness of channel utilization. Here the principal shortcomings of the CTI system, as thus far demonstrated, are the impracticability of using dot interlace and the poor small-area flicker performance. If dot interlace were attempted, while the resolution would be doubled, the complete color-sequence rate would be lowered to 5 per second, thus greatly accentuating the small-area flicker effects.

The nature of the compromise necessary to fit the CTI system into the 6-megacycle channel can now be stated. The resolution and the large-area flicker performance are maintained, so far as scanning is concerned, at the values of the black-and-white system, but to secure this performance in color it is necessary to lower the rate of the complete scanning cycle to 10 per second, one-third the value of the black-and-white system. Accompanying the lower scanning cycle rate are small-area flicker effects, notably interline flicker and line crawl.

The fifth performance characteristic is color fidelity, section 14 (e). On the assumption that proper fluorescent materials and color filter are used in the picture tube and proper color filters are used in the camera, the large-area color fidelity of the system suffers no limitation. Lack of registration, noted below, may affect adversely the color fidelity in small areas, particularly in the fine details and along the edges of brightly colored objects.

Superposition defects, section 14 (f), are limited to improper registration, since color breakup and color fringing are confined to the depth of one or two scanning lines. Faulty registration may appear in four independent ways: (1) Misadjustment of the camera optics may produce color images of different size, shape, or orientation on the sensitive plate of the camera tube; (2) the motion of the camera electron beam may not be uniform or not properly alined with the images; (3) the scanning at the receiver picture tube may not produce congruent and properly oriented images; and (4) the projection lenses of the receiver may not bring the images into correct superposition on the viewing screen.

Finally, the method of depicting fine detail, section 14 (d), in this system is to impose the fine detail on all three primary-color images. The mixed-highs system of transmitting fine detail only in shades of gray cannot be used in the line-sequential system for the reasons outlined in section 12.

19. Summary

The essential attributes of the CTI line-sequential system are as follows:

(*a*) It is a compatible system, employing the same number of lines per picture and the same number of fields per second as the black-and-white system. This permits a black-and-white version of the color image to be reproduced on standard black-and-white receivers, without modification of the receiver.

(*b*) It achieves resolution and large-area flicker performance equivalent to the black-and-white system, but is deficient in apparent vertical resolution and small-area flicker performance.

(*c*) It is subject to registration difficulties.

(*d*) It does not employ the channel width effectively, since neither the dot-interlace nor the mixed-highs principle are employed.

CHAPTER 4

THE CBS FIELD-SEQUENTIAL SYSTEM

20. Introduction

The information in this chapter is based on the testimony submitted by the Columbia Broadcasting System to the FCC during the color-television hearing, and on demonstrations of the CBS system viewed by members of the committee prior to May 1, 1950.

21. The CBS scanning pattern

Figures 2 and 3 illustrate the manner in which the CBS field-sequential color-television image is scanned. In figure 2 is shown the conventional line-interlaced version of the system. Each picture consists of 405 lines, divided into two fields of 202½ lines each. The fields are scanned at a rate of 144 fields per second. As shown in the figure, all the lines in one field are scanned in blue, the next in green, and so on in the sequence red, blue, green.

After six successive fields have been scanned, every dot in the image has been scanned in all three primary colors. Consequently, the whole scanning sequence occurs at a rate one-sixth as great as the field-scanning rate, that is, $144/6 = 24$ complete scanning cycles per second. The complete scanning cycle is termed a "color picture." The color-picture rate of the CBS system is, accordingly, 24 per second.

In the dot-interlaced version of the CBS system (fig. 3) each line is broken up into dots, all of the same primary color, with blank spaces of equal size between the dots. These blank spaces are filled in with dots of another primary color, on the next successive scanning of that line. Consequently, a given dot in the image is scanned in all three colors only after 12 consecutive fields have been scanned, and the complete scanning cycle occurs at a rate of $144/12 = 12$ color pictures per second. The corresponding color picture rate of the CTI line-sequential system (sec. 16) is 10 per second, and that of the RCA dot-sequential system (sec. 26) is 15 per second.

22. Essential apparatus of the CBS system

The CBS color camera employs one image orthicon camera tube and one lens. Between the lens and the sensitive plate of the camera tube is located a filter disk containing six transparent filter segments, two for each of the three primary colors. The disk rotates at 1,440 revolutions per minute, so the filter segments move past the sensitive plate at a rate of 144 segments per second. The disk rotation is synchronized with the 144-per-second field-scanning rate of the camera. In this manner all the lines in one field are illuminated in red light, the lines of the next field in blue, and the lines of the third field in green, and so on, in the sequence red, blue, green.

These elements of the CBS camera are the same in the line-interlaced and dot-interlaced versions of the system. In the dot-interlaced version, the electrical output of the camera is rapidly switched on and off. The camera is thus effectively connected to the circuit during the scanning of a particular dot, and is disconnected during the scanning of the adjacent blank space, then reconnected for the next dot, and so on. The rate of connecting and disconnecting the camera is about 9,000,000 per second (9 megacycles).

Two types of receiver have been demonstrated by CBS. In the first a rotating filter disk, similar to that used in the camera, is positioned before the screen of the picture tube. This disk carries six filter segments, two in each of the three primary colors. The disk rotates at 1,440 r. p. m. and is synchronized with the 144-per-second field-scanning rate of the receiver. The image formed on the screen of the picture tube is displayed in white light, and this light, passing through the colored filters, takes on successively the three primary colors. Thus, the light emerging from the receiver is red on one field, blue on the next successive field, and green on the third, and so on. By means of synchronizing impulses, the position of the receiver filter disk is controlled so that red light is produced by the receiver only when the red filter is positioned before the camera tube at the transmitter, and similarly for the other two colors.

The system thus comprises two filter disks rotating in rigid synchronism, so positioned that the filters before the camera and the picture screen always have the same color at any instant.

It is not considered feasible to use a rotating disk with picture tubes exceeding about 12½ inches in diameter because of the physical size of the disk involved.

The second type of receiver is very similar to that used in the CTI system, described in section 17, chapter 3. A single picture tube is used, but three separate images are formed on the screen, one above the other, one in each of the primary colors. The blue-colored image is formed only during the fields scanned in blue by the camera, and similarly for the images in the other two colors. An optical system comprising three lenses projects the three images so that they fall, one on top of the other, on a common viewing screen. The scanning of the images, and the choice of lenses and positioning of the lenses with respect to the image, must be precisely controlled to preserve registration between the projected images. By using a green phosphor of comparatively long decay time this type of receiver eliminates practically all flicker and color break-up.

Both types of receiver may be used with the line-interlaced as well as the dot-interlaced version of the system. For dot-interlaced reception, additional circuits are required which effectively connect and disconnect the picture tube in synchronism with the corresponding connections and disconnections of the camera tube, described above.

23. Performance characteristics of the CBS system

We proceed now to examine the performance characteristics of the CBS system, in accordance with the outline of sections 13 and 14, chapter 2.

Resolution (sec. 14 (a)).—For reasons given below (under flicker-brightness relationship), the field-scanning rate of the CBS system must be chosen substantially higher than that of the black-and-white system. The rate used in the CBS demonstrations is 144 fields per second.

In section 7 it was explained that the standard black-and-white television system has a geometric resolution of approximately 200,000 picture elements per frame (two interlaced fields). This corresponds to a field repetition rate of 60 per second and a video band width of 4 megacycles (6-megacyle radio channel).

In the CBS line-interlaced system the geometric resolution is also determined by the number of picture elements in two interlaced fields, but the field repetition rate is now increased to 144 per second. It was explained in section 9 that for a given band width the number of picture elements in a frame is inversely proportional to the field-scanning rate. The geometric resolution of the CBS line-interlaced system is therefore 200,000 times 60/144 or 83,000 picture elements. Thus the higher field-repetition rate decreases the geometric resolution of the CBS line-interlaced system to 60/144 or 42 percent of that of the standard black-and-white system.

In the dot-interlaced version of the CBS system, the resolution is doubled in theory, and very nearly doubled in practice. Thus the resolution of the dot-interlaced CBS color image is about $2 \times 83,000 = 166,000$ picture elements or 83 percent of the resolution of the standard black-and-white image.

Flicker-brightness relationship (sec. 14 (b)).—In a field-sequential color system, such as the CBS system, flicker is a much more difficult problem than in a line-sequential or dot-sequential color system. This follows from the fact that the eye is more sensitive to large-area flicker than to small-area flicker and from the fact that in the field-sequential system, the interruption of the image in changing from color to color occurs over the whole picture area.

To counteract the prominence of large-area flicker, it is necessary to increase the field-scanning rate by a substantial amount. Experience has indicated that, for equal flicker-brightness performance under all conditions, the field-scanning rate of a field-sequential system should be about 3 times that of a black-and-white system. Actually, in the CBS system, the field rate has been increased by the ratio 144/60=2.4 times, rather than 3 times. The lower value was chosen to preserve as much geometric resolution as possible within the confines of the 6-megacycle radio channel.

It follows that large-area flicker is more prominent in the CBS system than in the black-and-white system. The comparable flicker rates in the two systems are 48 per second in the CBS color system, (twice the complete picture rate) and 60 per second in the black-and-white system (the field-scanning rate). The difference in the rates is 12 per second. According to the Ferry-Porter flicker law, this difference in flicker rate would allow the black-and-white image to be about 9 times as bright as the color image, for equal visibility of flicker.

Corresponding to these theoretical values are various practical values quoted in the testimony given at the FCC hearing. It was reported that flicker can be held within tolerable levels if the high-light brightness of the CBS color image is not greater than about 25 foot-lamberts, whereas the corresponding limit for the standard black-and-white image is well above 100 foot-lamberts. The 25-foot-lambert figure was quoted for the filter-disk type CBS receiver. In the projection-type CBS receiver, using a long-persistence phosphor in the green image, higher brightnesses were attained within the tolerable limit of flicker.

At the request of the committee, tests of large-area flicker were made by the National Bureau of Standards. The results are given in annex D of this report.

So far as small-area flicker is concerned, the CBS line-interlaced system is not substantially different from that of the black-and-white system, and it may be somewhat superior when the colors transmitted are not too close to saturated red, green, or blue. The dot-interlace version of the CBS system is, on the other hand, somewhat inferior in this respect to the black-and-white system. Small areas (dimensions of the order of a picture element) are scanned at a color-picture rate of 12 per second in the dot-interlaced CBS system.

Interline flicker should be somewhat more pronounced in the CBS system, when colors in the scene approximate the primary colors, because adjacent lines are then laid down at intervals of one-forty-eighth second, compared to one-sixtieth second in the black-and-white system. However, when the colors comprise components of all three primaries in roughly equal amounts (and this is likely to be the case in bright—e. g., white—portions of the scene), adjacent lines are laid down at intervals of one one-forty-fourth second, and the interline flicker is then less noticeable than in the other systems.

Continuity of motion (sec. 14 (c)).—Continuity of motion, like flicker, is affected in the CBS system by the composition of the colors transmitted. If the object in motion is displayed in one of the primary colors, the other two primaries being substantially absent, then that portion of the image is illuminated only one-third of the time, and the motion may appear jerky. If two or three primary-color components are present, the illumination is more nearly continuous and the discontinuity is not so pronounced.

In either event, motion is portrayed with sufficient smoothness to satisfy the eye, at color-picture rates in excess of 10 per second. So far as large areas are concerned, this requirement is met by both the line-interlaced and the dot-interlaced versions of the CBS system. In small areas, notably the detail of vertical and horizontal edges of objects, the dot-interlaced version of the system may display ragged edges on an object in rapid motion.

Noncompatible and convertible nature of the CBS system (sec. 13).—The fact that the field-scanning rate of the CBS system must be substantially higher than that of the black-and-white system leads to a most important difference in receivers designed for the two systems. In the black-and-white system, the vertical (field) scanning occurs at a rate of 60 per second, and the horizontal scanning at a rate of 15,750 per second (30 frames per second, each having 525 lines). In the CBS field-sequential system, the vertical (field) scanning occurs at 144 per second, and the horizontal scanning at 29,160 per second (72 frames per second, each of 405 lines).

The respective values in the two systems are so different that receivers built for black-and-white reception cannot be adjusted to scan at the higher rates required for the CBS color system, unless modifications are made in the receiver scanning circuits. This fact is the root of the "compatibility" argument. The cost of modifying existing receivers to make them operative on both sets of scanning standards may be substantial, and no reliable data have been submitted as to what this cost would be. However, by modifications of the circuits and the addition of a rotating disk, existing sets with picture tubes less than 12½ inches diameter can be converted to color reception. Thus the CBS system is convertible but not compatible.

Effectiveness of channel utilization (sec. 14 (d)).—We have previously noted that both the line-interlaced and the dot-interlaced CBS systems have a flicker-brightness performance somewhat lower than that of the black-and-white system. The line-interlaced version displays resolution which is substantially lower than the black-and-white value. The dot-interlace version has poorer performance so far as small-area flicker and small-area continuity are concerned, but achieves resolution not markedly below that of the black-and-white system. The dot-interlace system makes substantially more effective use of the channel and is to be preferred, on this account, to the line-interlaced version of the system.

The nature of the compromise, adopted to fit the CBS system into the 6-megacycle channel, is determined principally by the large-area flicker effect. Since the color sequence is introduced by changing the color of the whole image at once, it is necessary to increase the field rate by a substantial amount, relative to the black-and-white system, and to lower the geometric resolution in proportion.

It may be argued, therefore, that the field-sequential scheme is less effective in channel utilization, because it devotes a disproportionately large amount of spectrum space to the reduction of flicker, at the expense of a substantial loss in resolution. Stated in another way, the use of the field-sequential technique, with dot-interlace, results in a picture having less geometric resolution (about 83 percent of the black-and-white value) and lower large-area flicker-brightness performance (brightness at flicker threshold about one-ninth the black-and-white value, for a given phosphor decay characteristic). Finally, the fact that the mixed-highs technique cannot be used in the field-sequential system has the effect of lowering the channel utilization, relative to that of a dot-sequential color system using mixed highs.

Color fidelity (sec. 14 (e)).—There is, as noted previously, no basic difference in the color fidelity of the three color systems. This statement assumes a proper choice of filters, phosphors, and light sources, proper color balance and gradation, and freedom from superposition defects. In practice, as the systems were demonstrated to the committee, the CBS system displayed superior color fidelity to the other two systems when filter-disk receivers were employed. This superiority is explained by better color balance (the same area is scanned in all three primary colors in the CBS camera, and in the filter-disk type CBS receiver as noted below), and by more accurate registration between the primary color images.

At the request of the committee, tests on the fidelity of color reproduction by both the CBS and the RCA systems were undertaken by the National Bureau of Standards. Results are given in annex E of this report.

Superposition defects (sec. 14–f).—A noteworthy characteristic of the field-sequential system is the fact that the color sequence occurs at a slow rate (144 per second), compared with the CTI line-sequential system (15,750 per second) and the RCA dot-sequential system (10,800,000 per second). The slow color sequence, while making the flicker problem comparatively serious, has the compensating advantage of allowing the color sequence to be introduced mechanically by the rotating filter-disk method. Since, in this method, filter segments are placed successively in front of the camera tube and picture tube, it is necessary to employ only one scanned surface for all three primary colors. The CTI and RCA systems require in the camera a separate image for each of the three primary colors, and similar images in the receiver.

Since only one scanned surface is used in the CBS filter-disk system, maintenance of proper registration between the primary color images is a simple matter. The optical elements are common to all three images, so optical misregistration cannot occur. Electrical registration is assured if the scanning pattern of each field is precisely congruent to those preceding and following it, and this requirement is readily met, provided only that the scanning system is adequately protected from stray magnetic and electric fields. The absence of registration defects is a noteworthy characteristic of the CBS system, compared with the present state of development of equipment in the two other color systems.

The other types of superposition defects are, however, more pronounced in the CBS system than in the others, due to the inherent nature of the scanning process. Color break-up and color fringing are detectable when either the eye or the image are in rapid motion.

Depiction of fine detail.—The CBS system cannot, by virtue of the nature of the scanning method used, take advantage of the mixed-highs principle. In compensation for this fact, and to improve the resolution, a circuit technique known as "crispening" has been developed by CBS. This is a method of causing the vertical edges of objects to appear more sharply defined. This technique is not unique to the CBS system, but may be used in any system to achieve the same result. It is believed, therefore, that the use of the crispening technique is not a significant difference between systems.

24. Summary

The essential characteristics of the CBS field-sequential system are as follows:

(a) The CBS system scanning standards are not compatible with the black-and-white scanning standards. This requires modification of existing black-and-white receivers, and additional complication in receivers of the future, to permit reception on both sets of scanning standards.

(*b*) The line-interlaced version of the CBS system has substantially poorer resolution than the black-and-white system. The dot-interlaced version has slightly poorer resolution than the black-and-white system. The crispening technique, applied to the CBS system, improves its resolution. However, this technique, applied to other systems, would improve their apparent resolution also.

(*c*) The large-area flicker-brightness performance of the CBS system is inferior to that of the black-and-white system. This means that CBS color image cannot be as bright, by a factor of 5 to 10 times, as the black-and-white image, for equal freedom from flicker. The dot-interlaced version of the CBS system, operating at the low color-picture rate of 12 per second, has a small-area flicker performance (inter-dot flicker) not as good as the black-and-white system.

(*d*) The color fidelity of the CBS system, as demonstrated, is superior to that of the other color systems. This superiority is due to the maintenance of better color balance and more accurate registration, both of which are implicit in the use of but one scanned surface in the camera and one in the receiver. Much of this advantage is lost in the electronic version of the CBS receiver, since three surfaces are necessary at the receiver.

(*e*) The effectiveness of channel utilization is satisfactory in the line-interlaced version, and is good in the dot-interlaced version. The impossibility of employing the mixed-highs technique lowers the channel utilization with respect to the dot-sequential color system.

(*f*) Existing receivers with picture tubes of 12½ inches and smaller diameter can be converted to color reception, but at an appreciable cost.

CHAPTER 5

THE RCA DOT-SEQUENTIAL SYSTEM

25. *Introduction*

The information in this chapter is based on the testimony submitted by the Radio Corp. of America to the FCC during the color-television hearing, and on demonstrations of the RCA system witnessed by members of the committee prior to May 1, 1950.

26. *The RCA scanning pattern*

Figure 4 shows the manner in which the RCA dot-sequential color television image is scanned. The basic scanning pattern is identical to that of the standard black-and-white system, i. e., the image consists of 525 lines, scanned at a rate of 60 fields per second. About 490 lines of the image are active, and about one-eighteenth of a second is available for the active scanning of all the picture elements in a single field.

Each line of any one field in the image consists of dots in the three primary colors. The dots are arranged from left to right in the sequence red, blue, green. The space between two dots of the same color—e. g., green—is equal to the width of the dots; consequently the dots tend to overlap each other.

On successive scannings of the same line, the dots are shifted, so that the position of a dot of given color falls midway between the positions of two dots of the other two colors, scanned on the preceding frame. Consequently at the end of two frames (four fields), every point on each line has been scanned in all three primary colors. The color picture rate is accordingly $60/4 = 15$ color pictures per second.

The positions of the dots on adjacent lines, scanned on successive fields, is shifted so that a dot of one color falls midway between the dots of the other two colors on the adjacent line. Consequently the whole area of the image, after four fields have been scanned, is covered with a uniform distribution of dots in the three primary colors.

The scanning of the RCA dot-sequential image is of the dot-interlaced variety, as may be appreciated by considering dots of one color only, e. g., green. As noted above, two green dots on one line are separated by a blank space, in which dots of red and blue are fitted, with some overlap. On the next scanning of that line, the space midway between two green dots is filled in by a green dot. The same sequence applies, on successive scanning of any given line, in respect to the red and blue colors. As a consequence of this dot-interlaced technique, the resolution of the RCA image is approximately twice as great as it would be if the interlacing was confined to the lines alone. The dots of any one color are laid down along each line at a rate of 3.58 million per second.

27. Essential apparatus of the RCA system

The camera, used in the demonstrations of the RCA system, employs three image orthicon camera tubes, one lens and a set of color selective mirrors which separate the light from the scene into three colors. Red light enters one camera tube, blue light the second tube, and green light the third tube. The sensitive plate of each camera tube is scanned in identical fashion, at the normal scanning rates of the standard black-and-white system, i. e., 525 lines per picture, 60 fields per second. In this fashion three complete images are televised, one for each of the primary colors. The optical and electrical adjustment of the camera must be such that each of these images is precisely congruent to, and properly oriented with, the others.

When the camera views a scene having fine detail, the output signal of each camera tube contains signal components up to 4 megacycles (actually components of higher frequency may be present but are not transmitted through the system). To take advantage of the mixed-highs principle, the signal from each camera tube is divided into two groups of frequencies. The components of frequencies above 2 megacycles, representing the finest detail in the image, are combined at the outputs of the three camera tubes. This mixed signal represents the finest details of the picture in tones of gray.

The signal components lower than 2 megacycles, corresponding to the respective primary colors, and representing all details of larger size, are transmitted separately. The structure of the image is depicted in color except for the smallest details, which are shown in tones of gray.

The three color signals are transmitted in interspersed fashion by means of a switch which connects and disconnects each camera tube in sequence to the transmitter. This switch (which operates electronically since no mechanical switch could operate at the high speed required), makes and breaks the connection to each camera in rotation at a rate of 3.58 million times a second. Each time a camera tube is connected, it generates a dot of the respective color. When disconnected, that camera is inactive, leaving a blank space in that color. As the switching progresses, the blank spaces between dots of one color are filled in by dots of the other two colors.

The net result is a sequence of overlapping dots along each line, in the sequence red, blue, green, each dot being somewhat larger than a picture element. Superimposed on the colored dot signal is the mixed-high signal, including details from the size of one picture element to several picture elements, in tones of gray.

Two types of receiver have been demonstrated by RCA. In the first type, three picture tubes are employed, one for each of the primary colors. By means of a high-speed electronic switch, like that at the transmitter, each tube is connected and disconnected from the receiver. This switch operates in strict synchronism with the transmitter switch. So the green tube, for example, is connected to the receiver only while green dots are being generated and is disconnected while the red and blue dots are generated. Consequently, on the face of the green tube, a dot-interlaced image appears which represents the image picked up by the camera tube which scans the scene in green. This image does not contain the finest detail of the picture, but the mixed-highs signal is also applied to each picture tube, through the switch, so that the fine detail is in fact present on the face of the green tube. The same arrangement is provided for the red and blue tubes, so that they reproduce images representative of those picked up by the red and blue camera tubes respectively, together with the mixed-highs component, derived from all the camera tubes.

The three primary-color images are combined by viewing them through a system of color-selective mirrors, which reflect light of a given primary color while transmitting light of the other two colors. Care must be taken to assure that the images on the three picture tubes are precisely the same size, have the proper orientation with respect to one another, and are congruent throughout. If these requirements are met, the primary-color images combine in register before the eye of the observer. The fine detail of the combined image, being present in equal amount in all three primary-color images, appears in tones of gray.

The second type of receiver employs but one picture tube, which is viewed directly. The viewing screen of this tube is composed of a very large number of small precisely alined areas, each area consisting of a cluster of three types of phosphor, which glow in the three primary colors. Each cluster represents a picture element which may be made to glow in any one of the primary colors. In one type of the tube demonstrated, three electron guns are used, one gun for each primary color. The guns are so positioned that the electron beams strike the screen at slightly different angles, having passed through perforations in a metal plate parallel to, and just behind, the screen. The angle of each beam is such as to cause it to fall on the phosphor of each cluster which glows in the color assigned to that beam. Thus each picture element in the image may be made to assume any primary color, by activating one gun as it passes that particular cluster, the other two beams remaining inactive during that interval.

To re-create the color image in the single-tube receiver, a high-speed switch, like those previously described, applies the picture signal to the three electron guns in sequence. The timing of the switch is such that the gun associated with one color becomes active at the instant corresponding to the time the camera tube of the same color is connected at the transmitter, and similarly for the other two colors. In this manner the clusters along each line in the image are caused to assume the color and intensity associated with the sequence of red, blue, and green dots transmitted over the system.

The single-tube receiver employs but one scanned surface, so the optical and electrical requirements for proper registration are considerably simpler than in the three-tube type of receiver. Moreover, the electrical and optical components of the single-tube receiver are substantially simpler.

28. Performance characteristics of the RCA system

The performance characteristics of the RCA system, based on the outline of sections 13 and 14, chapter 2, are as follows:

Resolution (sec. 14 (a)).—The resolution of the system must be considered in two categories: the mixed-highs component and the color components. In the mixed-highs component, the maximum picture-signal frequency is 4 megacycles, the same as that of the black-and-white system. Since the time for scanning the active portion of each field is also the same, one eightieth of a second, the number of picture elements per field is the same, about 100,000, and the total resolution (contained in two successive fields) is 200,000 picture elements. This fine structure is, of course, depicted in tones of gray.

The color components, considered individually, each have a maximum picture signal frequency of about 2 megacycles. In the dot-interlace type of transmission each cycle produces one picture element. Moreover, in accordance with the dot-interlace technique, all the dots in any one color are laid down in four consecutive fields, or in one-fifteenth of a second. When account is taken of the portion of the image blanked off, this time is reduced to one-twentieth of a second. Consequently 2,000,000 green dots are scanned per second, or 100,000 green dots during the complete color picture period. Thus, nominally, the resolution in each color is one-half that of the black-and-white image.

Actually the resolution in the individual primary colors is not as high as 100,000 dots because there is a certain amount of dilution of each color by the other two colors. This dilution occurs because the signal corresponding to one color dot overlaps that corresponding to the adjacent color dot by about 50 percent. This phenomenon, known as "cross-talk," has the effect of causing a part of the color values to combine into shades of gray, much in the manner of the mixed-highs portion of the image. The net effect is that details of width from one to eight picture elements are reproduced in shades of gray, whereas all larger portions of the image are reproduced in their component colors.

As indicated in section 12, chapter 2, the superimposed fine detail in the mixed-highs method of transmission provides a substantial economy in the use of the channel, without appreciable degradation of the color or tonal values of the image. In theory, therefore, the resolution of the RCA system is equal to that of the black-and-white system. It should be noted, however, that the tricolor tubes demonstrated had a resolution of 117,000 picture elements, rather than the 200,000 elements of which the system is theoretically capable. This limitation was imposed by the number of phosphor clusters on the screen and perforations in the metal plate which could be accommodated in the tube. Refinements in the design and construction of the tricolor tube may remove this limitation in the future.

In passing it may be mentioned that the tubes as demonstrated were laboratory models of a special design which may involve considerable difficulty in adapting to factory production. At present one of the most urgent needs of all color television systems is for a three-color receiver tube adaptable to quantity production. Besides RCA, a number of others are known to be actively engaged in seeking solutions to this important problem, notably Dr. E. O. Lawrence, of Berkeley, Calif., and Dr. C. W. Geer, of Los Angeles, Calif.

Flicker-brightness relationship (sec. 14 (b)).—The large-area flicker-brightness performance of the RCA system is equal to that of the black-and-white system, since the systems employ the same field rate, 60 per second. The small-area performance is inferior to that of the black-and-white system, however, since a given picture element is scanned in all colors at the comparatively slow rate of 15 per second. Accordingly interdot and interline flicker are present at lower light levels than are the corresponding small-area flicker effects in the black-and-white image.

In early demonstrations of the RCA system a prominent form of dot crawl was evident along vertical or nearly vertical boundaries in the image. In later demonstrations, the geometry of the dot scanning had been altered to minimize this effect, and dot crawl was not then evident.

Continuity of motion (sec. 14 (c)).—Since the field scanning rate of the RCA system is equal to that of the black-and-white system, the continuity of large objects in motion is the same. The continuity of small objects (of the dimensions of a few picture elements) is adversely affected by the low color-picture rate of 15 per second. This short-coming is inherent in the dot-interlace system, and is parallel to the small-area effect noted in section 23, chapter 4, as applying to the dot-interlaced version of the CBS system.

Compatible nature of RCA system (sec. 13).—Since the line and field scanning rates of the RCA color system are identical to those of the black-and-white system, the two systems are compatible so far as scanning goes. Consequently a black-and-white rendition of RCA color transmission can be received on existing and future sets designed for black-and-white reception only, without change in the scanning circuits of these receivers. Moreover, the presence of the mixed-highs component in the color transmission assures high resolution in the black-and-white rendition. The black-and-white rendition of the RCA color transmission has higher resolution and better flicker-brightness performance than do the black-and-white renditions of the CTI and CBS systems.

Effectivenes of channel utilization (sec. 14 (d)).—The RCA system makes highly effective use of the channel because it employs both the principal spectrum-saving techniques—dot-interlace scanning and mixed-highs transmission.

Color fidelity (sec. 14 (e)).—As noted elsewhere in this report, proper choice of mirrors, filters, and phosphors permits the RCA system to achieve satisfactory color fidelity. However, if color balance and accurate superposition of the primary-color images are not maintained, the color fidelity suffers. The color fidelity demonstrated in the RCA system was considered by the committee to be not as satisfactory as that of the CBS system. The larger colored areas in the RCA images were not always uniform in hue and saturation. This may have been caused by differences in the spectral responses of the three camera tubes. Color distortions noted in small areas are explained by overlapping and cross-talk between the color signals, described above. In the early demonstrations of the RCA system, gradual shifting of colors with time was observed, due to uncontrolled shifts in the relative positions of the interspersed color dots along each line. In the later demonstrations, these shifts were controlled by improvements in the synchronization of the high-speed switch of the receiver, and the colors were then found to be free of such variations with time. (See annex E for results of National Bureau of Standards tests on color fidelity of the RCA system.)

Superposition defects (*sec. 14* (*f*)).—Of the three principal superposition defects—color break-up, color fringing, and faulty registration—only the last is present in the RCA system. Registration is more difficult to maintain in the RCA system than in the other systems. This follows from the fact that three separate camera tubes are used, introducing the possibility of optical and electrical errors in the size, orientation, and congruency of the primary images as transmitted. In the three-tube type of receiver, these possibilities of improper registration are present also in the receiver. In the single-tube receiver, faulty registration may occur between the scanning of the three electron guns, but optical misregistration does not occur.

Depiction of fine detail.—The dot-sequential color system, alone of all sequential systems, can use the mixed-highs method of depicting fine detail. In the RCA dot-sequential system, no color information is transmitted at frequencies above 2 megacycles whereas the fine detail, transmitted by the signal from 2 to 4 megacycles, is shown in shades of gray.

29. Summary

The essential performance characteristics of the RCA system are as follows:

(*a*) The RCA system scanning standards are compatible with the black-and-white scanning standards. Consequently a black-and-white rendition of the RCA color transmission can be received on receivers built for black-and-white reception, without modification of their scanning circuits. Moreover, the characteristics of the RCA color system are such that the quality of the black-and-white rendition may be equal to that of standard black-and-white reception, in resolution and large-area flicker-brightness performance.

(*b*) The RCA color image has an over-all resolution approximately equal to that of the black-and-white system. The finest details are depicted in shades of gray, while larger details are rendered in color. The color transmission has sufficiently fine detail that, when the gray-tone detail is added to it, the apparent resolution of the image as a whole is approximately 200,000 picture elements.

(*c*) The large area flicker-brightness and continuity performance of the RCA system is equal to that of the black-and-white system. The small-area performance in these respects is somewhat inferior, due to the fact that the color-picture rate is 15 per second, half the corresponding rate in the black-and-white system.

(*d*) The color fidelity of the RCA system suffers to a certain extent from uneven color balance in large areas. Overlap and cross-talk between the color components, and faulty registration, affect the color fidelity in small areas.

(e) The effectiveness of channel utilization of the RCA color system
is the highest of all the systems discussed in this report.

(f) Existing receivers cannot be converted to color reception in the
RCA system, except at a substantial cost.

CHAPTER 6

COMPARISON OF SYSTEMS AND CONCLUSIONS

30. Introduction

To avoid confusion, each of the foregoing three chapters has been
confined to a discussion of one of the proposed color systems, with a
minimum of comparative comment. The plan of the discussion in
each chapter follows the same pattern, however, so it is possible to
bring together comparably the data and conclusions on the performance of the three systems. This comparison has been set forth in
chart form, in the accompanying "Tabular summary of performance
characteristics." Explanatory comments are given below.

31. Comments on the tabular summary

The committee is of the opinion that the essential differences among
the three proposed color systems are embodied in nine categories,
listed alphabetically at the left of the tabular summary, and defined
in sections 13 and 14, chapter 2, as follows: Adaptability, color fidelity,
compatibility, continuity of motion, convertibility, effectiveness of
channel utilization, flicker-brightness performance, geometric resolution, and superposition defects.

This list purposely omits consideration of certain peculiarities of apparatus such as mechanical versus electronic operation of the receiver
color-sequence device, limitation of size of image, and limitation of
angle of view. These matters once loomed large in the competitive
consideration of the systems, but they have become progressively less
prominent as the development of the systems has proceeded. It
appears, in fact, that all of the systems may use a tri-color tube to
advantage, and this fact puts all three systems on a par with respect
to all-electronic receiver operation, size of image, and angle of view.
Moreover, such differences are not fundamental, either in the transition stage during which color service is introduced to the public or in
the long run as the color service consolidates its position.

The performance characteristics listed in the tabular summary, on
the other hand, are believed by the committee to be fundamental,
either because they reside in the nature of the scanning process, or
because (as in the case of adaptability, compatibility, and convertibility) they are matters of importance during the transition from
black-and-white service to color service.

Under some of the main characteristics are listed a number of subdivisions. These subdivisions are not necessarily of equal importance;
they merely represent items on which system performance displays a
significant difference. For example, under geometric resolution, the
total number of picture elements per frame is more fundamental than
either the vertical or horizontal values of resolution considered separately. To aid the reader, the subdivision believed by the committee
to have outstanding importance within each main category is marked
with an asterisk (*).

No attempt has been made to place relative emphasis on the main categories, which are listed alphabetically to avoid any connotation of relative importance. The emphasis on main categories must be assigned at the highest level of administrative decision, taking into account the economic, political, and sociological factors, as well as the technical factors, involved.

The difficulty of placing this emphasis can be well illustrated by such questions as: Is compatibility (preservation of existing investment) more important than convertibility (converting existing investment)? How do each of these compare with effectiveness of channel utilization (conservation of the public domain) or geometric resolution (providing the maximum flexibility to program producers in choice of subject matter, range of action, and field of view)? Answers to these vexing questions must be found but they are not properly the concern of technical specialists.

So much for the basis of the listings. Opposite each performance characteristic, the committee has placed a verbal or numerical index to the relative performance of the three color systems. These indices represent technical judgments, based either on evident fact, well-established theory, or on the subjective reactions of the committee members to the demonstrations. For the most part, the basis of the committee's judgments will be found in the preceding chapters of this report. But the subjective reactions are difficult to analyze, and the terms "excellent," "good," "satisfactory," "fair," "poor" are, in the last analysis, merely words on which the committee was able to agree as being most indicative of relative performance. The final column in the table indicates the system whose performance is, in the opinion of the committee, superior in each category or subdivision. Where two systems share a superior position, both are listed in alphabetical order.

It is the belief of the committee (1) that this table, with the accompanying text of the report, provides a sound basis for a technical decision among the three systems, and (2) that the only missing element is the relative weight to be accorded each main category. When such weights are assigned, a preponderance of advantage for one system over the others can be found.

The main conclusions reached by the committee have been stated at the outset, in chapter 1. These favor a color service based on the six-megacycle channel, the service to be limited to one of the sequential systems (dot, line, or field).

32. Comments on Possibilities of Future Developments

This report would not be complete without one additional observation, namely that all the systems are subject to improvement as a result of further technical and operational development. The process of improvement will go on in each system until the decision between them is handed down, so long as the proponents and other members of the industry continue to expend manpower and resources on their development.

However, the prospect for future improvement is not of equal magnitude in each system. This is a matter of evident importance in setting standards, since the standards may be expected to be in use for a long time after their full potential has been realized. The net long-term good to the public is thus greatest in that system which can be expected to reach the highest pitch of performance during the next few years. Such technical advances, presuming a choice of one system in the immediate future, will be limited to those matters capable of improvement within the framework of the then-established standards.

It is the opinion of the committee that the CBS system has progressed furthest toward full realization of its potentialities, within the confines of the field-sequential system. It is not likely, for example, that the color fidelity will improve beyond the highly satisfactory state now achieved. Equally, the CBS system is not likely to improve substantially its channel utilization beyond that achieved in the dot-interlaced version of the system. Nor is the flicker-brightness performance capable of substantial improvement, except by methods equally available to other systems, once the picture rate is established at 24 color pictures per second.

The CTI system, being less fully developed, has somewhat greater possibility of future improvement, particularly with respect to correction of faulty registration and small area color distortions, and the development of convertible receiver circuits using a tricolor tube. But in other respects the CTI system cannot reasonably be expected to overcome certain inherent limitations imposed by the choice of scanning method. These include the difficulty of avoiding interline flicker and the impracticability of using dot interlace (at a color picture rate of 5 per second, which is too low for satisfactory rendition of small areas and sharp edges).

The RCA system also has considerable opportunity for improvement within the confines of the scanning standards proposed for this system. The registration of the color images, and the balance of the color values in both large and small areas can be expected to improve substantially with advances in camera design. Convertible circuits, to convert existing sets to color, using the tricolor tube and auxiliary components, can be developed.

The systems discussed above are confined to those developed and demonstrated by their proponents, CTI, CBS, and RCA. An additional demonstration of a dot-sequential system was viewed by the committee. The Hazeltine Electronics Corp. demonstrated a technique known as constant-luminance sampling, which considerably reduces the visible effect of noise and interference in a dot-sequential color image. This demonstration also provided conclusive proof of the efficacy of the mixed-highs technique, in that a video channel of 4 megacycles, carrying a mixed-highs, dot-sequential transmission was found to offer substantially the same quality of image as a 12-megacycle channel carrying an equivalent simultaneous color transmission. The committee concludes that the Hazeltine developments are an important contribution to the dot-sequential system.

The present state of development of each system has been reached through the efforts of single organizations working in competition. Once the decision is reached among the systems, all that effort, plus additional effort from other quarters, can be applied to the one system then chosen. It may then be found that the real limit to future progress is that imposed by the nature of the scanning standards, not by present equipment limitations or present relative costs.

On this account, the final conclusion of the committee is that principal importance should be attached to those fundamental capabilities and limitations which relate to the choice of scanning method. These fundamentals have been discussed at length in this report and listed in detail in the tabular summary. Other factors, relating to the present performance and costs of apparatus, deserve consideration, but, in the opinion of the committee, such matters should take second place in the technical assessment of the systems.

Respectfully submitted.

E. U. CONDON, *Chairman*.
S. L. BAILEY.
W. L. EVERITT.
D. G. FINK.
NEWBERN SMITH.

Tabular summary of performance characteristics

Performance characteristic	Standard black-and-white	CTI color	CBS color		RCA color	Superior system
			Line interlaced	Dot interlaced		
Adaptability		Not needed	Adaptable	Adaptable	Not needed	CTI; RCA.
Color fidelity:						
*Large areas		Satisfactory	Excellent	Excellent	Satisfactory	CBS.
Small areas and edges of objects		Fair	do	do	Fair	CBS.
Compatibility: Quality of image rendered on existing sets		do	Not compatible	Not compatible	Excellent	RCA.
Continuity of motion:						
*Large objects	Excellent	Good	Good	Good	Good	(All comparable)
Small objects	Good	Fair	do	Fair	do	CBS (line); RCA.
Convertibility		Not easily convertible at present.	Convertible 12½-inch tube diameter, maximum.	Convertible 12½-inch tube diameter, maximum.	Not easily convertible at present.	CBS.
Effectiveness of channel utilization	Good	Good	Satisfactory	Good	Excellent	RCA.
Flicker-brightness relationship:						
*Large areas	Excellent	Excellent	Good	do	do	CTI; RCA.
Small areas	Good	Fair	do	Satisfactory	Good	CBS (line); RCA.
Interdot flicker	Absent	Absent	Absent	Fair	Fair	CTI; CBS (line).
Interline flicker	Good	Poor	Good	Good	Good	CBS; RCA.
Geometric resolution:						
*Number of picture elements per color picture	200,000	200,000	83,000	166,000	200,000	CTI; RCA.
Vertical resolution	490 lines	490 lines [1]	378 lines	378 lines	490 lines	RCA.
Horizontal resolution	320 lines	320 lines	185 lines	370 lines	320 lines	CBS (dot).
Superposition performance:						
Registration		Fair	Excellent	Excellent	Fair	CBS.
Color breakup		Excellent	Satisfactory	Satisfactory	Excellent	CTI; RCA.
Color fringing		do	do	do	do	CTI; RCA.

*See explanation on p. 36.

NOTE 1.—This is the geometric resolution; the apparent vertical resolution is considerably less, due to interline flicker.

HISTORICAL STUDIES IN TELECOMMUNICATIONS

An Arno Press Collection

The Electric Telegraph: An Historical Anthology. Edited, with an Introduction by George Shiers. 1977

The Telephone: An Historical Anthology. Edited by George Shiers with an Introduction by Elliot N. Sivowitch. 1977

The Development of Wireless To 1920. Edited by George Shiers with an Introduction by Elliot N. Sivowitch. 1977

Technical Development of Television. Edited, with an Introduction by George Shiers. 1977

Documents in American Telecommunications Policy. Edited, with an Introduction by John M. Kittross. Two volumes. 1977